姚建铨 ○ 编著

追光

从激光到太赫兹的科学探索之路

ZHUIGUANG
CONG JIGUANG DAO TAIHEZI DE
KEXUE TANSUO ZHI LU

天津大学出版社

图书在版编目（CIP）数据

追光：从激光到太赫兹的科学探索之路 / 姚建铨著. -- 天津：天津大学出版社, 2021.5
　ISBN 978-7-5618-6895-9

　Ⅰ.①追… Ⅱ.①姚… Ⅲ.①激光技术 Ⅳ.①TN24

中国版本图书馆CIP数据核字(2021)第063367号

策划编辑　田　园
责任编辑　李　源
装帧设计　谷英卉

出版发行	天津大学出版社
地　　址	天津市卫津路92号天津大学内（邮编：300072）
电　　话	发行部：022—27403647
网　　址	www.tjupress.com.cn
印　　刷	北京盛通印刷股份有限公司
经　　销	全国各地新华书店
开　　本	185mm×260mm
印　　张	46.5
字　　数	1161千
版　　次	2021年5月第1版
印　　次	2021年5月第1次
定　　价	298.00元

序

随着现代科学技术的飞速发展和社会信息化进程的不断加快，各种新技术、新思想及交叉学科不断涌现。激光技术、光电子技术、太赫兹技术、微纳光电子技术以及海洋技术等成为国内外学术界热门的研究领域，它们在通信、雷达、能源、环境、生物、健康、安全与国防等领域都具有广阔的应用前景。但是也应当看到，这些技术的实际应用潜力还远未被充分发掘，还有许多理论及工程技术问题有待突破。信息技术展示了巨大的发展前景。

天津大学姚建铨院士和他领导的研究团队长期致力于激光、光电子、非线性光学、太赫兹及海洋光学等领域的研究，尤其在激光与非线性光学频率变换技术、高功率倍频激光器、光子学太赫兹源及激光致声技术等方面取得了很多具有创造性的重要成果。在本书中，姚建铨院士以研究团队优秀论文的汇编及本人专门撰写的科研创新总结的方式向读者提供了丰富的资料，系统地总结了在激光类高斯光束理论，激光谐振腔理论，光学参量振荡器，全固态激光技术，激光可调谐技术，光子学太赫兹源、调控、成像和检测技术，微纳光电子技术及海洋光学等方面的研究成果。本书既是对姚建铨院士及研究团队近四十年来在这些领域内研究成果的总结和升华，又是他在这些领域内的学术思想、技术水平的具体体现，读者或许能从中吸取一些营养。本书确是我国这一领域内很有特色的科技专著。相信本书的出版将会有力地促进我国激光、太赫兹、信息技术的发展，使其在国民经济和国防建设中发挥更大的作用。

清华大学教授
中国科学院院士
2020年5月于清华园

前言

  我已年过八十，在天津大学学习、工作和生活了近六十三年，最近我重读了五年前在从教五十周年时写的《给我的学生的一封公开信》。回首往事，思绪万千，感慨颇多，我想以此信作为我此书的前言。

2020年5月

愉快而又艰辛的岁月——76个春秋略记

  正值天津大学（北洋大学）迎来建校120周年之际，敝人也度过了愉快而又艰辛的岁月——76个春秋及从教50周年，思绪万千，浮想联翩；回首往事，感慨颇多。

  我是一个幸运者。

  我赶上了一个空前动荡而变革的时代，赶上了中华民族历经深重灾难，而今又走上独立、民主、富强道路的新时代；

  我经历了陡门桥小学、胶南初级中学、苏州中学、天津大学这几所人杰地灵、英才辈出的好学校，其中又有一批关心我、教育我的好师长；

  我遇见了一批在科研道路上与我并肩披荆斩棘、奋力拼搏的同事们，尤其有一批刻苦钻研、勇于创新的学生们；

  我又赶上了正在改革创新、走向努力实现中华民族伟大复兴"中国梦"的新时代；

  我也幸运地获得中国科技界最高学术称号（终身荣誉）——中国科学院院士，并成为北大、清华等高等学府的兼职教授。

  我曾记得在上海金陵西路孝和里、无锡太湖湖畔、锡澄运河边陡门桥镇度过的愉快的童年，自己还曾不知天高地厚地树立了成为"科学家"的理想。

  还记得在苏州三元坊、状元巷对面，孔庙旁的苏州中学里，在碧霞池边、道山亭上的晨读；记得在天津大学青年湖里游泳，在图书馆里苦读。天津大学三年研究生学习为我夯实了科研根基。

我难以忘却"文革"时期的乱象，更清楚地记得改革开放后首批被派往美国进修的场景。记得导师西格曼（A. E. Siegman）教授（美国光学学会主席）对我耐心地讲解，还带我去太浩湖（Lake Tahoe），一路上不仅业务上收获颇丰，还领略了美国西部的美景。在XMR，公司工作时，同事给我起了"Green"（绿光）的绰号，我又尝到了在世界顶尖杂志上发表4篇论文时的喜悦的味道。在美国这段经历，使我的研究工作拓宽到一个崭新的国际化领域。

我也记得回国后在仓库中做倍频激光器的艰辛。在不到5年内取得了第一阶段的科研成果：高功率绿光激光器达到国际领先水平，获国家发明二等奖、布鲁塞尔尤里卡世界发明博览会金奖，个人获比利时一级"骑士勋章"等。

科研道路也并不平坦。我适时地开展可调谐激光及全固态激光技术的研究，取得了第二阶段的科研成果。在学生们的协助下，我撰写了3部专著，其中2部在科学出版社出版（1995年，2007年），1部在美国施普林格出版集团出版（2012年）。这些著作是我和科研团队的师生科研的结晶，也算是对光学领域的贡献吧！

山重水复疑无路，柳暗花明又一村。我不失时机地在我国较早开展太赫兹技术的研究，现又指导研究所开展微纳结构材料以及物联网—智慧城市—智慧海洋方面的研究。

我永远不会忘记一些师长对我的教育、指导、关怀，从陡门桥小学的姚绥英，胶南中学的陈龙海、胡琛，到苏州中学的许楠英，再到天大的李维铮、孙祖宝等。其中，我不得不提到王大珩院士对我从事激光光电子研究的支持，他亲自参加光电子中心的成立大会，在去俄罗斯访问的一路上对我谆谆教诲。此外还有斯坦福大学的西格曼（Siegman）教授、柏林工大的韦伯（Weber）教授、法国的齐斯（Zyss）教授等也在我的科研生涯中起到了举足轻重的作用，把我的科研水平逐步提升到国际领先水平。

我也永远不会忘记我的可爱的研究生群体。我们共同钻研，切磋业务，甚

前言

至热烈争论，还在一起下棋、打球。我还亲自组织了去水上公园的英语派对，以及到黄崖关的团建活动。我永远珍惜50年来集体的温暖及师生之情。种种事例，不胜枚举；感激之情，无以言表。我的所有成绩都是我的团队及学生共同奋斗的结晶。还有天津大学、天津市各部门的关怀支持，我不会忘记天大老校长李曙森及市委书记张立昌对我的支持及关怀。1997年我当选天津市唯一的院士。当想到自己是天津大学培养出的第一个院士时，我一共买了99支玫瑰花，分别来到李曙森、吴咏诗、李光泉3位校长的家中，与他们分享喜悦，感谢天大对我8年的培育之恩。每逢节日，我都能收到近千封祝贺的信件，其中一半来自我的学生，这是我最大的欣慰吧！

我最喜欢的名言是：

  在科学上没有平坦的大道，只有不畏劳苦沿着陡峭山路攀登的人，才有希望达到光辉的顶点。（马克思）

  人的一生应当是这样度过的：当他回忆往事的时候，他不至于因为虚度年华而悔恨，也不至于因为过去的碌碌无为而羞愧。（奥斯特洛夫斯基）

世界的历史，人类的历史，中国的历史翻开了新的一页，我所领导的激光与光电子研究所也将进入新的发展期，形势无限好！任重而道远，我虽年迈，但不算衰老，老骥伏枥，志在千里。我还可以为祖国的未来，为中华民族的腾飞，为中国的教育和科技事业发挥余热！祝大家前程无量！

谦逊　勤奋　求实　创新

姚建铨

2015年7月于天大

目录

姚建铨专著与教材	10
姚建铨主持香山科学会议一览表	11
姚建铨近20年完成的国家自然科学基金及重大科研项目一览表	11
天津大学激光与电子研究所重要论文	13
姚建铨学术生涯图片回顾	16
第一部分　追光	23
第一章　江南少年	24
第二章　北洋求学	30
第三章　海外进修	31
第四章　回国创新	34
第五章　新的起点	41
第六章　怀念恩师	46
第七章　耕耘不辍	50
第八章　科研回溯	54
附访谈录　打破学科的藩篱	61
第二部分　姚建铨学术团队论文集	65
论文集目录	67
第一章　非线性光学频率变换及固体激光技术	109
第二章　光电子及光纤技术	329
第三章　太赫兹技术	518
后记	742

姚建铨专著与教材

专著

[1] 姚建铨. 非线性光学频率变换及激光调谐技术. 北京: 科学出版社, 1995.

[2] 姚建铨. 奇异的光——激光. 北京: 清华大学出版社, 广州: 暨南大学出版社, 2000.

[3] 姚建铨. 全固态激光及非线性光学频率变换技术 // 红外与激光工程编辑部. 现代光学与光子学的进展. 天津: 天津科学技术出版社, 2002.

[4] 姚建铨, 徐德刚. 全固态激光及非线性光学频率变换技术. 北京: 科学出版社, 2007.

[5] Jianquan Yao, Yuye Wang. Nonlinear optics and solid-state lasers, advanced concepts, tuning-fundamentals and applications. New York: Springer Publishing Company, 2012.

章节

[1] 姚建铨. 第2章: 谐波产生及光学参量 // 叶佩弦. 振荡非线性光学. 北京: 中国科学技术出版社, 1999: 22-51.

[2] 姚建铨, 王杰. 第6章: 激光分子流场检测技术 // 范洁川, 等. 近代空气动力学丛书. 北京: 国防工业出版社, 2002.

[3] Jianquan Yao, Ran Wang, Haixia Cui, Jingli Wang. Chapter 16: Atmospheric propagation of terahertz radiation // Boris Escalante-Ramírez. Remote sensing. London: InTech Open Access Publisher, 2012.

教材

[1] 姚建铨. 第3章: 锁模技术 // 激光技术. 长沙: 湖南科技出版社, 1979.

[2] 姚建铨, 于意仲. 光电子技术. 北京: 高等教育出版社, 2006.

姚建铨主持香山科学会议一览表

	时间	会议顺次	会议主题名称	执行主席	姚建铨报告题目
1	2003-09-17	210	激光制造与未来技术产业的发展	左铁钏教授　王大珩院士 姚建铨院士　周炳琨院士	我国在先进制造业中激光技术领域的发展现状、差距及对策
2	2005-11-13	270	太赫兹科学与技术	刘盛纲院士　姚建铨院士 张杰院士　　封松林教授	基于光子学原理的THz辐射源
3	2012-06-20	427	老年健康信息化服务的科学问题与前沿技术	姚建铨院士　陈可冀院士 俞梦孙院士　王志良教授	老年健康信息化服务的科学问题与前沿技术
4	2014-04-08	528	太赫兹波在生物医学应用中的科学问题与前沿技术	姚建铨院士 杜祥琬院士　孔祥复院士 王正国院士　刘仓理研究员	太赫兹波在生物医学应用中的科学问题与前沿技术
5	2015-09-24	541	激光制造科学与工程前沿	姜澜教授　　姚建铨院士 陆永枫教授　李琳教授 肖荣诗教授	时代、机遇及挑战
6	2016-10-20	573	空间碎片监测移除前沿技术与系统发展	孙家栋院士 包为民院士　姚建铨院士 李明教授　　严俊教授	激光及太赫兹技术在碎片监测移除应用中的探索

姚建铨近20年完成的国家自然科学基金及重大科研项目一览表

序号	课题名称	项目类型	负责人	项目起止时间	总经费（万元）	项目状态
1	受激拉曼激光感应电子荧光（RELIEF）方法进气道内流场诊断装置研究	航空部预研	姚建铨	1998年1月至1999年	85	通过验收
2	紫外及序列脉冲激光瑞利散射用于非稳空气流场	国家自然科学基金项目	姚建铨	1998年至2000年12月	10	已结项
3	宽带输出的Ti:Al_2O_3-PPLN光学参量振荡器	国家自然科学基金项目	姚建铨	2003年1月至2005年12月	24	已结项
4	双波长运转全固态激光器泵浦准相位匹配差频THz波辐射源的研究	国家自然科学基金项目	姚建铨	2005年1月至2007年12月	29	已结项
5	THz波在金属镀层空芯波导中传输的研究	国家自然科学基金项目	姚建铨	2007年1月至2009年12月	30	已结项
6	超大功率集成式多芯微结构光纤激光器的研究	国家自然科学基金项目	姚建铨	2007年1月至2010年12月	57/200	已结项

续表

序号	课题名称	项目类型	负责人	项目起止时间	总经费(万元)	项目状态
7	基于非对称量子阱非线性光学差频产生THz波的研究	国家自然科学基金项目	姚建铨	2013年1月至2016年12月	88	已结项
8	柔性无机-光电场效应非易失性浮栅型存储	国家自然科学基金项目	姚建铨	2017年1月至2020年12月	60	已结项
9	太赫兹调控的新材料、新器件及关键技术研究	国家自然科学基金重点项目	姚建铨	2018年1月至2022年12月	290	在研
10	高功率、高重复频率全固态绿光激光器样机研究	天津市科技发展计划项目	姚建铨	2000年1月至2002年11月	150	已结项
11	High Power SHG Laser (532 nm)	天津市光电子联合科学研究中心项目	姚建铨	2001年至2004年	150	已结项
12	高功率红、绿、蓝全固态激光器及其激光显示系统的关键技术研究	天津市光电子联合科学研究中心项目	姚建铨	2002年1月至2004年11月	200	已结项
13	基于非线性效应超宽带光通信系统中关键技术研究	教育部南开大学天津大学联合研究院项目	姚建铨	2002年1月至2004年11月	320	已结项
14	全固态激光大面积全色显示关键技术研究	国家新技术研究发展计划（863计划）	姚建铨	2003年1月至2005年12月	130	已结项
15	大模场光纤波导特性研究	国家重点基础研究发展计划（973计划）"光纤激光器技术基础研究"专题	姚建铨	2006年10至2009年10月	85/100	已结项
16	光子学与非线性光学产生太赫兹辐射源的基础研究	国家重点基础研究发展计划（973计划）"太赫兹重要辐射源、探测及应用的基础研究"专题	姚建铨	2007年7月至2011年8月	376/513	已结项
17	新型光子晶体光纤传感器的基础研究	国家重点基础研究发展计划（973计划）"新一代光纤智能传感网与关键器件基础研究"专题	姚建铨 陆 颖	2011年1月至2015年6月	504	已结项
18	全固态激光器	国家高技术研究发展计划（863计划）	姚建铨	2011年12月至2013年12月	100	已结项
19	基于数字编码超构材料的电磁信息处理	国家重点研发计划	姚建铨	2018年4月至2023年4月	271	在研

天津大学激光与电子研究所重要论文

研究所早期发表的关于双轴晶体最佳相位匹配计算、激光类高斯光束理论、高转换效率准连续泵浦内腔倍频YAG激光器、可调谐激光技术体系、RELIEF方法测量高速空气流场理论、太赫兹技术、超材料及微纳光电子材料与器件等方面的若干重要论文目录如下。

1 J. Q. Yao, Theodore S. Fahlen. Calculations of optimum phase match parameters for the biaxial crystal $KTiOPO_4$. Journal of Applied Physics, 1984, Vol. 55, No. 1: 65-68.

2 Yao Jianquan, Zhou Guosheng, A. E. Siegman. Large-signal results for degenerate four-wave mixing and phase conjugate resonators. Appl. Phys. B, 1983, 30: 11-18.

3 姚建铨,李昱,薛彬,T.S.Fahlen. 使用$KTiOPO_4$的高功率内腔倍频YAG激光器, 中国激光, 1983, Vol. 10, No. Z1: 101.

4 Jianquan Yao, Bin Xue. Optimum parameters of a high-conversion efficiency intracavity frequency-doubled laser with gaussian-like beam. CLEO'84, THC6, Anaheim, California, U.S.A.: June19-22, 1984.

5 姚建铨,薛彬. 类高斯分布的理论及其应用. 量子电子学, 1984, Vol. 1, No. 2: 133-142.

6 姚建铨,薛彬. 高转换效率下具有高斯及类高斯光束的内腔倍频. 光学学报, 1985, Vol. 5, No. 2: 142-150.

7 J. Q. Yao, W. Q. Shi, J. E. Millerd, G. F. Xu, E. Garmire, M. Birnbaum. Room-temperature 1.06-0.53-μm second-harmonic generation with $MgO: LiNbO_3$. Optics Letters, 1990, Vol. 15, No. 23: 1339-1341.

8 Jianquan Yao, Weidong Sheng, Weiqiang Shi. Accurate calculation of the optimum phase-matching parameters in three-wave interactions with biaxial nonlinear-optical crystals. Journal of the Optical Society of America B, 1992, Vol. 9, No. 6: 891-902.

9 Yao Jianquan, Liu Hang, Ashok Puri. Femtosecond pulse-second and third harmonic generation with BBO, Acta Optica Sinica, 1995, Vol. 15, No. 6: 641-647.

10 J. Q. Yao, Y. Cui, X. W. Sun, M. H. Dunn, W. Sibbett. Mixed mode coefficient and gaussian-like distribution, Acta Optica Sinica, 1995, Vol. 15, No. 12: 1633-1640.

11 王杰,姚建铨,于意仲,张帆,王涛. 基于混频效应的宽带激光谐波转换理论. 物理学报, 2001, Vol. 50, No. 6: 1092-1096.

12 施翔春,王杰,肖绪辉,王晓勇,姚建铨. 拉曼激发激光诱导电子荧光流场测量系统中标记过程的研究, 光学学报, 2001, Vol. 21, No. 2: 206-210.

13 姚建铨. 非线性光学频率变换及准相位匹配技术. 人工晶体学报, 2002, Vol. 31, No. 3: 201-207.

14 Yao Jianquan, Zhang Baigang, Lu Yang, Ding Xing, Xu Degang. Wavelength tunable optical parametric oscillator based on periodically poled lithium niobate. Journal of Synthetic Crystals, 2004, Vol. 33, No. 4: 465–470.

15 Zhang Baigang, Yao Jianquan, Ding Xin, Zhang Hao, Xu Degang, Yu Guojun, Zhang Fan. Low-threshold, high-efficiency, high-repetition-rate optical parametric generator based on periodically poled $LiNbO_3$. Chinese Physics, 2004, Vol. 13, No. 3: 364–368.

16 孙博, 姚建铨. 基于光学方法的太赫兹辐射源. 中国激光, 2006, Vol.33, No.10: 1349–1359.

17 孙博, 姚建铨, 王卓. 利用各向同性半导体晶体差频产生可调谐THz辐射的理论研究. 物理学报, 2007, Vol.56, No.3: 1390–1396.

18 姚建铨, 迟楠, 杨鹏飞, 崔海霞, 汪静丽, 李九生, 徐德刚, 丁欣. 太赫兹通信技术的研究与展望, 中国激光, 2009, Vol.36, No.9: 2213–2233. (被下载5000余次, 被引用240余次, 被光学学会评为"高被引论文"奖)

19 钟凯, 姚建铨, 徐德刚, 张会云. 级联差频产生太赫兹辐射的理论研究. 物理学报, 2011, 60(3): 291–298.

20 Kai Zhong, Jialin Mei, Yang Liu, Hongzhan Qiao, Kefei Liu, Degang Xu, Jianquan Yao. Widely tunable eye-safe optical parametric oscillator with noncollinear phase-matching in a ring cavity. Optics Express, 2019, Vol. 27, NO. 8: 10449–10455.

21 Wu L, Du T, Xu N, Ding C, Li H, Sheng Q, Liu M, Yao J, Wang Z, Lou X, and Zhang W. Metamaterials: a new $Ba_{0.6}Sr_{0.4}TiO_3$-Silicon hybrid metamaterial device in terahertz regime, Small, 2016, Vol.12, NO.19: 2609–2615.

22 Xin Yan, Maosheng Yang, Zhang Zhang, Lanju Liang, Dequan Wei, Meng Wang, Mengjin Zhang, Tao Wang, Longhai Liu, Jianhua Xie, Jianquan Yao. The terahertz electromagnetically induced transparency-like metamaterials for sensitive biosensors in the detection of cancer cells. Biosensors, Bioelectronics, 2019, Vol.126: 485–492.

23 Zhang Zhang, Ju Gao, Maosheng Yang, Xin Yan, Yuying Lu, Liang Wu, Jining Li, Dequan Wei, Longhai Liu, Jianhua Xie, Lanju Liang, Jianquan Yao. Microfluidic integrated metamaterials for active terahertz photonics. Photonics Research, 2019, Vol. 7, NO.12: 1400–1406.

24 Luan N, Ding C, Yao J. A refractive index and temperature sensor based on surface plasmon resonance in an exposed-core microstructured optical fiber. IEEE Photonics Journal, 2016, Vol. 8, NO. 2: 1–8.

25 Luan N, Yao J. High refractive index surface plasmon resonance sensor based on a silver wire filled hollow fiber. IEEE Photonics Journal, 2016, Vol. 8, NO.1: 1–9.

26 Dexian Yan, Yuye Wang, Degang Xu, Pengxiang Liu, Jianquan Yao, et al. High-average-power, high-repetition-rate tunable terahertz difference frequency generation with

GaSe crystal pumped by 2 μm dual-wavelength intracavity KTP optical parametric oscillator. Photonics Research, 2017, 5(2): 82-87.（影响因子为5.242，被引13次）

27 Pengxiang Liu, Xinyuan Zhang, Chao Yan, Degang Xu, Jianquan Yao, et al. Widely tunable and monochromatic terahertz difference frequency generation with organic crystal 2-(3-(4-hydroxystyryl)-5, 5-dime-thylcyclohex-2-enylidene) malononitrile. Applied Physics Letters, 2016, 108(1): 011104.

28 Yuye Wang, Zhinan Jiang, Degang Xu, Tunan Chen, Beike Chen, Shi Wang, Ning Mu, Hua Feng, Jianquan Yao. Study of the dielectric characteristics of living glial-like cells using terahertz ATR spectroscopy. Biomedical Optics Express, 2019, 10(10): 5351-5361.

29 王与烨, 孙忠成, 徐德刚, 姜智南, 穆宁, 杨川艳, 陈图南, 冯华, 姚建铨. 基于太赫兹时域光谱系统的脑缺血检测. 光学学报, 2020, 40(4): 0430001-1-0430001-7.（封面文章）

30 Yu Y, Zhang Y, Song X, Zhang H, Cao M, Che Y, Dai H, Yang J, Zhang H, Yao J. PbS-Decorated WS_2 phototransistors with fast response. ACS Photonics, 2017. Vol. 4, NO. 4: 950-956.（当年被该杂志评为亚洲地区高被引论文榜第5位，2019年被SCI评为高被引论文，被引54次）

31 Yu Y, Zhang Y, Zhang Z, Zhang H, Song X, Cao M, Che Y, Dai H, Yang J, Wang J, Zhang H, Yao J. Broadband phototransistor based on $CH_3NH_3PbI_3$ Perovskite and PbSe quantum dot heterojunction. The Journal of Physical Chemistry Letters, 2017. Vol. 8, NO.2: 445-451.（单篇最高影响因子为9.353，被引48次）

32 Li Y, Zhang Y, Yu Y, Chen Z, Li Q, Li T, Li J, Zhao H, Sheng Q, Yan F, Ge Z, Ren Y, Chen Y, Yao J. Ultraviolet-to-microwave room-temperature photodetectors based on three-dimensional graphene foams. Photonics Research, 2020, Vol. 8: 368-374.（制备并获得了响应波长跨越4个数量级的超宽谱探测器，探测范围从紫外线直至毫米波）

33 Chen Z, Yu Y, Jin L, Li Y, Li Q, Li T, Li J, Zhao H, Zhang Y, Dai H, Yao J. Broadband photoelectric tunable quantum dot based resistive random access memory. J. Mater. Chem. C, 2020, Vol. 8: 2178—2185.（制备并获得可多波长操作和宽谱光操作的阻变存储器）

姚建铨学术生涯图片回顾

▶▶▶ **与诺贝尔奖得主及世界知名教授合影**

1989年诺贝尔物理学奖获得者杨振宁教授在访问姚建铨教授实验室时与姚建铨合影

1988年1月姚建铨与诺贝尔物理学奖获得者霍尔（Hall）教授的合影

1995年姚建铨与王大珩先生访俄时与诺贝尔物理学奖获得者普罗霍洛夫（Prokhorov）教授的合影

2003年9月姚建铨与诺贝尔物理学奖获得者高锟教授的合影

1993年姚建铨与诺贝尔物理学奖获得者汤斯（Townes）教授的合影

姚建铨在斯坦福大学进修时与导师西格曼（Siegman）的合影

1995年姚建铨作为中国光学代表团成员与王大珩先生访问俄罗斯时的合影

2000年姚建铨与师昌绪院士的合影

2003年姚建铨与德国韦伯（Weber）教授的合影

2011年6月姚建铨与贝尔实验室原光通信部主任厉鼎毅的合影

2015年"两弹一星"元勋孙家栋院士会见姚建铨

2010年姚建铨与日本伊藤教授在北洋纪念亭的合影

2010年姚建铨与刘国治院士的合影

2004年姚建铨与法国齐斯（Zyss）教授的合影

2012年姚建铨与程津培院士的合影

姚建铨与香港科技大学校长朱经武教授的合影

2008年姚建铨与赵伊君院士的合影

出席重要会议及国内外学术交流活动

2000年姚建铨访问英国

2000年姚建铨访问法国

2005年4月姚建铨在天津市参加光学物理年全球点灯仪式

2005年11月姚建铨作为召集人之一组织召开第270次香山科学会议研讨太赫兹科学技术

2007年姚建铨在中国国际光电博览会（CIOE）上发言

姚建铨参加中俄激光物理研讨会（摄于俄罗斯新西伯利亚）

2007年姚建铨在深圳举行的国际光电展会上与王兴龙博士及友人交谈

2010年姚建铨参加中俄2010年激光物理研讨会并做特邀报告

2010年4月姚建铨在中国电子技术年会上做报告

2011年9月姚建铨参加第十三届中国科协年会
——天津光谷产业发展高峰论坛并做主题报告

2011年姚建铨在国际智能物联网会议上发言

2010年8月姚建铨在内蒙古自治区参加项目论证会时的留影

2008年姚建铨在武汉参加第一届光子与光电子学会议（POEM 2008）时与中、日、美等国教授合影

2011年10月姚建铨在美国硅谷高科技创新·创业高峰会（SVIEF）上做主题发言

2019年姚建铨在第五届中国智慧城市博览会上发言

参加全国政协、高校和社会团体活动

2003年姚建铨参加全国政协十届一次会议

2007年姚建铨参加全国政协十届五次会议

2003年姚建铨在华中科技大学给研究生做学术报告

2002年姚建铨在北京青少年科技俱乐部科学家系列讲座上与青少年代表合影

2009年7月姚建铨被聘为南京大学客座教授

2008年8月姚建铨获聘香港科技大学兼职教授

2008年姚建铨访问香港理工大学

2009年1月姚建铨作为天津市专家协会会长与欧美同学会会长出席两会二〇〇九新春年会

潜心育人，桃李满天下

2011年5月姚建铨参加李忠洋等博士论文答辩会

2013年5月姚建铨参加盛泉等博士论文答辩会

2011年姚建铨与硕士毕业生合影

姚建铨带领学生进行学术交流

2011年姚建铨赴俄罗斯参加中俄激光会议时与助手钟凯合影

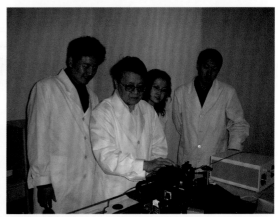

2002年姚建铨在实验室指导学生

姚建铨与自己指导的博士毕业生合影

第一部分

追光

第一章　江南少年
第二章　北洋求学
第三章　海外进修
第四章　回国创新
第五章　新的起点
第六章　怀念恩师
第七章　耕耘不辍
第八章　科研回溯
附访谈录　打破学科的藩篱

第一章　江南少年

从大上海回到江南故乡

　　1939年，我出生在上海市金陵西路孝和里的一个普通家庭。彼时的中国风雨飘摇，时局维艰。我赶上的是一个空前动荡而又瞬息万变的时代。上海这个"冒险家的乐园"在我的记忆中并不熟悉，只是定格着一些那个时代特有的场景：马路上戴船形帽的美国大兵，街中央值勤的印度警察，霓虹闪烁的舞厅酒楼……当然也有与表弟一起在孝和里里弄中追逐打斗的温馨画面。

　　上学没多久，我们就举家迁回了家乡。这是父亲的主张，一是要照看祖宅，更深一层的考虑是让子女接受传统教育并感受家乡氛围。记得那是小学二年级，我从上海回到了老家——无锡太湖湖畔、锡澄运河旁的陡门桥镇，并在这里度过了愉快的童年。

　　无锡山清水秀，人杰地灵，是江南文明的发源地之一，有文字记载的历史可追溯到商朝末年。历史上无锡曾属常州专区和苏州专区管辖，近代以山水名城、工商名城著称于世，素

图1-1　姚建铨（前排中）童年时的全家合影

有"布码头""钱码头""窑码头""丝都""米市""小上海"之称。这里山水秀美、人文景观众多,被誉为"太湖明珠"。无锡朴实清静,古意盎然,文脉绵延,俊彦如林,英杰辈出,历史文化积淀深厚。独特的自然资源和劳动者的创造禀赋,不仅造就了繁荣发达的经济社会,也使崇文尚教之风勃兴,各类英才辈出。东晋文学家、画家顾恺之,元代画家、诗人倪瓒,明代地理学家徐霞客,清代外交家薛福成,都曾在历史的星空中熠熠生辉;近代革命家秦邦宪,科学家周培源、钱伟长,经济学家孙治方,现代作家、文学研究家钱钟书,画家徐悲鸿及音乐家华彦钧皆为无锡人;从无锡还走出了许许多多的院士、大学校长、教授学者。他们是无锡儿女的杰出代表,也是这座城市的骄傲,为这座城市增添了无穷的魅力。有人云:江南佳绝地,尚有桃花源。而无锡是江南中的江南,桃源里的桃源。

一枚印章,一生座右铭

我的老家陡门桥镇是无锡北门外的一个小镇,当地周边几十里范围内有十多个村庄,六七条河浜。在这几条河浜中唯有陡门桥旁边的河是与锡澄运河及南北大运河相通的,我们称之为"活水河"。当地的小孩都喜欢在"活水河"中游泳,我也经常与小朋友下河游泳,在陡门桥的桥墩上玩跳水。

小时候,我是一个调皮的孩子。当地有一风俗,中秋节后,由于水凉,不能游泳了,大人就会警告小孩"中秋节后不能下河!"。而我正是既淘气又逆反的年龄,偏不听,还下河,从河道的东岸游到西岸,又从西岸游到东岸。我母亲知道后,跑来催我上岸,我还是不听,母亲就拿着竹竿轰我上岸,我游得快,竹竿够不着,直到母亲最后要把我的裤子拿走,我才上了岸。

我知道自己聪明,学习不费劲,但同时我也很警醒,知道自己有两个毛病——有点傲气,不够刻苦。大约是小学五六年级时,我立志要克服自己的毛病,做到勤奋、好学、谦逊、认真。在我的提议下,我和哥哥决定给自己刻个言志的章,就刻"勤奋谦逊"四个字。我们准备了一块圆卵石及刻刀,不会写反字,就先在纸上写好这四个字,然后反过来贴在圆石上,刻出来就是四个正着的字了。从此以后这四个字就成为我的座右铭,我时常以此来告诫自己,警醒自己。

我提醒自己要勤奋,勤奋,再勤奋!我总感到时间不够用,想学的东西,想干的研究工作很多。我的想法很简单:抓紧一切时间多学、多干,积少成多总能成功。以后的几十年间我总是早起晚睡。记得在天大工作时,我早上起来买早点的时候,早点铺的师傅还在生火点炉子,我基本上是每天第一个顾客。烧饼、茶鸡蛋,是我的早点,冬天买完后就裹在棉衣里,有时晚上回家时才发现烧饼还没啃完。我一般早上6点就到办公室了,有时不到7点就给学生、朋友发电子邮件,他们收到后,总是来电话说:"姚老师,您几点就给我发电子邮件啊?"

我不管去哪儿出差都会带着一个笔记本和一台相机,走到哪儿就学到哪儿,回来整理学到的东西。记得在"文化大革命"后期我到上海出差,没正式的旅店住,就住在防空洞的地道里。地道里面很湿，抽风机的"嗡嗡"声太大，根本无法入睡。于是我刚过5点就从地下室里跑出来，坐在马路边的石头上看资料，那时清洁工才刚刚开始工作。

回首往事，我发现自己的很多志向都是小时候立下的，然后用一点一滴的努力去实现。大概是在我上小学五六年级时，北京召开了全国政治协商会议，我看到新闻后暗下决心，总有一天我也要去北京参会，协商国家大事。后来这个愿望实现了，1988年至2007年，我当选第七至第十届共四届全国政协委员。

在我小的时候，电风扇尚未普及，当地人们在夏夜解暑的土办法就是躺在门板上乘凉。那时，我躺在门板上仰望着星空，总是好奇地想，比星星更远的地方有什么？我们所在的地球在宇宙中到底占据怎样的位置？后来，读到牛顿、居里夫人等著名科学家的故事，我的人生志向渐渐明朗，自己将来一定要当科学家。

后来，我总是从科技与学术界一些有突出贡献的人物身上吸取正能量，将其作为自己教学科研中的宝贵经验。我认为，科学研究最吸引人的地方在于探究未知：从开始的一无所知，然后逐步深入，就像在森林里探宝一样，经过艰难困苦最后找到科学宝藏的时候，那真是一种莫大的享受。这种乐趣，是对自己最大的回馈。熟悉我的朋友及学生经常说，能感受到我身上充满着一种力量，这种力量就是我探究未知的喜悦，可能这也是我对知识、对社会、对人生保持着高度兴趣和热情的原因吧。

苦中作乐的快乐少年

1951年至1954年，我在无锡市北乡胶南中学读初中。胶南中学在堰桥镇胡家渡村北端，占地2万多平方米。1920年胡家渡村胡氏祠堂"思饴堂"内设小学，取名胶南小学；1944年，在小学基础上增设初中。校园内古树参天，移步换景，奇花异草，争奇斗艳。记得校歌中有一句"西胶山南胡家渡，一乡先觉乐树人"。胶南中学从1944年建校，到1997年并入堰桥中学，前后办学50余载，培养了大批人才，桃李满园，在20世纪五六十年代是锡澄地区名声远播的好学校。现在胶南中学没有了，胶南厚文圣贤之地、孜孜求学之景已俱往矣，我还是感到非常遗憾。

中华人民共和国成立初期的胶南中学，校长范学农与一批优秀的教师艰辛办学的精神让人心生敬意，而学生求学的刻苦也值得纪念。当时的生活条件很差，冬天宿舍里既没有炉子，晚上也没有开水供应，我们就穿着棉裤直接钻进被窝。早上起来时，毛巾结着冰，硬邦邦的，没法洗脸，我们就用茶杯去开水房打一点开水擦擦眼睛，然后去食堂吃早点，去教室读书。当时没有电灯，到了夏天，为了上晚自习，就在教室点燃一盏煤油打气的"汽灯"，可供两三百人一起看书。灯很亮，飞蛾、蚊子也一起来了，我们有自己的土办法，把长裤的裤脚捆住穿在脚上，防止蚊子叮，买一种"防蚊油"涂在手臂及背上。天气又闷又热，我们一边看书一边出汗，手臂下降温用的毛巾也湿透了。尽管生活条件十分艰苦，但没有一个学生叫苦，大家苦中作乐、乐在其中。我们一个年级两个班共约190人，其中我与"二尤"——

尤永坚、尤永章三个人学习成绩总是排在年级前三名，我们之间的关系也特别好。

那时正值抗美援朝期间，我还参加了学校的文娱晚会，登场扮演了一个抗美机智小英雄。剧情大致是这样的：一位同学演美国空降兵，另一位男同学和我分别演父女二人，我们用机智的办法在家里将空降兵抓住了。我当时读初一，个头较小，还长得十分俊俏，几个高年级的女同学给我搽脂抹粉，把我打扮成小丫头，从此我有了一个外号——"小丫头"。初中时，我还发生过一件糗事。我当时是少先队大队长，酷爱下象棋，一天下午四五点钟，我正与同学下棋下到紧张之际，辅导员陈龙海老师通知我："明天下午4：30开大队委会议，要通知所有大队委员。"我当时一口答应，可转头就忘了。到第二天辅导员等着开会时，谁都没去。事后他把我狠狠批评了一顿。

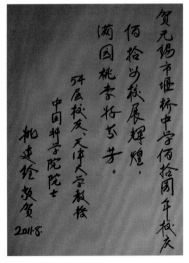

图1-2　姚建铨为初中母校胶南一堰桥中学110周年校庆题写的贺词

启蒙老师点燃了我的科学梦想，给少年的我留下了刻骨的印记，并深深地影响着我的一生。

初中三年中我印象最深的有两位老师，分别是甲、乙两班的班主任陈龙海及胡琛老师。陈龙海老师教语文，他授课时采取的启发式、哲理式的教学方式令人印象深刻，从古典诗词到现代文学作品，他的讲解深入浅出，让人受益匪浅。胡琛老师教数学，他年轻聪慧，讲课逻辑性强、效率高，从代数、几何到三角，每门课都讲得清晰到位。由于他的精心教导，帮助我打下了坚实的数学基础。我从初中升高中，高中升大学，数学都是满分。胡琛老师后来调到"天一中学"去当数学老师，成为无锡市、江苏省的模范教师。我在天大读研究生时，他来北京参加全国劳模会议，我还接待过他。

陡门桥小学与胶南中学（初中）虽然地处农村，环境很艰苦，但因为有好老师，成为远近闻名的好学校。陡门桥小学的姚绥英老师，胶南中学的陈龙海、胡琛、陈渊成、沈松炯、毛君白、庄祖耕等优秀老师，他们有丰富的知识、精湛的讲授方法和一颗爱孩子的心。这些老师对学生谆谆教诲，悉心培养，不仅给了我知识，还让我幼小的心灵得到了启发与激励，让我终身受益。

心系道山，苏中求学

我是不幸的，在祖国风雨飘摇的时候出生；我也是幸运的，在新中国的生机勃发中成长。

苏州是一座有着2500多年历史的文化名城，自古就有崇文尚教的优良传统，教育一直非常发达。苏州籍两院院士总共有100余名，数量始终居全国各大城市之首。苏州中学是一所千年名校，其历史可追溯至北宋景祐二年（1035年）范仲淹创建的苏州府学。当年，北宋政治家、文学家范仲淹在今苏州中学校址上创办苏州府学，首开东南兴学之风。据记载："苏州府学是宋代历史上规模最大的官办地方学府，号称东南学宫之首。"尽管苏州中学的占地面积已无法与当年相比，但历经千年，办学历史未断，校址始终未变，格局基本保存，这在中外教育史上十分罕见。

从1035年到2020年，苏州中学走过了近千年，千年漫漫，风起云涌，在姑苏城南这片钟灵毓秀的土地上，苏州中学弦歌不断、薪火相传、桃李芬芳，在苏州乃至中国的教育史上留下了光辉的一页。苏州中学环境优美、校风淳朴、师资一流，是孕育优秀人才的摇篮。处处洋溢着宁静、自觉、奋进气息的苏州中学素以名师众多、人才辈出而著称。

名相范仲淹创建苏州府学以来，胡瑗、沈德潜、俞樾等一批名师在此传道授业、衣钵相传。20世纪上半叶，苏州中学可谓群星璀璨，熠熠生辉。自1904年改办新学以来，国学大师罗振玉、王国维、钱穆，语言学大师吕叔湘，美术大师颜文樑，史学大师吕思勉，地理学家胡焕庸，教育家孙起孟等都曾在苏中执教，蔡元培、胡适等名家亦曾到苏中讲学授课。苏州中学培养了一大批牵动社会历史发展走向，构建人文、科学世界的时代精英。他们中有博古（原名秦邦宪）等中共早期领导人，有叶圣陶、顾颉刚、胡绳、匡亚明、吴作人、陆文夫等大师名家，还有诺贝尔物理学奖获得者李政道博士和钱伟长院士等科学家。一流学校素以培养造就一流人才为己任，苏州中学培养的杰出人才遍布各地，蜚声中外。

历经近千年风雨飘摇而屹立不倒，苏州中学是一所充满传奇色彩而又富有活力的学校。1954年，我有幸进入这所全国著名的中学学习，亲身感受到苏州中学优渥的学习氛围和基于悠久历史传统所形成的教学规范。

苏州中学在苏州三元坊、状元巷对面，孔庙旁。我爱苏中的学习环境，苏中的校园并不算大，除明亮的教学主楼和室外运动场外，在校园的一角有一座小山，名字叫道山，还有一湾碧水，人们常称之为碧霞池、春雨池，道山上古色古香的道山亭悠然矗立，那是我特别钟爱的地方。中学时，我经常在道山亭上及湖边晨读,诵读诗歌、课文，背诵英语，也常和三五同窗在树荫小径上嬉戏切磋。

遥想近千年之前，如果不是因为道山孕育的风水气象灵秀宜人，就不会有当年的范仲淹办学，就不会有后来绵延千年的苏中文脉，也就更加不会有现在苏中引以为傲的一切。纵观近千年的办学历程，我坚信凡在此教过书或者读过书的

图1-3 2011年6月2日姚建铨在母校苏州中学科学馆前的范仲淹像前留影

人，无人不曾登过道山。不同时代的人有着不同的故事、不同的情感、不同的记忆，那么唯独道山才是所有在此求学的人共同的故事、共同的情感、共同的记忆。我想，如果道山有眼，如果道山有灵，唯有他真正目睹和见证了这所学校在漫漫历史长河中所发生的一切。

苏中是道山，苏中是大学。

我想对每一个苏中的校友说，每当你离开或者回到母校的时候，你一定要去看望一位"老人"——道山，他不仅陪伴了你三年的中学生活，还和你的母校相依为命走过了近千年的风风雨雨。道山，在我们每个苏中学子的心中，他重于泰山。三年中，他一定陪过你追逐游戏、探险寻宝、读书解题、谈天论地。他是道义的象征、真理的象征、智慧的象征、信念的象征。

和无数先贤骄子一样，

也许你曾"道山寻道"，发现道在脚下走；

也许你曾"道山思道"，发现道在脑中悟；

也许你曾"道山论道"，让你知晓天道、懂得人道；

也许你曾"道山叛道"，让你理解老子的"道可道，非常道"。

道山之高是无数先贤思想的堆积，道山之深是无数骄子勤奋的挖掘，道山之道是近千年苏中学子走出来的成功之道。

最近看到一位小校友在网上议论母校："在我眼里，这是一所像大学的高中。校园像大学，学生像大学生，整体氛围也如大学一般。如果你不知道，走在校园里，大概很难想象这是一座中学。"

其实，90年前，这所刚刚成为省立高中的首任校长汪懋祖先生在建校伊始就提出了"学术高中"的概念，要求"以教师的学术去引领学生的学业"，他所带来的钱穆那样的一批学者、大家，现在连北大、清华这些高校都难以望其项背。文脉绵延，前人所遗传的大学文化基因，在苏州中学的校园里依然随处可见。

"先天下之忧而忧，后天下之乐而乐。"北宋名相范仲淹这句千古名句已经成为中华知识分子的道德训诫、精神追求。范仲淹创办苏州府学时就将"家国天下"的精神基因融入这所学校，并一脉相承，融入每一个在此求学的学子的血脉中。

苏州中学首任校长汪懋祖，少时赴美国哥伦比亚大学攻读教育学位。为求索教育救国之路，实现亲自办学、普及

图1-4 2004年姚建铨与苏州中学同年级校友张钟华院士、何鸣元院士的合影

教育的理想，他毅然辞去大学校长及江苏督学等职位，回故乡组建苏州中学。他舍近求远、舍高就低，为的是兴学强国。汪懋祖校长当年所提出的苏中精神——"有转移环境之能力，而不为不良环境所屈服"，完全是以天下为己任的精神。敏而好学，刻苦勤奋，精勤求学，敦笃励志，字字清晰，一丝不苟，苏州中学的老师讲课深入浅出，引人入胜，令学生受益终生。无论我们在哪里，我们都会心系道山，我们都会胸怀天下！

苏中是天下！自古至今，从苏州中学这块土地上出来的人都有"以天下为己任"的高尚精神，都有救国报国的博大情怀。

我在苏州中学求学的三年里，校长陈六中，教导主任许楠英，班主任、生物老师朱伯尼，物理老师姚昌学、吴保让，语文老师范烟桥、夏蕴文，数学老师陈浣华以及芮和师、毛礼垣、张道中、马成烈等优秀老师，他们以丰富的知识、循循善诱的教育方法、精湛的讲授方法，给我打好了各方面的基础，培养出我坚韧不拔、刻苦钻研的学风。也正是有了这样一批师德高尚、学养深厚的老师，才培育出了弃文从理、科学救国的钱伟长，生物力学之父冯元桢，物理学家冯端等无数立志救国的优秀人才。先校长汪懋祖先生曾说："要恢复民族精神，先要恢复教育精神，因为民族精神的传递与发扬全是教育的责任。"苏州中学不但有令人尊敬、学识渊博的名师大家，而且素以教学有方、治学严谨著称，形成了一整套教学科学方法和严肃的科学学风。苏州中学的三年对我一生的成长起到了关键作用，让我时常怀念。苏中的精神永远激励我前行！

第二章　北洋求学

1957年我考入天津大学，这是中国第一所现代大学，其前身是北洋大学，创建于1895年，是一所百年名校。

如果在胶南中学和苏州中学的学习对我一生产生的重大作用是开启了科学探索之梦并打下坚实的数理基础，那在天津大学的五年本科及三年研究生阶段的学习就是为我全面夯实了基础，并培养了我的各种能力。

图2-1　20世纪50年代天津大学七里台校门

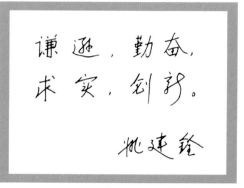

图2-2　姚建铨手书的座右铭"谦逊、勤奋、求实、创新"

记得天大老校长张国藩曾在我们1962级研究生入学的第一课讲了一句至理名言："三年研究生学习的真正意义不是学习某种专业知识，而是培养学习能力，包括自学能力、外语能力、查阅资料向国际同行学习的能力，最重要的是分析问题和解决问题的能力。"

回首我在天大八年的学习生涯，最大的收获是锻炼了能力。我研究生从事的科研方向是振动测量，之后搞激光测距仪、固体激光器、非线性光学频率变换技术、可调谐激光技术，又组织研究太赫兹技术，60多岁时转向物联网技术、智慧养老等智能科技，到70多岁后又转到智慧海洋技术和宽带太赫兹通信技术研究，过了80岁又筹划参与新型病毒检测技术的探索研究。我做的这些研究很多都是跨学科的，有的跨度还很大，之所以能深入下去，与我有比较宽的交叉学科知识面、能触类旁通有关系。

五年的本科学习期间我只回过老家三次，寒暑假大多在学校学习，在图书馆里苦读。三年研究生学习更是夯实了我的科研根基。求学期间，我遵循自己谦逊、勤奋的诺言，踏踏实实，尽量多学多做。在刻苦学习的同时，我还加强锻炼身体，在班里学习成绩一直保持在前三名。

"实事求是"是天津大学的校训，另外考虑到"创新是民族的灵魂"，我在自己的座右铭中又加入"求实创新"四个字。从此，我就将"谦逊、勤奋、求实、创新"作为我做人做事的准绳及奋斗目标。

在天大这八年中我遇到了很多好老师，他们对我的成长起到了重大作用。著名教授王守融、刘豹、蔡其恕，研究生导师吴又芝，还有孙祖宝、周昌震、陈林才、齐植兰、李维铮等老师对我的科研工作给予了悉心指导及重要帮助。

研究生毕业后，我留校工作至今。在此期间，1997年我也幸运地获得中国科技界最高学术称号（终身荣誉）——中国科学院院士，还成为北大、清华等著名高校的兼职教授。

回首我的求学之路，从陡门桥小学到胶南初级中学，从苏州中学再到天津大学，我所经历的都是人杰地灵、英才辈出的好学校，遇到的都是一批批关心我、爱护我的好师长。回首科研生活，我遇到的是一批批在科研道路上与我并肩披荆斩棘、奋力拼搏的同事们，还有一批刻苦钻研、勇于创新的学生们。而我又何其有幸，赶上了中国改革创新、努力实现中华民族伟大复兴"中国梦"的新时代。

第三章　海外进修

从斯坦福大学到XMR公司

1965年，在我研究生毕业的时候，正好赶上"四清"运动，后来爆发了"文化大革命"，学校的教学科研秩序被破坏，我们这些教师和科技工作者也失去了十年的宝贵光阴。"文化大革命"结束后我一直在想，要把失去的时间补回来。学校复课后，机会终于来了。1980年，天津大学根据中央精神，决定派我作为访问学者前往美国斯坦福大学、加利福尼亚大学进修及研究激光技术。出国前，我在天津大学激光教研室开展激光领域的学习和研究已有近十年之久，学术基础还是不错的。

1980年5月在出国前，我作为当时中国在激光领域最年轻的学者出席了在美国召开的两个国际会议，并且在"国际微微秒现象会议"上宣读了论文。这是中国学者第一次在该领域的会议上宣读论文。在得知自己的论文被录用为口头报告论文后，我的心里一直惴惴不安。我担心自己的英语口语比较差，怕在会上的提问环节听不懂专家所提的问题。当我愁眉不展的时候，前一天晚上遇到的美国巴尔的摩大学华裔教授黄文广对我说："你放心，我可以当你的翻译。"在我做完报告之后，黄教授主动站出来帮助我翻译，帮我过了一大关。在这次会议上我还结识了斯坦福大学的西格曼（A. E. Siegman）教授，他提出邀请我去斯坦福当访问学者。1980年年底，我在美国的加利福尼亚大学戴维斯分校和伯克利分校短期访问。当Siegman教授得知我已到了美国后，就正式邀请我去斯坦福大学，我开始了近两年的访问学者生涯。

初到美国，"语言关"是我前进道路上的一道坎。

初中三年，我们的外文课学的是"英语"。考入苏州高中后，由于那个时代全国上下都在学苏联，因此我高中三年外语课学的是俄语，大学学的也是俄语，研究生第一外语还是俄语，只在第二外语课堂上学了一点英语。所以我的英语底子较差，尤其是口语。到美国以后，问题就暴露出来。为了补习英语，听懂老师讲课的内容，我想了很多办法。在斯坦福大学听Siegman和哈里斯（Harris）教授讲课时，我就早早去，坐在第一排，用录音机全程录音，回去放录音，一句一句反复听，反复练习。当地政府有一种为来自非英语国家（比如日本、印度、墨西哥等国家）的人开的免费英语补习班，我报了名，很快从初级班升入中级班，一个月后升到高级班。因为参加英语培训的学员来自不同的国家，所以大家把学习班称为"联合国"。培训班课程中间休息的时间很长，我就利用这段时间多与学员聊天，这对提升英语水平也大有好处。记得有一次傍晚，天降大雨，我骑着自行车，打着雨伞去上课，在斯坦福大学校门前，因为修路，道路泥泞，我一不小心连人带自行车、伞及书包摔入泥坑里。我从坑里爬出来，不顾一身狼狈，还是准时赶到了补习班。其他学员看到我的样子，都很惊讶，但我还是坚持听完了课。经过了几个月的补习，我也算"毕业"了。斯坦福大学还有一项专门提升来自非英语国家的研究生、访问学者英语水平的制度——"英语伙伴"（English Partner），即让我们这样的访问学者与一位美国学生结对子，建立练习英语的合作关系，这对双方有益。一位来自夏威夷的学

图3-1 1982年月1日姚建铨在美国斯坦福大学进修时于胡佛塔前留影

生，每周跟我见面，为我讲解英语日常词汇，我们也成了朋友。这样的伙伴关系一直持续了近两年，我在英语的口语方面有了不少提高和长进。

在斯坦福大学期间，Siegman教授对我特别有耐心，总是不厌其烦地为我讲解业务。我一边听Siegman教授的"激光"、Harris教授的"非线性光学"等课程，一边参加学术活动。

我还克服了使用电脑及软件上的困难，经常一天在实验室里工作10多个小时，有时回住所时怕因为累而导致骑自行车时摔倒，我不敢骑车，只能推着走。功夫不负有心人，我终于在"四波混频与相位共轭谐振腔的理论"项目的研究中得到好的结果，撰写的论文发表在 *Applied Physics B* 上。此文章的数据及结论对于采用四波混频相位共轭谐振腔的设计及制作具有指导意义。该文章也是对Siegman教授激光谐振腔理论的发展。同时我把飞秒（fs）激光超短腔最佳运转参数的论文发表在 *Applied Physics Letters* 上，此论文对于超短腔染料激光器的研究具有理论指导意义。

在斯坦福大学工作一段时间后，经Siegman教授介绍，我转去XMR公司从事激光倍频的研究工作。在XMR公司，我的主要任务是研究一种能够部分取代大功率氩离子激光器的新型高功率倍频绿光固体激光器。因为传统的氩离子激光器输出功率虽能达到10~20 W，但是运行电流大，预热工作时间长，维护困难。那时倍频的固体激光功率只有1~2 W，甚至更低。当时国际上由美国杜邦公司刚研究出来一种新型晶体钛氧磷酸钾（KTiOPO$_4$，简称KTP）。是否有可能用KTP晶体作为倍频器，用固体激光器来获取高功率绿光激光？这是当时的研究方向，也是世界级难题。经Siegman教授推荐，我去该公司兼职研发，同时每周还回斯坦福大学参加Siegman教授的组会（Group Meeting）。

与杜邦公司较量中小试牛刀

我确定了采用带有声光Q开关的KTP晶体作为内腔倍频器的Nd:YAG激光器的总体方案。接下来需要面对一系列难题：研究KTP晶体的非线性光学特性及如何实现相位匹配？怎么进行设计计算？怎么切割？如何运用？首要的问题是对KTP晶体的设计计算。我从杜邦公司了解到KTP晶体的非线性参数，又在斯坦福大学图书馆查阅了大量文献资料。因为KTP是一种双轴晶体，我经过反复思考，从双轴晶体的折射率椭球及光率体椭球，分析出KTP晶体的双轴晶体的特性不能做解析解，只能做数值解，通过对有效非线性系统、走离角的分析计算，最终解决了它的最佳非线性相位匹配的问题，求得最佳相位匹配角度为：$\theta=90°$，$\phi=21.3°$。我还请公司副总裁、光学博士泰德（Dr. Ted）配合我做了软件的设计。

此时KTP晶体已由杜邦公司的专家设计并切割后转到公司，在实验中我发现他们切割的角度与我计算的不同，有很大的偏离。经联系后，杜邦公司的专家不承认他们的设计有误，认为测试方法有误。我请专家从纽约飞到加州XMR公司，一起做了实验，经过反复验证确认实验无误。我坚持认为问题出在计算方法上。杜邦将我的计算公式、方法带走，核对一周后终于来信承认是他们的方法没考虑走离效应。经过这一场小小的测试，我的最佳相位匹配的计算理论及方法经受住了考验。在XMR我解决了设计、计算、稳功率、晶体冷却等一系列问题。投出去的论文后于1984年被 Journal of Applied Physics 接收发表。此文为多国学者引用数百次，是双轴晶体最佳相位匹配的原创性论文。

获得绰号"Green"

完成最佳相位匹配的设计计算之后，我在实验室中得到的倍频输出逐步提高，功率从 0.5 W→1 W→3 W→5 W→9.1 W，达到当时绿光输出的国际最高水平。这近10 W的绿光输出功率惊动了XMR公司。公司上下均来实验室现场参观，还给我起了一个绰号"Green"（绿光）。在研究过程中还发生了一个KTP晶体差点丢失的小故事。KTP晶体在当时是十分珍贵的，价格畸高不说，还属于"非卖品"。我早期用的晶体的尺寸是 3 mm × 3 mm × 5 mm，一次因为不小心，KTP晶体被我掉落在光学平台上，平台上有一系列间隙为20 mm、直径10 mm的螺纹孔，小晶体有可能掉入螺纹孔。当发现一小块KTP晶体不见时，我很紧张，因为我是外国人，如果找不到，就说不清楚了，会被怀疑是故意拿走的，有拿回国作为"晶种"的嫌疑。当时公司找来几个人把光学平台反过来，用木榔头敲，搞得沸沸扬扬，最后还是泰德博士仔细，是他在实验室地板角落里找到了这块KTP晶体，洗清了我的不白之冤。这件事，也让我暗下决心，一定要让我们国家有自己的KTP晶体。

与XMR公司的工作合同结束时，XMR公司的正副总裁与一家使用该绿光激光器作为医疗仪器的公司的正副总裁给我开了一个告别会。会上他们希望我能留在美国工作，我说我是中国人，要回去报效祖国，他们也没有强留。我记得我在会上发言，感谢公司给我这次研发高功率绿光激光器的机会，我祝愿中国人民与美国人民的友谊，像喜马拉雅山和洛基山那么高，同长江和密西西比河一样源远流长。公司还为我保留了一个职位——高级研究员，我在任何时候来均可以。数年后我研发的高功率绿光激光器，经斯坦福大学的一位博士后开发为高档医疗仪器，在国际市场有很大的销售量。

第四章 回国创新

归国，做自己的KTP

看到KTP晶体对高功率固体倍频激光器的作用，我在美国时就暗下决心，回国后要帮助祖国做出我们自己的KTP晶体，用我们自己生成的KTP晶体来从事高功率固体倍频激光器的研究。我在公开发表的论文中搜集了很多相关科研资料，整理并保存下来。回国后，我没回家，在到达北京的第二天，我就跑去见了北京人工晶体研究所的沈德忠教授，给他看美国生成KTP晶体的相关资料，沈教授后来也开始了对新型晶体的研发。

此后不久我在南开大学又见到了山东大学晶体所的蒋民华教授。他曾在沈元壤教授的建议下开展过用水热法生长晶体的研究。但该方法需要高压,由于不熟悉工艺而发生了爆炸,这项研究被终止了。我建议蒋教授采用溶剂法生长晶体,并联合申请了国家自然科学基金,获得批准,从此开始了两年多的合作研究。后来我把在天大申请到的天津科委给我们的3.6万元的经费也投入项目研究中。开始时,由于晶体太小没法磨,我们把生长出来的小晶体粘在一根筷子的头部,用手拿着,对准激光器的激光束,一边转,一边观察。近半年后,当看到有绿光出现时,在场的人高呼了起来,这是在中国采用KTP晶体倍频所看到的第一缕绿光。以后生长出的晶体慢慢变大,绿光慢慢变强。再后来,山东大学研究生长晶体的老师索性让研磨晶体的师傅一起进驻天大,一边试验,一边打磨晶体的角度,经过一年多的反复试验,在天津大学和山东大学的共同努力下,终于成功得到了高质量的KTP晶体及倍频激光器。

1985年2月5日,在我回国两年后,由天津科委主持,我们与山东大学晶体所在天津联合召开了成果鉴定会,当时我们的绿光输出功率达到8.7 W这一国际先进水平,比我在美国做到的9.1 W仅差0.4 W。参加鉴定的专家包括10余名教授、副教授、高级工程师,大家给出的鉴定意见是该器件"达到国际1982年水平",在准连续泵浦方面有"创造性,国际上未见报道"。

百瓦功率输出,国际领先水平

项目成功结项后,后续的研发工作一直在继续。天津大学从倍频晶体的最佳设计、倍频激光器的设计、谐振腔的结构、倍频晶体的热效应分析及功率稳定性等方面开展研究;山东大学从晶体生长工艺改进等方面继续研究。在双轴晶体最佳相位匹配理论及类高斯分布理论的指导下,我们就声光Q开关KTP内腔倍频Nd:YAG激光器,进行试验研究,有几点创新:

①倍频晶体的最佳相位匹配的理论设计及方法,倍频激光器的最佳腔形结构设计;
②分析解决了KTP晶体的热效应及冷却技术;
③提出了输出绿光的最佳耦合;
④通过对声光Q开关调制频率的反馈控制,提高了输出功率稳定性。

在研究中,张大鹏博士做了大量精细的试验及技术方面的工作。

图4-1 高效倍频Nd:YAG激光器钛氧磷酸钾(KTP)晶体鉴定会现场

图4-2 高效倍频Nd:YAG激光器装置

图4-3 准连续泵浦内腔倍频钇铝石榴石激光器获得国家技术发明二等奖（左图）和布鲁塞尔尤里卡世界发明博览会金奖（右图）

图4-4 姚建铨获比利时国王颁发的最高荣誉——一级"骑士勋章"

后来，在1987年组织的科研成果鉴定会上，我们已做到绿光（532 nm）的输出功率达到31.4 W的国际领先水平。当时新华社及全国10多家报刊均报道了这一项重大成果。该项目先后获得国家技术发明二等奖、国家教委科技进步二等奖及天津市世界进步二等奖共四次，又获得尤里卡世界发明博览会金奖，我个人获比利时国王颁发的一级"骑士勋章"。就这样中国的KTP晶体研发成功了，打破了国际垄断，杜邦公司反过来买中国的晶体，还发了杜邦奖，KTP晶体从火箭价变成了白菜价。

那时试验条件有限，除了在技术上攻克了一道道难关，师生们还克服了很多实际困难。高功率激光器要冷却，采用内外循环水，内循环可以用"泵"，外循环要用自来水，我们在一个破旧的仓库内做试验，室内没有水源，隔壁是女厕所，我们只能在女厕所内接一个水龙头，用塑料管引过来。因为怕有人使用厕所时关了水龙头阀门，所以在做激光试验时，总要派一个学生坐在厕所门外守着，以免水源被切断。我们就是在这么艰难的条件下，获得了国际领先水平的研究成果。

沈德忠教授曾在一篇论文中指出：KTP晶体在中国的研制成功，姚建铨是"功不可没"的。后来蒋民华教授谈起此事也常说，在KTP晶体的制备过程中，姚先生功不可没，他是参与者、探索者，也是应用者。这件事在我们光学及非线性光学晶体材料界被传为佳话。我和这两位非线性光学晶体专家——蒋民华及沈德忠均先后成为院士，我们被评为院士与该晶体的成功研制有很大关系。这既说明研制中国品牌的晶体及激光器属于国际前沿，更说明我们正是在做着前人没有做到的有意义的工作。

20世纪80年代，我从倍频理论出发，对国际上有关激光倍频的理论进行了总结及分析，提出了在不同情况下的八个公式，发展了倍频的理论，此外对于皮秒（ps）、飞秒（fs）等超短脉冲激光的倍频也有创新的分析及发展，在激光倍频理论的发展方面做出了若干贡献。

提出并发展了"类高斯光束理论"

在发展高功率倍频激光器的同时，我提出并发展了激光谐振腔的类高斯光束理论。该理论是怎么提出的呢？如果只考虑倍频效率，单基模光束（TEM_{00}）的高斯光束倍频有最高的效率，但实际的激光器都运转在高阶模或有很多模混合的状态，称为混合模。采用新型双轴晶体KTP做内腔倍频激光器的研究时，KTP放入腔内对模式有何影响？如何计算及考虑混合模倍频的效率问题？我们有必要对谐振腔的模式进行仔细的研究。1984年以

图4-5 姚建铨与学术同行交流高效倍频$Nd: YAG$激光器研制经验

前，以Siegman教授为主要代表的科学家提出的谐振理论仅有某一高阶模的模场分布公式，谐振腔理论中只有对某一特定高阶模（q,m,n）的光斑描述。我与研究生薛斌从谐振腔理论中的多模混合模（Mixed Modes）开始，在数学教授齐植兰的帮助下，从厄米特-高斯及拉盖尔-高斯两种分布函数的递推公式及正交归一化条件出发，求得混合模的分布特性。该特性需要用很复杂的公式来表示，经简化后可获得混合模的光斑半径w_{mn}为基模光斑半径w_{00}的M倍（M为混合模系数），即$w_{mn}=Mw_{00}$。光束在传输过程中的任何空间位置（z）都可看成一种新型的高斯光束，称为"类高斯光束"或"类高斯分布（Gaussian like Distribution）"，这就是我提出的"类高斯光束理论"概念及要点。我们的理论及定义的多模系数M与Siegman教授用二阶矩算出的M^2是相当的，相关论文发表在我国《量子电子学》1985年第一卷第二期首页。在发表前我曾将此文译成英文发给斯坦福大学的Siegman教授，请他指正。有趣的是，四年后的1989年，Siegman教授采用二阶矩的方法也求出了混合模的M^2公式，与我们的结论相似，企业界也发展了M^2测量仪器。我认为同一个科学问题可以用不同的方法解决，所谓殊途同归，我与Siegman教授在类高斯光束研究方面的结论有异曲同工之妙，真可谓英雄所见略同吧。

图4-6 姚建铨科研团队研制的20 W高稳定性的KTP内腔倍频N_d: YAG激光器

图4-7 姚建铨科研团队研制的100 W高功率、高稳定性的KTP内腔倍频N_d: YAG激光器

我们在此理论基础上又提出了类高斯光束的理论模型，发展了类高斯光束在均匀介质中的传播与变换理论，完善了类高斯光束的聚焦、准直、透镜变换以及类高斯光束的倍频与光纤耦合技术。对于混合模系数M，当$M=1$时即为单模TEM_{00}模，M数值越大则高阶模越多。我们给出了混合模系数M的测量方法，该方法很简单，只要在类高斯光束传输的两个位置（z_1、z_2）上测出两处的光斑半径及功率，根据测得的四个指数就可用我们的方法算出类高斯光束的基模光斑半径w_{00}及混合模系数M，这是很实用的技术。我们提出的上述方法是M及M^2测量仪的理论基础。

类高斯分布理论只是激光谐振腔理论的一种发展，它本质上与腔内各个模之间的耦合、相干性均有关系。对此，专家有两种不同意见：一种认为各个模式之间完全相干，另一种认为完全不相干。我们后来的发展是在原有的基础上引入一个"相干系数"——$K(m, n)$，当$K(m, n)=0$时完全不相干，当$K(m, n)=1$时完全相干。通过引入从0到1之间变化的相干系数K，我们的类高斯分布理论得到更好的理解及更完整的表述。在该理论基础上，我们创新性地研发成功了由高功率绿光激光器泵浦的两种可调谐激光器——染料（Dye）可调谐激光器和钛宝石（Ti: Sapphire）可调谐激光器，建立了一个新的技术体系——准连续泵浦-可调谐激光技术体系。

1991年至1993年间，我们直接用高功率绿光激光器泵浦研制成功了两种可调谐激光器。染料可调谐激光器方面，采用染料双调Q技术，激光输出脉宽范围覆盖10～30ns和100～300 ns，脉冲重复频率1～10 kHz。该可调谐激光器用于国家重大项目——高空机载遥感实用系统中"激光三维测距系统"，获中科院科技进步奖特等奖（1993年）。钛宝石可调谐激光器方面，激光脉冲重复频率5 kHz，波长在750～870 nm范围内可调，输出功率2.58 W，转换效率21%。该可调谐激光器在激光致盲研究中获得军队科技进步奖一等奖（1996年）。

由此，我创新性地建立了双轴晶体最佳相位匹配—类高斯光束理论—准连续泵浦高功率内腔倍频激光器—高功率绿光激光器泵浦的染料及钛宝石可调谐激光系统这一崭新的激光技术体系。

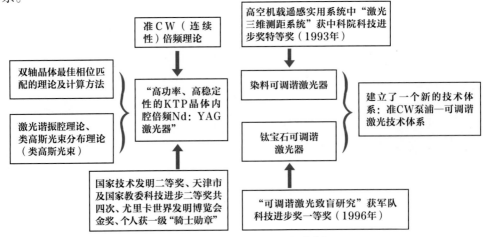

图4-8 姚建铨团队建立的创新性激光技术体系

高功率绿光激光器还有两方面的应用：一是利用血红蛋白对绿光的敏感吸收，开发出以绿光激光器作为光源的胃出血观察仪；二是由于绿光波段在海水中传输能力强，即衰减系数较低，绿光激光器可以应用于海洋中的通信、检测及成像等。

建设一个能战斗的集体

任何重大科研成果的取得都不可能是一个人战斗的结果，我致力于将我领导的研究所建设成为一个分工明确、各负其责、相互协调、具有浓厚学术气氛的讲团结、能奋战的科研集体。

1988年，由我提议，天津大学批准建立了"激光与光电子研究所"。从1988年到1998年近十年的时间里，我除了曾去美国宾夕法尼亚大学和普林斯顿大学开展合作研究两年多，以及参加过几次国际会议外，

图4-9 1990年姚建铨在普林斯顿大学开展学术研究

我把绝大部分时间都放在了研究所科研项目的组织和协调工作上。当时研究所承担的科研项目较多，如千瓦级YAG激光器、多波长钛宝石可调谐激光器、光学参量振荡器、超高速流场测速仪、周期极化晶体频率变换、固体电源等。我的团队中博士研究生和硕士研究生人数也很多，怎么管理才高效？我想出了分工负责的管理方法。我以"项目"为单位建立一个大组，将教师、博士、硕士分入不同小组。大组中由我决定总体方案，各小组分别具体研制，小组一起向大组汇报，共同讨论决定解决办法，总结及鉴定工作由大组负责。这样可以发挥

图4-10 姚建铨科研团队的师生在天津市蓟县（现蓟州区）黄崖关长城开展团建活动时的合影

图4-11 2009年6月姚建铨重返普林斯顿大学与学术同行合影留念

每位科研人员的特点，既分工又合作，建设成了一个既有浓厚学术气氛，又有奋发向上精神的战斗集体，使科研团队承担的每项科研任务都能圆满完成。如我们很快在两种可调谐激光的技术上取得突破，钛宝石及染料可调谐激光在激光致盲研究及机载激光工作中得到了重大应用，分别获得军队科技进步奖一等奖（1996年）及中科院科技进步奖特等奖（1993年）。

除了抓科研工作外，我还十分关心师生的身体健康及生活娱乐。研究所经常组织师生打球、下棋、唱歌等，如组织过两次去水上公园的游园活动，其中一次是"English Party"，提高了大家的英语水平；还组织大家到天津市蓟县（现蓟州区）黄崖关长城开展团队建设活动。

用姚方法测量不稳定流场

20世纪90年代初，我曾访问美国著名的斯坦福大学和普林斯顿大学。在访问普林斯顿大学航天学院迈尔斯（Miles）教授的实验室时，我看到一种国际上最新的不用外加荧光指示剂的激光测量高速空气流场的方法，即拉曼激发激光感应电子荧光法，简称RELIEF方法，其是对风洞流场进行诊断的光学方法。我在参观时发现了一个大问题：实验中所用荧光标记是相隔100 ms延迟时间的两次不同的流场测量，对于稳定流场还算可以，但对于不稳定流场是不行的，该项目原理是有问题的。当时我就找Miles教授提出了疑问。他听了，就问有办法吗？我说可以用序列脉冲激光来解决此问题。我提出用间隔为15 μs的序列脉冲测量流场。我的方案得到了Miles教授的肯定，他决定让我开展用RELIEF方法测流场的研究。就这样，我在普林斯顿大学做了近一年半的实验。

回国后，从1995年至1996年，由中国航天工业总公司支持，我的科研团队也进行了RELIEF高速流场测量的实验研究。由于采用序列脉冲测量高速流场的方法是对Miles方法的发展，学生们把我创新提出的测量方法称为YAO-RELIEF，简称Y-RELIEF方法。我们研制完成了简易的风洞实验装置，完善了三套激光系统及ICCD（增强型电荷耦合器件）测量装置，抓到了流场中的氧分子流，完成了原理试验样机。在此基础上，科研团队又发展了利用激光瑞利散射技术测量空气流场技术等。

准相位匹配＋RGB光源→激光电视技术

2001年至2003年，国际上出现了一种新型匹配技术——准相位匹配技术，它是利用调节光学超晶格（也称为声学超晶格或微米超晶格）的周期，弥补由于折射率色散而产生的波矢失配，可以拓宽非线性晶体的应用范围，大大提高了非线性光学转换效率。我的科研团队提出了在PPLN-OPO（周期极化铌酸锂晶体—中红外光参量震荡技术）中四波混频（FWM）的产生过程，初步建立了PPLN-OPO-FWM的理论体系，对实验得到的750 nm波长光谱谱线给予了解释与分析，发表了多篇高质量的论文。

同时科研团队也为实现红绿蓝三原色电视做了不少工作，取得了重大突破。将非线性频率变化技术及固体激光技术有机结合，所用晶体从KTP到LBO，再从BBO到

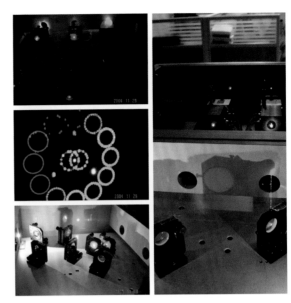

图4-12 姚建铨科研团队开展红绿蓝三原色激光电视研发

LiNbO$_3$，激光波长从1 064 nm（倍频后又增加532 nm绿光）到946 nm（倍频后又增加473 nm蓝光），再到1 320 nm（倍频后又增加660 nm红光），激光波长扩展到红、绿、蓝。在此基础上，又扩展到研究红绿蓝三色激光电视。我们采用瓦级以上的红、绿、蓝全固态激光器作为光源，合理计算三基色的功能配比，采用多面转鼓和振镜分别实现行、场扫描，制成了扫描面积为5 m^{2}的激光电视样机，显示的彩色视频图像亮度高、色彩鲜明。

一直到20世纪80年代末，激光与光电子研究所还在致力于提高绿光功率。到了2002年和2003年，经过泵浦KTP冷却、声光Q开关等技术的改进，我们的高功率绿光激光已达到百瓦输出，最高可达150 W，进入国际先进行列。

第五章　新的起点

聚焦前沿，锁定太赫兹

1997年10月，我当选为中国科学院院士。我把当选院士看作国家对我全面的肯定，但更重要的是我把它作为我在科研道路上继续奋进的新起点。

激光与光电子研究所在科研方面已取得长足的进步。其中最擅长的高功率绿光激光器已到百瓦量级；在长脉冲高功率Nd:YAG激光器方面，我们利用三根棒串接，也达到了1000W以上；从一般的角度相位匹配发展到周期极化晶体的准相位匹配；重大应用领域扩展到了高速流场的检测技术及红绿蓝三色激光电视等。

虽然在熟悉的领域已经取得了一定成绩，但结合国际科技发展趋势，考虑到发展新的电磁波段的战略意义，我毅然做出了全所大力投入研究太赫兹技术的决定。2004年开始太赫兹

的研究，这在全国都是较早的。

太赫兹波（THz Wave）是指频率在0.1~10 THz范围内的电磁波，介于微波和红外波波段之间。从电磁波看它是亚毫米波，从光波看它是远红外波，因此太赫兹波本质上是处于宏观电子学到微观光子学之间的频段，兼有微波和光波的特性。太赫兹波具有很多优点，如低量子能量、大带宽、良好的穿透性等，是大容量、超宽带、实时无线传输最有效的技术手段。太赫兹通信与微波通信相比，带宽大、信息容量高、载流频率高、能有效穿透等离子体鞘套、波长短、易于实现小型化。与激光通信相比，太赫兹波束宽度适中，对平台稳定度和跟瞄要求低。大气对太赫兹波的吸收较强，有利于实现空间保密通信。太赫兹是人类最后一块没有得到充分认识及应用并未开垦的处女地，已引起了各国科学家与工程技术人员的高度关注。当时欧美国家及日本等已经把太赫兹技术列为国家科技的发展重点。我们将研究的重点放在太赫兹波的产生、调制、放大、滤波、探测及天线设计，并努力实现太赫兹波关键器件的小型化、集成化、芯片化及产业化。

在研究中我逐步发现由于太赫兹波具有特殊的性质，除了在检测、科研中可以有广泛的应用外，其在通信、雷达等方向也有很多潜在的应用前景。2009年我与迟楠教授等撰写了一篇论文《太赫兹通信技术的研究与展望》（《中国激光》，Vol.36，No.9，2009），在文章中我曾预言太赫兹技术有望在空间及地面通信中得到应用。这篇关于太赫兹通信预测的文章被下载5 000余次，被引用约240余次，被光学学会授予"高被引论文"奖。

我通过研究太赫兹技术的特点、发展现状及前景，结合我们团队的特色，决定从研究太赫兹源技术着手。太赫兹源有电子学（真空电子学、固态电子学等）及光子学的源，还有用X-ray（X射线）和自由电子激光来产生太赫兹辐射等。我决定从光子学源起步，即将采用非线性光学差频方法产生太赫兹源作为重点研究方向。我们可以用两台激光器或一台激光器泵浦光学参量振荡器发出的信号光及空闲光在非线性晶体中差频来产生太赫兹辐射，所以其中最重要的是需要合适的激光器及非线性晶体材料。

因为预见了太赫兹的广阔应用前景，2005年11月底，我们联合了10多位院士及数十位代表，在北京召开了关于"太赫兹科学技术"的第270次香山会议。该会议是我国开展太赫兹科学技术研究的新起点。我作为会议执行主席之一，在会议上做了主题为"用光子学方法产生太赫兹源"的报告。我们的团队在太赫兹光子源领域做了大量工作，先后完成了国家重点研发计划、国家自然科学基金项目等10余项。我又联合上海大学及东南大学的科研人员共同申请到关于太赫兹调控方面的国家自然科学重点基金项目，还完成了小型化光子太赫兹源的实用化样机。

我在学术上，总能在适当的时候提出一些超前的思想。比如在2008年，申报国内移动通信项目时还有一个小故事。我与复旦大学迟楠教授联合申报了太赫兹用于移动通信的项目。答辩时专家组组长说："姚院士，你们怎么在'指南'中未提到太赫兹通信的情况下来报项目？"我回答说："我是'醉翁之意不在酒'，我知道'指南'中没有提到的项目不能申请,我只是想让你们专家组知道,在中国有一批搞太赫兹的人,很可能在未来某一天,太赫兹就会成为移动通信的主要频段!10余年后的今天,国际上所有从事高速通信(5G和6G)的国家都把太赫兹通信作为首选方案。

2010年,由第三军医大学及中国工程物理研究院发起召开了太赫兹中物理与激光交叉学科学术会议。第一次会议在绵阳召开,我应邀参加,在会上我提出一个构想,把太赫兹技术与生物医学相互融合,两者结合起来开展研究。这个想法得到与会代表的赞同,以后每年一次的交叉学科会变成专门讨论太赫兹技术与生物医学相结合问题的会议。到2013年,我建议在全国以此为主题召开香山会议,由我联合几位院士共同发起会议申报。经过不懈的努力,2014年4月8日在北京召开了"太赫兹在生物医学应用中的科学问题与前沿技术"的香山会议,我作为首席执行主席,在会上做了主题报告。该会议受到全国学术界的重视,我又不失时机地提出联合申请与此主题有关的"973"项目。经过多年努力,由第三军医大学、中国工程物理研究院、天津大学和吉林大学等单位共同申报的"973"项目于2017年获得批准立项,在我国兴起了太赫兹在生物医学方面应用研究的热潮。此后,天津大学又在太赫兹脑科学研究领域申请到另一项"973"项目。我还帮助枣庄学院开展了将太赫兹高灵敏度超材料传感器应用于癌细胞凋亡率检测技术的研究,取得重要成果,发表了数篇重要论文。

中国有关资源库对世界范围内的学者在太赫兹方向发表的论文进行了统计。2015年至2019年,全球太赫兹科学技术成果分析中,发表论文共13 904篇,分布在25个国家,分别为:中国4 933篇,占42%;美国2 650篇,占17%;以下依次是德国、日本、俄罗斯、英国、法国、韩国、印度和伊朗。以研究机构发表相关论文数量排名依次为:中国科学院、俄罗斯科学院、法国国家科学研究中心、电子科技大学、美国能源部、加州大学、天津大学、德国亥姆霍兹联合会、马普学会、中国科学院大学。以学者个人发表论文情况看,我发表的论文连续五年在每年发表论文数量Top10榜单中排首位,五年发表论文总数亦排在首位。这个数据从侧面证明我领导的团队及助手、学生们在论文发表方面还是取得了很好的成绩。不过虽然论文数量多,但在顶级期刊发表文章数量及引用次数还有待提高。

"空—天—地—海"一体化构建智慧海洋

2015年初,国家网信办在北京召开了有关我国国家网络基础设施规划的座谈会,在会上,我提出了我国应规划构建"空—天—地—海"一体化网络系统的建议,受到国家网信办领导及参会专家的一致好评。我提出的理由有四个:一是地球上海洋面积占地球总面积的71%;二是中国是海洋大国,海洋面积300多万km^2;三是中国在东海、南海与多国存在争端,这些国家背后均有美国的支持;四是习近平主席提出的"丝绸之路"构想中也包括"海上丝绸之路",维护海洋权益及开发海洋是我国的战略决策。而其中陆上的互联网已很发达,空间也有微波及激光通信等,恰恰是海上和海下的网络很少,也很不成熟,现在的网络都没有覆盖沙漠、高山及海洋。所以未来要实现万物互联,构建包括"空—天—地—海"一体化网络系统是我国网络基础设施规划的重要方面。

当晚,国家网信办主任给我打电话,告知与会专家一致同意我的意见。网信办决定由我牵头,联合一些单位成立专题调研组开启进一步的论证。我接受了此任务,很快成立了由天津大学、北京邮电大学、中国海洋大学、南开大学、航天九院十三所、航天十一院、中国计量学院(现中国计量大学)、山东科技大学等八家单位组成的调研组,分别在天津、北京、青岛等地多次召开座谈会,于2015年12月21日公布了题为"海洋网络空间基础设施研究"的调研报告,并报国家网信办及中央有关部门。现在,构建"空—天—地—海"一体化网络系统的概念及构想已在全国达成共识了。

推动中国物联网技术及智慧城市健康发展

2013年10月,我应邀担任了中国智慧城市建设投资联盟和深圳市智慧城市研究会荣誉主席、首席科学家,这是在深圳成立的全国最早的智慧城市建设投资联盟。

我那时就提出智慧城市必须要做好顶层规划,积极推动中国智慧城市健康发展。2013年初,从住建部发布第一批智慧城市试点城市开始,中国智慧城市建设拉开了帷幕。在我国政府、企业以及相关机构都在探索智慧城市如何建设发展的大背景下,社会上逐渐出现了拿智慧城市理念和题材做政府项目的现象。在利益的驱使下,有些地方政府被动地启动建设了一批与本地需求不相匹配的数据库基础设施和部门信息化建设项目,成为新的"信息孤岛"。建设智慧城市本应该做到互联互通、信息共享和避免重复建设。可这种盲目上马的项目适得其反,陆续出现了诸多浪费现象,智慧城市建设受到了政府和社会的质疑。针对当时这种情况,我首次提出智慧城市建设必须要做好需求调研的顶层规划。在中国智慧城市建设投资联盟和深圳市智慧城市研究会理事会上我特别强调,联盟要积极按照国家发改委等八部委下发的《关于印发〈促进智慧城市健康发展的指导意见〉的通知》(发改高技〔2014〕1770号)文件要求,健康有序地推进智慧城市建设。

我还顶住各方面压力,力挺深圳市编制全国首部地方团体标准《智慧城市系列标准》,并积极参加评审和推广工作。2013年,当时中国的智慧城市建设还处在探索初期,没有相关

图5-1 2014年10月18日,《智慧城市系列标准》国家级专家评审会在深圳举行

图5-2 2015年11月15日，《智慧城市系列标准》信息发布会在深圳举行

图5-3 2015年12月13日，《智慧城市系列标准》出版发行会在北京举行

的标准和规范。以我为首的联盟专家委员会，积极支持联盟在企业自愿参与的情况下牵头编制（制定）智慧城市联盟标准。为此，时任深圳市智慧城市研究会会长的李林教授和执行会长张晓新教授积极筹备，于2013年12月开始带领106家各行各业的企业和科研机构联合编制（制定）了深圳团体标准《智慧城市系列标准》，该标准第一卷共包括54个子系统。在2014年10月召开的国家级专家评审会上，专家们对《智慧城市系列标准》提出了多项不同意见，辩论异常激烈，本次评审岌岌可危。在此情况下，我力排众议，提出联盟《智慧城市系列标准》是自下而上的由企业自愿编制的核心专利标准，填补了中国智慧城市目前没有标准可依的空白，它至少有四大创新，即标准体系创新、编制内容创新、编制模式创新、产业发展创新。我在评审会议上强调，哪怕只是联盟标准，这也是深圳市社会组织和企业家本着家国情怀，愿意为中国智慧城市建设探索分担责任，愿意为国家智慧城市标准的出台先行先试所贡献的经验。在我的积极支持下，联盟《智慧城市系列标准》于2015年11月和12月分别在深圳和北京两地隆重发布，工信部中国电子信息行业联合会领导莅临现场指导并给予高度评价。

对中国智慧城市建设的探索已经从智慧城市到新型智慧城市，再到数字政府、智慧民生和智慧产业。在《智慧城市系列标准》专家评审会上，我从智慧城市的复杂性、系统性、科学性等情况出发，对深圳智慧城市建设投资联盟编制的联盟《智慧城市系列标准》提出了四个"W"的评价，即：Why, When, Where, Who。

Why——我从一开始就见证和指导、参与了深圳智慧城市建设投资联盟的发展，并及时、科学地指出《智慧城市系列标准》就是智慧城市规划与建设的"顶"。没有智慧城市标准体系的指导性、规范性和约束性，就不可能建成绿色低碳、和谐美好、可持续发展的智慧城市。编制（制定）《智慧城市系列标准》的重大意义不言而喻。

When——2014年7月，国家发改委等八部委联合下发《关于促进智慧城市健康发展的指导意见》，呼唤健康发展的标准体系。《智慧城市系列标准》在这个时间节点上发布试行，为智慧城市规划与建设提供了共性先立、急用先行的指导性、规范性和约束性的标准，可有效消除"信息孤岛"和避免重复建设，为智慧城市健康有序发展奠定了坚实的基础。

Where——深圳是我国改革开放的前沿，聚集了一大批敢为天下先、具有创新精神的高科技优秀企业。因此只有在深圳才有条件创造出这样的奇迹。机会总是给有准备的人，2015年国务院发布了智慧城市规划与建设的标准政策，极大地鼓励和推动了团体标准的编制（制定）。

Who——《智慧城市系列标准》是深圳市智慧城市研究会自2013年3月成立以来，带领106家以深圳为主体、有全国各地企业会员用核心专利技术、成功案例和实践经验，采取以政府引导、企业为主自下而上的方式制定的，这是从企业联盟的《智慧城市系列标准》到深圳地方团体标准的《智慧城市系列标准》。这也是我国标准编制（制定）模式的一次创新。

我在全国物联网及智慧城市建设的群众性学会、联会等机构中任职，应邀参加过50多次各种会议，在全国各大中城市发表过30多份报告。对于智慧城市、物联网等新一代信息技术，我不断学习，从不懂到懂，从学习到宣讲，对我国物联网及智慧城市建设，起到了一定的推动作用。

第六章　怀念恩师

我在科研的道路上走了几十年，一路上遇到无数良师益友。在这里回忆与恩师的二三事，向一中一西两位恩师致敬。

王大珩院士

王大珩，祖籍江苏苏州，著名光学家。他1915年2月26日生于日本东京，1936年毕业于清华大学，1955年当选中国科学院技术科学部学部委员（院士），中国科学院长春光学精密机械与物理研究所研究员、所长、名誉所长，是我国光学事业奠基人之一。王大珩在国防现代化所需要的各种大型光学观测设备研制领域做出了突出贡献，对我国的光学事业及计量科学的发展起了重要作用。20世纪50年代，他创办了中国科学院仪器馆，该馆以后发展成为长春光学精密机械研究所。他领导该所研制出我国第一锅光学玻璃、第一台电子显微镜、第一台激光器，并使它成为国际知名的从事应用光学和光学工程的研究开发基地。1986年，他和王淦昌、陈芳允、杨嘉墀联名，提出发展高技术的建议（"863"计划），还与王淦昌联名倡议，促成了激光核聚变重大装备的建设。他提倡并组织学部委员主动为国家重大科技问题进行专题咨询，颇有成效。1992年，他与其他五位学部委员倡议并促成中国工程院的成立。1999年，他荣获"两弹一星功勋奖章"（获得该勋章的有：王淦昌、赵九章、郭永怀、钱学森、钱三强、王大珩、彭桓武、任新民、陈芳允、黄纬禄、屠守锷、吴自良、钱骥、程开甲、杨嘉墀、王希季、姚桐斌、陈能宽、邓稼先、朱光亚、于敏、孙家栋、周光召）。

图6-1 王大珩先生为姚建铨书写的题词

我与王大珩院士第一次较长时间的接触是在1993年6月到7月。当时应俄罗斯光学学会的邀请，中国光学学会组成了由王大珩院士任团长的中国光学代表团，团员有长春光机所王乃弘研究员和我。我们一行三人于1993年6月由北京飞赴莫斯科，第一站访问了俄罗斯科学院物理研究所。我们受到了诺贝尔物理学奖得主、物理研究所所长普罗霍洛夫院士的接待，交谈中双方商定中俄光学学科要开展全面深入的合作。在莫斯科我们还参观了宏大雄伟的红场及莫斯科大学，并在莫斯科大学物理学院看望了几位来自中国的博士生。第二站，我们来到圣彼得堡，圣彼得堡是俄罗斯第二大城，在苏联时期的名字为"列宁格勒"。圣彼得堡是我们访问的重点。我们去了俄罗斯科学院圣彼得堡约飞物理技术研究所，在那里还看到了中国黄昆院士当时在约飞物理技术研究所写的论文。我们还参观了激光研究所，了解了该所"军转民"机制的运行情况。闲暇时，我们还参观了冬宫及郊区的夏宫（也叫彼得宫）。如果拿圣彼得堡与北京做比较，中国的故宫相当于俄罗斯的冬宫；中国的颐和园相当于俄罗斯的夏宫。从冬宫到夏宫，我们看到了伟大的俄罗斯民族悠久的历史和深厚的文化底蕴。

苏联解体后，经济受到很大影响，约飞物理技术研究所楼内的地板坑洼不平，楼外的马路上遍布水坑，路上行驶的汽车的表面满是灰尘，以致在与王先生过马路及上下楼梯时我需要经常搀扶他。不过我们均认为俄罗斯的这种状况是暂时的，由于俄罗斯基础扎实，有深厚的科学功底，他们赶上世界先进水平指日可待。在俄罗斯，我们还召开过一次宴会，俄罗斯光学学会的多位领导赴宴，还自带了伏特加酒。我还记得，在会上我赞扬中俄人民友谊像喜马拉雅山和乌拉尔山脉那样高，像长江及伏尔加河那样源远流长。在俄罗斯访问期间，王大珩院士经常对我说："光学问题我和王乃弘接待，激光问题你来接待"。

图6-2 2004年4月28日姚建铨与王大珩先生共同组织香山会议时的合影

1995年，天津大学（北洋大学）建校100周年，我曾邀请王大珩先生来天大参加校庆纪念活动，他愉快地接受了邀请，出席了隆重的天大庆祝建校百年的大会。

2000年，天津市在天津科委支持下成立了"天津市光电子联合研究中心"，南开大学母国光院士为中心学术委员会主任，我任中心主任。王大珩先生应我之邀专程从北京赴天津主持了中心成立的论证会，对中心的发展做出了指示。2002年在中国科学院大会间隙，王大珩先生为我书写了题词，让我做到"传承辟新，寻优勇进"，这是对我的极大鞭策及鼓励。

2003年9月17日，我与王大珩先生共同作为香山会议执行主席主持了以"激光制造与未来技术产业的发展"为题的香山会议，他在会议上发表了"激光材料加工对未来产业发展意义"的重要讲话。

2004年2月26日，为祝贺王大珩先生90大寿，我和我老伴专程赴北京中关村王大珩先生家中表示祝贺。当天下午由中国光学学会举办了一个祝贺王大珩先生90大寿的庆寿会，王大珩先生愉快地接受了大家的祝福。在王大珩先生90大寿前夜，我给王大珩先生撰写了90大寿的祝贺信，信中既表达了对王大珩先生在中国光学事业所做重大贡献的敬佩之意，更表达了我对王大珩先生多年关爱的谢意。2004年4月27日，在北京国际光电子博览会开幕式上，王大珩先生和我陪同德国光学界的专家到会表示祝贺，当天下午，几位到会的院士王大珩、周立伟、许祖彦、周寿恒、沈德忠和我共同合影留念。

2005年11月10日，在中国科学院大会信息学部会议期间，91岁高龄的王大珩先生、师母与学部的全体院士合影。2011年7月21日，王大珩先生不幸因病去世，终年96岁。光学界为了永久纪念王大珩先生的伟业，在长春光机所内建了王大珩纪念馆，并在馆前建了王大珩先生的塑像。我在2018年7月专程赴长春光机所参观王大珩纪念馆，并在王大珩先生塑像前留影，以表哀悼之情。虽然我并不是王大珩先生的学生，但他对我无微不至的关怀，让我体会到一位有崇高威望的老科学家的高尚胸怀，对祖国光学事业的无私奉献，我从他身上吸取了无穷的动力。王大珩院士永远是激励我前进的良师，他对我的关怀、教导将永远激励我前进。

A. E. Siegman教授

A. E. Siegman是美国斯坦福大学电子系及应用物理系教授、应用物理系及金兹顿（Ginzton）实验室主任、美国工程院院士、美国光学学会主席。他是我1980年到1982年在美国斯坦福大学进修时的指导教授。

我第一次见Siegman教授是在1980年5月的"国际微微秒现象会议"上。会议在美国马萨诸塞州的海边城市Cape Code（科德角）镇召开，在我做会议报告结束后，Siegman教授问我是否愿意到他那里去进修，还给了我一份斯坦福大学的地图。我告诉他，如果国家派遣，我就可以去斯坦福大学。1980年底，我先去了加利福尼亚大学戴维斯分校，在那里我接到了他的邀请函，于是很快就转到斯坦福大学。在那里，Siegman教授给我安排了工作和要用的计算机网络。我的第一项工作是关于"四波混频与相位共轭谐振腔的研究"。在斯坦福大学除了听Siegman教授、Harris教授的"激光"及"非线性光学"课程外，我主要是在实验室从事研究。我的英文水平较差，Siegman教授经常很耐心地给我解答问题，一遍没说明白，就再解释一遍，还给我很多资料，帮助我学习。每周五上午研究组有一个组会，到了组会的时候大家都带着午餐来，一边开会一边吃饭。组会内容很丰富，他会谈研究组方向，有时他也会讲他出去开会的收获，更多的时候是让博士生讲自己的读书报告、研究进度及总结报告，他会一一给予点评、仔细指导，这让我受益匪浅。

Siegman教授有时会让我到他家去玩。一进门他家的小狗就会冲出来，Siegman就把一个网球丢得远远的，小狗马上飞快地去捡球。在家里，他一改在学校时那种非常"绅士"的教授风度，而是表现出一种非常美国式的放松状态。一个周末，他开车带我和他女儿去Lake Tour的度假住处。一路上他十分亲切地和我谈了很多业务及美国名牌大学中博士生如何做研究的事。在他的指导下，我完成了有关"相位共轭谐振腔大信号理论研究"的论文，论文被发表在美国的 *Appl. Phys. B* 上（1983）。我自己还完成了一篇有关超短腔的论文，论文被发表在 *Appl. Phys. Lett.* 上（1982）。

我在Ginzton实验室工作了一年多，后来Siegman教授把我介绍到一个叫XMR的公司，去从事倍频固体激光器的研究。不过我经常回到斯坦福大学来参加他们的组会。

在我离开美国前夕，Siegman教授曾把他在斯坦福大学的"激光"课程讲稿给我，他有意将此书翻译成中文，让我帮忙在中国出版，后来由于版权的问题就放弃了。这本书 *Lasers* 于1984年在牛津大学出版社出版，后来被中国作为优秀国外教材引进到中国。该书是国际上最优秀的关于激光的教材之一，全书近2000页，它是基于经典模型即可对激光物理进行完整、详细和准确的处理，还包括很多习题，无论是理科还是工科师生均十分欢迎。

我回国后有两件事与Siegman教授有关。

第一件事，在从事倍频固体激光器的研究中，我提出了激光谐振腔的类光束理论，推导出了多模系数 M，论文《类高斯分布的理论及其应用》被发表在1984年的《量子电子学》

图6-3 姚建铨在斯坦福大学做访问学者时与导师A. E. Siegman合影

上。之后Siegman教授在1989年提出了谐振腔多模理论，与我们方法不同，但结论相似，可谓殊途同归。

Siegman教授的*Lasers*一书是出现较早的全面阐述激光原理的经典教材，该书让读者从量子电子学的神秘感中解脱出来，从而能轻松地阅读全书。该书没有从量子电子学的角度去论述和进行数学推导，而是从经典的量子电子学的角度论述，这样工科学生学习起来容易一些，但学生需要有本科生的光学和经典电子学的基础。Siegman认为激光物理中的大部分基本概念及所有的应用原理均属于经典理论的范围，因此该书的论述方法主要基于电子振荡理论模型，并适当引入原子、分子的能级结构图和能级跃迁这些物理学者所熟知的量子力学概念，这样就不用要求学生熟悉深奥的量子力学理论，只要掌握了电磁理论、电动力学方法和基础光学知识就能很容易地理解掌握全书所阐述的基本理论。另外，该书物理概念论述清楚、系统，图示和数学表述容易理解，内容丰富，并且该书每一节的后面都有参考文献和思考题，便于学生查阅参考和思考。

第二件事，2005年我承担高功率光纤激光器项目时，采用了Siegman教授2003年提出的GG-IAG大模场光纤激光器，即大模场单模光纤的构想及理论。此外，我还在一个"973"项目中用过他的这个理论。当时国内很多激光项目都应用了他的理论。他的"大模场理论"为我国高功率光纤激光器的发展做出了重大贡献。

功率增益导引技术是Siegman教授于2003年提出的获得更大模场直径的单模光纤的新方法。根据Siegman的理论分析，采用增益导引可以在单模传输的前提下将光纤纤芯提高到几百微米，这个数字对于大功率光纤激光器来讲，是极具诱惑力的。增益导引早已被应用于半导体激光器中，用于产生单模高功率激光。人们曾经对增益导引在半导体激光器和表面发射激光中的应用以及表现出来的现象进行过研究，但是将增益导引技术应用于光纤激光器中还是首次提出。目前在国内外还很少有人对增益导引技术在单模光纤激光器中的应用展开研究，其理论分析体系还不够完善，这给实际应用造成了很大障碍。

Siegman教授作为我的导师把我带入了一个更高、更深的激光领域，令我终生受益。

第七章　耕耘不辍

勤勉为师

我研究生毕业后留校，先后担任了教研室副主任、教研室主任等职。我十分重视本科教学，先后讲解"激光技术"课程，参编《激光技术》教材，后又主编全国统编教材《光电子技术》，亲自做视频公开课；主持教学改革，制定了"工科偏理"的本科培养目标，并在课程中加入四大力学，使本专业毕业生有很强的基础。我带领团队申报硕士点、博士点，大量引进人才。天津大学精仪学院最早的长江学者、许多学科交叉人才如王瑞康等都是我帮助引进的。我还十分注重学科体系及实验室建设，先后把天津激光技术、光电子技术、电子科学与技术、物理电子学等大学本科专业建设成了国内先进学科。我牵线引进的史伟教授，已成为我国高功率光纤激光器方面的学术及产业领域内的领军人才。

作为老师，最大的成就就是培养了一批又一批学生。从教50多年来，我培养了20多名博士后、120多名博士生、近200名硕士生。青出于蓝而胜于蓝，我的很多研究生已成为各单位业务骨干。看到他们不断成长成才是我感到最欣慰的事情。

作为科技工作者我也十分注重科普。我在全国数十次科普会议上做了科普讲座，内容涵盖激光光电子技术、太赫兹技术、物联网技术、智慧城市、智慧养老、智慧海洋、太赫兹在生物医学领域中的应用、太赫兹通信等，受到普遍好评。我还出版了科普书籍《奇异的光——激光》。

图7-1 姚建铨撰写的科普书籍《奇异的光——激光》

恪尽职守

我1997年担任中国民主促进会（简称民进）天津市委会主委。多年来，我恪尽职守、殚精竭虑，在科研工作和社会活动占去很大一部分精力的情况下，对民进市委会的工作仍然是精心布置，努力做到细致入微。我曾在天津接待民进中央主席、全国人大常委会副委员长许嘉璐同志，同他一起讨论民进成员如何在我国政治生活及科技、教育、文化方面做出更多贡献。他曾书写"深藏若虚、盛德若愚"赠送给我，我深感有愧，只能作为自己的努力方向吧。我在天津还多次接待民进中央副主席张怀西及九三学社中央主席韩启德院士。

2002年民进天津市委会换届后，我非常重视新一届领导班子的开局工作，撰文提出三点要求，并在《民进之声》报上发表，供天津的民进组织和广大会员审阅并监督。我要求：一要加强学习，学习政治理论，关心国家大事和民进发展，增强政党意识、责任感和使命感，树立大局意识；二要培养谦虚谨慎的工作作风及原则，要加强服务意识和创新意识；三要将自身建设、议政调研、人才培养和机关工作作为今后工作的重点。我还挤

图7-2 2006年姚建铨担任民进天津市委会主委时主持全会并讲话

图7-3 民进中央主席、全国人大常委会副委员长许嘉璐为姚建铨书写的题词

图7-4 姚建铨在天津接待许嘉璐（中）、张怀西（左）和韩启德（右）

出时间参加了市委会年终述职会，以科学家高度务实的思维方式指出今后的工作思路，并一再强调市委会机关建设的重要性，说机关是个大本营、大家庭，要让大家在这里都能得到锻炼。我重视对年轻干部的培养，做出安排分批培训新来的同志，提高其写作能力及理论研究能力。

2002年11月，在我的倡议下，民进天津市委会和天津市教育科学研究院联合召开了"高等教育发展与改革（天津）研讨会"。会议得到了民进中央与教育部的高度重视，国家有关领导，教育部领导，部分院士，国内外部分著名高校的校长、教育专家学者出席，共同探讨在新形势下我国高等教育的发展方向及改革措施。我在会上做了以"我国创建世界先进水平大学的基本对策"为题的报告。这次活动使民进天津市委会在全市开高层次参政议政之先河，充分调动起民进高教界会员的积极性。民进中央主席许嘉璐用"巩固老阵地、开辟新领域的天津经验"高度评价我们的创新举措。

2003年"非典"期间，我专门到民进天津市委会召开全体干部会，向大家通报疫情的发展情况，慰问并嘱咐大家要做好自我防护、及时消毒，要求党派工作不能停摆，要利用这段时间多学习，提高自己的业务能力。我出席了民进天津科技界会员向市卫生局捐赠医疗设备和消毒剂的仪式，到在抗击"非典"一线战斗的医务界会员家中慰问，使会员们深受感动。

图7-5 2007年3月3日姚建铨院士参加全国政协会议时留影

建言献策

自1987年我当选为全国政协委员（第七届至第十届）以来，我在参政议政这个全新的领域里，一如既往地辛苦工作，每年都要精心准备提案，大多是关于促进科技及教育发展方面的。

2003年是我第16次参加全国政协会，这一届全国政协委员中有107位院士，既反映了党对知识分子特别是高级知识分子的高度重视，又反映出院士群体在我国经济、科技发展及国家政治生活中的作用和地位。作为我国自然科学界及工程科技界最高层次知识分子的代表，院士应该在国家民主生活中发挥怎样的作用？我认为，院士群体拥有精深的科学知识和实践经验，可为国家提供有科学依据的决策咨询意见；但院士也有不足，非专业领域的一般知识有限，所以应正确对待自己，严于律己，脚踏实地，力争为社会进步及国家强盛做出更大贡献。对我来说，首先要把院士的科研工作做好，同时还应担负起参政议政的责任。要密切联系群众，反映群众的呼声。要在自己较熟悉的科技、教育等领域深入分析、研究，提出中肯的建议或提案。

作为科技组委员（第七届至第九届），我多次在政协会议上提出高质量的提案。我撰写了《改革旨在增强宏观调控能力》的提案，指出根据中国科学院发布的《中国可持续发展战略报告》，我国的可持续发展综合国力排第七，经济力排第三，科技力则排在倒数第二的第12位。我认为这说明我们国家近几年经济增长中投资、出口、消费"三架马车"拉动力最为强劲，科技拉动力还不够，反映出科技对经济增长的贡献率还是偏低。要实现党的十六大提出的全面建设小康社会的战略目标，就一定要在科技上下更大的功夫，力争有更多的新增长点。

2003年3月12日，我与政协科技组的几位委员碰在一起，谈到目前国家科研项目在年龄限制上的"一刀切"问题。有许多55岁以上的科学家既有充沛的精力和体能，又有丰富的经验和学识，科学家本人也愿意负责承担较大的科研课题，却因为55岁这道门槛而被拒之于国家科研项目带头人门外。我赶在政协会提案截止日前提交了一份《关于取消55岁以上科技人员不能作为项目负责人申请科研项目的建议》。我在建议中写道："我国科研队伍目前还存在青黄不接现象，55岁左右的科学家正是承上启下的一支骨干力量，弃之不用是很可惜的。比较可行的做法是，既发挥年轻人精力充沛、思想活跃的优势，营造青年科学家不断脱颖而出的有利条件，又发挥中年科学家经验学识丰富的优势，让他们充分展示自己的学识与经验，这样才能保持我们科研队伍的活力和原创性。"

一位成功的科学家，也应是一位政治家、社会活动家，我以花甲之年在科学的崎岖道路上不畏劳苦地攀登，在民主党派这块园地里不知疲倦地辛勤耕耘，永无止境地为社会奉献着我的智慧和才情。2009年，在中华人民共和国成立60周年之际，我荣幸地入选"60位感动天津人物——海河骄子"。

图7-6 2009年姚建铨获得"海河骄子"荣誉称号

第八章　科研回溯

与"光"结下了不解之缘

我大学本科（1957—1962年）学的是自动控制仪表，研究生时期（1962—1965年）是搞精密仪器中的振动测量的，真正接触光学是在1969年。从那时起我与"光"结下了不解之缘，从开始搞激光研究，到现在50余年了，对光学及激光从不熟悉到熟悉，从不懂到懂了一些，半个世纪的磨合，让我对"光"产生了深厚的感情，而且痴迷至今。当然就光学学科这棵参天大树而言，我懂的仅仅是一小片树叶而已。我自己总想，作为一个从事光学事业的教师、科技工作者，应该为光学研究贡献更多的东西。但我已经老了，很多事力不从心了，现在我将自己在光学探索之路走过的50年科研教学工作做一个总结，如果读者能从中汲取一些营养，特别是在科研精神、创新思维、奋发心态等方面找到一些对自己有益的东西，我就满足了。

有时我总问自己：未来光学、激光将向何处发展？

光学领域是前途无限的，其重要性从近几年诺贝尔物理学奖获奖情况就能窥见一斑。2014年诺贝尔物理学奖颁给了日、美科学家，表彰他们发明的高效环保光源——蓝光LED。2017年诺贝尔物理学奖颁给了美国科学家，表彰他们为"激光干涉引力波天文台"（LIGO）项目和发现引力波所做的贡献。2018年诺贝尔物理学奖颁给了美、法、加的三位科学家，表彰他们在激光物理学领域的突破性贡献，其中一半颁给光镊技术（Optical Tweezers），一半颁给啁啾脉冲放大技术（Chirped Pulse Amplification）；前者可用激光来操纵粒子、原子和分子，后者为创造最短、最强激光脉冲铺平了道路。再结合早年被授予诺贝尔物理学奖的科学家，有很多人的研究领域是激光理论、激光器件、激光冷却、多种光学及光子学的器件及技术。

所以，我认为光学（再加上激光、光子学、光电子学）既是人类认识世界的起点，又是改变世界的重要手段及技术。我特别盼望我国从事光学领域的师生们，共同为光学领域的发展做出创新性的贡献。

我从事激光技术科研工作几十年，对激光很有感情，激光器可以产生不同功率大小的激光源以适应于不同的领域。在半导体材料中产生的，能量只有几个或几十个微瓦的激光，可以放大后做出能量巨大的激光武器，可以摧毁卫星、飞机、导弹。这是因为激光可以由放大器来放大，有时用低功率或低能量

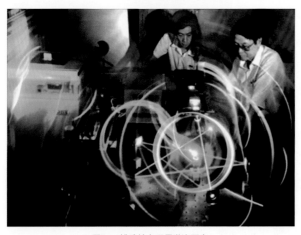

图8-1　姚建铨在开展激光研究

的激光器来保证其质量（光束质量发散角、谱线宽度等），用放大器来放大功率或能量，最后获得既能保证好的光束质量，又能有很大功率或能量的激光。

太赫兹辐射与激光类似，但是一般功率及能量均较小，还达不到一般应用的要求。目前还缺少有效的太赫兹放大器，无法满足很多要求高功率或高能量的太赫兹技术应用的需求。我和研究所的师生们试图参照激光的情况，考虑了几种太赫兹放大器的技术路径。

一是采用四波混频的方法，借助于两个泵浦波把能量转移到太赫兹信号波上，但要找到对太赫兹波三阶非线性系数大的材料很难，也曾想过用高压气体等来实现，困难也很大。

图8-2 姚建铨工作照

二是采用激光振荡加放大的多级结构来实现，但由于本身太赫兹的源很少，其放大介质也不好找，仍然没有办法。

三是索性可以考虑把长波长激光器的能量、功率转换到太赫兹源上，但也有非常大的困难。

放大器恐怕是太赫兹技术发展及应用中的一个极大的瓶颈。目前用光子学方法差频产生太赫兹的方法中只能用两束很强的激光束来差频产生太赫兹波，但由于两束大功率、大能量的激光本身不易获得，再加上差频晶体容易损坏，也只能运转在较低的太赫兹功率及能量水平上。

我想应该另起炉灶，另辟蹊径，从太赫兹辐射的机理出发，探索出颠覆性的新技术，才有攻克大功率太赫兹技术难关的希望。找不到放大器技术路径而引起的另一个问题是对太赫兹的非线性光学效应也很难开展研究。因为非线性效应就是一种强场效应，本身功率、能量较低时，这种非线性效应必然低。这就是太赫兹波的放大为什么很难的根源所在。不过，可以预料，太赫兹放大技术的解决，必将推动太赫兹技术应用的发展，也将对太赫兹波的非线性效应提供推动力及创造力。

智慧海洋——海洋技术问题

海洋是人类文明的起源，是人类生命的摇篮。地球上有71%的面积是海洋，我国有300多万km^2的海洋，还有1.8万km的大陆海岸线和1.4万km的岛屿岸线。我们要深入贯彻落实习总书记提出的海洋强国战略思想，努力担负起建设海洋强国的历史使命。

我认为贯彻海洋强国战略需要解决四个方面的问题，一是科学问题，二是技术问题，三是工程问题，四是权益及管理问题。我主要研究技术方面的问题。我较关心的海洋技术有：通信（海上、海面下及空—海界面通信）、检测、定位、导航及水下成像等。我想用各种科学技术手段来解决海洋中的技术难题，希望能对海洋资源利用及开发有益，对渔业、养殖业、石油天然气开采等人类活动有益。我关注海洋的起点是较熟悉的光学、光电子技术、激光技术及太赫兹技术等领域，从这些领域出发寻找与机械、电磁学、微波、声学等领域的结合点，将上述技术融合在一起，为海洋技术服务。目前我们科研团队正在以下这些领域开展研究与探索。

· 蓝绿激光水下通信

蓝绿激光水下通信利用波长为450～530 nm的蓝绿色光作为通信载波，具有海水穿透能力强、数据传输速率快、方向性好、设备轻巧且抗截获和抗核辐射影响能力好的优点，相当于给水下游弋的潜艇戴上了数据通信的"助听器"，因而得到快速发展和广泛研究。那么，如何研发实用化、小型化、高功率的绿光激光器？如何适应水下环境及海洋环境？如何发射？用什么接收？采用内部探测还是外部探测？如何选择波长？蓝或者绿？是否要可调谐？近海和远海都能适用吗？通信系统将装备岸基还是放船上？军事应用如何选？潜艇、战舰、浮标？海水下通信与海面上通信甚至与飞机、卫星的通信如何延接？

· 激光致声通信

激光致声通信方式是在空气中利用激光、在水中利用声波，通过机载激光对水的作用实现光波到声波的转换，把两种介质中最佳的信道物理场结合起来，形成强大的技术优势，为空中与水下载体间的通信开辟一条新的技术途径。发展激光致声通信是建立空—天—地—海通信网络的重要一环，可形成具有不对称性、颠覆性的技术优势，为打赢海上战争增加重要保障。激光致声通信有许多难题待解，如是机载、汽艇载、卫星载还是无人机载？如何克服海洋风浪？接收平台放在何处？能否调制？是否能实现简单的编码？能传输信号、数据、图像吗？与水下设备如何延接？如何实现海—空跨界面的有效传输？

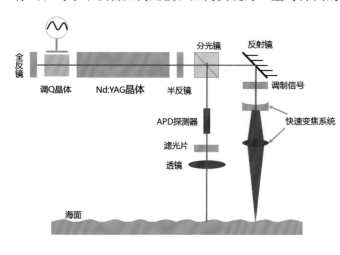

图8-2 激光致声下行通信示意图

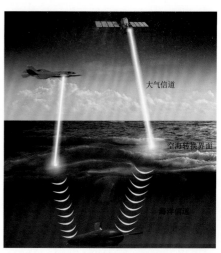

图8-3 激光致声通信模拟图

现代科技目前已明确，中、高频段的无线电波不能从海面上透过海—空界面后向下传播，声波可在海水下传播而一般不能透射到海面上，所以如何解决海洋中跨海—空界面的传输是海洋中面临的重大理论及实际问题。我们有一个不是很完善的成果——激光致声效应研究。具体描述如下。

被调制后的激光从空间机载平台由上而下发射，激光的能量较高，"聚焦"在海面上，会产生声波，这种声波就携带了信号传入海水。我们在实验室1～2 m的高度，将经脉冲编码后的激光射入水面，可以听到"啪、啪、啪"的响声。入射点处得到声波后，它以点源的状态、球面波的形状向下向多方位传播出去，经厦门大学水声通信与海洋信息技术教育部重点实验室工作人员进行声源级测试及计算，在现有实验条件下信息传输距离可达到1～1.4 km。该成果经鉴定会评价为国际先进水平。

这中间有几个问题，在实际条件下海面上风浪较大时，激光聚焦点不易产生声波。虽然飞机的波动加波浪高低问题可以用激光测距仪加反馈补偿发射波得到充分的补偿，但较大的风浪仍会导致"激光致声"的失败，对于这个问题，我们还在进一步分析研究中。另外我们还有另一个考虑，能否利用海水下声波在海面产生的波纹，再在海面上采用激光及太赫兹波，通过分析这种波的状态从而充分获取声波的信息，判断海面下是否有物体在移动。我们也许可以采用多普勒效应＋微波多普勒效应来测出海面下移动的物体的类别。

- **激光与太赫兹波在空间碎片监测及成像技术中的应用**

激光与太赫兹技术在空间领域有通信、雷达、遥感、成像等方面的很多应用。比如我曾经研究的空间碎片检测原理及应用。

空间碎片又称轨道碎片，是指宇宙空间中除正常工作的飞行器外的所有人造物体，包括飞行着的各种残骸和碎片，大到废弃的卫星、运载火箭，小到构件、固体火箭发动机燃烧后的氧化铝小颗粒或从航天器上剥落下来的漆片等。这些人造物体长期运行在空间轨道上，并随着人类航天活动的扩展日益增多，早晚会成为人类航天探索的"麻烦制造者"。就好像在地面的高速公路上如果有很多废物垃圾，汽车就无法在公路上正常行驶一样，空间轨道的碎片也会影响飞行器在空间中的正常运行。自1957年以来，现在空间轨道上分布的大大小小的空间碎片总数已超过44万个，总质量已达数百万kg，其中大于10 cm的空间碎片每年增加200个，现在总量已超过1万个。我们的研究是利用激光及太赫兹技术来检测出碎片（>1 cm以上）的大小、形状、质心及运行速度等，然后再用激光对准其质心进行推动，将其推碎后送至人类不再到达的地方或推动碎片从地球轨道进入大气层后摩擦、烧掉。用太赫兹检测激光推离是多种方法之一。2014年3月4—5日我在天津首次组织召开了空间碎片移除技术研讨会，此后，我连续参加了在北京、苏州等地举办的空间碎片专题研讨会，并作为大会主席或特邀专家发表了自己的见解。

提出"空—天—地—海"一体化网络系统的构想

2015年，国家网信办召集科技界和互联网界有关人士共同讨论未来的国家空间网络系统发展。我是青岛海洋科学与技术国家实验室学术委员会委员,对海洋网络的重要性和现状有着深刻认识。在这次座谈会上,我提出了应规划构建"空—天—地—海"一体化网络系统的建议,受到国家网信办领导及参会专家的一致好评,并牵头起草国家海基空间网络基础设施规划。这项工作有四个任务:一是了解国内外情况,指出中国在相关领域的薄弱环节;二是提出中国在未来海洋空间网络系统的发展战略和重大项目;三是相关的信息安全和应急措施;四是规划和实施相关示范工程。这项工作原本计划六个月完成,后来要求缩短到两个月,可见国家对此高度重视。2015年12月21日,由天津大学、北京邮电大学、中国海洋大学、南开大学、航天九院十三所、航天十一院、中国计量学院(现中国计量大学)、山东科技大学等八家单位共同完成的"海洋网络空间基础设施研究"调研报告上报国家网信办及中央有关部门。在调研报告起草过程中,国家又要求进行一个相关的海洋专项研究,由国家和民间共同承担,投资巨大。这也成为我当时的重点工作之一。

光声结合解决海洋通信关键问题

前面提到的激光致声下行通信,只解决了部分特别需要从80～100 m的空间向海洋内实现编码通信的问题。结合我从事的工作,要解决海洋内部与通信问题,应用光声结合的技术应该是解决问题的关键环节。

光声结合(光声融合)主要研究以下内容:一是激光致声下行通信;二是应用激光多普勒效应+微波多普勒效应,实现"空—水下"通信及水下检测;三是利用LED(发光二极管)+LD(激光二极管)+Fiber(光纤)技术+浮标技术,实现水下多种设备与水面上的联系及沟通;四是利用蓝绿激光+各种先进激光光源对海洋中各参数进行测量控制等。

由于海洋环境与地面差异太大,我们要采用多种先进技术进行研究,如光波通信、声纳通信、蓝绿激光通信、LED通信、LD通信、卫星通信、量子通信、中微子通信等。我提出要从环境,应用对象,技术的成熟性、可塑性多方面来选择研究,我认为"海洋"对全球而言十分重要,如何去认识海洋、关注海洋,真正在海洋强国战略指导下深入开展海洋通信领域的科学研究是大有作为的。

学科交叉培育太赫兹调控技术

2017年7月,我作为项目负责人代表天津大学、东南大学及上海大学联合申请关于太赫兹调控方面的一个国家自然科学基金重点项目。项目题目为"太赫兹调控的新材料、新器件

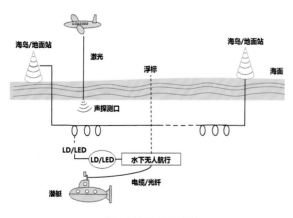

图8-4 海洋通信示意图

及关键技术研究"。我与另外两个大学的主要负责人多次商议，查阅了大量文献，发现当时国内外进行太赫兹调控的研究组织及研究人员很多，大家从不同角度、不同方面、不同应用对象对太赫兹进行调控研究，千头万绪，并没有形成非常清晰的主流研究方向。我首先分析了太赫兹需要调控的参数，如频率（波长）、功率（能量）、偏振、相位、时间、空间、角动量，以及调控需要实现的功能，如开关、吸收、调制、解调、偏振、滤波、耦合等。通过分析，我更加认识到对太赫兹辐

图8-5 太赫兹调控的新材料、新器件及关键技术研究项目树示意图

射的"灵活调控"是太赫兹技术在国民经济、国际等领域应用的核心问题。现有的太赫兹波特性调控能力包括频率调控、宽带偏振调控及空间光调制器，这些调控均具有局限性、单一性及孤立性。我们急需从原理上、研究思路上进行重大创新，也急需从材料、器件和技术上进行重大突破。于是，我针对克服局限性、单一性及孤立性提出了要解决的科学问题，通过归纳创造性地提出了"一级调控（源）""二级调控（器件）"及复合、联动太赫兹调控的新概念。我们确定在新型可调太赫兹源、外加可调控器件及调控器件等共七个研究方向进行分析阐述，最后的联动调控是根据人工智能的原则，按需求分析对调控的源和器件分别进行检测，反馈控制有望得到全面、准确而快速的调控。我们通过一棵树来说明该项目的总体思路，受到评委们的一致好评，最终项目申请全票通过。通过这次项目申报我自己也得到了很大的提高，一方面对太赫兹调控的原理、器件、技术等方面有了深刻的认识，另一方面分析认识水平得到了升华。

该树表示该项目的基础（树根）为自旋流相干调控、光学Cherenkov相位匹配、界面调控、泄漏模波导结构及超构材料；通过建立太赫兹新机理、寻找新材料、研究新器件、采用新结构及技术，实现太赫兹的频率、功率、偏振、相位、时间、空间的调控；对于若干太赫兹功能，能调控开关、吸收、调制、解调、滤波及耦合，最终适用于太赫兹成像、光谱、通信、雷达、安检、天文、国防、生物医疗及科学研究等方面的应用。这棵树状的结构很直观，也体现了项目研究内容间的联系，形成一个整体，在项目答辩时受到评委的赞许，可作为大家在申报项目及分析课题研究时的参考。

研发太赫兹超构材料光谱法用于新型病毒检测的探索

2020年新型冠状病毒肺炎疫情全球大爆发，中国采取了最严格的疫情防控措施，已取得抗击疫情的重大胜利，但全球疫情依然严重。几个月来，我每天了解疫情的消息，一直在思

考如何为彻底赢得对新冠肺炎疫情的战斗而尽一份力。

抗击新冠肺炎疫情在技术层面有三大问题：一是病毒的检测方法、手段，二是疫苗的研发及使用，三是针对病毒的特效药的研究。后两个问题很重要，但完全是生物、化学的问题，我所掌握的知识对此无能为力，但在病毒检测方面我还是能做些工作，出些力的。

关于病毒检测，我了解到现有的作为金标准的办法是"核酸检测—荧光定量PCR方法"，这是一种很有效的方法，原理是以病毒独特的基因序列为检测靶标，通过PCR扩增，使我们选择的这段靶标DNA序列指数级增加，每一个扩增出来的DNA序列，都可与我们预先加入的一段荧光标记探针结合，产生荧光信号，扩增出来的靶基因越多，累计的荧光效果越强。所以核酸检测试剂盒（多重荧光RT-PCR法）能够实现一小时快速精准诊断。但我查阅现有的资料，显示这种方法也有很多问题，如假阳性问题，染料的饱和、漂白、闪烁问题，以及光毒性及光漂白等问题。当然也有其他检测方法，如激光二次谐波法、抗原抗体法、光学干涉反射成像、质谱分析法等，各种方法均有自身的特点及适用范围。面对病毒不断产生的变异，研发多种原理、多种技术的检测手段十分重要。

我一直搞太赫兹技术，太赫兹有很多特点，其中有一条就是它对大分子结构的物质具有特殊的光谱。我觉得这一特性也许能在病毒检测中发挥作用，于是我们科研团队多次开会研讨，提出了太赫兹-光谱检测法，技术要点如下：一是基本方案采用太赫兹光谱技术；二是采用超材料谐振法的太赫兹光谱法；三是采用分子特异性结合的精准检测方法，提高了灵敏度；四是与人工智能结合；五是最终实现芯片化。

2020年5月中旬，我发给国家自然科学基金委员会的项目建议书中提出的项目名称为"基于分子特异性结合的太赫兹超材料谐振器光谱法用于新型病毒检测技术的探索"。该项目提出的检测方法与以往的普通光谱测量法相比，最重要的创新点为采用了"特异性结合"的方式，让谐振器件自身与活化分子形成共价键，从而具有了"抓捕"指定病毒的能力。实现生物芯片与待测病毒的"一对一"精准结合，排除了其他杂质对谐振特性变化的影响，让检测结果更加真实可靠。在"特异性结合"方式的前提下，我们又充分利用其优势，创造性地在待测物体和测量方案上进行了优化设计，拟以冠状病毒的特征蛋白刺突蛋白（Spike Proteins）为判据，大大消除"假阴性"对社会造成的隐患，预防多病毒交叉流行。

该项目是一个典型的学科交叉项目。项目包括了太赫兹技术、生物技术、分子特异性结合技术、人工智能技术、芯片技术，最终提供的研发成果是芯片，被检测者将血液、尿液或分泌物等放在芯片上，经一定处理后即可测出是阳性还是阴性。

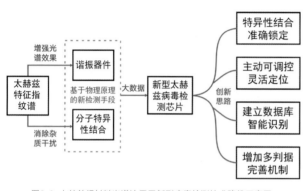

图8-6 太赫兹超材料光谱法用于新型病毒检测技术路线示意图

附访谈录

打破学科的藩篱
——姚建铨谈科研

天津大学党委宣传部 赵晖

问：姚先生您好，您让中国的激光技术站在了世界前沿，还把KTP晶体从"火箭价"拉回到"白菜价"，1982年在美国时您被业内誉为"绿光"。这个称呼是怎样得到的？您怎样看待这个称呼？

答：1960年激光器就已经出现了，激光的应用前景也备受关注。但当时的激光器预热时间长、电流大、输出功率小，这些缺点让科研工作者不得不另辟蹊径去寻找更好的方法做出理想中的激光器。能不能用固体激光器内腔倍频，做出高功率绿光输出的激光器是我思考的一个问题。当时我在斯坦福大学做访问学者。我的导师，美国著名学者A. E. Siegman教授，推荐我去硅谷的XMR公司从事激光倍频研究工作。当时国际上由美国杜邦公司刚研究出来一种新型晶体——$KTiOPO_4$（简称KTP）晶体。能不能用KTP作为倍频，用固体激光器来获取高功率绿光激光器？晶体是新的，用固体激光器来获取高功率也是新的（研究方向）。我就从研究KTP晶体的特性入手，对这个新的晶体进行深入研究。在KTP晶体出现以前都是单轴晶体，其相位匹配计算可以用简单的解析解来解决，而KTP晶体是双轴晶体结构，不能用解析解，只能用数值解。双轴晶体十分复杂，它的折射率椭球是复杂的三维立体图。怎样研究？我首先从杜邦公司拿到了几个公开的KTP晶体基本数据，但到底怎样得到相位匹配角？我克服了重重困难，进行了大量的研究。当时没有资料，我就在斯坦福大学的图书馆常常一待就是一整天，终于研究出了双轴晶体相位匹配的计算方法。我从双轴晶体的折射率椭球及光率体椭球，分析了KTP晶体的双轴晶体的特性，通过有效非线性系统、走离角的分析计算，最终解决了它的最佳非线性相位匹配的问题，求得最佳相位匹配角度。在XMR我解决了设计、计算、稳功率、晶体冷却等一系列问题。投出去的论文于1984年被 *Journal of Applied Physics* 接收、发表。此文被多国学者引用几百次，是双轴晶体最佳相位匹配的原创性工作。因为我做出了高功率绿光激光器，所以大家送了我"绿光"的绰号。

问：您在美国为高功率绿光激光器研制做了开创性的工作，回国后您是如何继续放大这项工作的"功率"的？

答：可以这么说吧。那时我国还没有KTP晶体，因为我本身不是做材料的，回国后我就去拜访材料界的专家沈德忠、蒋民华两位院士，一起研究能不能做出中国自己的KTP晶体。我是做激光器的，最了解激光器需要的晶体特性，他们是做晶体的，最了解人工晶体的生长方法，我们共同研究材料，反复交流，反复磨合，终于做出了中国自己的KTP晶体。晶体从一开始的米粒大小，越长越大。有了晶体，我们搞的高功率固体倍频激光器也成功了。在1987年的鉴定会上，绿光（532 nm）的输出功率达到31.4 W的国际领先水平。这个项目先后获得国家技术发明二等奖、国家教委科技进步二等奖及天津市科技进步二等奖等奖项，又获得尤里卡世界发明博览会金奖，我个人获比利时国王颁发的一级"骑士勋章"。

问："绿光"给您最大的收获是什么？

答：任何科技领域，即便是再小的分支，只要你找对方向，深入下去，切中要害进行科学创新都能取得成功。

我并不是学激光的，但我一辈子都围绕着激光进行研究。我大学时学的是控制仪表，研究生时研究振动，工作后因为要面向国家战略需求，我的研究内容才开始进入激光领域。

在把高功率倍频激光器做到世界领先之后，我们没有就此止步，而是继续深入研究。高功率绿光倍频激光器本身的用途就很广泛，在医疗、海洋、测量等领域都有很多用处，值得进一步研发。此外我还想到一个新的方向，把绿光激光器作为泵浦源，应用于其他激光器，这个方向也取得了成功，其中两个项目还分别得了中科院及军队的大奖。

在发展高功率倍频激光器的同时，我还注意不断发展理论，提出并发展了激光谐振腔的类高斯光束理论；提出了准连续泵浦的理论；又从倍频的角度提出了在不同情况下有关倍频的八个公式，全面发展了倍频的理论。经过几十年的研究，我们逐步创新性地建立了一个可调谐激光技术体系。

回首我的科技工作历程，我最大的心得就是不能轻视科技领域中任何一个小的分支。只要你深入地做下去，就能豁然开朗，找到一片广阔的创新天地。

问：高功率绿光倍频激光器也是学科之间成功交叉融合的结果吗？

答：这正是我想说的另外一个问题。高功率倍频激光器的研究是一个多学科综合的研究，其中不仅包括激光方面的研究，还包括电子学、物理学、材料学、热力学，是一个系统的综合。其实科技发展到今天，任何一个科技项目都不可能是单一学科，而是多学科的综合研究，一定要打破学科的藩篱，把目光放开阔一些，更开阔一些。

问：从激光到太赫兹，再到物联网、智慧养老、智慧城市、智慧海洋、太赫兹通信，您的科学探索之路从未停下脚步。业界认为您的科学嗅觉特别灵敏，总能先人一步找到下一个科研热点。您的科学敏锐性是怎样炼成的？

答：我的大部分研究都在激光领域，正是对激光的深入研究让我注意到了太赫兹波。太赫兹波又称远红外波，波长在0.03~3 mm之间，比微波更短，该波长范围两侧的微波与红外线均已有了广泛的应用，它是电磁波段中最后一段未被人类充分认识和应用的波段，故而这一频段有个外号叫做"太赫兹鸿沟"。正是因为其特殊性，让其具有频率高、脉冲短、穿透性强，且能量很小、对物质与人体的破坏较小等特质。太赫兹曾被评为"改变未来世界的十大技术"之一。我对太赫兹的关注也是基于对激光的研究。正是看到发展新的电磁波段的战略意义，我带领科研团队全面投入太赫兹技术的研究。我是2004年开始太赫兹研究的，这在全国是较早的。2005年11月底，我联合了十多位院士及数十位相关代表在北京召开了关于"太赫兹科学技术"的香山会议。我作为会议执行主席之一，在会议上做了用光子学方法产生太赫兹源的报告。有人说正是该会议开创了我国开展太赫兹科学技术研究新领域的先河，具有里程碑意义。

太赫兹波所处的特殊电磁波谱的位置，使它有很多优越的特性，在天体物理学、等离子体物理学、光谱学、材料学、生物学、医学成像、环境科学、信息科学等领域都有着广阔的科研前景。其应用前景更是广阔，在生物组织的活体检查、高精度保密雷达、卫星间宽带通信、违禁物品反恐检查等方面都大有作为。这将又是一片科研的海洋、创新的乐园。找到可以创造未来的科研方向，作为一个科研人员能不兴奋吗？！

我后来提出的智慧海洋、智慧城市也是从对"光"的研究衍生出来的。海洋面积占地球总面积的71%，海洋本来就是人类生命的起源地。海洋本身又十分奇特，大部分电磁波不能进入海洋，而声波又出不了海洋。为了研究海洋，特别是从海洋技术（通信、探测、成像等）出发，我考虑可采用光—声相结合的办法来解决空—海界面的传输，这恐怕是建设智慧海洋要解决的最大问题。

现在有一种说法，要面向未来培养"T型人才"，我对这个说法很赞同。"—"表示有广博的知识面，"|"表示知识的深度。既有较深的专业知识，又有广博的知识面的"T型人才"，能够在科学的道路上走得更远。打牢基础很重要，科研就像一颗树，根扎得深，枝蔓才能茂盛。

问：教育的根本任务是培养担当民族复兴大任的时代新人，如何培养？您至今已培养了几十名博士后，120多名博士生及更多的硕士生，对学生的学业及科研工作的培养最主要应该抓住哪些方面？您对青年学子评价如何？您认为他们最主要应该从哪些方面努力？

答：充分信任，大胆使用，挖掘并发挥学生最大的潜能，这就是我"教"学生的方法。在科研的舞台上，学生是主角，老师是搭台的人。搭好舞台，给予必要的指导，让学生尽情地发挥，这是老师应该做的。在很长一段时间里我的项目多，研究生数量也很多，我就以"项目"为单位，将教师、博士、硕士分入不同小组，由我决定总体方案，跟踪进度，各小组分别具体研制，小组一起向我汇报，共同讨论决定解决办法。这样学生不仅能在各自领域负责，又能在更高层面交叉合作，对于他们的科研创新是十分有好处的。有人说，姚老师对学生是"放羊"，我觉得这样的放羊效果不错。很多学生出来都能独当一面，现在我的学生有很多都是大企业大集团的业务骨干、管理骨干，也有在知名院校当校长的。

科研团队建设，氛围是最重要的。氛围营造靠导师的言传身教。在身体允许的情况下，我是团队中来得最早、走得最晚的。我这没有下班时间，学生有问题，可以随时来问。科研第一，攻关时更是可以不眠不休。天大17教学楼晚上11点锁门，往往一个问题研究讨论告一段落，已是凌晨，楼门落锁出不来怎么办？我们就开辟了一条"厕所通道"，学生先从一楼厕所跳窗出去，在窗下面垫上几块石头，再扶我出去。

科研团队是学生成长的精神家园，工作之余我经常组织学生开展团队建设活动，一起打球，一起郊游，一起旅行，一起聚餐，一起奋斗。记得一次去水上公园，我要求进入园区就不能说中文了，大家玩得不亦乐乎。我想让同学们认同的是：我们是一个严肃的科研集体，也是志同道合的朋友集体。

任何科研成果的取得都不是一个人的战斗，要建设一个分工负责、相互协调，具有浓厚学术氛围的团结的战斗集体。这点很重要，也是老师的重要职责。

问：未来的竞争将是创新能力的竞争，您是如何培养学生的创新思维，如何引导学生的创新行为的？

答：既要低头做事又要抬头看路。学校的入学教育、科学夏令营、培训班只要请我，我从来没有拒绝过。我愿意和同学们交往，愿意和同学们交朋友，愿意给他们讲讲我的科研心得。我希望学生都给自己定一个长远的奋斗目标。做学问也好，做老师也好，做社会工作也好，都需要一个目标。这个目标怎么定呢？量才而定！根据自己的情况、兴趣，定一个跳一跳能够得着的奋斗目标。不能好高骛远，也不能碌碌无为。同学们切记，不能贪图享乐，不经过奋斗、不经过努力而得到的，都是不真实的。既要有奋斗目标又要踏踏实实地奋斗，有家国情怀、德才兼备、脚踏实地的学生最可爱。"谦虚勤奋"是我一生的座右铭，在此与同学们共勉。

作为科研人员，科技强国是我们的历史使命。跟踪世界科技前沿，聚焦国家重大需求，就能找到科研方向，进而实现自己的人生价值。比如我现在关注的智慧海洋，就是从国家重大需求出发，试图打通海面上、海面下的空—海界面通信，进行更有效的检测、定位、导航及水下成像等，为海洋开发提供技术支持。

创新是一个系统工程，需要创新环境、创新土壤、创新氛围等，其中最重要的就是创新思维。我现在希望能在培养学生的创新思维方面多做一些工作。

第二部分

姚建铨学术团队论文集

第一章 非线性光学频率变换及固体激光技术
第二章 光电子及光纤技术
第三章 太赫兹技术

论文集目录

第一章 非线性光学频率变换及固体激光技术

1. Simultaneous active AM plus passive FM with Q-controlling mode-locked high power pulsed laser ……110
2. Optimum operational parameters of the ultrashort cavity laser ……110
3. Large-signal results for degenerate four-wave mixing and phase conjugate resonators ……111
4. 使用 $KTiOPO_4$ 的高功率内腔倍频 YAG 激光器 ……111
5. Calculations of optimum phase match parameters for the biaxial crystal $KTiOPO_4$ ……112
6. Optimum parameters of a high-conversion efficiency intracavity frequency-doubled laser with Gaussian-like beam ……112
7. 类高斯分布的理论及其应用 ……113
8. 高转换效率下具有高斯及类高斯光束的内腔倍频 ……113
9. Phase conjugation in iron-doped Lithium Niobate: study and application ……114
10. Intracavity frequency-doubling of quasi-CW pumped YAG laser ……114
11. The theoretical research on the FWM effect of F_2-centers in LiF crystals ……115
12. 准连续泵浦 KTP 内腔倍频 YAG 激光器及其热效应分析 ……115
13. KTP 内腔倍频 YAG 激光器 ……116
14. 大电流脉冲方波激光电源 ……116
15. 可饱和吸收体在激光腔内的调 Q 及四波混频效应 ……117
16. Analyses of BBO and KTP crystals for optical parametric oscillation ……117
17. 双轴晶体 $KTiOPO_4$ 的多频率变换 ……118
18. High power green laser by intracavity frequency doubling with KTP crystal ……118
19. Applications and features of a new nonlinear crystal: Lithiun Triborate (LBO) ……119
20. 用 $KNbO_3$ 晶体实现角度调谐频率下转换的计算与分析 ……119
21. Room-temperature 1.06-0.53-μm second-harmonic generation with $MgO:LiNbO_3$ ……120
22. 光折变二波混频双稳态 ……120
23. 高功率脉冲 YAG 激光器 ……121
24. 高功率脉冲 Nd:YAG 激光器实验研究 ……121
25. Room temperature 1.06-0.53 μm second harmonic generation with $MgO:LiNbO_3$ ……122
26. Q-switched Nd:YAG laser and color-center laser tunable over the 1.1-1.26 μm range operating simultaneously ……122

27. Accurate calculation of the optimum phase-matching parameters in three-wave interactions with biaxial nonlinear-optical crystals ·················123

28. Transmitted and tuning characteristics of birefringent filters ·················123

29. Transverse parametric intensity distribution of optical parametric oscillation in MgO:LiNbO$_3$ crystal ·················124

30. 温度调谐 MgO:LiNbO$_3$ 晶体光参量振荡器 ·················124

31. 声光 - 染料双调 Q 激光器的理论及实验研究 ·················125

32. Numerical method of astigmatic compensation and stability regions of folded or ring cavity ·················125

33. Quasi-continuous wave laser operation of Titanium-doped sapphire ·················126

34. BSO 晶体光电特性及光调制器的研究 ·················126

35. Azimuthal and incident angle dependences in the second-harmonic generation from aluminum ·················127

36. Mechanism of parametric pulse duration change of a temperature-tuned MgO:LiNbO$_3$ optical parametric oscillator ·················127

37. 光折变二波耦合增益振荡 ·················128

38. 脉冲激光泵浦的钛宝石激光器 ·················128

39. Tunable 5-watts Ti:Al$_2$O$_3$ laser oscillating with a quasi-continuous wave pumping ·················129

40. Phase matching of optical parametric oscillator in KTP pumped by doubled YAG laser in the quasi-degenerate region ·················129

41. 厄米特—高斯混合模的混合模系数 ·················130

42. Second and third harmonic generation in BBO by femtosecond Ti:sapphire laser pulses ·················130

43. Numerical analysis of beam parameters and stability regions in a folded or ring cavity ·················131

44. 一种新的多波长数字散斑干涉技术 ·················131

45. Theoretical analysis and experimental investigation of pulsed tunable forsterite laser ·················132

46. 连续波飞秒锁模激光器中的三阶色散计算 ·················132

47. 可用于高功率脉冲激光的面阵 CCD 模式测量装置 ·················133

48. The characteristics of sum-frequency-mixing by biaxial crystal LiB$_3$O$_5$ and KTiOPO$_4$ ·················133

49. Femtosecond Pulse—Second and Third Harmonic Generation with BBO ·················134

50. Mixed mode coefficient and Gaussian-like distribution ·················134

51. 调 Q 倍频 Nd:YAG 激光泵浦氧气和氧 - 氩混合气体的受激拉曼散射 ·················135

52. 两个激光二极管端面泵浦的基横模 Nd:YAG 激光器 ·················135

53. 激光二极管端面泵浦的调 Q 内腔倍频 Nd:YAG 激光器 ·················136

54. 倍频 YAG 激光泵浦钛宝石激光器 ·················136

55. High power pulsed laser power supply based on IGBT ·················137

56. 脉冲钛宝石激光器对兔眼损伤效应的研究 ·················137

57. Study on the output stability of Ti:Sapphire laser ·················138

58. Single frequency operation of diode-pumped intracavity frequency doubled Nd:YAG laser with tilted KTP crystal: theoretical study ………………………………………………………………138
59. 外腔锁模多量子阱二极管激光器 ……………………………………………………………139
60. 激光二极管泵浦的高重复频率 Nd:YAG 激光器 ……………………………………………139
61. 用修正的 RELIEF 方法测量高速空气流瞬时速度的理论研究 ……………………………140
62. 双脉冲数字散斑干涉系统的实验研究 ………………………………………………………140
63. 高功率准连续激光二极管泵浦的 Nd:YAG 激光器 …………………………………………141
64. 高功率开关型脉冲激光电源逆变器的理论分析 ……………………………………………141
65. 钛宝石晶体色散特性的理论研究 ……………………………………………………………142
66. 脉冲钛宝石激光器动力学特性的理论及实验研究 …………………………………………142
67. Analysis of simultaneous Q-switching and frequency doubling in KTP ……………………143
68. The temperature distribution in longitudinally pumped Ti:sapphire lasers …………………143
69. 双波长脉冲激光器的速率方程理论及其数值计算 …………………………………………144
70. 脉冲钛宝石激光器的双波长运转 ……………………………………………………………144
71. 激光二极管端面泵浦的单频固体激光器的理论研究 ………………………………………145
72. 脉冲泵浦 KTP 单谐振光学参量振荡器 ………………………………………………………145
73. LD 泵浦的内腔倍频激光器单频运转的理论研究 …………………………………………146
74. ArF 准分子激光器的窄线宽可调谐运转及注入放大 ………………………………………146
75. ArF 准分子激光振荡放大系统的研究 ………………………………………………………147
76. 序列脉冲倍频 YAG 激光器泵浦的氧气受激拉曼散射研究 ………………………………147
77. 利用双折射效应设计的窄带滤光系统 ………………………………………………………148
78. 激光应用的新领域——激光清洗 ……………………………………………………………148
79. 半导体激光直接倍频的蓝紫光激光器 ………………………………………………………149
80. 脉冲可调谐钛宝石激光器的实验 ……………………………………………………………149
81. 瑞利散射/激光诱导荧光技术用于空气、O_2 流场的二维瞬态测量 ………………………150
82. Theoretic analyzing of optical parametric oscillator with intracavity sum frequency generation ……150
83. 激光二极管面阵列泵浦固体激光器的理论分析 ……………………………………………151
84. 分子标记示踪用于氧气/空气流场速度测量 …………………………………………………151
85. 单片机在可调谐固体激光器中的应用 ………………………………………………………152
86. VISAR 测速中的信号丢失及丢失条纹数的确定 ……………………………………………152
87. Blue-violet light second harmonic generation with CMTC crystals …………………………153
88. Blue-violet light by direct frequency doubling of laser diode ………………………………153
89. 激光二极管端面抽运的 1 342 nmNd:YVO_4 激光器 …………………………………………154
90. 瑞利散射用于分子流场多参数测量 …………………………………………………………154
91. 泵浦波长调谐 KTP-OPO 及其倍频、和频的理论与实验研究 ……………………………155

69

92. 激光表面改性技术的研究与发展 ··155
93. The characteristics of sum-frequency-mixing by biaxial crystal LiB$_3$O$_5$ and KTiOPO$_4$ ·············156
94. Pump-tuning optical parametric oscillation and sum-frequency mixing with KTP pumped by a Ti:sapphire laser ···········156
95. Rate-equation theory and experimental research on dual-wavelength operation of a Ti:sapphire laser ··············157
96. Nearly-noncritical phase matching in MgO:LiNbO$_3$ optical parametric oscillators ··············157
97. Thermal effect in KTP crystals during high power laser operation ···········158
98. Pump-tuning KTP optical parametric oscillator with continuous output wavelength pumped by a pulsed tunable Ti:sapphire laser ···········158
99. Tunable sum frequency mixing of a Ti:sapphire laser and a Nd:YAG laser ···········159
100. A micro blue-violet laser by frequency doubling of semiconductor laser ···········159
101. 基于混频效应的宽带激光谐波转换理论 ··160
102. 准相位匹配 PPLN、PPKTP、PPRTA 光学参量振荡器及其应用 ··············160
103. 高功率准连续激光二极管抽运的 Q 开关内腔倍频固体激光器的研究 ··············161
104. 非线性晶体三波互作用允许参量的修正计算 ··161
105. 拉曼激发激光诱导电子荧光流场测量系统中标记过程的研究 ··············162
106. 宽波段温度调谐 MgO:LiNbO$_3$ 光学参量振荡器 ··162
107. 可调谐钛宝石激光抽运的 KTP 单谐振光学参量振荡器的研究 ··············163
108. LD 抽运 Nd:YAG 激光器声光调 Q 高效内腔谐波转换 ··163
109. LD 抽运的内腔倍频 Nd:YAG 激光器输出倍频绿光达 40 W ··164
110. 瑞利散射用于气体流场二维瞬态密度测量 ··164
111. 非线性光学频率变换及准相位匹配技术 ··165
112. 含预反馈的激光自混合干涉型位移测量结构 ··165
113. 掺铜 KNSBN 光折变晶体的光致吸收特性 ··166
114. 多模激光自混合干涉实验与理论分析 ··166
115. 准连续 660 nm Nd:YAG 内腔倍频激光器的研究 ··167
116. LD 抽运的高效率内腔倍频绿光激光器的研究 ··167
117. LD 抽运的全固态三倍频紫外激光器 ··168
118. 高功率全固态绿光激光器的设计与研究 ··168
119. 红光 (660 nm) 准连续 Nd:YAG 内腔倍频激光器 ··169
120. 半导体激光自混合干涉系统的稳态分析 ··169
121. 光互连中两微腔激光器耦合机制研究 ··170
122. 激光动态识别 Line 元件三维形状及动态的研究 ··170
123. (1 064/1 319,532/659 nm)Nd:YAG 激光治疗仪的实验研究 ··171

124. Study on high power laser diode pumped Nd:YAG laser and its frequency conversion ……171
125. Different configurations of continuous-wave single resonant cavity and design of periodically poled LiNbO$_3$ ……172
126. All solid-state wide beam flat-top laser ……172
127. High power high efficiency CW operation of a LD-pumped Nd:YAG rod laser ……173
128. A new method to measure the ultrashort thermal focal length of high-power solid-state laser ……173
129. The study on high average power all-solid-state green laser ……174
130. The method research of used laser to processes the slot pipe applied to petroleum exploitation ……174
131. Analysis of self-mixing interference signals in LD pumped solid-state laser using fast Fourier transform technique ……175
132. Studies on the relation between temperature tuning value and grating period deviation of periodically poled crystal for SHG ……175
133. View for the development of theory on the self-mixing interference and the general model of the displacement measurement ……176
134. Research on the laser disturbing and damage of TV tracking system ……176
135. 从 LASERS' 2000 看激光技术发展趋向 ……177
136. Analysis for the self-mixing interference effects in a laser diode at high optical feedback levels ……177
137. Temperature tunable infrared optical parametric oscillator with periodically poled LiNbO$_3$ ……178
138. High power diode single-end-pumped Nd:YVO$_4$ laser ……178
139. Study on 660 nm quasi-contionuous-wave intracavity frequency-doubled Nd:YAG laser ……179
140. Angle-tuned signal-resonated optical parametric oscillator based on periodically poled lithium niobate ……179
141. Study on CW Nd:YAG infrared laser at 1 319 nm ……180
142. LD pumped high efficient intracavity frequency-doubled green laser ……180
143. Self-mixing interference effects in LD pumped multi-mode solid-state Laser ……181
144. 激光自混合干涉位移测量系统的稳态解 ……181
145. 104 W 全固态 532 nm Nd: YAG 激光器 ……182
146. 波长为 1 319 nm 的连续输出 Nd:YAG 激光器的研究 ……182
147. 85 W 绿光激光器中的 KTP 晶体热效应研究 ……183
148. 激光在人头部立体形状建模中的应用研究 ……183
149. 送粉激光熔覆结合界面组织结构与特性 ……184
150. 波长 1 319 nm 连续 Nd:YAG 激光器的研究 ……184
151. 天津大学精仪学院激光与光电子研究所研制出全固态绿光激光器已突破百瓦 ……185
152. LD 纵向泵浦的 946 nm Nd:YAG 准三能级激光系统的研究 ……185
153. 激光微加工中运动控制系统的设计 ……186

154. 片式电阻激光调阻机检测系统的设计 ·· 186
155. 送粉激光熔敷耐磨涂层制备工艺与组织结构 ······································ 187
156. 用偏振光学相干层析技术分析研究组织偏振特性 ······························· 187
157. 天津大学激光与光电子研究所研制成功 104 W 全固态 532 nm 绿光激光器 ················ 188
158. Low-threshold, high-efficiency, high-repetition-rate optical parametric generator based on periodically poled LiNbO$_3$ ·· 188
159. Wavelength tunable optical parametric oscillator based on periodically poled lithium niobate ··· 189
160. 在 PPLN-OPO 中伴生四波混频的研究 ·· 189
161. 绿光输出达 85 W 的全固态绿光激光器谐振腔研究 ·································· 190
162. 8.1 W 全固态准连续红光 Nd:YAG 激光器 ··· 190
163. 高功率、高效率、全固态准连续钛宝石激光器 ·· 191
164. 104 W 内腔倍频全固态 Nd:YAG 绿光激光器 ·· 191
165. 1 319 nm 与 660 nm 双波长 Nd:YAG 激光器的研究 ······························ 192
166. Self-mixing interferences effects in LD pumped multi-mode solid-state laser ············ 192
167. 强激光对星载光电探测系统的干扰与破坏研究 ······································ 193
168. 利用 QPM-OPO 的调谐特性精确测定晶体的极化周期 ···························· 193
169. LD 侧泵激光器抽运光和温度分布数值研究 ·· 194
170. SESAM mode-locked YAG laser with LD sidepumping module ······················ 194
171. Study on CW Nd:YAG laser at 1 319 nm and 660 nm ······························ 195
172. The influence of KTP's thermal effects on the stabilization of a high-power diode-side-pumped all-solid-state QCW intracavity-frequency-doubled Nd:YAG laser and the compensating methods ······ 195
173. Phase mismatch compensation of second harmonic generation with controlling boundary temperature of type-II KTP crystal in high power green laser ···························· 196
174. LD pumped Q-CW Nd:YAG/KTP red laser ··· 197
175. The study of stability on a laser-diode-pumped high-power high-repetition-rate intracavity frequency doubled 532 nm laser ·· 198
176. High power high efficiency all-solid-state quasi-continuous-wave tunable Ti:sapphire laser system ·· 199
177. New study on self-mixing interference effects in LD pumped laser ···················· 200
178. Study on self-mixing interference effects in LD pumped laser with compensatory cavity ········ 200
179. Quantum mechanism method for self-mixing interference effects in LD pumped laser ········ 201
180. Diode end-pumped passively mode-locked Nd:YVO$_4$ picosecond laser with SESAM ······ 201
181. The study on laser scanning three-dimensional modeling of human head ·············· 202
182. LD pumped Nd:YAG/KTP intracavity frequency-doubled 1.2 W CW red laser ·········· 202
183. Research on the correlation of the signal and output in the stimulated Raman scattering ······ 203

184. Ultra-broad band wavelength tuning from 400 nm to 1 700 nm with near-NCPM OPO and frequency doubling ·················203
185. Total internal reflection noncollinear quasi-phase-matching PPLN OPO ···············204
186. All-solid-state quasi-continuous-wave tunable Ti:Sapphire laser and pump-tuning PPLN-OPO 205
187. High conversion efficiency continuous wave quasi-phase-matched second harmonic generation in MgO doped stoichiometric lithium tantalite ···············206
188. Continuous-wave operation at 1 386 nm in a diode-end-pumped Nd:YVO$_4$ laser ···············206
189. 8.3 W diode-end-pumped continuous-wave Nd:YAG laser operating at 946-nm ···············207
190. Multi-point laser Doppler velocimeter ···············207
191. 110 W high stability green laser using type II phase matching KTiOPO$_4$ (KTP) crystal with boundary temperature control ···············208
192. Broadening angular acceptance bandwidth of second-harmonic generation by using nearly ideal temperature quasi-phase-matching ···············209
193. High-power continuous-wave Nd:YAG laser at 946 nm laser and intracavity frequency-doubling with a compact three-element cavity ···············210
194. High-efficiency single-pass cw quasi-phase-matched frequency doubling based on PP-MgO:SLT ···············210
195. High power diode-end-pumped Nd:YAG 946-nm laser and its efficient frequency doubling ······211
196. High-average-power nanosecond quasi-phase-matched single–pass optical parametric generator in periodically poled lithium niobate ···············211
197. Experimental 511 W composite Nd: YAG ceramic laser ···············212
198. Influence of the KTP crystal boundary temperature on conversion efficiency in high power green laser ···············212
199. High power all-solid-state quasi-continuous-wave tunable Ti: sapphire laser system ···············213
200. Diode end-pumped 1 123-nm Nd: YAG laser with 2.6-W output power ···············213
201. 5.3-W Nd:YVO$_4$ passively mode-locked laser by a novel semiconductor saturable absorber mirror 214
202. Efficient stable simultaneous CW dual-wavelength diode-end-pumped Nd: YAG laser operating at 1.319 and 1.338 μm ···············214
203. High power output quasi-continuous-wave nanosecond optical parametric generator based on periodically poled lithium niobate ···············215
204. 脉冲激光对类金刚石(DLC)薄膜的热冲击效应研究 ···············215
205. LD泵浦准连续Nd:YAG/KTP 12 W红光激光器 ···············216
206. 高功率全固态准连续钛宝石激光器的实验研究 ···············216
207. LD抽运Nd:YAG/KTP腔内倍频连续波1.2 W红光激光器 ···············217
208. 大功率连续锁模皮秒激光器单端输出超过5 W ···············217

209. 半导体可饱和吸收镜连续被动锁模端面抽运 Nd:YVO$_4$ 激光器 ·············218
210. 准相位匹配周期极化高掺镁铌酸锂 532 nm 倍频准连续输出研究 ·············218
211. 高功率绿光激光器热稳定性的研究 ·············219
212. 外差椭偏测量技术及其混频误差分析 ·············219
213. 纳米厚度薄膜外差椭偏测量技术研究 ·············220
214. 透射式外差椭偏测量及非线性误差分析 ·············220
215. A laser-diode-pumped 7.36 W continuous-wave Nd:YVO$_4$ laser at 1 342 nm ·············221
216. A low-pump-threshold high-repetition-rate intracavity optical parametric generator based on periodically poled lithium niobate ·············221
217. The study on the high-power LD-pumped Nd:YAG cw dual-wavelength laser ·············222
218. High-power continuous-wave diode-end-pumped intracavity frequency doubled Nd:YVO$_4$ laser at 671 nm with a compact three-element cavity ·············222
219. A tunable optical parametric generator by using a quasi-phase-matched crystal with different wedge angles ·············223
220. All-solid-state quasi-continuous-wave high power dispersion cavity tunable Ti:sapphire laser ···223
221. Precise calculation of the KTP crystal used as both an intracavity electro-optic Q-switch and a second harmonic generator ·············224
222. Efficient and high-power laser-diode single-end-pumped Nd:YVO$_4$ continuous wave laser at 1 342 nm ·············224
223. Dual-signal-wavelength optical parametric generator based on ppr-PP-MgO:LN ·············225
224. Theoretical analysis of broadband source by using retracing behavior of collinear quasi-phase-matching optical parametric generator ·············225
225. All-solid-state high power dual-wavelength Ti: sapphire laser ·············226
226. Theoretical and experimental research on the birefringent filter applied for the high power tunable Ti: sapphire laser ·············226
227. Continuous-wave, 15.2 W diode-end-pumped Nd:YAG laser operating at 946 nm ·············227
228. Theoretical study of the electro-optic effect of aperiodically poled lithium niobate in a Q-switched dual-wavelength laser ·············227
229. Design of a broadband source by using the retracing behavior of a collinear quasi-phase-matching optical parametric generator ·············228
230. Experimental study on a high conversion efficiency, low threshold, high-repetition-rate periodically poled lithium niobate optical parametric generator ·············228
231. LD 端泵浦 1 319 nm 3.11 W-CW Nd :YAG 激光器 ·············229
232. 基于复合晶体的全固态 1 064 nm 激光器的实验研究 ·············229
233. 激光二极管端面抽运 Nd:YVO$_4$ 实现 1 386 nm 连续波激光输出 ·············230

234. 高稳定 LD 泵浦腔内倍频 Nd:YVO$_4$/KTP 连续绿光激光器 ……………………230

235. 激光在机械制造技术中的应用 …………………………………………………231

236. 抽运光角度调谐准相位匹配光学参量振荡器的研究 …………………………231

237. 高功率激光二极管抽运 Nd:YAG 连续双波长激光器 …………………………232

238. The study on the high-power LD-pumped Nd:YAG cw dual-wavelength laser …………232

239. High-power continuous-wave diode-end-pumped intracavity frequency doubled Nd:YVO$_4$ laser at 671 nm with a compact three-element cavity ………………………………………233

240. 104 W high stability green laser generation by using diode laser pumped intracavity frequency-doubling Q-switched composite ceramic Nd:YAG laser ………………………………233

241. Widely tunable, high-repetition-rate, dual signal-wave optical parametric oscillator by using two periodically poled crystals ……………………………………………………234

242. Tunable quasi-phase-matched optical parametric generator by translating a segmented crystal …234

243. Tunable dual-signal PPLN optical parametric generator by using an acousto-optic beam splitter 235

244. High-repetition-rate dual-signal intracavity optical parametric generator based on periodically-phase-reversal PPMgLN ……………………………………………………235

245. An all-solid-state high power quasi-continuous-wave tunable dual-wavelength Ti:sapphire laser system using birefringence filter ………………………………………………236

246. A bidirectional diode-pumped, passively mode-locked Nd:YVO$_4$ ring laser with a low-temperature-grown semiconductor saturable absorber mirror …………………………………236

247. An all-solid-state tunable dual-wavelength Ti: sapphire laser with quasi-continuous-wave outputs ……………………………………………………………………………237

248. High-power diode-side-pumped intracavity-frequency-doubled continuous wave 532 nm laser …237

249. Efficient nanosecond dual-signal optical parametric generator with a periodically phase reversed PPMgLN ……………………………………………………………………238

250. Analysis of frequency mixing error on heterodyne interferometric ellipsometry ……………238

251. High-peak-power, high-repetition-rate intracavity optical parametric oscillator at 1.57 μm ……239

252. 端面抽运全固态皮秒被动锁模激光器 …………………………………………239

253. 端面抽运高功率连续单频 1 064 nm Nd:YVO$_4$ 环形腔激光器 …………………240

254. 高稳定 LD 端面泵浦腔内倍频 Nd:YVO$_4$/LBO 连续红光激光器 ………………240

255. New Progress on all solid state laser technology in China ………………………241

256. 15.2 W LD 端面抽运 946 nm 连续波 Nd:YAG 激光器 ………………………241

257. 激光电视的研究进展及趋势分析 ………………………………………………242

258. 氧化锆掺杂激光熔覆涂层成形、结构与增韧机制 ……………………………242

259. 1 064 nm 泵浦温度调谐 PPLN 光学参量振荡器 ………………………………243

260. 高功率全固态内腔倍频调 Q 激光器中相位失配的理论研究 …………………243

261. 基于周期极化铌酸锂的可调谐内腔光参量产生的实验研究 ··· 244
262. 基于掺氧化镁周期性极化铌酸锂的光学参量产生器 ··· 244
263. PPLN 准相位匹配内腔光学参量振荡器的实验研究 ·· 245
264. 三镜折叠腔内腔倍频高功率绿光激光器 ·· 245
265. High-efficiency direct-pumped Nd:YVO$_4$ laser operating at 1.34 μm ··························· 246
266. Generation of 3.5 W high efficiency blue-violet laser by intracavity frequency-doubling of an all-solid-state tunable Ti:sapphire laser ·· 246
267. 26.3-W continuous-wave, diode-double-end-pumped all-solid-state Nd:GdVO$_4$ laser operating at 1.34 μm ··· 247
268. 1-W, high-repetition-rate room temperature operation of mid-infrared optical parametric oscillator based on periodically poled Mgo-doped LiNbO$_3$ ··· 247
269. Compact efficient optical parametric generator internal to a Q-switched Nd:YVO$_4$ laser with periodically poled MgO:LiNbO$_3$ ··· 248
270. Continuous-wave mid-infrared intracavity singly resonant optical parametric oscillator based on periodically poled lithium niobate ··· 248
271. Performance of gain-switched all-solid-state quasi-continuous-wave tunable Ti:sapphire laser system ·· 249
272. Numerical modelling of QCW-pumped passively q-switched Nd:YAG lasers with Cr^{4+}:YAG as saturable absorber ·· 249
273. Passively Q-switched quasi-continuous-wave diode-pumped intracavity optical parametric oscillator at 1.57 μm ··· 250
274. Threshold characteristic of intracavity optical parametric oscillator pumped by all-solid-state Q-switched laser ··· 250
275. LD 端面抽运 Nd:YAG 1 319 nm/1 338 nm 双波长激光器研究 ································· 251
276. 高效高功率侧面抽运腔内倍频连续绿光激光器 ·· 251
277. 全固态激光电视的研究 ·· 252
278. 准连续 LD 侧面泵浦的 Nd:YAG 双制式脉冲激光器 ··· 252
279. 电光调 Q 1 064 nm/532 nm 脉冲激光器 ·· 253
280. Ar^{3+} 激光器泵浦的连续波 KTP 光学参量振荡器的研究 ·· 253
281. 双棒 Nd:YAG 激光器双折射补偿未对准灵敏度的研究 ··· 254
282. 514.5 nm 泵浦 KTP 晶体的连续光学参量振荡器的研究 ··· 254
283. Ar$^+$ 激光器泵浦的连续光学参量振荡器的理论设计 ·· 255
284. Efficient electro-optic Q-switched eye-safe optical parametric oscillator based on KTiAsO$_4$ ······ 255
285. High-energy pulsed laser of twin wavelengths from KTP intracavity optical parametric oscillator 256

286. Optimum design for 160-Gb/s all-optical time-domain demultiplexing based on cascaded second-order nonlinearities of SHG and DFG ···················256
287. A high-power continuous-wave diffraction-limited Nd:GdVO$_4$ laser operating at 1.34 μm ······257
288. Efficient dual signal output from a Nd:YVO$_4$ laser-pumped PPMgLN OPO ···················257
289. High efficiency continuous-wave tunable signal output of an intracavity singly resonant optical parametric oscillator based on periodically poled lithium niobate ···················258
290. High power continuous-wave diode-end-pumped 1.34-μm Nd:GdVO$_4$ laser ···················258
291. Green-beam generation of 40 W by use of an intracavity frequency-doubled diode side-pumped Nd:YAG laser ···················259
292. All-solid-state Nd:YAG laser operating at 1 064 nm and 1 319 nm under 885 nm thermally boosted pumping ···················259
293. Low threshold and high conversion efficiency nanosecond mid-infrared KTA OPO ···················260
294. 高效高稳定高光束质量声光调 Q 绿光激光器的研究 ···················260
295. 36 W 侧面抽运腔内倍频 Nd:YAG/KTP 连续绿光激光器 ···················261
296. 准连续 TEM$_{00}$ 模被动调 Q Nd:YAG 激光器的研究 ···················261
297. 双棒串接 Nd:YAG 激光器的稳区分析和实验研究 ···················262
298. 大功率衍射极限准连续 1.34 μm 激光器的研究 ···················262
299. Image processing in spectral domain optical coherence tomography with phase shifting interferometry ···················263
301. High efficiency generation of 355 nm radiation by extra-cavity frequency conversion ···················264
302. High-power continuous wave green beam generation by use of simple linear cavity with side-pumped module ···················264
303. Using relaxation time approximation for collisions in laser-cluster interaction simulation ···················265
304. Multi-wavelength generation based on cascaded Raman scattering and self-frequency doubling in KTA ···················265
305. Compact and tunable mid-infrared source based on a 2 μm dual-wavelength KTiOPO$_4$ optical parametric oscillator ···················266
306. High-average-power, high-repetition-rate dual signal optical parametric oscillator based on PPMgLN ···················266
307. 基于常规组合光谱仪下的 LIBS 快速铅检测研究 ···················267
308. 角度调谐双信号光运转准相位匹配光学参量的产生 ···················267
309. 径向、切向热透镜效应对高亮度激光器输出特性的影响 ···················268
310. 连续波可调谐内腔光学参量振荡器及橙红光源 ···················268
311. 正向抽运脉冲染料激光放大时间特性研究 ···················269

312. Theoretical study of tunable mid-infrared radiation based on sum-frequency generation in nonresonant fresnel phase matching ZnSe and GaP crystal ·········269
313. LDA 侧面泵浦 1 319 nm / 1 338 nm 双波长激光器的研究 ·········270
314. 高效率紧凑紫外 355 nm 激光器 ·········270
315. 高峰值 266 nm 紫外激光器 ·········271
316. Tunable and coherent nanosecond 7.2-12.2 μm mid-infrared generation based on difference frequency mixing in ZnGeP$_2$ crystal ·········271
317. 基于 GaSe 差频产生 8~19 μm 可调中红外辐射 ·········272
318. 165 W high stability green laser based on composite ceramic Nd:YAG crystal ·········272
319. Generation of 1 178 nm based on cascaded stimulated Raman scattering in KTA crystal ·········273
320. High power widely tunable all-solid-state pulsed titanium-doped sapphire laser ·········273
321. Generation of tunable coherent nanosecond 8-12 μm mid-infrared pulses based on difference frequency generation in GaSe and ZnGeP$_2$ ·········274
322. A low-threshold efficient KTA OPO by a fiber-coupled diode-end-pumped Nd:YVO$_4$ laser ·········274
323. Efficient diode-end-pumped dual-wavelength Nd, Gd:YSGG laser ·········275
324. High efficiency 1 342 nm Nd:YVO$_4$ laser in-band pumped at 914 nm ·········275
325. Effects of crystal parameters on power and efficiency of Nd:YVO$_4$ laser in-band pumped by 914 nm laser ·········276
326. A continuous-wave tunable orange-red source based on sum-frequency generation in an intra-cavity periodically poled LiNbO$_3$ singly resonant optical parametric oscillator cavity ·········276
327. Comparison of eye-safe KTA OPOs pumped by Nd:YVO$_4$ and Nd:YLF lasers ·········277
328. High power widely tunable narrow linewidth all-solid-state pulsed titanium-doped sapphire laser 277
329. 飞秒抽运随机激光输出波形的可控性研究 ·········278
330. 激光二极管抽运共轴双晶体黄光激光器 ·········278
331. 铅的单脉冲 LIBS 定量光谱检测与比较 ·········279
332. 常温条件下 KTP 晶体应用于 1 319 nm 激光三倍频相位匹配角的测量 ·········279
333. 高效腔外频率变换紫外激光器 ·········280
334. 激光二极管侧面泵浦高功率 266 nm 紫外激光 ·········280
335. 强光电离团簇并行数值模拟 ·········281
336. 基于序列脉冲激光法高速空气流场速度的研究 ·········281
337. 利用通用多媒体技术的激光光斑测量方法的研究 ·········282
338. Continuous-wave mid-infrared intra-cavity singly resonant PPLN-OPO under 880 nm in-band pumping ·········282
339. Continuous-wave intra-cavity singly resonant optical parametric oscillator with resonant wave output coupling ·········283

340. Continuous-wave Nd:GYSGG laser around 1.3 μm ·········284
341. Stimulated emission cross section of the $^4F_{3/2} \rightarrow {}^4I_{11/2}$ transition of Nd:GYSGG ·········284
342. High-power high-stability Q-switched green laser with intracavity frequency doubling of diode-pumped composite ceramic Nd:YAG laser ·········285
343. Efficient continuous-wave eye-safe region signal output from intra-cavity singly resonant optical parametric oscillator ·········285
344. 双电子 CO_2 模型及其高次谐波模拟 ·········286
345. H_2^+ 和 H_3^+ 系统的电子频谱与高次谐波谱 ·········286
346. Dynamic Stark effect on XUV-laser-generated photoelectron spectra: Numerical experiment on atomic hydrogen ·········287
347. Dynamic Stark effect and interference photoelectron spectra of H_2^+ ·········288
348. Continuous-wave Nd:GYSGG laser properties in 1.3 and 1.4 μm regions based on $^4F_{3/2}$ to $^4I_{13/2}$ transition ·········289
349. Time-dependent Born–Oppenheimer approximation approach for Schrödinger equation: Application to H_2^+ ·········289
350. Diode-pumped continuous-wave quasi-three-level Nd:GYSGG laser at 937 nm ·········290
351. The double-ended 750 nm and 532 nm laser output from PPLN-FWM ·········290
352. 双电子 H_3^+ 体系的高次谐波谱的数值模拟 ·········291
353. 914 nm 基态高斯塔克能级共振抽运的高效率 Nd:YVO$_4$ 激光器 ·········291
354. 880 nm 共振抽运连续波内腔单谐振光学参量振荡器及其逆转换 ·········292
355. 提高全固态内腔倍频绿光激光器功率稳定性的一种新方法 ·········292
356. High-power picosecond 355 nm laser based on $La_2CaB_{10}O_{19}$ crystal ·········293
357. 5.2-W high-repetition-rate eye-safe laser at 1 525 nm generated by Nd:YVO$_4$–YVO$_4$ stimulated Raman conversion ·········293
358. A 7.81 W 355 nm ultraviolet picosecond laser using $La_2CaB_{10}O_{19}$ as a nonlinear optical crystal ·········294
359. 12.45 W wavelength-locked 878.6 nm laser diode in-band pumped multisegmented Nd:YVO$_4$ laser operating at 1 342 nm ·········294
360. Efficient Nd:YVO$_4$ self-Raman laser in-band pumped by wavelength-locked laser diode at 878.7 nm ·········295
361. A novel CW yellow light generated by a diode-end-pumped intra-cavity frequency mixed Nd:YVO$_4$ laser ·········295
362. 28.2 W 波长锁定 878.6 nm 激光二极管共振抽运双晶体 1 064 nm 激光器 ·········296
363. 波长锁定 878.6 nm 激光二极管抽运 Nd:YVO$_4$ 1 064 nm 激光器 ·········296
364. Efficient eye-safe Nd:YVO$_4$ selfRaman laser in-band pumped at 914 nm ·········297

365. A non-critically phase matched KTA optical parametric oscillator intracavity pumped by an actively Q-switched Nd:GYSGG laser with dual signal wavelengths ·············297

366. High-order Stokes generation in a KTP Raman laser pumped by a passively Q-switched ND:YLF laser ·············298

367. 飞秒激光等离子体在高超声速飞行器减阻中的应用 ·············298

368. 激光二极管抽运 Nd:YAG/Nd:YVO$_4$ 共轴双晶体 Cr:YAG 被动调 Q 激光器 ·············299

369. 矢量多高斯 - 谢尔模型光束在大气湍流中上行链路中的传输特性 ·············299

370. Random lasing in a colloidal quantum dot-doped disordered polymer ·············300

371. Simultaneous dual-wavelength eye-safe KTP OPO intracavity pumped by a Nd:GYSGG laser 300

372. Efficient 914-nm Nd:YVO$_4$ laser under double-end polarized pumping ·············301

373. High efficiency actively Q-switched Nd:YVO$_4$ self-Raman laser under 880 nm in-band pumping 301

374. Dynamic interference photoelectron spectra in double ionization: numerical simulation of 1D helium ·············302

375. 轨道电子的三重动量相关：氖原子的从头计算 ·············302

376. Optical parametric oscillation in a random polycrystalline medium ·············303

377. 9.80-W and 0.54-mJ actively Q-switched Nd:YAG/Nd:YVO$_4$ hybrid gain intracavity raman laser at 1 176 nm ·············303

378. 16.7 W 885 nm diode-side-pumped actively Q-switched Nd:YAG/YVO$_4$ intracavity raman laser at 1 176 nm ·············304

379. High power, widely tunable dual-wavelength 2 μm laser based on intracavity KTP optical parametric oscillator ·············304

380. Thermal management of Nd:YVO$_4$ laser by 808-/880-nm dual-wavelength pumping ·············305

381. Compact and flexible dual-wavelength laser generation in coaxial diode-end-pumped configuration ·············305

382. Characterizing carrier envelope phase of an isolated attosecond pulse with annular photoionization momentum spectra ·············306

383. Electron vortices in photoionization by a pair of elliptically polarized attosecond pulses ·············306

384. High-pulse-energy mid-infrared optical parametric oscillator based on BaGa$_4$Se$_7$ crystal pumped at 1.064 μm ·············307

385. Ultrastable, high efficiency picosecond green light generation using $K_3B_6O_{10}Br$ series nonlinear optical crystals ·············307

386. Power-ratio tunable dual-band Nd:GYSGG laser at 0.94 μm and 1.06 μm ·············308

387. High efficiency actively Q-switched Nd:YVO$_4$ self-Raman laser under 914 nm in-band pumping 308

388. Generation of high-output yellow light by intracavity doubling Nd:YAG-Nd:YVO$_4$ hybrid gain Raman laser ·············309

389. Efficient yellow-light generation based on a Q-switched frequencydoubled self-Raman laser……309
390. High-performance second-Stokes generation of a Nd:YVO$_4$/YVO$_4$ Raman laser based on a folded coupled cavity ……310
391. Dual-signal-resonant optical parametric oscillator intracavity driven by a coaxially end-pumped laser with compound gain media ……311
392. Wavelength tuning and power enhancement of an intracavity Nd:GdVO$_4$-BaWO$_4$ Raman laser using an etalon ……312
393. Dynamic Stark induced vortex momentum of hydrogen in circular fields ……312
394. Symmetric electron vortices of hydrogen ionized by orthogonal elliptical Fields. ……313
395. Odd-fold-symmetric spiral momentum distributions and their Stark distortions in hydrogen ……313
396. Stable, high power, high efficiency picosecond ultraviolet generation at 355 nm in K$_3$B$_6$O$_{10}$Br crystal ……314
397. Compact, efficient and widely tunable 2-μm high-repetition-rate optical parametric oscillators ……315
398. 10.3-W actively Q-switched Nd: YVO$_4$/YVO$_4$ folded coupled-cavity Raman laser at 1 176 nm……316
399. Dual-wavelength eye-safe optical parametric oscillator intracavity driven by a coaxially end pumped laser ……317
400. Compact low-energy high-repetition rate laser-induced breakdown spectroscopy for lead element detection ……318
401. 基于相机式激光光束参数测量精度的影响因素分析……318
402. 航天器用内螺纹的气体湍流激光多普勒式检测……319
403. 基于面阵探测器的激光光束参数测量精度的影响因素分析……319
404. 基于小波阈值法的激光雷达回波信号去噪研究……320
405. 四波长激光准六自由度激光增材制造异形永磁件……320
406. Dual-wavelength intracavity Raman laser driven by a coaxially pumped dual-crystal fundamental laser ……321
407. Widely tunable eye-safe optical parametric oscillator with noncollinear phase-matching in a ring cavity ……321
408. High energy and tunable mid-infrared source based on BaGa$_4$Se$_7$ crystal by single-pass difference-frequency generation ……322
409. A single-frequency intracavity Raman laser ……322
410. Efficient and tunable 1.6-μm MgO:PPLN optical parametric oscillator pumped by Nd: YVO$_4$/YVO$_4$ Raman laser ……323
411. Intracavity-pumped, mid-infrared tandem optical parametric oscillator based on BaGa$_4$Se$_7$ crystal ……323
412. AC Stark effect on vortex spectra generated by circularly polarized pulses ……324

413. Laser performance of neodymium-and erbium-doped GYSGG crystals ……………325
414. Evolution of the Raman beam quality in a folded-coupled-cavity Nd: YVO$_4$/YVO$_4$ Raman laser 326
415. Carrier envelope phase description for an isolated attosecond pulse by momentum vortices ……326
416. 由动态 Stark 效应诱导的氢原子涡旋动量分布 ………………………………………327
417. 基于 Nd: YAG/LBO 倍频蓝光的全固态激光综合实验系统 ……………………………327
418. 基于 TDLAS 技术的人体呼气末二氧化碳在线检测 ……………………………………328

第二章　光电子及光纤技术

1. 掺铒光纤激光放大的研究 ……………………………………………………………330
2. 四能级系统光纤激光放大器的理论分析 ………………………………………………330
3. 光学广义载波条纹图的计算机辅助分析 ………………………………………………331
4. 1 477 nm LD 泵浦掺铒光纤放大器的研究 ……………………………………………331
5. 电极处势垒对新结构器件电子输运的影响 ……………………………………………332
6. Multipoint velocity interferometer system for any reflector ……………………………332
7. Optimal conditions of coupling between the propagating mode in a tapered fiber and the given WG mode in a high-Q microsphere ………………………………………………………333
8. An investigation of a tapered fiber-microsphere coupling system with gain and evanescent-field sensing device ………………………………………………………………………333
9. 基于微球与锥形光纤耦合系统的窄带信道下载滤波器 …………………………………334
10. Novel alternating current electroluminescent devices with an asymmetric structure based on a polymer heterojunction ………………………………………………………………334
11. 基于微球-锥形光纤耦合系统的窄带信道下载滤波器 …………………………………335
12. 大气污染监测中的 DOAS 技术 ……………………………………………………335
13. Analysis of speckle in optical coherence tomography …………………………………336
14. Formulation of beam propagating through the organized tissues with polarization sensitive OCT 336
15. Tissue clearing of bio-tissues for optical coherence tomography ………………………337
16. Analyses on propagation and imaging properties of GRIN lens ………………………337
17. Research on parameters option of a pressure fiber sensor ………………………………338
18. The amplified model of erbium-doped fiber amplifier (EDFA) with cross-phaser modulation (XPM) ……………………………………………………………………………338
19. 紫外差分光学吸收法测量污染气体的实验研究 ………………………………………339
20. 新型单片式开关电源的电磁干扰及其抑制 ……………………………………………339
21. 长光程差分吸收光谱法 ………………………………………………………………340
22. Power distribution in Yb^{3+}-doped double-cladding fiber laser ……………………340

23. Absorption and emission of ErNbO$_4$ powder ··············341
24. 一种光纤压力传感器的设计理论分析 ··············342
25. Image distortion of optical coherence tomography ··············342
26. A monte carlo model of light propagation in nontransparent tissue ··············343
27. 2.2 W 掺 Yb^{3+} 双包层光子晶体光纤激光器 ··············343
28. 15 W 光子晶体光纤激光器的研究 ··············344
29. Supercontinuum generation at 1.6 μm region using a polarization-maintaining photonic crystal fiber ··············344
30. Theoretical study of double microcavity resonators system with absorption or gain ··············345
31. Conceptual design of LD side pumped high power disk fiber laser ··············345
32. High-speed spectral domain optical coherence tomography for imaging of biological tissues ··············346
33. Photoacoustic tomography imaging of biological tissues ··············346
34. Fourier domain optical coherence tomography for imaging of biological tissues ··············347
35. Multi-channel FBG sensing system using a dense wavelength demultiplexing module ··············348
36. Mobile on-line DOAS trace-gases monitoring system with fiber spectroscopy ··············349
37. Optical bistability and differential amplification in coupled nonlinear microcavity resonators ··············350
38. Optical phase shifting with acousto-optic devices ··············350
39. Tunable asymmetrical Fano resonance and bistability in a microcavity-resonator-coupled Mach-Zehnder interferometer ··············351
40. Multi-frequency and multiple phase-shift sinusoidal fringe projection for 3D profilometry ··············351
41. Crosstalk characteristics of resonant dispersion EDFAs in WDM systems ··············352
42. A photoacoustic tomography system for imaging of biological tissues ··············352
43. Analysis of cross-phase modulation in WDM systems ··············353
44. Small-signal analysis of cross-phase modulation instability in lossy fibres ··············353
45. Supercontinuum generation using a polarization-maintaining photonic crystal fibre by a regeneratively amplified Ti:sapphire laser ··············354
46. Spectral optical coherence tomography using two-phase shifting method ··············354
47. Spectral broadening in the 1.3 μm region using a 1.8-m-long photonic crystal fiber by femtosecond pulses from an optical parametric amplifier ··············355
48. Analysis and simulation of XPM intensity modulation ··············355
49. A method of simulating intensity modulation-direct detection WDM systems ··············356
50. 时域光声技术及其在生物组织检测中的应用 ··············356
51. Bandgap extension of disordered 1D ternary photonic crystals ··············357
52. 泵浦波长对光子晶体光纤产生超连续谱的影响 ··············357
53. 使用混合高渗制剂提高 OCT 的探测深度和清晰度 ··············358

54. 光纤激光器透镜耦合系统的优化设计 ···358
55. 高功率光子晶体光纤激光器及关键技术 ··359
56. 基于周期极化晶体的 Solc 型滤波器透射谱的研究 ·······························359
57. 大功率激光光纤透镜耦合系统设计 ···360
58. 光声技术在医疗成像中的应用 ···360
59. Two-dimensional photoacoustic imaging of blood vessel networks within biological tissues ·········361
60. Omnidirectional zero-ñ gap in symmetrical Fibonacci sequences composed of positive and negative refractive index materials ···361
61. Dispersion compensation methods for ultrahigh-resolution optical coherence tomography ·········362
62. Fiber-optic bending sensor for cochlear implantation ·································363
63. Reconstruction algorithm in photoacoustic tomography ······························364
64. Photoacoustic imaging of blood-vessel networks of biotissue ······················364
65. Spectral optical coherence tomography using three-phase shifting method ·······365
66. Improving dynamic response of a temperature only FBG sensor ·················365
67. Double-wavelength Fano resonance and enhanced coupled-resonator-induced transparency in a double-microcavity resonator system···366
68. Photoacoustic imaging: its current status and future development ···············367
69. Simulation study on sensitive detection of small absorbers in photoacoustic tomography ············368
70. Arbitrary three-phase shifting algorithm for achieving full range spectral optical coherence tomography ···369
71. Simulation on sensitive detection of small absorber in photoacoustic tomography ·······369
72. 基于 DPS 的双通道动平衡仪研制与应用 ··370
73. 含负折射率材料一维三元光子晶体的特性研究 ····································370
74. 液晶红外磁控双折射效应的研究 ··371
75. 光纤传输的脉冲展宽研究 ···371
76. 液晶的磁旋光特性 ···372
77. 三维光子晶体典型结构完全禁带的最佳参数理论分析 ··························372
78. 光声成像技术 ···373
79. Frequency response in photonic heterostructures consisting of single-negative materials ············373
80. Analysis of square-structured photonic crystal fibers using localized orthogonal function algorithm 374
81. A new type ultraflattened dispersion photonic crystal fiber with low confinement loss ···········374
82. Omnidirectional single-negative gap and in fibonacci sequences composed of single-negative materials ···375
83. 液晶磁致旋光的研究 ···375
84. 掺钕保偏光纤激光器的研究 ··376

85. 基于六角结构二维光子晶体绝对带隙的优化设计研究 ·········376
86. The pulse broadening study of Gauss-chirped pulse in optical fibers ·········377
87. Feasibility of photoacoustic tomography for ophthalmology ·········377
88. Low-cost high-resolution wavelength demodulator for multi-channel dynamic FBG sensing ······378
89. 保偏光纤激光器的实验研究 ·········378
90. 液晶磁控偏光特性的研究 ·········379
91. 含负折射率材料一维光子晶体的全方位带隙和缺陷模 ·········379
92. 波片和旋转器复合退偏的矩阵研究 ·········380
93. 一种激光打标控制系统的软件研究 ·········380
94. 基于阵列探测方式的时域光声成像系统 ·········381
95. CMI 编译码电路的设计 ·········381
96. 微环共振器的开关特性分析 ·········382
97. Spectral domain polarization sensitive optical coherence tomography based on the two phase method ·········382
98. Birefringence imaging of biological tissue by spectral domain polarization sensitive optical coherence tomography ·········383
99. Signal processing using wavelet transform in photoacoustic tomography ·········384
100. Proposal to produce coupled resonator-induced transparency and bistability using microresonator enhanced Mach-Zehnder interferometer ·········384
101. Tunable zero-phase-shift omnidirectional filter consisting of single-negative materials ·········385
102. Characterization of bent large-mode-area photonic crystal fiber ·········385
103. Effect of thiol on the holographic properties of TiO_2 nanoparticle dispersed acrylate photopolymer films ·········386
104. Bend-induced distortion in large mode area holey fibre ·········386
105. Structure and optical damage resistance of near-stoichiometric Zn:Fe:$LiNbO_3$ crystals ·········387
106. Performance comparisons between 10 Gb s^{-1} hybrid TDM/WDM and WDM systems ·········387
107. 增益导引和折射率导引在大模场单模光纤设计中的应用 ·········388
108. 八角格子光子晶体光纤的传输特性 ·········388
109. Tailoring optical transmission via the arrangement of compound subwavelength hole arrays ······389
110. A method to design transmission resonances through subwavelength apertures based on designed surface plasmons ·········389
111. The leakage current mechanisms in the Schottky diode with a thin Al layer insertion between $Al_{0.245}Ga_{0.755}N$/GaN heterostructure and Ni/Au Schottky contact ·········390
112. Thermal annealing behaviour of Al/Ni/Au multilayer on n-GaN Schottky contacts ·········391
113. Study of ultraflattende dispersion square-lattice photonic crystal fiber with low confinement loss 391

114. Influence of lattice symmetry and hole shape on the light enhanced transmission through the subwavelength hole arrays··················392
115. 调 Q 脉冲保偏光纤激光器的研究··················392
116. 掺铷保偏光纤放大器的研究··················393
117. 四方格子全固光子带隙光纤带隙特性研究··················393
118. 增益导引折射率反导引光纤激光特性的研究··················394
119. Analyses of the performances of 10 Gb s⁻¹ time-division multiplexing and wavelength- division multiplexing signals in single-mode fibers and non-zero dispersion-shifted fibers··················394
120. 通信波段液晶光电特性的实验研究··················395
121. 亚波长环形电磁结构的光学特性研究··················395
122. Loss properties of all-solid photonic band gap fibers with an array of rings··················396
123. Dynamical analysis of evanescent field lose based fiber laser sensing··················396
124. Study on the supermode and in-phase locking in multicore fiber lasers··················397
125. Supermode analysis in multi-core photonic crystal fiber laser··················397
126. Microstructured-core photonic-crystal fiber for ultra-sensitive refractive index sensing··················398
127. Fano resonance and spectral compression in a ring resonator drop filter with feedback··················398
128. 双端泵浦保偏光纤激光器··················399
129. Photonic crystal fiber SERS sensors··················399
130. Study on HDPE-PCF evanescent wave sensor··················400
131. 表面等离子体共振类熊猫型光子晶体光纤传感器··················400
132. 纳米银基底表面增强拉曼散射效应仿真及优化··················401
133. 亚波长周期性排列空气孔的传输特性研究··················401
134. 物联网产业的形成及智慧城市建设··················402
135. Transmission and group delay in a double microring resonator reflector··················402
136. Grapefruit fiber filled with silver nanowires surface plasmon resonance sensor in aqueous environments··················403
137. Characteristics of bend sensor based on two-notch Mach–Zehnder fiber interferometer··················403
138. Interference effect in a dual microresonator-coupled Mach–Zehnder interferometer··················404
139. A surface-plasmon-resonance sensor based on photonic-crystal-fiber with large size microfluidic channels··················404
140. A photonic crystal fiber based on surface Plasmon resonance temperature sensor with liquid core 405
141. Estimation of the fourth-order dispersion coefficient β_4··················405
142. Three-dimensional thermal analysis of 18-core photonic crystal fiber lasers··················406
143. 液晶光子晶体光纤电场传感的模式特性··················406
144. Supermode analysis of the 18-core photonic crystal filber laser··················407

145. Tunable thermo-optic switch based on fluid-filled photonic crystal fibers ············407
146. Research on the transverse mode competition in a Yb-doped 18-core photonic crystal fiber laser 408
147. Theoretical study on modulating group velocity of light in photonic crystal coupled cavity optical waveguide ············408
148. Theoretical and experimental researches on a PCF-based SPR sensor ············409
149. Microstructured polymer optical fiber-based surface plasmon resonance sensor ············409
150. Surface plasmon resonance sensor based on a novel grapefruit photonic crystal fiber ············410
151. Surface plasmon resonance sensor based on grapefruit fiber filled with silver nanowires ············410
152. 用非球面透镜制作光纤约 1:1 空间耦合器 ············411
153. Surface plasmon resonance refractive index sensor based on active photonic crystal fiber ············411
154. Ferrofluid-infiltrated microstructured optical fiber long-period grating ············412
155. 11 mJ all-fiber-based actively Q-switched fiber master oscillator power amplifier ············412
156. Agarose gel-coated LPG based on two sensing mechanisms for relative humidity measurement 413
157. 230 W average-power all-fiber-based actively Q-switched fiber master oscillator-power amplifier 414
158. Surface plasmon resonance sensor based on polymer photonic crystal fibers with metal nanolayers ············414
159. A reflective photonic crystal fiber temperature sensor probe based on infiltration with liquid mixtures ············415
160. The use of a dual-wavelength erbium-doped fiber laser for intra-cavity sensing ············416
161. Magneto-optical tunability of magnetic fluid infiltrated microstructured optical fiber ············416
162. Thermo-optic characteristics of micro-structured optical fiber infiltrated with mixture liquids ···417
163. Effects of heterogeneity on the surface plasmon resonance biosensor based on three-hole photonic crystal fiber ············418
164. Hollow-core photonic crystal fiber based on C_2H_2 and NH_3 gas sensor ············418
165. Numerical investigation of the microstructured optical fiber-based surface plasmon resonance sensor with silver nanolayer ············419
166. A surface plasmon resonance sensor based on a multi-core photonic crystal fiber ············419
167. 基于液体填充微结构光纤的新型光子功能器件 ············420
168. Intra-cavity absorption sensor based on erbium doped fiber laser ············420
169. 1.55 μm 波长处具有高非线性低限制损耗的八边形实心光子晶体光纤 ············421
170. 光纤光栅温度应变同时测量传感技术研究进展 ············421
171. 光纤气体传感器及其组网技术综述 ············422
172. 物联网与智慧城市的关系 ············422
173. 物联网是工具手段　智慧城市是目标 ············423

174. Low temperature sensitive intensity-interrogated magnetic field sensor based on modal interference in thin-core fiber and magnetic fluid ……423

175. Intracavity absorption multiplexed sensor network based on dense wavelength division multiplexing filter ……424

176. 700-kW-peak-power monolithic nanosecond pulsed fiber laser ……424

177. 978 nm single frequency actively Q-switched all fiber laser ……425

178. Dual-direction magnetic field sensor based on core-offset microfiber and ferrofluid ……425

179. Temperature sensing using photonic crystal fiber filled with silver nanowires and liquid ……426

180. Lensed water-core teflon-amorphous fluoroplastics optical fiber ……427

181. Magnetic field tunability of square tapered no-core fibers based on magnetic fluid ……428

182. Surface plasmon resonance temperature sensor based on photonic crystal fibers randomly filled with silver nanowires ……429

183. Fiber lasers and their applications [Invited] ……430

184. Temperature-insensitive optical fiber refractometer based on multimode interference in two cascaded no-core square fibers ……430

185. Green function method for the time domain simulation of pulse propagation ……431

186. Magnetic-field sensor based on core-offset tapered optical fiber and magnetic fluid ……431

187. 300-W-average-power monolithic actively Q-switched fiber laser at 1 064 nm ……432

188. Simulation of surface plasmon resonance temperature sensor based on liquid mixture-filling microstructured optical fiber ……432

189. A photonic crystal fiber sensor based on differential optical absorption spectroscopy for mixed gases detection ……433

190. 光谱频移的炸药熔铸温度网络监测系统研究……433

191. 基于两种光纤介质模型的长周期光纤光栅传感特性分析……434

192. 基于亚波长悬浮芯光纤的高灵敏度气体传感器……434

193. Simultaneous measurement of temperature and magnetic field based on a long period grating concatenated with multimode fiber ……435

194. Single-frequency fiber laser at 1 950 nm based on thulium-doped silica fiber ……436

195. Low-temperature cross-talk magnetic-field sensor based on tapered all-solid waveguide-array fiber and magnetic fluids ……437

196. Passive Q-switching of an all-fiber laser induced by the Kerr effect of multimode interference ……438

197. Surface plasmon resonance sensor based on D-shaped microstructured optical fiber with hollow core ……438

198. High-power all-fiber single-frequency erbium-ytterbium co-doped fiber master oscillator power amplifier ……439

199. An exposed-core grapefruit fibers based surface plasmon resonance sensor······439
200. 2 μm actively Q-switched all fiber laser based on stress-induced birefringence and commercial Tm-doped silica fiber ······440
201. Fiber ring laser sensor based on hollow-core photonic crystal fiber ······440
202. Simultaneous measurement of displacement and temperature based on thin-core fiber modal interferometer ······441
203. Surface plasmon resonance sensor based on exposed-core microstructured optical fibres ······441
204. Surface plasmon resonance sensor based on hollow-core PCFs filled with silver nanowires ······442
205. Simulation analysis of a temperature sensor based on photonic crystal fiber filled with different shapes of nanowires ······442
206. 基于银纳米颗粒的HCPCF SERS传感系统优化设计 ······443
207. All-fiber passively Q-switched fiber laser based on the multimode interference effect ······443
208. Multipoint hollow core photonic crystal fiber sensor network based on intracavity absorption spectroscopy ······444
209. Linearly polarized narrow linewidth single mode fiber laser and nonlinear phenomena ······444
210. 基于长周期光纤光栅和ZigBee组网技术的无线溶液折射率传感网络 ······445
211. 从工业物联网到智慧城市 ······445
212. Dual-wavelength fiber laser operating above 2 μm based on cascaded single-mode-multimode single-mode fiber structures ······446
213. ASE suppression in backward-pumped Er/Yb double-cladding fiber amplifier via cladding feedback ······446
214. A refractive index and temperature sensor based on surface plasmon resonance in an exposed-core microstructured optical fiber ······447
215. High refractive index surface plasmon resonance sensor based on a silver wire filled hollow fiber 447
216. Surface plasmon resonance sensor based on exposed-core microstructured optical fiber placed with A silver wire ······448
217. Remote gas pressure sensor based on fiber ring laser embedded with fabry-pérot interferometer and sagnac loop ······448
218. Remote magnetic field sensor based on intracavity absorption of evanescent field ······449
219. Temperature sensor based on photonic crystal fiber filled with liquid and silver nanowires ······449
220. Experimental investigation on spectral linewidth and relative intensity noise of high-power single-frequency polarization-maintained thulium-doped fiber amplifier ······450
221. Analysis of graphene-based photonic crystal fiber sensor using birefringence and surface plasmon resonance ······450

222. Platinum-scatterer-based random lasers from dye-doped polymer-dispersed liquid crystals in capillary tubes ·············451
223. Simultaneous magnetic field and temperature measurement based on no-core fiber coated with magnetic fluid ·············452
224. Analysis of hollow fiber temperature sensor filled with graphene-Ag composite nanowire and liquid ·············453
225. Temperature sensor based on fiber ring laser with sagnac loop ·············454
226. SPR sensor based on exposed-core grapefruit fiber with bimetallic structure ·············454
227. A dual-parameter sensor using a long-period grating concatenated with polarization maintaining fiber in sagnac loop ·············455
228. Low-temperature cross-sensitivity refractive index sensor based on single-mode fiber with periodically modulated taper ·············456
229. Low-temperature-sensitive relative humidity sensor based on tapered square no-core fiber coated with SiO_2 nanoparticles ·············457
230. Amplified spontaneous emission in distributed feedback active microcavities fabricated by the sol-gel dip-coating method ·············458
231. Backward-pumped 1 550 nm EYDF amplifier with ASE suppression by cladding feedback ······458
232. Temperature distribution of double-cladding high-power thulium-dope fiber amplifier ·············459
233. Dual-point automatic switching intracavity-absorption photonic crystal fiber gas sensor based on mode competition ·············459
234. Multidimensional microstructured photonic device based on all-solid waveguide array fiber and magnetic fluid ·············460
235. Compact CNT mode-locked Ho^{3+}-doped fluoride fiber laser at 1.2 μm ·············461
236. Simulation of LSPR sensor based on exposed-core grapefruit fiber with a silver nanoshell ·············462
237. Humidity sensor based on fabry-perot interferometer and intracavity sensing of fiber laser ······463
238. Relative humidity sensor based on no-core fiber coated by agarose-gel film ·············464
239. Improved numerical calculation of the single-mode-no-core-single-mode fiber structure using the fields far from cutoff approximation ·············465
240. Humidity sensor based on intracavity sensing of fiber ring laser ·············466
241. General description and understanding of the nonlinear dynamics of mode-locked fiber lasers ···467
242. Fiber ring laser temperature sensor based on liquid-filled photonic crystal fiber ·············468
243. Compact hundred-mW 2 μm single-frequency Thulium-doped silica fiber laser ·············468
244. Temperature self-compensation high-resolution refractive index sensor based on fiber ring Laser 469
245. Dynamic propagation of initially chirped airy pulses in a quintic nonlinear fiber ·············470

246. High sensitivity hollow fiber temperature sensor based on surface plasmon resonance and liquid filling ···470
247. Design of a tunable single-polarization photonic crystal fiber filter with silver-coated and liquid-filled air holes ···471
248. 5 kW near-diffraction-limited and 8 kW high-brightness monolithic continuous wave fiber lasers directly pumped by laser diodes···472
249. Linewidth-narrowed, linear-polarized single-frequency Thulium-doped fiber laser based on stimulated Brillouin scattering effect···473
250. A hollow-core photonic crystal fiber-based SPR sensor with large detection range ·············473
251. Polarization characteristics of high-birefringence photonic crystal fiber selectively coated with silver layers ···474
252. Tunable surface plasmon resonance sensor based on photonic crystal fiber filled with gold nanoshells ···474
253. Generation of 2.5 μm and 4.6 μm dispersive waves in kagome photonic crystal fiber with plasma production ···475
254. Fiber lasers and their applications: introduction···475
255. Review of recent progress on single-frequency fiber lasers [invited] ·································476
256. Simultaneous measurement of refractive index and temperature based on SPR in D-shaped MOF ···476
257. Hollow-fiber-based surface plasmon resonance sensor with large refractive index detection range and high linearity ···477
258. Surface plasmon resonance sensor based on photonic crystal fiber filled with gold-silica-gold multilayer nanoshells ···478
259. Megawatt-peak-power picosecond all-fiber-based laser in MOPA using highly Yb^{3+}-doped LMA phosphate fiber···479
260. Tunable polarization filter based on high-birefringence photonic crystal fiber filled with silver wires ···479
261. 97-μJ single frequency linearly polarized nanosecond pulsed laser at 775 nm using frequency doubling of a high-energy fiber laser system ···480
262. A novel variable baseline visibility detection system and its measurement method ···········480
263. 高功率双包层掺铒光纤放大器温度分布特性···481
264. 激光雷达用高性能光纤激光器··481
265. High power all fiber-based ultrafast lasers ···482
266. Automatic channel-switched intracavity-absorption acetylene sensor based on mode-competition via Sagnac loop filter ···482

267. High-resolution temperature sensor through measuring the frequency shift of single-frequency Erbium-doped fiber ring laser ············483

268. Investigation of ASE and SRS effects on 1 018 nm short-wavelength Yb^{3+}-doped fiber laser ······483

269. Linewidth-narrowed 2-μm single-frequency fiber laser based on stimulated Brillouin scattering effect ············484

270. Extended linear cavity 2 μm single-frequency fiber laser using Tm-doped fiber saturable absorber 484

271. 空心 Kagome 光子晶体光纤中等离子诱导产生的色散波 ············485

272. 金属 - 电介质微盘阵列红外吸收器的光学特性分析 ············485

273. 基于保偏光纤和 LPFG 的 Sagnac 环温度及环境折射率双参量光纤传感器研究 ············486

274. Enhancement and modulation of photonic spin Hall effect by defect modes in photonic crystal with grapheme ············486

275. Realization of tunable Goos-Hänchen effect with magneto-optical effect in graphene ············487

276. Compact CNT mode-locked Ho^{3+}-doped fluoride fiber laser at 1.2 μm ············487

277. A highly sensitive magnetic field sensor based on a tapered microfiber ············488

278. High-resolution temperature sensor based on single-frequency ring fiber laser via optical heterodyne spectroscopy technology ············488

279. Megawatt-peak-power picosecond all-fiber-based laser in MOPA using highly Yb^{3+}-doped LMA phosphate fiber ············489

280. Numerical simulation of reflective infrared absorber based on metal and dielectric nanorings ······489

281. Liquid crystal-modulated tunable filter based on coupling between plasmon-induced transparency and cavity mode ············490

282. The research on the design and performance of 7 × 1 pump combiners ············490

283. Influence of seed power and gain fiber temperature on output linewidth in single-frequency Er^{3+}/Yb^{3+} co-doped fiber amplifier ············491

284. High-energy, 100-ns, single-frequency all-fiber laser at 1 064 nm ············491

285. Dual-wavelength noise-like pulse generation in passively mode-locked all-fiber laser based on MMI effect ············492

286. Theoretical study and design of third-order random fiber laser ············492

287. Switchable and tunable dual-wavelength Er-doped fiber ring laser with single-frequency lasing wavelengths ············493

288. 1-kW monolithic narrow linewidth linear-polarized fiber laser at 1 030 nm ············493

289. Compact bi-direction pumped hybrid double-cladding EYDF amplifier ············494

290. 基于 915 nm 半导体激光单端前向抽运的单纤准单模 2 kW 全光纤激光振荡器 ············494

291. 基于激光器内腔调制的低探测极限折射率传感系统 ············495

292. 基于 DOAS 的消防应急救援多气体快速遥感仪 ············495

293. 有机金属卤化物钙钛矿薄膜中的光诱导载流子动力学和动态带重整效应 ……………496
294. Highly sensitive dual-wavelength fiber ring laser sensor for the low concentration gas detection   497
295. High-resolution temperature sensor based on intracavity sensing of fiber ring laser ……………498
296. Ultrasharp LSPR temperature sensor based on grapefruit fiber filled with a silver nanoshell and liquid ……………………………………………………………………………………………499
297. Simultaneous measurement of temperature and relative humidity based on a microfiber Sagnac loop and MoS_2 ………………………………………………………………………………………500
298. Multi-direction bending sensor based on supermodes of multicore PCF laser ……………500
299. All-fiber seawater salinity sensor based on fiber laser intracavity loss modulation with low detection limit ……………………………………………………………………………………………501
300. Ultrafine frequency linearly tunable single-frequency fiber laser based on intracavity active tuning ………………………………………………………………………………………………501
301. A highly sensitive magnetic field sensor based on a tapered microfiber ……………502
302. Simultaneous measurement of relative humidity and temperature using a microfiber coupler coated with molybdenum disulfide nanosheets ……………………………………………………503
303. Opening up dual-core microstructured optical fiber-based plasmonic sensor with large detection range and linear sensitivity ……………………………………………………………………504
304. High-sensitivity magnetic field sensor based on a dual-core photonic crystal fiber ……………504
305. Microfiber coupler with a Sagnac loop for water pollution detection ……………505
306. Square-lattice alcohol-filled photonic crystal fiber temperature sensor based on a Sagnac interferometer ……………………………………………………………………………………505
307. Tunability of Hi-Bi photonic crystal fiber integrated with selectively filled magnetic fluid and microfluidic manipulation ………………………………………………………………………506
308. 1.7-μm thulium fiber laser with all-fiber ring cavity ……………………………507
309. All-fiberized single-frequency silica fiber laser operating above 2 μm based on SMS fiber devices   507
310. Hundred-watts-level monolithic narrow linewidth linearly-polarized fiber laser at 1 018 nm ……508
311. Highly sensitive chloride ion concentration measurement based on a multitaper modulated fiber   509
312. A novel photonic crystal fiber sensor with three d-shaped holes based on surface plasmon resonance ………………………………………………………………………………………………509
313. Analysis of a photonic crystal fiber sensor with reuleaux triangle ……………510
314. Research of laser-induced underwater communication zoom optical system ……………511
315. Numerical investigation of high efficiency random fiber lasers at 1.5 μm ……………512
316. Thulium doped silica fiber laser operating in single-longitudinal-mode at a wavelength above 2 μm ………………………………………………………………………………………………513

317. Tunable CW all-fiber optical parametric oscillator based on the cascaded single-mode-multimode-single-mode fiber structures ········· 513

318. Single-frequency, ultra-narrow linewidth hybrid brillouin-thulium fiber laser based on In-band pumping ········· 514

319. An FBG-based high-resolution temperature sensor through measuring the beat frequency of single-frequency ring fiber laser ········· 514

320. 有源光纤中稀土离子激光上能级寿命测量的研究 ········· 515

321. 重复频率 1.2GHz 皮秒脉冲全光纤掺镱激光器 ········· 515

322. 基于 HCPCF SERS 传感器的吡啶痕量检测 ········· 516

323. 1mJ 窄线宽掺镱脉冲光纤放大器 ········· 516

324. 基于 MMI 滤波器的可调谐连续光全光纤 OPO ········· 517

第三章　太赫兹技术

第一节　太赫兹辐射源

1. THz 辐射的研究和应用新进展（New research progress of THz radiation）········· 519

2. Analysis of surface-emitted terahertz-wave difference frequency generation in slant-stripe-type MgO-doped periodically poled lithium niobate ········· 519

3. Study of tunable terahertz-wave generation in isotropic semiconductor crystals based on dual-wavelength KTP-OPO operating near degenerate point ········· 520

4. Theoretical investigation of dual-wavelength terahertz wave generation based on slant-stripe-type periodic poled lithium niobate crystal ········· 521

5. Theoretical study of dual-wavelength PPKTP-OPO as a source of DFG THz-wave ········· 522

6. Widely tunable, dual-signal-wave optical parametric oscillator for terahertz generation by using two periodically poled crystals ········· 522

7. 基于光学方法的太赫兹辐射源 ········· 523

8. 利用各向同性半导体晶体差频产生可调谐 THz 辐射的理论研究 ········· 523

9. 光泵重水气体产生 THz 激光的半经典理论分析 ········· 524

10. 抽运光强度对光学抽运重水气体产生 THz 激光的影响分析 ········· 524

11. 紧凑型超辐射光泵重水气体 THz 激光器的研制 ········· 525

12. The numerical calculation and analyze of the pulse-laser pumped D_2O gas Terahertz laser ········· 525

13. Effects of pump source on spectra of optically pumped sub-millimeter wave laser ········· 526

14. Theoretical study of phase-matching properties for tunable terahertz-wave generation in isotropic nonlinear crystals ········· 526

15. $ZnGeP_2$ 晶体差频产生 THz 波的研究 ········· 527

16. Study of optimal gas pressure in optically pumped D_2O gas terahertz laser ·················527
17. Tera-Hz radiation source by deference frequency generation (DFG) and TPO with all solid state lasers ·················528
18. 基于 GaSe 和 $ZnGeP_2$ 晶体差频产生可调谐太赫兹辐射的理论研究 ·················529
19. 产生太赫兹辐射源的 Nd:YAG 双波长准连续激光器 ·················529
20. 紧凑型超辐射光泵重水气体 THz 激光器的研制 ·················530
21. Collinear phase-matching study of terahertz-wave generation via difference frequency mixed in GaAs and Inp ·················530
22. The numerical calculation and analyze of the pulse-laser pumped D_2O Gas Terahertz laser ·········531
23. Simultaneous all-solid-state multi-wavelength lasers-a promising pump source for generating highly coherent terahertz waves ·················531
24. Investigation of pump-wavelength dependence of terahertz-wave parametric oscillator based on $LiNbO_3$ ·················532
25. 基于钽酸锂晶体的太赫兹波参量振荡器运转特性的研究 ·················532
26. High-energy, continuously tunable intracavity terahertz-wave parametric oscillator ·················533
27. The generation of THz frequency comb via surface-emitted optical rectification of fs-pulses in periodically poled lithium niobate ·················533
28. Compact and widely tunable terahertz source based on a dual-wavelength intracavity optical parametric oscillation ·················534
29. Enhancement of terahertz wave difference generation based on a compact walk-off compensated KTP OPO ·················534
30. Threshold analysis of THz-wave parametric oscillator ·················535
31. 基于闪锌矿晶体中受激电磁耦子产生可调谐太赫兹波的理论研究 ·················535
32. 周期结构 GaAs 晶体 ps 脉冲差频产生窄带 THz 辐射的研究 ·················536
33. Study on the generation of high-power terahertz wave from surface-emitted THz-wave parametric oscillator with $MgO:LiNbO_3$ crystal ·················536
34. Terahertz difference frequency generation in GaSe based on a doubly-resonant walk-off compensated KTP OPO ·················537
35. Theoretical study on the generation of THz sub-comb via surface-emitted optical rectification of ultra-short pulse in periodically poled lithium niobate ·················537
36. Terahertz-wave parametric oscillator with a misalignment-resistant tuning cavity ·················538
37. p-polarized Cherenkov THz wave radiation generated by optical rectification for a Brewster-cut $LiNbO_3$ crystal ·················538
38. High-power terahertz radiation from surface-emitted THz-wave parametric oscillator ·················539
39. High-power terahertz radiation based on a compact eudipleural THz-wave parametric oscillator   539

40. Output enhancement of a THz wave based on a surface-emitted THz-wave parametric oscillator 540

41. 级联差频产生太赫兹辐射的理论研究 ··················540

42. 铌酸锂晶体中参量振荡产生高功率可调谐太赫兹波的实验研究 ··········541

43. Theoretical analysis and numerical simulation of terahertz wave generation and modulation based on GaAs:O ··················541

44. Analysis on characteristic and application of THz frequency comb and THz sub-comb ··········542

45. Terahertz generation by optical rectification in zincblende crystals with arbitrary crystal-oriention 542

46. 周期极化 GaAs 晶体中差频产生太赫兹辐射的研究 ··········543

47. 太赫兹波及其常用源··················543

48. Theory of monochromatic terahertz generation via Cherenkov phase-matched difference frequency generation in $LiNbO_3$ crystal ··················544

49. High-powered tunable terahertz source based on a surface-emitted terahertz-wave parametric oscillator ··················544

50. THz-wave difference frequency generation by phase-matching in $GaAs/Al_xGa_{1-x}$ as asymmetric quantum well ··················545

51. THz source based on optical Cherenkov radiation ··········545

52. 硅棱镜耦合输出 THz 波参量振荡器的实验研究··················546

53. Design and threshold analysis for a novel intracavity THz-wave parametric oscillator ··········546

54. Design of $GaAs/Al_xGa_{1-x}$ As asymmetric quantum wells for THz-wave by difference frequency generation ··················547

55. Investigation on phase matching in a THz-wave parametric oscillator ··········547

56. Coupled-mode theory for Cherenkov-type guided-wave terahertz generation via cascaded difference frequency generation··················548

57. Widely tunable, monochromatic THz generation via Cherenkov-type difference frequency generation in an asymmetric waveguide ··················549

58. Efficient continuous-wave 1053-nm Nd:GYSGG laser with passively Q-switched dual-wavelength operation for terahertz generation··················549

59. A high-energy, low-threshold tunable intracavity terahertz-wave parametric oscillator with surface-emitted configuration ··················550

60. Frequency tuning characteristics of a THz-wave parametric oscillator ··········550

61. Intersubband absorption properties of $GaAs/Al_xGa_{1-x}As$ asymmetric quantum well based on optical difference frequency ··················551

63. High-power tunable terahertz generation from a surface-emitted THz-wave parametric oscillator based on two $MgO:LiNbO_3$ crystals··················552

64. High energy terahertz parametric oscillator based on surface-emitted configuration ··········552

65. 级联参量振荡产生太赫兹辐射的理论研究 ···553
66. 小型化外腔可调谐 THz 参量振荡器 ··553
67. 表面出射太赫兹波参量振荡器的设计与增强输出 ·······························554
68. The widely tunable THz generation in QPM-GaAs crystal pumped by a near-degenerate dual-wavelength KTP OPO at around 2.127 μm ··554
69. High-energy, tunable intracavity terahertz-wave parametric oscillator with surface-emitted configuration ···555
70. Monochromatic Cherenkov THz source pumped by a singly resonant optical parametric oscillator 555
71. Widely tunable and monochromatic terahertz difference frequency generation with organic crystal DSTMS ··556
72. A study of the multi-mode pumping of terahertz parametric oscillators ················556
73. Energy scaling of a tunable terahertz parametric oscillator with a surface emitted configuration ···557
74. Investigation of a terahertz-wave parametric oscillator using LiTaO₃ with the pump-wavelength tuning method ···557
75. 太赫兹波光学参量效应放大特性的理论研究 ·······································558
76. Study on terahertz parametric oscillator pumped by multi-transverse-mode lasers ········558
77. 太赫兹参量振荡器研究进展 ···559
78. Terahertz wave parametric oscillations at polariton resonance using a MgO:LiNbO₃ crystal ······559
79. Investigation on terahertz generation at polariton resonance of MgO:LiNbO₃ by difference frequency generation ···560
80. Investigation on terahertz parametric oscillators using GaP crystal with a noncollinear phase-matching scheme ··560
81. Investigation of terahertz generation from bulk and periodically poled LiTaO₃ crystal with a cherenkov phase matching scheme ···561
82. Theoretical analysis of terahertz generation in periodically inverted nonlinear crystals based on cascaded difference frequency generation process ··561
83. Investigation on frequency mixing effects in terahertz parametricoscillator with a noncollinear phase-matching scheme ···562
84. 非共线相位匹配太赫兹波参量振荡器级联参量过程的研究 ···················562
85. Terahertz fiber laser based on a novel crystal fiber converter ····························563
86. 高能量、快速可调谐太赫兹参量振荡器 ···563
87. High-power high-brightness terahertz source based on nonlinear optical crystal fiber ·········564
88. Widely tunable and monochromatic terahertz difference frequency generation with organic crystal 2-(3-(4-hydroxystyryl)-5, 5-dime-thylcyclohex-2-enylidene) malononitrile ·················565
89. Green laser induced terahertz tuning range expanding in KTiOPO₄ terahertz parametric oscillator 565

90. Molecular design on isoxazolone-based derivatives with large second-order harmonic generation effect and terahertz wave generation ……566

91. High-energy terahertz wave parametric oscillator with a surface-emitted ring-cavity configuration 567

92. Efficient phase-matching for difference frequency generation with pump of Bessel laser beams ……567

93. Widely-tunable high-repetition-rate terahertz generation in GaSe with a compact dualwavelength KTP OPO around 2 μm ……568

94. Numerical study of compact terahertz gas laser based on photonic crystal fiber cavity ……569

95. Compact high-repetition-rate monochromatic terahertz source based on difference frequency generation from a dual-wavelength Nd:YAG laser and DAST crystal ……570

96. High-repetition-rate terahertz generation in QPM GaAs with a compact efficient 2-μm KTP OPO ……571

97. Compact high-repetition-rate terahertz source based on difference frequency generation from an efficient 2-μm dual-wavelength KTP OPO ……572

98. Enhanced stimulated polariton scattering in $KTiOPO_4$ terahertz parametric oscillator based on green laser pumping ……573

99. Widely-tunable terahertz parametric oscillator based on MgO-doped near-stoichiometric $LiNbO_3$ crystal ……573

100. High-repetition-rate, widely tunable terahertz generation in GaSe pumped by a dual-wavelength KTP-OPO ……574

101. THz wave parametric oscillator with a surface-emitted ring-cavity configuration ……575

102. 小型化可调谐太赫兹波参量振荡器的研究……576

103. High-average-power, high-repetition-rate tunable terahertz difference frequency generation with GaSe crystal pumped by 2 μm dual-wavelength intracavity KTP optical parametric oscillator ……577

104. Energy scaling and extended tunability of terahertz wave parametric oscillator with MgO-doped near-stoichiometric $LiNbO_3$ crystal ……578

105. Compact and stable high-repetition-rate terahertz generation based on an efficient coaxially pumped dual-wavelength laser ……579

106. Compact high-repetition-rate monochromatic terahertz source based on difference frequency generation from a dual-wavelength Nd:YAG laser and DAST crystal ……580

107. Optically pumped terahertz sources ……581

108. Bursts of efficient terahertz radiation with saturation effect from metal-based ferromagnetic heterostructures. ……582

109. Low-threshold terahertz-wave generation based on a cavity phase-matched parametric process in a Fabry–Perot microresonator. ……582

110. Investigation of stimulated polariton scattering from the B_1-symmetry modes of the $KNbO_3$ crystal ·····583

111. High-energy and ultra-wideband tunable terahertz source with DAST crystal via difference frequency generation ·····584

112. Theoretical study of organic crystal-based terahertz-wave difference frequency generation and up-conversion detection ·····585

113. Energy scaling and extended tunability of a ring cavity terahertz parametric oscillator based on $KTiOPO_4$ crystal ·····586

114. Conduction-band nonparabolicity effect on refractive index and phase match in asymmetric quantum wells pumped by two infrared beams ·····587

115. Stimulated polariton scattering in β-BTM crystal ·····588

116. Synchronous dual-wavelength pulse generation in coaxial pumping scheme and its application in terahertz difference frequency generation ·····589

117. THz radiation modulated by confinement of transient current based on patterned CoFeB/Pt heterostructures ·····590

118. 基于 MgO:SLN 晶体的环形腔太赫兹参量振荡器 ·····590

119. 铁磁异质结构中的超快自旋流调制实现相干太赫兹辐射 ·····591

120. Rational structural design of benzothiazolium-based crystal HDB-T with high nonlinearity and efficient terahertz-wave generation. ·····591

121. Injection pulse-seeded terahertz-wave parametric generator with gain enhancement in wide frequency range ·····592

122. Active multifunctional terahertz modulator based on plasmonic metasurface ·····593

123. Tunable dual-color terahertz wave parametric oscillator based on KTP crystal ·····594

124. Enhanced terahertz wave generation via stokes wave recycling in non-synchronously picosecond pulse pumped terahertz source ·····595

125. Efficient ring-cavity terahertz parametric oscillator with pump recycling technique ·····596

126. Efficient terahertz generation via GaAs hybrid ridge waveguides ·····597

127. Optically pumped gas terahertz fiber laser based on gold-coated quartz hollow-core fiber ·····597

128. Two parallel polarized terahertz waves generation from quasi-phase-matching stimulated polariton scattering with periodically-inverted GaAs ·····598

129. Theoretical investigation on collinear phase matching stimulated polariton scattering generating THz waves with a KTP crystal ·····599

130. Simultaneous generation of two THz waves with bulk $LiNbO_3$ and four THz waves with PPLN by coupled optical parametric generation. ·····600

131. High-energy and ultra-wideband tunable monochromatic terahertz source and frequency domain system based on DAST crystal ·······601
132. A gain-boosted terahertz-wave parametric generator in high frequency tuning range via pulse-seed injection ·······602
133. 新型有机晶体及超宽带太赫兹辐射源研究进展 ·······602
134. 基于 DAST 晶体的高能量超宽带可调谐小型化差频 THz 辐射源研究 ·······603
135. 基于 DAST 晶体差频的可调谐 THz 辐射源 ·······603
136. 基于负曲率空芯光纤的光泵太赫兹光纤激光器的理论研究 ·······604
137. 基于超快电子自旋动力学的太赫兹辐射研究进展 ·······604

第二节　太赫兹传输及功能器件

1. Propagation characteristics of two-dimensional photonic crystals in the terahertz range ·······605
2. Characteristics of photonic band gaps in woodpile three-dimensional terahertz photonic crystals ···605
3. Design of terahertz photonic crystal fibers by finite difference frequency domain method ·······606
4. A novel woodpile three-dimensional terahertz photonic crystal ·······606
5. Transmission loss and dispersion in plastic terahertz photonic band-gap fibers ·······607
6. Novel optical controllable terahertz wave switch ·······607
7. Controllable terahertz wave attenuator ·······608
8. Low loss plastic Terahertz photonic band-gap fibres ·······608
9. 基于法布里 - 珀罗干涉仪的太赫兹波波长测试方法 ·······609
10. 基于法布里 - 珀罗干涉仪的太赫兹波长测试仪 ·······609
11. Terahertz liquid crystal tunable filter ·······610
12. Fe-doped polycrystalline CeO_2 as terahertz optical material ·······610
13. 太赫兹波在有限电导率金属空芯波导中的传输特性 ·······611
14. The propagation characteristics of THz radiation in hollow circular waveguides coated with different material films ·······611
15. Propagation characteristics of THz radiation in hollow elliptical waveguide ·······612
16. Time-dependent theoretical model for terahertz wave detector using a parametric process ·······612
17. Compact terahertz wave polarizing beam splitter ·······613
18. Proposal of an electrically controlled terahertz switch based on liquid-crystal-filled dual-metallic grating structures ·······613
19. Steady-state theoretical model for terahertz wave detector using a parametric process ·······614
20. Four-wave mixing model solutions for polarization control of terahertz pulse generated by a two-color laser field in air ·······614
21. THz 辐射大气传输研究和展望 ·······615

22. A ferroelectric polyvinylidene fluoride-coated porous fiber based surface-plasmon- resonance-like gas sensor in the terahertz region ············615
23. 基于非对称量子阱的太赫兹波调制器 ············616
24. The study on THz wave propagation feature in atmosphere ············616
25. Parameter selection and design considerations with MPOF evanescent wave sensor in the THz wavelength range ············617
26. Low-loss and birefringent terahertz polymer elliptical-tube waveguides ············617
27. A THz modulator use the photo-carrier surface plasma effect ············618
28. The guidance mechanism and numerical simulation of THz polymer hollow-core photonic crystal fiber ············618
29. Analysis on characteristic and application of THz frequency comb and THz sub-comb ············619
30. The physical theory and propagation model of THz atmospheric propagation ············619
31. Dielectric behavior of $CaCu_3Ti_4O_{12}$ ceramics in the terahertz range ············620
32. Directional terahertz beams realized by depth-modulated metallic surface grating structures ············620
33. Far-infrared dispersion of complex dielectric constant in the ferroelectric near-stoichiometric $LiNbO_3$:Fe ············621
34. Far-infrared dispersion of the complex dielectric constant in ferroelectric near-stoichiometric $LiNbO_3$:Ce ············621
35. Ultrahigh birefringent polymer terahertz fiber based on a near-tie unit ············622
36. Simulation of continuous terahertz wave transient state thermal effects on static water ············622
37. A simple birefringent terahertz waveguide based on polymer elliptical tube ············623
38. Single mode condition and power fraction of air-cladding total refractive guided porous polymer terahertz fibers ············623
39. 高双折射的混合格子太赫兹光子晶体光纤的设计与研究 ············624
40. THz 波在金属镀层空芯波导中传输的理论和实验研究 ············624
41. THz modulator based on the Drude model ············625
42. Propagation characteristics of THz radiation in hollow rectangle metal waveguide ············625
43. The study on THz wave propagation feature in atmosphere ············626
44. Electrically controlled broadband THz switch based on liquid-crystal-filled multi-layer metallic grating structures ············626
45. 基于液晶的可调谐太赫兹双折射滤波器的设计 ············627
46. 谐振环左手材料设计参数对太赫兹传输的影响 ············627
47. Terahertz photonic states in semiconductor-graphene cylinder structures ············628
48. The study of negative THz conductivity of graphene under the phonon scattering mechanism ············628
49. The dielectric behaviour of doped near-stoichiometric lithium niobate in the terahertz range ············629

50. 基于平行金属双柱的太赫兹波二维左手材料 ······629

51. Optical tuning of dielectric properties of $Ba_{0.6}Sr_{0.4}TiO_3$-$La(Mg_{0.5}Ti_{0.5})O_3$ ceramics in the terahertz range ······630

52. Optical tuning of dielectric properties of $SrTiO_3$:Fe in the terahertz range ······630

53. Hyperfine spectrum measurement of an optically pumped far-infrared laser with a Michelson interferometer ······631

54. Real propagation speed of the ultraslow plasmonic THz waveguide ······631

55. Effect of optical pumping on the momentum relaxation time of graphene in the terahertz range ···632

56. Resonance mode-switching in terahertz metamaterials based on varying gallium arsenide conductivity ······632

57. Optical control of terahertz nested split-ring resonators ······633

58. Optical tuning of dielectric properties of $LiNbO_3$: Mg in the terahertz range ······633

59. 太赫兹频段开口环谐振器的可调谐振模式转换 ······634

60. 基于太赫兹目标散射特性测试系统的设计与应用 ······634

61. 铁磁材料在太赫兹波段的研究进展 ······635

62. Modulation of dielectric properties of $KTaO_3$ in terahertz region via 532 nm continuous-wave laser ······635

63. Effect of an optical pump on the absorption coefficient of $Ba_{0.6}Sr_{0.4}TiO_3$-$La(Mg_{0.5}Ti_{0.5})O_3$ ceramics in the terahertz range ······636

64. Dynamic trapping of terahertz waves by silicon-filled metallic grating structure ······636

65. Guided modes in asymmetric negative-zero-positive index metamaterial waveguide in the terahertz regime ······637

66. Effect of optical pump on the dielectric properties of $SrTiO_3$ in terahertz range ······637

67. Low-loss terahertz waveguide with InAs-graphene-SiC structure ······638

68. Tunable ultra-wideband terahertz filter based on three-dimensional arrays of H-shaped plasmonic crystals ······639

69. Reflection-type electromagnetically induced transparency analogue in terahertz metamaterials ···640

70. Design of the novel steering-wheel micro-structured optical fibers sensor based on evanescent wave of terahertz wave band ······640

71. Thermal tunability and sensitivity of bandgap photonic crystal fiber of teraherz wave ······641

72. Voltage influence on propagation characteristics of liquid crystal photonic crystal fiber of terahertz wave ······641

73. Multiband metamaterial absorber at terahertz frequencies ······642

74. Propagation speed calculation of a plasmonic THz wave trapping system ······642

75. The study of the fundamental nature and electromagnetic parameter retrieval of reverse nested Split-ring resonators ·········643
76. Mechanisms of THz trapping devices based on plasmonic grating ·········643
77. Design of a tunable multiband terahertz waves absorber ·········644
78. Plasmon-induced transparency in metamaterial based on graphene and split-ring resonators ······644
79. Dual-band ultrasensitive THz sensing utilizing high quality Fano and quadrupole resonances in metamaterials ·········645
80. Graphene metamaterial for multiband and broadband terahertz absorber ·········646
81. Photoexited switchable metamaterial absorber at terahertz frequencies ·········647
82. Stable terahertz toroidal dipolar resonance in a planar metamaterial ·········648
83. Resonant conversion based on GaAs-metal metamaterials within terahertz range ·········649
84. Direct thermal tuning of the terahertz plasmonic response of semiconductor metasurface ·········649
85. 基于编码超表面的太赫兹宽频段雷达散射截面缩减的研究 ·········650
86. 太赫兹波在沙尘中衰减特性 ·········650
87. Study of the properties of $BaGa_4Se_7$ crystal in the terahertz region ·········651
88. 棋盘型结构在太赫兹宽频段RCS缩减中应用研究 ·········651
89. A New $Ba_{0.6}Sr_{0.4}TiO_3$–Silicon Hybrid Metamaterial Device in Terahertz Regime ·········652
90. Graphene-based tunable terahertz plasmon-induced transparency metamaterial ·········653
91. Tunable plasmon-induced transparency in a grating-coupled double-layer graphene hybrid system at far-infrared frequencies ·········653
92. Investigation of optical pump on dielectric tunability in PZT/PT thin film by THz spectroscopy 654
93. Dual-band tunable perfect metamaterial absorber in the THz range ·········654
94. Reconfigurable hybrid metamaterial waveguide system at terahertz regime ·········655
95. Effect of electric field on the dielectric properties of the Barium Strontium Titanate film ·········655
96. Dynamically tunable graphene plasmon-induced transparency in the terahertz region ·········656
97. Dynamically electrically tunable broadband absorber based on graphene analog of electromagnetically induced transparency ·········656
98. Optical-induced absorption tunability of Barium Strontium Titanate film ·········657
99. Slowing and trapping THz waves system based on plasmonic graded period grating ·········657
100. Triple-band high Q factor Fano resonances in bilayer THz metamaterials ·········658
101. Terahertz wavemeter based on scanning Fabry–Perot interferometer: accuracy and optimum designation ·········658
102. 太赫兹波段表面等离子体波传播距离的调控 ·········659
103. Extracting dielectric parameter based on multiple beam interference principle and FTIR system in terahertz range ·········659

104. Analyzing terahertz time-domain transmission spectra with multibeam interference principle ⋯660

105. Flexible manipulation of Terahertz wave reflection using polarization insensitive coding metasurfaces ⋯660

106. Optically tuned dielectric property of ferroelectric PZT/STO/PT superlattice by THz spectroscopy ⋯661

107. Linear optical properties of $ZnGeP_2$ in the terahertz range ⋯661

108. Active $KTaO_3$ hybrid terahertz metamaterial ⋯662

109. Terahertz wavefront manipulating by double-layer graphene ribbons metasurface ⋯662

110. Theoretical and experimental study on broadband terahertz atmospheric transmission characteristics ⋯663

111. Optical coefficients extraction from terahertz time-domain transmission spectra based on multibeam interference principle ⋯663

112. Characteristic analysis of a photoexcited metamaterial perfect absorber at terahertz frequencies ⋯664

113. Magneto-optical modulation of photonic spin hall effect of graphene in terahertz region ⋯664

114. The novel hybrid metal-graphene metasurfaces for broadband focusing and beam-steering in farfield at the terahertz frequencies ⋯665

115. Electrical terahertz modulator based on photo-excited ferroelectric superlattice ⋯666

116. Electrically tuned transmission and dielectric properties of illuminated and non-illuminated barium titanate thin film in terahertz regime ⋯666

117. Terahertz magnon and crystal-field transition manipulated by R^{3+}-Fe^{3+} interaction in $Sm_{0.5}Pr_{0.5}FeO_3$ ⋯667

118. Terahertz optical properties of nonlinear optical CdSe crystals. ⋯667

119. The effect of optical pump on the absorption coefficient of $0.65CaTiO_3$ -$0.35NdAlO_3$, ceramics in terahertz range. ⋯668

120. A broadband metamaterial absorber based on multi-layer graphene in the terahertz region ⋯668

121. Effect of optical pumping on the dielectric properties of $0.6CaTiO_3$-$0.4NdAlO_3$ ceramics in the terahertz range ⋯669

122. Active optical modulator based on a metasurface in the terahertz region ⋯669

123. Optical modulation of BST/STO thin films in the terahertz range ⋯670

124. Dynamically tunable terahertz passband filter based on metamaterials integrated with a graphene middle layer ⋯671

125. Highly sensitive sensors of fluid detection based on magneto-optical optical Tamm state ⋯672

126. Optically tuned optical properties of ferroelectric superlattice by THz spectroscopy ⋯672

127. 基于石墨烯编码超构材料的太赫兹波束多功能动态调控 ⋯673

128. Manipulation of terahertz wave using coding pancharatnam–berry phase metasurface ⋯673

129. Active control of terahertz plasmon-induced transparency in the hybrid metamaterial/monolayer MoS$_2$/Si structure ···················674
130. Amplitude modulation of anomalously reflected terahertz beams using all-optical active Pancharatnam-Berry coding metasurfaces ···················675
131. Microfluidic integrated metamaterials for active terahertz photonics ···················676
132. Characteristic analysis of a photoexcited tunable metamaterial absorber for terahertz waves ······677
133. Effect of optical pump on Pb$_{0.52}$Zr$_{0.48}$TiO$_3$ ultrathin film on LaNiO$_3$/Si substrate in the terahertz region ···················677
134. BaTeMo$_2$O$_9$ crystals: optical properties and applications in the terahertz range ···················678
135. Tunable characteristics of the SWCNTs thin film modulator in the THz region ···················679
136. Modulation of terahertz electromagnetically induced absorption analogue in a hybrid metamaterial/graphene structure ···················680
137. Optical-induced dielectric tunability properties of DAST crystal in THz range ···················680
138. Effect of optical pumping on the dielectric properties of 0.55SrTiO$_3$-0.45NdAlO$_3$ ceramics in terahertz range ···················681
139. Photoexcited blueshift and redshift switchable metamaterial absorber at terahertz frequencies ···681
140. Temperature-dependent dielectric characterization of magneto-optical Tb$_3$Sc$_2$Al$_3$O$_{12}$ crystal investigated by terahertz time-domain spectroscopy ···················682
141. Ultrafast carrier dynamics and terahertz photoconductivity of mixed-cation and lead mixed-halide hybrid perovskites···················682
142. All-optical switchable vanadium dioxide integrated coding metasurfaces for wavefront and polarization manipulation of terahertz beams ···················683
143. Investigation of optical tuning on the dielectric properties of 0.3 Ba$_{0.4}$Sr$_{0.6}$TiO$_3$–0.7NdAlO$_3$ ceramics in terahertz range ···················684
144. A novel terahertz beam splitter using ultrathin flexible transmission-type coding metasurface ···685
145. Nested anti-resonant hollow core fiber for terahertz propagation ···················686
146. Terahertz wave transmission and reflection characteristics in plasma ···················687
147. Photo active control of plasmon-induced reflection in complementary terahertz metamaterials ···688
148. 基于超材料的可调谐的太赫兹波宽频吸收器···················688
149. 稀土正铁氧体中THz自旋波的相干调控与强耦合研究进展···················689
150. 可调控的太赫兹多频带吸波器特性研究···················689
151. Ultra-wideband low-loss control of Terahertz scatterings via an all-dielectric coding metasurface 690
152. Position-guided Fano resonance and amended GaussAmp model for the control of slow light in hybrid graphene–silicon metamaterials ···················691

153. Plasmon-induced reflection metasurface with dual-mode modulation for multi-functional THz devices ··692

154. Frequency-switchable VO$_2$-based coding metasurfaces at the terahertz band ···············692

155. Metal-graphene hybrid active chiral metasurfaces for dynamic terahertz wavefront modulation and near field imaging ··693

156. Active controllable bandwidth of THz metamaterial bandpass filter based on vanadium dioxide   694

157. Active controllable dual broadband terahertz absorber based on hybrid metamaterials with vanadium dioxide ··695

158. All-optical switchable terahertz spin-photonic devices based on vanadium dioxide integrated metasurfaces ··696

第三节　太赫兹技术的应用

1. Highly precise determination of optical constants olive oil in terahertz time-domain spectroscopy···697
2. Terahertz spectrum analysis of leather at room temperature ···698
3. Terahertz time-domain spectroscopy and application on peanut oils ···································699
4. Terahertz-wave Spectrum of Cotton ···700
5. 太赫兹通信技术的研究与展望 ···700
6. High-quality continuous-wave imaging with a 2.53 THz optical pumped terahertz laser and a pyroelectric detector ··701
7. High quality continuous-wave imaging at 2.53 THz ···701
8. Terahertz imaging technique and application in large scale integrated circuit failure inspection ·······702
9. Review of explosive detection using terahertz spectroscopy technique ·································703
10. 粗糙表面对雷达目标散射截面的影响 ···703
11. 球型目标在不同波段的雷达散射截面 ···704
12. 太赫兹成像技术在无损检测中的实验研究 ··704
13. 太赫兹光谱技术在气体检测中的应用 ···705
14. 爆炸物太赫兹光谱探测技术研究进展 ···705
15. 太赫兹波段纳米颗粒表面增强拉曼散射的研究 ··706
16. THz 光谱技术检测 DNAN 炸药含量的研究 ··706
17. 粗糙铜表面对低频太赫兹波的散射实验 ···707
18. Numerical simulation of the thermal response of continuous-wave terahertz irradiated skin ········707
19. 太赫兹空间应用研究与展望 ···708
20. 太赫兹波段超材料在生物传感器的应用研究进展 ···708
21. 太赫兹雷达散射截面测量中定标体的确定 ··709
22. THz 波连续波透射式逐点扫描快速成像实验研究 ···709
23. 低频太赫兹标准目标雷达散射截面的实验研究 ··710

24. 基于相干层析的太赫兹成像技术研究 ···710
25. 超高速太赫兹通信系统中调制方式的探讨 ···711
26. Single pixel imaging with tunable terahertz parametric oscillator ································711
27. Characterizing the oil and water distribution in low permeability core by reconstruction of terahertz images ···712
28. 太赫兹波近场成像综述 ··713
29. Attenuated total internal reflection imaging with continuous terahertz wave ···············713
30. Biomedical diagnosis of cerebral ischemia with continuous-wave THz imaging ···········714
31. Fast terahertz imaging based on compressive sensing ··714
32. 基于太赫兹参量振荡器的太赫兹压缩感知成像研究 ···715
33. Terahertz imaging based on morphological reconstruction ······································716
34. High-sensitivity attenuated total internal reflection continuous-wave terahertz imaging ··········717
35. Label-free and reagentless bacterial detection and assessment by continuous-wave terahertz imaging ··717
36. Label-free bacterial colony detection and viability assessment by continuous-wave terahertz transmission imaging ···718
37. Optimization for vertically scanning terahertz attenuated total reflection imaging ···········719
38. Automatic evaluation of traumatic brain injury based on terahertz imaging with machine learning 720
39. Terahertz computed tomography of high-refractive-index objects based on refractive index matching ··721
40. Interference elimination in terahertz imaging based on inverse image processing ···········722
41. High-sensitivity terahertz imaging of traumatic brain injury in a rat model ···············723
42. Terahertz two-pixel imaging based on complementary compressive sensing ···············724
43. Broadband terahertz dielectric measurement based on multi-beam interference and Fourier transform infrared spectrometer ···724
44. Terahertz computed tomography of high-refractive-index object based on a novel experimental procedure ···725
45. Terahertz reflectometry imaging of traumatic brain injury ······································725
46. 3.11 THz 标准体雷达散射截面测量 ··726
47. 迈克尔逊干涉法精确测量太赫兹频谱及目标速度 ···726
48. 基于衰减全反射式太赫兹时域光谱技术的食用油光谱特性研究 ·······································727
49. The terahertz electromagnetically induced transparency-like metamaterials for sensitive biosensors in the detection of cancer cells ···728
50. Multiple modes integrated biosensor based on higher order Fano metamaterials ···········729
51. Study of the dielectric characteristics of living glial-like cells using terahertz ATR spectroscopy ···730

52. Study of *in vivo* brain glioma in a mouse model using continuous-wave terahertz reflection imaging ·· 730
53. Electromagnetically induced transparency-like metamaterials for detection of lung cancer cells ··· 731
54. Feasibility of terahertz imaging for discrimination of human hepatocellular carcinoma ············ 732
55. Super-resolution reconstruction for terahertz imaging based on sub-pixel gradient field transform 733
56. Dual-wavelength terahertz sensing based on anisotropic Fano resonance metamaterials ············ 734
57. Sensitive detection of the concentrations for normal epithelial cells based on Fano resonance metamaterial biosensors in terahertz range ··· 735
58. The biosensing of liver cancer cells based on the terahertz plasmonic metamaterials ············ 736
59. Biosensor platforms of the polarization-dependent metamaterials for the detection of cancer-cell concentration ·· 737
60. 太赫兹波三维成像技术研究进展 ··· 738
61. 基于太赫兹时域光谱系统的脑缺血检测 ·· 738
62. 基于分块压缩感知理论的太赫兹波宽光束成像技术 ·· 739
63. 基于太赫兹波成像的鼠脑创伤三维重构 ·· 739
64. 基于梯度变换的太赫兹图像超分辨率重建 ··· 740
65. The antibody-free recognition of cancer cells using plasmonic biosensor platforms with the anisotropic resonant metasurfaces ··· 741

第一章 非线性光学频率变换及固体激光技术

1. Simultaneous active AM plus passive FM with Q-controlling mode-locked high power pulsed laser

Yao Jianquan

Picosecond Phenomena II, Springer Berlin Heidelberg, 1980: 54−58.

Abstract: After 1973 there were some new developments on the research of mode-locked solid-state lasers. In order to obtain a stable and high-power ultrashort pulse a new mode-locked method is presented. We intend to combine active AM with passive FM and prelasing.
Key words: Mode-locked laser, Active AM, Passive FM

2. Optimum operational parameters of the ultrashort cavity laser

J.Q. Yao

Applied Physics Letters, 1982, 41(2): 136−138.

Abstract: The influence of input parameters and cavity parameters of the ultrashort cavity laser is discussed in terms of numerical calculation of the rate equations of this laser. The optimum parameters of the ultrashort cavity laser are obtained. The results of this letter will provide some theoretical guides for designing and adjusting picosecond tunable dye lasers with ultrashort cavities.
Key words: Tunable dye laser, Ultrashort cavity, Rate equation

3. Large-signal results for degenerate four-wave mixing and phase conjugate resonators

Yao Jianquan, Zhou Guosheng, A. E. Siegman
Applied Physics B-Lasers and Optics, 1983, 30: 11–18.

Abstract: A numerical calculation procedure and various large-signal numerical solutions are presented for degenerate four-wave mixing in optical Kerr media, and for phase conjugate resonators using degenerate four-wave mixing. The solutions presented take full account of nonlinear refractive index changes, pump depletion, signal saturation, distributed losses, and possible external mirrors with laser gain. We find that including the nonlinear index change generally causes little change in the reflectivity or power output of degenerate four-wave mixing devices, at least with symmetric pumping. The optimum power output from a phase conjugate resonator with and without a laser gain medium is calculated. The results provide some theoretical guidance for designing phase conjugate resonators.

Key words: Degenerate four-wave mixing, Phase conjugate, Large signal

4. 使用 KTiOPO$_4$ 的高功率内腔倍频 YAG 激光器

姚建铨，李昱，薛彬，T.S.Fahlen
中国激光，1983, 10(Z1): 101.

摘要： 讨论了双轴晶体的最佳相位匹配，即在可能的相位匹配方向中，有效非线性系数 d_{eff} 为最大值时的匹配方向，导出了 d_{eff} 的精确表示式。分析了在高转换效率（即应考虑基波在倍频晶体中的功率衰减效应）下的激光器的速率方程，利用数值解讨论了激光工作介质的增益、腔内损耗及晶体参数的影响，得到了最佳运转参数。并用类高斯分布处理具有低阶混合模运转的光束特性，找到了最佳腔长及热稳定腔的参数。利用 L 型腔，在声光 Q 开关的重复频率为 2.5 kHz 时，绿光输出平均功率为 11 W，峰值功率 22 kW，这就远远超过同类器件到目前为止所能达到的水平。

关键词： 双轴晶体，最佳相位匹配，内腔倍频，高功率 YAG 激光器。

5. Calculations of optimum phase match parameters for the biaxial crystal KTiOPO$_4$

J. Q. Yao, Theodore S. Fahlen

Journal of Applied Physics, 1984, 55(1): 65−68.

Abstract: The expression for the optimum phase matching angle for the biaxial crystal KTiOPO$_4$ (KTP) is presented and numerically calculated. The theoretical effective nonlinear coefficient, walk-off angle, and conversion efficiency are determined. The results provide some theoretical and practical guidance for optimum operation of biaxial crystals, with specific numerical calculations for KTP.

Key words: Optimum phase match, Biaxial crystal, KTiOPO$_4$

6. Optimum parameters of a high-conversion efficiency intracavity frequency-doubled laser with Gaussian-like beam

Jianquan Yao, Bin Xue

Conference on Lasers and Electro-Optics, THC6, Anaheim, California, USA, June19−22, 1984.

Abstract: Up to now, all analyses dealing with intracavity frequency-doubled laser devices are limited by plane wave and small signal approximations. There are few authors who give a full discussion of the operating states and the optimum parameters of a high conversion efficiency intracavity frequency-doubled laser with either plane wave beam or Gaussian beams. With high-quality nonlinear crystals, for example, KTP, coming into use and the technique of phase matching, it is necessary to investigate the operating states of high-conversion efficiency frequency-doubled lasers for some practical optimum parameters.

Key words: High-conversion efficiency, Intracavity frequency-doubling, Gaussian-like beam

7. 类高斯分布的理论及其应用

姚建铨,薛彬

量子电子学, 1984, 1(2): 133–142.

摘要： 当固体激光器运转在低阶混合模时,既可得到较高的功率输出,又可保证较好的光束质量。本文从引入混合模系数 M 出发,论证及提出了用类高斯分布对混合棋进行分析处理的理论及方法,然后讨论了类高斯光束在均匀介质中的传播及变换、通过非线性晶体时的二次谐波功率及与光纤耦合时的传输效率、最后给出了测量混合模系数 M 的实用方法。

关键词： 类高斯光束,混合模,固体激光器

8. 高转换效率下具有高斯及类高斯光束的内腔倍频

姚建铨,薛彬

光学学报, 1985, 5(2): 142–150.

摘要： 本文在考虑了振幅的横向分布和基波的衰减时,给出了一些关于倍频的有价值的结果。借助于求解三维耦合波方程,导出了二次谐波功率的最一般的表达式,讨论了基波功率和晶体长度的影响。作为一个特例,也给出了低转换效率下二次谐波功率的表达式。讨论了类高斯光束倍频的平面波近似处理方法,这里借助于类高斯光束的光线方程。作为小结,我们列出了八种不同情况下二次谐波功率的表达式,其中后五种来自本文的推导。最后,阐明了一种处理内腔倍频激光器的新方法。在我们的模型中,对于在腔内循环的基波功率而言,由于倍频的功率损耗,可视为一种可变的损耗。借助于速率方程的数值解,我们求得了激光腔参数和倍频晶体的最佳值。

关键词： 类高斯光束,高转换效率,内腔倍频

9. Phase conjugation in iron-doped Lithium Niobate: study and application

Yanming Liu, Yu Li, Jianquan Yao

International Quantum Electronics Conference, THGG19, Optical Society of America, 1986.

Abstract: Consider the three transport processes of charge carriers (typically due to $Fe^{2+} \leftrightarrow Fe^{3+}$); with the photovoltaic effect dominating in iron-doped lithium niobate, the photorefractive effects of one-beam incidence and two-wave mixing are analyzed and experiments made. The refractive-index spatial variation diverges the incident beams like a lens in the direction of the c axis, which is demonstrated by using a He–Ne laser beam.

Key words: Phase conjugation, Lithium niobate

10. Intracavity frequency-doubling of quasi-CW pumped YAG laser

Yao Jianquan, Li Yu, Liu Yanming

AIP Conference Proceedings, 1987, 160: 133–135.

Abstract: We designed a quasi-CW pumped intracavity frequency-doubling YAG laser, which has a higher average power and peak-power output, better beam quality and less thermal effect than that of the CW pumped laser.

Key words: Intracavity frequency-doubling, Quasi-CW, YAG laser

11. The theoretical research on the FWM effect of F_2^--centers in LiF crystals

Tao Zhang, Jianquan Yao, Liangfeng Wan, Yongfeng Run
Optics Communications, 1986, 60(5): 314–318.

Abstract: We report our theoretical research on the four wave mixing effect caused by F_2^- centers in LiF crystals. It is to our satisfaction that the calculation results agree with the experimental data very well.
Key words: Four wave mixing, LiF crystal

12. 准连续泵浦 KTP 内腔倍频 YAG 激光器及其热效应分析

姚建铨，李昱，薛彬，刘淑珍
光学学报，1986, 6(4): 326–331.

摘要： 由类高斯光束倍频理论出发，导出了类高斯光束在高转换效率下倍频时，谐波功率近似与基波功率平方成正比的结论。基于此提出了一种混合模内腔倍频的新方案——准连续泵浦。在占空比 1:1 时，理论和实验均证明可使倍频输出提高一倍多。由热传导方程出发，导出了准连续泵浦时 YAG 棒的光束传播参量比连续泵浦时低 22%（结果其他因素可低 30% 左右）的结论。实验中测出准连续泵浦时热焦距增长，发散角减小，与理论分析很好地吻合。
关键词： 准连续泵浦，内腔倍频，热效应，YAG 激光器，KTP。

13. KTP 内腔倍频 YAG 激光器

姚建铨, 李昱, 薛彬, 汪永一, 蒋民华, 谭忠恪, 列耀岗, 徐忠彬, 韩建儒

中国激光, 1986, 13(1): 48.

摘要:我们研制了一台 KTP 内腔倍频 YAG 激光器,输出平均功率达 8.9 W,脉宽 70 ns,发散角 5 mrad。

关键词:KTP 内腔倍频,YAG 激光器,高平均功率。

14. 大电流脉冲方波激光电源

薛彬, 汪永一, 姚建铨

中国激光, 1986, 13(2): 112–115.

摘要:采用可控硅开关元件和大功率三极管放大元件相结合的方案,得到稳定大电流脉冲方波激光电源,脉冲电流可达 80 A,平顶电流波动小于 2%。

关键词:激光电源,大电流,方波

15. 可饱和吸收体在激光腔内的调 Q 及四波混频效应

张涛, 姚建铨

光学学报, 1987, 7(11): 983–989.

摘要： 采用密度矩阵等方法, 对可饱和吸收体在激光腔内的调 Q 以及四波混频效应从理论上作出了统一的解释, 阐述了这些效应的物理起源。本文给出的理论计算结果与实验值的一致性令人满意。

关键词： 激光腔, Q 调制, 四波混频, 共振饱和吸收

16. Analyses of BBO and KTP crystals for optical parametric oscillation

J. Q. Yao, J. T. Lin, N. Due, Z. Liu

Conference on Lasers and Electro-Optics, TuJ5, Anaheim, April 25–29, 1988.

Abstract: The newly developed nonlinear crystals of $KTiOPO_4$ (KTP) and beta-barium-borate (BBO) provide the most efficient frequency conversion from the IR to visible and visible to UV, respectively. The unique features of BBO crystal and various potential applications including high-order harmonic generation and optical parametric oscillation (OPO) have been recently reported and compared with other crystals. On the other hand, KTP has been widely used as a frequency doubler. However, the phasematching condition limits the fundamental wavelength to not shorter than 1 μm. We present the new use of KTP for tunable laser sources, where the OPO processes, pumped by a doubled YAG (532 nm), generate output wavelength ranges of 0.7-1.9 μm. For comparison, we also present the OPO tuning ranges of BBO, which has a lower efficiency but wider tuning ranges (0.6-3.0 μm)

Key words: Optical parametric oscillation, BBO, KTP

17. 双轴晶体 KTiOPO$_4$ 的多频率变换

孙德才, 姚建铨

中国激光, 1988, 15(5): 315−317.

Abstract: With calculation of phase match parameters and effective nonlinear coefficients, several different nonlinear optical processes using one biaxial crystal KTiOPO$_4$ are considered. The nonlinear optical processes include (1) SHG (type Ⅱ) of 1.064~0.532 μm, (2) SFG of 1.064+0.6~0.383 μm, (3) DFG of 1.064~0.6~1.375 μm.

Key words: Biaxial crystal, Phase match parameters, Effective nonlinear coefficient

18. High power green laser by intracavity frequency doubling with KTP crystal

J. Q. Yao, Y. Li, D. P. Zhang

Proceedings of Society of Photo-Optical Instrumentation Engineers, 1988, 1021: 181−183.

Abstract: High power green laser has a lot of application. We have got 34.2-watt average power of laser at 532 nm from an intracavity frequency doubled YAG laser. KTP crystal was used as the frequency doubler. The advantage of Quasi-CW pumping for frequency doubling was taken, the output power of harmonic wave with Quasi-CW pumping is higher than CW pumping. The influence of the polarization property of YAG laser on frequency doubling efficiency was theoretically analysed, and the optimum orientational angle for frequency doubling crystal was found in addition to the traditional phase matching condition. Some experimental work, the parameters and properties of the laser was reported.

Key words: High power green laser, Intracavity frequency doubling, KTP

19. Applications and features of a new nonlinear crystal: Lithiun Triborate (LBO)

J. T. Lin, C. E. Huang, J. Q. Yao

Conference on Lasers and Electro-Optics, Anaheim, April 24–28, 1989.

Abstract: Lithium triborate (LBO) is a new nonlinear crystal recently discovered and developed by the Research Institute on the Structure of Matter (at Fujian, China) and the Research Institute for Synthetic Crystals (at Beijing, China). In some aspects, this LBO crystal is superior to another borate crystal, beta barium borate (BBO).

Key words: Lithium triborate, Nonlinear crystal

20. 用 $KNbO_3$ 晶体实现角度调谐频率下转换的计算与分析

张小洁，陈煦，姚建铨

量子电子学，1989, 6(2): 97–104.

摘要： 本文在与其他作者不同的坐标系下研究了用双轴晶体 $KNbO_3$ 实现角度调谐频率下转换的可能性，并对 1.06 μm 和 1.32 μm 泵浦的 xz 面运用以及 0.532 μm 泵浦的 xy 面运用情况重点进行了计算和分析，给出了对实验有指导意义的理论计算结果。

关键词： $KNbO_3$ 晶体，角度调谐，频率下转换

21. Room-temperature 1.06-0.53-μm second-harmonic generation with MgO:LiNbO$_3$

J. Q. Yao, W. Q. Shi, J. E. Millerd, G. F. Xu, E. Garmire, M. Birnbaum
Optics Letters, 1990, 15(23): 1339−1341.

Abstract: Room-temperature 1.06-0.53 μm second-harmonic generation (SHG) achieved with LiNbO$_3$ doped with 7 mol. % MgO has been studied. Phase matching was achieved with angle tuning. SHG conversion efficiency of 45% was obtained with a 12-mm-long crystal and a fundamental peak-power density of 140 MW/cm^2. SHG performance of MgO:LiNbO$_3$ is compared with that of KTP and LBO crystals. Various phase-matching parameters of MgO:LiNbO$_3$ were calculated as functions of the fundamental wavelength, using the experimentally determined Sellmeier equations. It was found that room-temperature, noncritically-phase-matched Type I SHG can be achieved in this crystal at 1.053 μm, where Nd:YLF lasers operate.

Key words: Second-harmonic generation, MgO:LiNbO$_3$, Phase matching

22. 光折变二波混频双稳态

王健，姚建铨
中国激光, 1990, 17(3): 179−182.

Abstract: Two-wave coupling in photorefractive media leads to intensity-dependent phase shift as well as energy transfer between the two waves without phase crosstalk. With this mechanism, the dispersive optical bistability can be established in an optical cavity. The effects of cavity length, linear absorption of photorefractive crystal and reflectivity of cavity mirrors on the nonlinear transmission characteristics are analysed.

Key words: Photorefractive crystal, Two-wave mixing, Optical bistability

23. 高功率脉冲 YAG 激光器

张锐

中国激光, 1990, 17(4): 216.

摘要: 天津大学精仪系研制的"高功率脉冲 YAG 激光器"于 1989 年 11 月 28 日通过了天津市科委技术鉴定。激光器主要技术指标有:器件采用两根 $\phi 8\times 106$ 中等质量 YAG 棒,$1.06\ \mu m$ 脉冲激光最高输出平均功率达 528.75 W,重复频率 400 次/s,单脉冲能量 23 J,脉冲宽度 0.3 ~ 5.6 ms,发散角小于 10 mrad,器件在高功率输出(高于 400 W)长时间稳定运转时不稳定性优于 2%。

关键词: 脉冲 YAG 激光器,高功率,大脉冲能量

24. 高功率脉冲 Nd:YAG 激光器实验研究

吴福顺,武星,姚建铨

天津大学学报, 1991, 4: 117–120.

摘要: 对高功率脉冲 YAG 激光器的输出镜的透过率、泵浦电压、器件的工作方式等进行实验研究,实现了平均输出功率 528 W YAG 脉冲激光器的运转。

关键词: 高功率,脉冲,激光器

25. Room temperature 1.06-0.53 μm second harmonic generation with MgO:LiNbO$_3$

Yao Jianquan, Shi Weiqiang, J. E. Millerd, Xu Guangfeng, E. Garmire, M. Birnbaum

Chinese Physics Letters, 1992, 8(6): 290–291.

Abstract: Second harmonic generation conversion efficiency of 45% was obtained with a 12 mm long specially grown and cut MgO:LiNbO$_3$ crystal and fundamental peak power density of 140 MW/cm^2.
Key words: Second harmonic generation, MgO:LiNbO$_3$ crystal

26. Q-switched Nd:YAG laser and color-center laser tunable over the 1.1-1.26 μm range operating simultaneously

Xing Wu, Fushun Wu, Hongping Li, Xifu Li

Proceedings of Society of Photo-Optical Instrumentation Engineers, 1992, 1627: 102–108.

Abstract: The opreating for the Q-switching of a Nd:YAG laser at 1.06 μm and for the tunable color-center laser over 1.1-1.26 μm has been obtained by using LiF:F$_2$ color center crystal both as the Q-switch for the Nd:YAG laser and the active medium for the color-center laser. The interactive affection of the two lasers has been analysed and calculated with the rate equation, we found the pulse duration of the YAG laser is compressed, the pulse duration of the color-center laser widened and the power is improved a lot significantly in this condition, they conformed to the experimental results.
Key words: Color-center laser, Q-switched Nd:YAG laser, Tunable laser

27. Accurate calculation of the optimum phase-matching parameters in three-wave interactions with biaxial nonlinear-optical crystals

Jianquan Yao, Weidong Sheng, Weiqiang Shi

Journal of the Optical Society of America B-Optical Physics, 1992, 9(6): 891–902.

Abstract: The effective nonlinear coefficients and the optimum phase-matching angles for three-wave mixing processes in biaxial crystals are calculated, and their analytical expressions are obtained. We carried out the calculations by taking into consideration the difference between the directions of the electric field vector **E** and the electric displacement vector **D** in biaxial crystals. Using the small-signal approximation, we also derived expressions for acceptance parameters and calculated their values for some typical three-wave interaction processes. In addition we obtained a theoretical curve for the energy conversion efficiency of three-wave interactions as a function of phase mismatch by numerically solving the coupled wave equations describing the processes. With this curve we are able to define and to compute the acceptance parameters for any type of three-wave mixing process. Finally, the numerical values of the phase-matching parameters are obtained for such biaxial crystals as $KTiOPO_4$ and LiB_3O_5, and the results are graphically presented.

Key words: Biaxial crystal, Optimal phase-matching, Three-wave interaction

28. Transmitted and tuning characteristics of birefringent filters

Xinglong Wang, Jianquan Yao

Applied Optics, 1992, 31(22): 4505–4508.

Abstract: Transmission formulas and transmission curves of birefringent filters when the optic axis is not in the plane of the filter plates are given and discussed in detail. The optimum parameters of birefringent filters, such as the most suitable ratio of thicknesses, tuning angles, and plate thicknesses, are obtained. As far as we know this is the first design of birefringent filters used in a tunable laser pumped by a quasi-cw source.

Key words: Birefringent filters, Tunable wavelength

29. Transverse parametric intensity distribution of optical parametric oscillation in MgO:LiNbO$_3$ crystal

Diankui Wang, Xiao Long, Feng Wu, Jianquan Yao

Conference on Lasers and Electro-Optics, CTuR6, Anaheim, May 10–15, 1992.

Abstract: Experimental results of the operation in our laboratory of singly and doubly resonant MgO:LiNbO$_3$ optical parametric oscillators tunable from 0.738 to 1.411 μm are given. The parametric wavelength of OPOs pumped by a frequency-doubled pulsed Q-switched Nd:YAG laser (0.532 μm) was tuned conveniently by tuning the temperature of MgO:LiNbO$_3$ crystal in the OPO cavity. A maximum total energy conversion efficiency of 10.4% at 5-Hz repetition rate was obtained using a 35-mm long MgO:LiNbO$_3$ crystal cut along θ = 90° (noncritical phase-matching) grown in Southwest institute of Technical Physics, Chengdu, China.

Key words: Optical parametric oscillation, MgO:LiNbO$_3$, Transverse parametric intensity

30. 温度调谐 MgO:LiNbO$_3$ 晶体光参量振荡器

王殿奎，龙晓，周定文，张锐，吴峰，姚建铨

光学学报，1992, 12(7): 611–615.

摘要： 采用调 Q Nd:YAG 激光倍频光（0.532 μm）泵浦温度调谐 MgO:LiNbO$_3$ 晶体单、双谐振光参量振荡器（OPO 包括 DRO、SRO）的实验结果。双谐振（DRO）调谐范围达 844.1~1 411.3 nm，最低泵浦阈值 0.22 mJ/pulse；单谐振（SRO）调谐范围达 738.9~1 032.2 nm，最低泵浦阈值 0.66 mJ/pulse。最大能量转换效率为 10.4%。

关键词： 光参量振荡器（OPO），双谐振（DRO），单谐振（SRO），MgO:LiNbO$_3$ 晶体

31. 声光 - 染料双调 Q 激光器的理论及实验研究

张小洁，杨杰，韩汝聪，姚建铨
中国激光，1992, 19(4): 241–246.

摘要：本文利用声光调 Q 和染料调 Q 各自的特点，提出了声光 - 染料双调 Q 理论。文中对双调 Q 激光器的输出特性随激光器各参量的变化关系进行了研究，为实现在合理的参数选择下的最佳输出提供了可靠的理论依据，并给出了与理论相符的实验结果。

关键词：声光 - 染料调 Q

32. Numerical method of astigmatic compensation and stability regions of folded or ring cavity

Xinglong Wang, Yu Li, Guojiang Hu, Jianquan Yao
Chinese Journal of lasers, 1992, 1(1): 37–42.

Abstract: Astigmatic compensation, stability regions and intracavity beam parameters of a folded or ring cavity are obtained numerically with a computer, based on the matrix transform theory of the propagation of fundamental Gaussian mode in cavity. This method is simple, precise and easy to operate.
Key words: Astigmatic compensation, Stability regions, ABCD law.

33. Quasi-continuous wave laser operation of Titanium-doped sapphire

Xinglong Wang, Changqing Wang, Xiaojun Fang, Weidong Sheng, Jianquan Yao

Chinese Journal of Lasers, 1992, 1(6): 548.

Abstract: The quasi-continuous wave Titanium-doped sapphire laser pumped by a frequency-doubled YAG laser developed in Tianjin University has been achieved new advance. A linear laser resonator consists of a reflecting mirror and an output coupler. The reflecting mirror had dielectric coating of 100% reflectivity, and the output coupler is with 7% transmissibility. The laser rod was 21 mm long and cut at Brewster's angle.

Key words: Titanium-doped sapphire laser, Quasi-continuous wave

34. BSO 晶体光电特性及光调制器的研究

宁继平，林华铁，张士勃，姚建铨

量子电子学，1992, 9(2): 199–203.

摘要： 本文对 BSO 晶体空间电光及光电导特性进行了理论分析，利用 532 nm、633 nm 波长的光进行了光电导及光调制的实验研究，实验结果与理论基本相符。

关键词： BSO 晶体，光电导效应，光调制

35. Azimuthal and incident angle dependences in the second-harmonic generation from aluminum

Z. C. Ying, J. Wang, G. Andronica, J. Q. Yao, E. W. Plummer

Journal of Vacuum Science & Technology A, 1993, 11(4): 2255–2259.

Abstract: Optical second-harmonic generation (SHG) from clean Al(111) surface has been investigated using a Nd:YAG laser at 1 064 nm. The second-harmonic signal exhibits both isotropic and anisotropic responses with respect to the rotation of the crystal azimuthal angle. The isotropic response can be largely accounted for by a free-electron contribution; the data agree with recent jellium-model calculations. On the other hand, the observation of anisotropic SHG suggests that the surface or bulk band structure of the material plays an important role in SHG and cannot be neglected in the description of SHG even for a simple-metal surface.

Key words: Second-harmonic generation, Aluminum, Azimuthal angle, Incident angle

36. Mechanism of parametric pulse duration change of a temperature-tuned MgO:LiNbO$_3$ optical parametric oscillator

Diankui Wang, Jianquan Yao, Xiao Long, Feng Wu

Conference on Lasers and Electro-Optics, CThS19, Baltimore, Maryland, United States May 2–7, 1993.

Abstract: The optical parametric oscillator (OPO) is an important means of producing tunable laser radiation. We have reported the operating characteristics of a temperature tuned OPO in MgO:LiNbO$_3$. In our experiments we found that the parametric pulse duration was generally narrower or wider than that of the pump pulse, and that the pulse shape was not symmetrical. Typically, parametric light (signal and idle) pulse duration in the experiment was ~8 ns when the MgO:LiNbO$_3$ OPO pumped by a doubled Q-switched Nd:YAG laser (532 nm) with pulse duration 10 ns. This phenomenon was also found by many researchers studying the properties of an OPO in a variety of crystals, for example, KTP, BBO and LBO, etc.

Key words: Optical parametric oscillator, MgO:LiNbO$_3$, Pulse duration

37. 光折变二波耦合增益振荡

宁继平,姚建铨,吴仲康
中国激光, 1993, 20(6): 478–480.

Abstract: Two-wave coupling in photorefractive media leads to intensity-dependent phase shift as well as energy transfer between the two waves without phase interdisturbance with this mechanism, the theoretical and experimental study on the optical gain oscillation has been done by using an optical ring cavity containing photorefractive media. The experimental results agree with the theoretical analysis.
Key words: Photorefractive crystal, Two-wave mixing, Gain oscillation

38. 脉冲激光泵浦的钛宝石激光器

宁继平,姚建铨,生卫东,王兴龙
中国激光, 1993, 20(9): 703–705.

Abstract: Starting from rate equations, output characteristics of titanium doped sapphire laser pumped by pulsed laser are studied by digital calculation. Titanium sapphire laser pumped by quasi-CW frequency doubled Nd:YAG laser is studied experimentally. The experimental results agree with the theoretical calculation.
Key words: Titanium doped sapphire laser, Pulsed laser, Quasi-continual wave

39. Tunable 5-watts Ti:Al$_2$O$_3$ laser oscillating with a quasi-continuous wave pumping

Xinglong Wang, Changqing Wang, Jianquan Yao

Chinese Journal of Lasers, 1993, B2(6): 485–487.

Abstract: An investigation of the output characteristics of Ti:Al$_2$O$_3$ laser pumped by the second harmonic generation from an acoustooptic Q-switched Nd:YAG laser is made. An average output power of 5 watts has been obtained when the pump power is 14.5 watts. The slope efficiency is 55.5%, and the quantum efficiency is approximately 80%.

Key words: Ti:Al$_2$O$_3$ laser, Quasi-continuous wave pump

40. Phase matching of optical parametric oscillator in KTP pumped by doubled YAG laser in the quasi-degenerate region

Wang Diankui, Long Xiao, Zhang Rui, Yao Jianquan

Journal of Tianjin University, 1993, 2: 30–35.

Abstract: We have realized the phase matching of optical parametric oscillator in KTP crystal pumped by doubled YAG laser in the quasi-degenerate region theoretically and obtained the condition of interactions of Type Ⅱ (A) and Type Ⅱ (B) in KTP in this region. The tuning region of wavelength, walk-off angle and effective nonlinear coefficient for Type Ⅱ (B) interaction are numerically calculated.

Key words: Phase matching, Optical parametric oscillator (OPO), Tuning plane, Quasi-degenerate region, Walk-off angle

41. 厄米特—高斯混合模的混合模系数

孙小卫，杨杰，于意仲，姚建铨

量子电子学, 1993, 10(1): 45–48.

摘要： 本文从厄米特 - 高斯近似的混合模出发，考虑了各模之间的相互作用项，引入了表征各模比例关系的振幅系数，对非对称光斑在 x，y 两个方向上计算了混合模系数，并对厄米特 - 高斯混合模进行了讨论。

关键词： 厄米特—高斯，混合模系数

42. Second and third harmonic generation in BBO by femtosecond Ti:sapphire laser pulses

Hang Liu, Jianquan Yao, Ashok Puri

Optics Communications, 1994, 109: 139–144.

Abstract: Harmonic generation in barium metaborate (BBO) by femtosecond Ti: sapphire laser pulses is studied, taking into account both the group velocity mismatch and the lowest and second order group velocity dispersion (GVD). The lowest and second order GVD is calculated as a function of wavelength in the BBO crystal. Second and third harmonic radiation using femtosecond Ti: sapphire laser pulses is computed numerically by solving the coupled wave equations. The effects of the lowest and second order GVD on the fundamental and the harmonic pulses are analyzed. The compensation of group velocity mismatch in sum frequency processes for third harmonic generation is considered.

Key words: Harmonic generation, BBO, Ti:Sapphire laser, Group velocity dispersion

43. Numerical analysis of beam parameters and stability regions in a folded or ring cavity

Xinglong Wang, Guojiang Hu, Yu Li, Jianquan Yao

Journal of the Optical Society of America A-Optics Image Science and Vision, 1994, 11(8): 2265–2270.

Abstract: A numerical method for analyzing a folded or ring cavity is presented. Astigmatic compensated parameters, stability regions, and beam parameters inside the cavity or within the gain medium can be obtained numerically as long as the intracavity beam propagation can be expressed by round-trip matrices, no matter how complicated the cavity is. This method is easy to follow, and the results are accurate.

Key words: Transfer matrix, ABCD law, Astigmatic compensation, Stability region, Beam parameter.

44. 一种新的多波长数字散斑干涉技术

彭翔, 姚建铨

光学学报, 1994, 14(7): 758–761.

摘要: 研究了一种新的多波长数字散斑干涉技术, 并证明了这种技术应用于分析复杂几何形貌表面的可能性。

关键词: 二极管激光器, 波长调谐, 散斑干涉, 表面形貌分析

45. Theoretical analysis and experimental investigation of pulsed tunable forsterite laser

Wu Fushun, Wu Xing, B. Hamilton, Yao Jianquan
Chinese Journal of Lasers, 1994, B3(4): 321–329.

Abstract: The theoretical analysis are presented using a $\chi^2_{(k)}$ function and propagation equations for a tunable chromium-doped forsterite laser. The emission cross section and the threshold changed with wavelength were given theoretically. Using the 1.06 μm of a pulsed Nd:YAG laser and its doubling-frequency 0.532 μm as the pump source, the laser action was achieved for a forsterite laser. A maximum energy of 2.5 mJ and a tuning range of 1.206~1.315 μm were obtained experimentally.
Key words: $Cr^{4+}:Mg_2SiO_4$ (chromium-doped forsterite) crystal, Tunable laser, Rate equations

46. 连续波飞秒锁模激光器中的三阶色散计算

王兴龙，肖绪辉，乔金元，吴小英，戴建明，王长青，姚建铨
中国激光，1994, 21(8): 624–630.

摘要：详细推导了连续波锁模激光器中各主要色散元件的二阶、三阶色散的表示式，并分析计算了这些元件的色散量和色散符号。
关键词：二阶色散，三阶色散，飞秒锁模激光器

47. 可用于高功率脉冲激光的面阵 CCD 模式测量装置

孙小卫，杨杰，于意仲，龙晓，张锐，史红民，段振广，王昊

中国激光，1994, 21(2): 96-101.

摘要： 本文介绍了一种用面阵 CCD 摄像机作探测器的激光光束测量装置。整个装置包括光束衰减器、单脉冲选取及脉冲同步、光束直径匹配、CCD 摄像机、图像卡、计算机及监示器以及系统软件。软件根据激光混合模的二阶矩阵理论编制，可测量高功率、高重复频率脉冲激光的空间能量分布，并可对所测结果进行模拟。由软件给出所测的光束参数，可对光束质量作出正确评价。

关键词： 面阵 CCD 摄像机，高功率脉冲激光，混合模，空间能量分布

48. The characteristics of sum-frequency-mixing by biaxial crystal LiB$_3$O$_5$ and KTiOPO$_4$

J.Q.Yao, Y.Zheng, Y.Tang, M.H.Dunn, W.Sibbett

The Pacific Rim Conference on Lasers and Electro-Optics, ThA5, Chiba, Japan, July 10-14, 1995.

Abstract: Some relative new type nonlinear optical crystals such as KTiOPO$_4$ (KTP), KTiOAsO$_4$ (KTA), LiB$_3$O$_5$ (LBO), β-BaB$_2$O$_4$ (BBO), KNbO$_3$ et al. have been widely used to main solid state laser (Nd:YAG, Nd:YLF, Ti:Al$_2$O$_3$ et al.), dye laser, excimer laser and diode laser et al. By frequency up conversion (sum frequency mixing) more short wavelength coherent radiation (even short to 190 nm) could be obtained. Using different nonlinear crystals and different technology a wide wavelength range (188 nm-459 nm) can be generated by sum frequency mixing. SFM with different crystals was summary in this paper.

Key words: Sum-frequency-mixing, Biaxial crystal, LiB$_3$O$_5$, KTiPO$_4$

49. Femtosecond Pulse—Second and Third Harmonic Generation with BBO

Yao Jianquan, Liu Hang, Ashok Puri

Acta Optica Sinica, 1995, 15(6): 641–647.

Abstract: Harmonic generation in BBO by femtosecond laser pulses is studied, taking into account both group velocity mismatch, lowest order group velocity dispersion (GVD) and second order GVD. The lowest and second order GVD of BBO as a function of wavelength is calculated. Second and third harmonic radiation with femtosecond laser pulses is computed by numerically solving the improved coupled wave equations. The effects of the lowest order GVD and second order GVD on fundamental and harmonic pulses are analyzed. The compensation of group velocity mismatch for third harmonic generation is considered.

Key words: Group velocity dispersion, Femtosecond pulse, Harmonic generation

50. Mixed mode coefficient and Gaussian-like distribution

J. Q. Yao, Y Cui, X. W. Sun, M. H. Dunn, W. Sibbett

Acta Optica Sinica, 1995, 15(12): 1633–1640.

Abstract: Starting from a sum of Hermite-Gaussian multimodes, the mixing mode coefficient is defined and the accurate expression of mixed mode coefficient M is derived. A Gaussian-like distribution (GLD) or Gaussian-like beam (GLB) is defined to describe and treat the mixed mode. The propagation and transformation of GLB in a homogeneous medium is discussed. Finally, a method for practical measurement of the coefficient, M, is given.

Key words: Mixed mode coefficient, Gaussian-like distribution, Gaussian-like beam

51. 调 Q 倍频 Nd:YAG 激光泵浦氧气和氧 - 氦混合气体的受激拉曼散射

郑义，姚建铨，朱少明，周定文，李艺
光学学报，1995, 15(11): 1594–1597.

摘要： 报道了调 Q 倍频 Nd:YAG 激光泵浦的高压工业用氧气和氧 - 氦混合气体的受激拉曼散射特性，研究了一阶、二阶斯托克斯（Stokes）光的能量转换效率与泵浦光的能量及气压的关系，并探讨了如何抑制氧气的二阶斯托克斯的产生等问题。

关键词： 调 Q 倍频 Nd:YAG 激光，氧气，氧 - 氦混合气体，受激拉曼散射

52. 两个激光二极管端面泵浦的基横模 Nd:YAG 激光器

生卫东，刘宏伟，宁继平，姚建铨，肖建伟
中国激光，1995, A22(1): 1–5.

摘要： 用两个功率为 1.5 W 的连续激光二极管偏振耦合端面泵浦及分别从双端同时泵浦 Nd:YAG 激光器，TEM_{00} 模 YAG 激光最大输出功率分别为 680 mW 和 492 mW，总的光 - 光转换效率分别为 26.4% 和 17.7%。从理论上对二极管单端面及双端面泵浦的 YAG 激光器进行了分析计算。

关键词： 激光二极管，端面泵浦，Nd:YAG 激光

53. 激光二极管端面泵浦的调 Q 内腔倍频 Nd:YAG 激光器

生卫东, 刘宏伟, 王鹏, 乔金元, 李昱, 姚建铨

光学学报, 1995, 15(9): 1195–1198.

摘要：报道了用两个 1.5 W 激光二极管偏振耦合端面泵浦的声光调 Q 内腔倍频 Nd:YAG 激光器。输出 532 nm 绿光重复频率 1 kHz 时, 最大峰值功率为 2.23 kW, 最窄脉宽为 18 ns, 平均功率 40 mW。最高重复频率 30 kHz。重复频率 15 kHz 时, 最高平均功率 128 mW。对声光调 Q 内腔倍频 Nd:YAG 激光器的动态特性进行了理论分析及计算。

关键词：激光二极管泵浦, 声光调 Q, 内腔倍频

54. 倍频 YAG 激光泵浦钛宝石激光器

赵长明, 陈安, 宋峰, 姚建铨

天津大学学报, 1995, 28(1): 118–122.

摘要：采用平 - 平腔结构, 实现了电光调 Q 倍频 Nd:YAG 激光泵浦的钛宝石激光运转。最大输出能量为 14.5 mJ, 最高能量转换效率为 35.6%, 波长调谐范围为 679~895 nm, 观察到脉宽展宽和延迟现象。

关键词：脉冲钛宝石激光器, 能量转换效率, 脉宽展宽

55. High power pulsed laser power supply based on IGBT

Wang Xueli, Yu Yizhong, Zhang Rui, Wu Feng, Yao Jianquan

Chinese Journal of Lasers, 1995, B4(5): 409–416.

Abstract: The operation process of a pulsed laser power supply which uses Insulated Gate Bipolar Transistor (IGBT) as a switching element is described with mathematical method. A new design model is proposed for the high power pulsed laser power supply. Based on the model four power supplies are designed, each output power is 8 kW, with parallel to each other a total 30 kW output power is given. The detailed circuit is given and the analysis is made.

Key words: Pulsed laser, Laser power supply, IGBT

56. 脉冲钛宝石激光器对兔眼损伤效应的研究

陈安，赵长明，姚建铨

激光杂志，1995, 16(3): 123–126.

摘要：用钛宝石脉冲可调谐激光器首次研究了单波长单脉冲；单波长重复频率；重复频率波长扫描三种状态下，波长在 680~900 nm 范围内钛宝石激光对兔眼照射所致损伤效应。报道了实验项目，损伤程度，得到了损伤规律。

关键词：可调激光器，兔眼损伤效应

57. Study on the output stability of Ti:Sapphire laser

Song Feng, Yao Jianquan, Chen Xiaobo, Hou Yanbin, Zhang Guanyin, Yang Jia
Proceedings of Society of Photo-Optical Instrumentation Engineers, 1996, 2889: 216–221.

Abstract: Stability of Ti:Sapphire laser is researched in the paper. The experimental results and theoretical analyzes show that the unstability of the output is contributed to the pump stability, the quality of the crystal, and the spontaneous emission photon numbers that enter into stimulated emission modes, which we believe is put forward for the first time.

Key words: Output stability, Ti:sapphire laser, Rate equations

58. Single frequency operation of diode-pumped intracavity frequency doubled Nd:YAG laser with tilted KTP crystal: theoretical study

Yao J. Q., Yang J. B., Sheng W. D., Zheng Y., Zheng H., Yang H. G.
Chinese Journal of Lasers, 1996, B5(3): 199–204.

Abstract: Combining a Brewster-plate and a birefringent crystal is one of the best methods to realize single-frequency coherent green light. Based on such an idea, a new method to optimize the output efficiency, which is to slightly tilt the nonlinear crystal is presented. The relationship between the tilting orientation angle β, its corresponding tilting angle α and crystal length d, and the influence upon the output are also analyzed.

Key words: Single-frequency Nd:YAG laser, Diode-pumped, Intracavity frequency doubling

59. 外腔锁模多量子阱二极管激光器

王兴龙，李昱，姚建铨

光学学报，1996, 16(12): 1829–1831.

摘要：报道了一种用作光通讯光源的外腔锁模多量子阱结构半导体激光器。其脉冲宽度为 2~5 ps，波长调谐范围为 1.52~1.57 μm，锁模频率 0.5~1.0 GHz，平均输出光功率为 1 mW。

关键词：半导体激光器，超短光脉冲，光通讯，外腔锁模，多量子阱

60. 激光二极管泵浦的高重复频率 Nd:YAG 激光器

生卫东，吴峰，刘宏伟，王鹏，姚建铨

光学学报，1996, 16(5): 595–597.

摘要：报道两个 1.5 W 连续激光二极管端面泵浦的声光调 Q Nd:YAG 激光器。输出激光脉冲的最高重复频率为 30 kHz，重复频率 1 kHz 时，最窄脉宽为 12 ns，最高峰值功率为 121 kW。

关键词：激光二极管，端面泵浦，声光调 Q，Nd:YAG 激光器

61. 用修正的 RELIEF 方法测量高速空气流瞬时速度的理论研究

郑义, 姚建铨, 吴峰, 房晓俊, 施祥春
光学学报, 1996, 16(8): 1148–1151.

摘要：修正了拉曼激发激光感应电子荧光（Raman excitation pulse laser-induced electronic fluorescence, 简称 RELIEF）方法，以实现对亚音速和超音速空气流多点瞬时速度的测量。测量精度优于 2%。
关键词：修正的拉曼激发激光感应电子荧光方法, 高速空气流, 瞬时速度, 标记线

62. 双脉冲数字散斑干涉系统的实验研究

彭翔, 朱绍明, 高志, 姚建铨
中国激光, 1996, 23(9): 847–851.

摘要：研究了一种基于脉冲 Nd:YAG 激光系统的数字散斑干涉技术, 设计并实现了物体与脉冲激光触发以及 CCD 图像获取的同步控制。这种双脉冲数字散斑干涉技术, 可记录亚毫秒时间间隔的物体状态变化并可作定量分析。
关键词：脉冲激光, 数字散斑干涉, 条纹自动分析

63. 高功率准连续激光二极管泵浦的 Nd:YAG 激光器

宁继平,肖绪辉,汤声书,熊英,姚建铨

中国激光, 1996, A23(6): 481−484.

摘要:对高功率准连续激光二极管侧面泵浦 Nd:YAG 固体激光器进行了理论和实验研究。当泵浦能量为 135 mJ 时,得到脉冲能量为 16 mJ 和 1 064 nm 激光,重复频率可达 400 Hz。

关键词:高功率准连续激光二极管,侧面泵浦,Nd:YAG 激光器

64. 高功率开关型脉冲激光电源逆变器的理论分析

于意仲,王学礼,张锐,吴峰,姚建铨

天津大学学报, 1996, 29(6): 971−975.

摘要:对高功率开关型脉冲激光电源逆变器的工作过程进行了理论分析,给出了一些说明开关电源逆变器电参数变化的方程,这些方程给出了电容充电时逆变回路电流、电压的变化情况以及逆变器输出功率与逆变器参数的关系。

关键词:高功率,开关型,脉冲激光电源,逆变器

65. 钛宝石晶体色散特性的理论研究

赵长明，史彦，姚建铨

激光技术，1996, 20(6): 338-341.

摘要： 本文采用激光经典理论中色散与吸收（增益）的关系，结合钛宝石吸收与荧光光谱资料，计算了钛宝石晶体的色散特性，得到了钛宝石在吸收区的反常色散和增益区正常色散曲线。结果表明，在一般条件下，由吸收和增益所造成的折射率变化可以忽略，在考虑折射率与波长的关系时，仍可按正常色散处理。

关键词： 钛宝石，折射率，色散特性

66. 脉冲钛宝石激光器动力学特性的理论及实验研究

赵长明，陈安，姚建铨

激光技术，1996, 20(4): 224-229.

摘要： 本文采用速率方程理论，比较全面详细地分析计算了脉冲钛宝石激光器的动力学特性。在实用参数范围内，给出了输出参数与泵浦参数的关系。采用这些结果可以统一解释以往文献报道的结果并与我们自己的实验数据相一致，证明了该方法的正确性和参数选择及近似处理的合理性。

关键词： 脉冲钛宝石激光器，动力学特性，激光速率方程。

67. Analysis of simultaneous Q-switching and frequency doubling in KTP

J. Q. Yao, X. W. Sun, H. S. Kwok
Journal of Modern Optics, 1997, 44(5): 997−1004.

Abstract: Simultaneous Q-switching and second harmonic generation in KTP is analyzed numerically. In this mode of operation, the phase matching condition for second harmonic generation and the half-wave voltage for Q-switching are interrelated. In this paper, this coupled numerical problem is solved iteratively until a self-consistent solution is obtained. The effective refractive index and effective electro-optical coefficient along the phase matching direction are defined and calculated in order to estimate the half-wave voltage for Q-switching with KTP. Additionally, the application of temperature and wavelength dispersion tuning to compensate for the constant retardation in this biaxial crystal is also studied. It is found that the operating temperature of the KTP crystal should be stable to within 0.1 ℃ for exact retardation compensation.
Key words: Second harmonic generation, KTP, Q-switching

68. The temperature distribution in longitudinally pumped Ti:sapphire lasers

Song Feng, Yao Jianquan, Yang Jia, Chen Xiaobo, Zhang Guanyin
Chinese journal of lasers, 1998, B6(1): 1−6.

Abstract: A simple model on the heat conduction of solid-state laser pumped longitudinally is put forward. The temperature distributions are numerically calculated and some useful conclusions are obtained. The laser media pumped longitudinally is easier to be damaged for the higher temperature gradient and the higher heat stress; coolant is need and two-end pump mode is much better.
Key words: Temperature distributions, Ti:sapphire lasers, Two-end pumping.

69. 双波长脉冲激光器的速率方程理论及其数值计算

宋峰,姚建铨,乔金元,陈晓波,张光寅
物理学报, 1997, 46(9): 1725-1730.

摘要：根据四能级脉冲激光器的速率方程理论,建立了激光器在双波长运转时的速率方程理论模型。以钛宝石激光器的有关数据进行了数值计算和分析,所得结果与实验相符。

关键词：钛宝石,激光器,双波长脉冲,速率方程

70. 脉冲钛宝石激光器的双波长运转

宋峰,周定文,李喜福,乔金元,姚建铨,陈晓波,张光寅
光学学报, 1997, 17(6): 661-666.

摘要：报道了脉冲钛宝石激光器的双波长可调谐运转,双波总能量达到 41.8 mJ,调谐范围大于 100 nm,并且研究了双波长运转时的竞争效应、增益损耗的影响、时域特性等。

关键词：脉冲,钛宝石激光器,双波长,可调谐

71. 激光二极管端面泵浦的单频固体激光器的理论研究

郑义,钱卫红,刘夫义,杨健柏,姚建铨

光学学报,1997, 17(7): 894–899.

摘要:给出了激光二极管端面泵浦的驻波激光器单频运转时的最大泵浦功率与腔参数和激光介质参数之间的简单函数关系,对于单频激光器的设计具有实际指导意义。

关键词:激光二极管,端面泵浦,固体激光器,单频

72. 脉冲泵浦 KTP 单谐振光学参量振荡器

施翔春,吴峰,于意仲,姚建铨,周定文

中国激光,1997, 24(1): 59–62.

摘要:报道了电光调 Q Nd:YAG 激光器的二次谐波泵浦的 KTP 单谐振光学参量振荡器的实验结果。实验获得了 1 203~1 399 nm 的单谐振参量光输出,总能量转换效率最大为 36%。对参量光与 1 064 nm 激光的和频进行了初步的实验,获得了 568~592 nm 的可见光输出。

关键词:$KTiOPO_4$ 晶体,光学参量振荡,光和频产生

73. LD 泵浦的内腔倍频激光器单频运转的理论研究

郑义，钱卫红，姚建铨
中国激光，1997, A24(8): 673–678.

摘要： 给出了激光二极管（LD）端面泵浦的内腔倍频激光器单频运转时的最大泵浦功率与腔参数和材料参数间的简单函数关系，并对 LD 泵浦的 $Nd:YVO_4$ 及 $Nd:YAG$ 内腔倍频激光器的单频运转进行了详细的分析。

关键词： 激光二极管，单频，内腔倍频激光器，$Nd:YVO_4$，$Nd:YAG$

74. ArF 准分子激光器的窄线宽可调谐运转及注入放大

王杰，吴峰，施翔春，肖绪辉，姚建铨
中国激光，1997, A24(9): 784–786.

摘要： 在一台快放电泵浦的 ArF 准分子激光振荡放大系统的振荡级上采用光栅、扩束镜、光阑等腔内元件，用组合输出镜获得了线宽小于 0.1 nm，调谐范围 ~1 nm 的激光输出。注入到非稳腔结构的放大级，注入后放大级效率提高了约 50%，获得了平均 30 mJ/脉冲的高光束质量的窄线宽可调谐激光，最大单脉冲能量 > 50 mJ，并进行了氧气的吸收光谱实验。

关键词： ArF 准分子激光器，窄线宽，可调谐，注入放大

75. ArF 准分子激光振荡放大系统的研究

赵震声,陈永荣,胡雪金,谷怀民,奚居雄,张虎,袁廷海,匡梅,姚建铨,吴峰,施翔春

光学学报, 1997, 17(2): 140–145.

摘要: 研究了 ArF 准分子激光振荡放大系统,中心波长 193 nm。用一只闸流管作为开关元件,通过主充放电网络匹配,有效地解决了两台器件的放电抖动,振荡放大倍数达到约 50 倍,系统单脉冲输出能量 50 mJ。光束发散角低于 0.20 mrad。

关键词: ArF 准分子激光,振荡放大

76. 序列脉冲倍频 YAG 激光器泵浦的氧气受激拉曼散射研究

吴峰,施翔春,王学礼,于意仲,郑义,姚建铨

光学学报, 1997, 17(3): 275–278.

摘要: 对序列脉冲倍频 YAG 激光器泵浦的氧气受激拉曼散射进行了研究。对氧气的受激拉曼散射池的热传导过程进行了数值模拟计算,实验研究了序列脉冲倍频 YAG 激光器泵浦情况下氧气受激拉曼散射的热效应问题。

关键词: 受激拉曼散射,热传导,序列脉冲

77. 利用双折射效应设计的窄带滤光系统

乔金元, 王兴龙, 王学礼, 李睿, 肖绪辉, 姚建铨

量子电子学报, 1997, 14(5): 453-457.

摘要：本文根据单轴晶体的双折射效应及偏振光干涉原理，设计制作了一种透射带宽大约为 0.05 nm 的窄带滤光系统。

关键词：滤光片, 双折射效应, 透射带宽

78. 激光应用的新领域——激光清洗

金杰, 房晓俊, 姚建铨, 焦强, 赵新

应用激光, 1997, 17(5): 228-230.

摘要：报导激光应用的新领域——激光清洗的原理及应用概况。介绍了两种激光清洗的实例：高重复频率 TEA-CO_2 激光高速清洗固体表面；和激光清除宇宙空间垃圾。

关键词：激光清洗, 表面污染

79. 半导体激光直接倍频的蓝紫光激光器

金杰,房晓俊,姚建铨,赵新,袁多荣,张光辉

中国激光, 1998, A25(11): 961-964.

摘要: 报道了用新型络合物非线性光学材料硫氰酸汞镉(CMTC)晶体实现激光二极管室温下直接倍频,产生蓝紫光激光输出。基频光功率为 1.98 W,波长为 808 nm,用长度为 4 mm 的 CMTC 晶体,获得 404 nm 波长的倍频光功率为 11.8 mW,SHG 非线性转换效率为 0.60%。

关键词: 激光二极管,CMTC 晶体,倍频,蓝紫光

80. 脉冲可调谐钛宝石激光器的实验

乔金元,肖绪辉,王学礼,李睿,宋峰,熊凯,姚建铨

天津大学学报, 1998, 31(4): 516-520.

摘要: 本文报道了一种小型化的脉冲钛宝石激光器,可输出 532 nm 绿光和可调谐钛宝石激光。钛宝石激光单脉冲输出最大能量为 77.5 mJ,重复率最大可达 20 Hz。由单片机控制波长调谐,调谐范围为 670~860 nm。整套器件的重量为 9.5 kg,置于一块 800 mm × 140 mm 的铝板上。使用该激光器进行了可调谐激光的生物效应的研究。

关键词: 脉冲钛宝石激光器,生物效应,小型化,可调谐

81. 瑞利散射/激光诱导荧光技术用于空气、O_2 流场的二维瞬态测量

王杰，施翔春，李喜福，肖绪辉，姚建铨

光学学报，1999, 19(10): 1375–1380.

摘要： 研究了空气分子的瑞利散射光强及真空紫外激光（波长 193 nm）诱导 O_2 荧光强度与温度、密度的关系，给出了一般表达式。指出了可用于测量高速空气、O_2 流场中的瞬态密度和温度二维分布。利用带像增强的 ICCD 照相机获得了室内空气及 O_2 喷流的瑞利散射及激光诱导氧气荧光流场图像，图像处理后获得了流场的二维分布参数。

关键词： 紫外脉冲激光，瑞利散射，激光诱导荧光，空气流场

82. Theoretic analyzing of optical parametric oscillator with intracavity sum frequency generation

Shi Xiangchun, Xiao Xuhui, Wang Jie, Ding Xin, Wang Xiaoyong, Yao Jianquan

Chinese Journal of Lasers, 1999, B8(5): 464–469.

Abstract: A new scheme of optical parametric oscillator with intracavity sum frequency generation is proposed, and the operation characters such as threshold, conversion efficiency, etc., are theoretically studied. The factor affecting conversion efficiency is discussed.

Key words: Optical parametric oscillator (OPO), Sum frequency generation (SFG), Nd: YAG laser

83. 激光二极管面阵列泵浦固体激光器的理论分析

宁继平, 汤声书, 熊英, 常志武, 姚建铨
光学学报, 1999, 19(9): 1183–1188.

摘要：对于激光二极管侧面泵浦的固体激光器, 泵浦光和振荡光场为空间分布函数, 利用四能级速率方程, 得到了激光器输入和输出参数之间的关系。分析了空间参数对激光器输出特性的影响, 找到了最佳参数。

关键词：激光二极管侧面泵浦, 空间分布函数, 输出特性, 最佳参数

84. 分子标记示踪用于氧气/空气流场速度测量

王杰, 施翔春, 王鹏, 姚建铨
实验力学, 1999, 14(4): 471–476.

摘要：利用受激喇曼激发加激光诱导电子荧光法(RELIEF)实现对氧气和空气中氧分子的标记示踪, 用于氧气和空气流场的速度的测量。实验装置测量氧气喷流的瞬时速度为 45 ± 5 m/s, 改进后精度可以优于1%, 并适合几个马赫的高速流场测量。

关键词：分子标记示踪, 空气流场, 氧气流场, 流场测速, RELIEF方法

85. 单片机在可调谐固体激光器中的应用

宁喜发, 姚建铨, 李喜福, 王鹏

激光杂志, 1999, 20(6): 64–66.

摘要: 本文阐述了脉冲钛宝石激光系统的光路安置和单片机自动控制的波长调谐系统。本文着重介绍了单片机的控制功能及其硬、软件系统。

关键词: 单片机控制, 波长调谐

86. VISAR 测速中的信号丢失及丢失条纹数的确定

李泽仁, 姚建铨

爆炸与冲击, 1999, 19(2): 182–187.

摘要: 论述了 VISAR 测速中信号频率与被测速度增量的关系和光电倍增管、数字示波器所能响应的最高速度增量。分析了信号丢失的原因, 给出了丢失条纹数的确定方法。最后对 VISAR 应用中如何正确选择条纹常数提出了建议。

关键词: 激光测速, 信号频率, 速度增量, 信号丢失, 条纹数

87. Blue-violet light second harmonic generation with CMTC crystals

Guanghui Zhang, Mingguo Liu, Dong Xu, Duorong Yuan, Weidong Sheng, Jianquan Yao

Journal of Materials Science Letters, 2000, 19: 1255–1257.

Abstract: In this work, the generation of blue-violet laser light by single-pass frequency doubling of a near-infrared GaAlAs diode laser with the large complex crystal cadmium mercury thiocyanate (CMTC) is reported. CMTC crystal is a metal-organic nonlinear crystal, which belongs to a tetragonal system with space group I4. The crystal is transparent at about 400 nm of blue spectral range and its measured optical damage threshold is about 200 MW cm^{-2}. This crystal is hard enough for crystal-polishing process. Large single crystals of CMTC were grown from aqueous solution using temperature lowing method.

Key words: Blue-violet laser, CMTC, Second harmonic generation

88. Blue-violet light by direct frequency doubling of laser diode

Jie Jin, Shuguang Guo, Fuyun Lu, Qiang Jiao, Jianquan Yao, Guangyin Zhang

Proceedings of Society of Photo-Optical Instrumentation Engineers, 2000, 3928: 228–231.

Abstract: In this letter we introduce a new nonlinear organornetallic complex CdHg(SCN)$_4$ crystal (CMTC), which is used to double the frequency of the 808 nm laser diodes. The blue-violet light output of 11.8 mW (λ=404 nm), and the SHG conversion efficiency of 0.60% are obtained.

Key words: Blue-violet light, Laser diode, Frequency doubling, CMTC

89. 激光二极管端面抽运的 1 342 nm Nd:YVO$_4$ 激光器

郑义，宋连科，徐德刚，王衍勇，王莉，尚连聚，吴福全，姚建铨

光学学报，2000, 20(8): 1141–1144.

摘要： 报道了用高功率线列阵光纤耦合激光二极管端面抽运的 1 342 nm Nd:YVO$_4$ 激光器，在 13.30 W 输入抽运功率下，1 342 nm 激光输出功率达到 3.22 W，斜效率为 25.5%。输出功率不稳定度小于 1%。

关键词： 激光二极管端面抽运，Nd:YVO$_4$ 激光器，抽运

90. 瑞利散射用于分子流场多参数测量

王杰，施祥春，肖绪辉，王晓勇，王鹏，姚建铨

天津大学学报，2000, 33(1): 21–24.

摘要： 研究了激光在分子流场的瑞利散射特性，包括散射光强、谱线加宽以及频率移动与流场中分子密度、温度和速度分布的关系；研究了实现瑞利散射测量的实验方法，利用脉冲深紫外 (193 nm) 及可见激光 (532 nm)，结合带像增强器 CCD 照像机，建立了脉冲为纳秒量级的流场测量装置，拍摄了清晰的空气瑞利散射图像。

关键词： 瑞利散射，流场检测，脉冲激光，多参数测量

91. 泵浦波长调谐 KTP-OPO 及其倍频、和频的理论与实验研究

丁欣,于意仲,王鹏,姚建铨
量子电子学报, 2000, 17(5): 474–475.

摘要：作为可见光到近红外波段的主要光学参量振荡器(OPO)，KTP-OPO 格外受到光学界重视。传统的 KTP 光学参量振荡器主要采用角度调谐方式,即通过改变泵浦光相对于晶体的入射角来实现参量光波长调谐,但随着晶体角度的改变,对于泵浦光来说已不是正入射,从而引入了损耗,提高了 OPO 的阈值。本实验创新之处是采用可调谐钛宝石激光(Ti:s)泵浦源,通过改变泵浦光波长作为参量振荡器的调谐方式,使得 OPO 中的 KTP 晶体不再需要转动角度,OPO 谐振腔相对比较稳定,同时由于钛宝石晶体具有很宽的荧光光谱(约 660~1 200 nm),用一块 KTP 晶体就可以得到很宽的连续可调谐参量光输出。

关键词：泵浦波长调谐,KTP-OPO,倍频,和频

92. 激光表面改性技术的研究与发展

赵新,金杰,姚建铨
光电子·激光, 2000, 11(3): 324–328.

摘要：本文综述了激光表面改性技术的研究现状及其最新进展,分析了多种工艺条件下的技术特点及其应用,并对未来的发展作了展望。

关键词：激光,表面改性,进展

93. The characteristics of sum-frequency-mixing by biaxial crystal LiB_3O_5 and $KTiOPO_4$

Yao Jianquan, Zhang Baigang, Wang Peng, Chen Jin, Wang Jie
Journal of Optoelectronics·Laser, 2000, 11(6): 652–662.

Abstract: Sum frequency mixing (SFM) in nonlinear crystals is reviewed. The phase matching (PM) region, main phase matching parameters: optimum phase matching angles θ and H, effective nonlinear coefficient d_{eff}, walk-off angles T and acceptance angles Δθ and ΔH for second harmonic generation (SHG)and SFM were numerically calculated and analyzed in LiB_3O_5 and $KTiOPO_4$. The polarization features of SFM on three principal planes are also discussed.
Key words: Biaxial crystals, Sum-frequency-mixing

94. Pump-tuning optical parametric oscillation and sum-frequency mixing with KTP pumped by a Ti:sapphire laser

J. Q. Yao, X. Ding, J. Y. Qiao, C .C. Yang, I. J. Hsu, C. W. Hsu
Optics Communications, 2001, 192: 407–416.

Abstract: We report the implementation of a KTP optical parametric oscillator pumped by a pulsed Ti:sapphire laser. Including signal and idler, it could be continuously tuned from 1.261 through 2.532 μm by varying the pump wavelength. Two major improvements were achieved, including the connection of the signal and idler tuning ranges and the high output pulse energy through the signal and idler tuning ranges. The first improvement was achieved by discovering two particular sets of phase-matching angles. The second improvement was realized by using five sets of resonator mirrors for oscillating either signal or idler. Theoretical calculations were also conducted for the effective nonlinear coefficient, walk-off angle, phase-matching acceptance angle, and output spectral widths for the concerned phase-matching angles. The second part of this paper presents the experimental and theoretical results of sum-frequency mixing (SFM) of a Ti:sapphire laser and a 1.064 μm Nd:YAG laser. By using two KTP crystals cut at θ= 76° and 85° (φ= 90° in both crystals), respectively, we have experimentally achieved SFM tuning range from 459.3 through 472.9 nm. The energy conversion efficiencies were reasonably high.
Key words: Optical parametric oscillator, Sum-frequency mixing, Ti:sapphire laser, KTP crystal

95. Rate-equation theory and experimental research on dual-wavelength operation of a Ti:sapphire laser

F. Song, J. Q. Yao, D. W. Zhou, J. Y. Qiao, G. Y. Zhang, J. G. Tian
Applied Physics B-Lasers and Optics, 2001, 72(5): 605–610.

Abstract: Rate-equation analysis and experiments on a pulsed dual-wavelength Ti:sapphire laser are reported in the paper. A four-energy-level system with two lower-energy levels is put forward and numerical calculations are given. A simple optical setup with two sub-resonators is constructed, which yields a total energy from dual-wavelength lasers up to 41.8 mJ. The largest tunable range is over 110 nm. Laser characteristics of dual-wavelength pulses are observed. Experimental results are in accord with rate-equation-theory results.
Key words: Rate-equation, Dual-wavelength, Ti:sapphire laser

96. Nearly-noncritical phase matching in MgO:LiNbO$_3$ optical parametric oscillators

Yao Jianquan, Yu Yizhong, Wang Peng, Wang Tao, Zhang Baigang, Ding Xin, Chen Jin, H. J. Peng, H. S. Kwok
Chinese Physics Letters, 2001, 18(9): 1214–1217.

Abstract: We have proposed and demonstrated a nearly-noncritical phase-matched (NCPM) optical parametric oscillation (OPO) in an MgO:LiNbO$_3$ crystal with 5 mol% MgO by temperature tuning. By giving up perfect NCPM, the practical tuning range for the OPO is increased by two times. For the crystal, the operated temperature decreases with the phase-matching angle at degeneracy. With a cutting angle of 82° instead of the noncritical case of 90°, the tuning range was increased. In order to obtain a sufficiently high output pulse energy for both signal and idler throughout the entire tuning range, five sets of mirrors were used for the resonator. The tuning range of the OPO was 800~1 700 nm with temperatures tuning from 83℃ to 224.2℃. The output energy was about 6.45 mJ with a conversion efficiency of nearly 13%. The bandwidth of the output was 1.0~1.1 nm.
Key words: Noncritical phase matching, MgO:LiNbO$_3$, Optical parametric oscillator (OPO)

97. Thermal effect in KTP crystals during high power laser operation

Yao Jianquan, Yu Yizhong, Chen Jin, Zhang Fan, Wang Peng, Wang Tao, Zhang Baigang

Chinese Physics Letters, 2001, 18(10): 1356–1359.

Abstract: We report on the theoretical and experimental studies of the thermal effect of the KTP crystal during high power operation. From the dependence of the refractive index temperature coefficients on wavelength, the dependence of the optimum phase-matching angles on temperature is derived. In the experiment, the angle of the frequency-doubled KTP crystal is tilted to compensate for the thermal effect and to obtain $\Delta\Phi = 0.7°$ when the green laser output power is 30 W and the KTP crystal temperature is about 80℃. We obtained the highest stable output power greater than 40 W with an L-shaped flat-flat intracavity frequency-doubled Nd:YAG laser. The experimental results are very consistent with the theoretical analysis.

Key words: Thermal effect, KTP crystals, High power laser

98. Pump-tuning KTP optical parametric oscillator with continuous output wavelength pumped by a pulsed tunable Ti:sapphire laser

Ding Xin, Yao Jianquan, Yu Yizhong, Yu Xuanyi, Xu Jingjun, Zhang Guangyin

Chinese Physics, 2001, 10(8): 725–729.

Abstract: We report on the implementation of a KTP optical parametric oscillator pumped by a pulsed tunable Ti:sapphire laser. Two major improvements were achieved, including the connection of the signal and idler tuning ranges and the high-output conversion efficiency through the signal and idler tuning ranges. Both in the signal and idler, the continuous output wavelength from 1.261 to 2.532 μm was obtained by varying the pump wavelength from 0.7 to 0.98 μm. The maximum output pulse energy was 27.2 mJ and the maximum conversion efficiency was 35.7% at 1.311 μm (signal).

Key words: Ti:sapphire laser, Pump-tuning, Optical parametric oscillator, Phase-matching condition

99. Tunable sum frequency mixing of a Ti:sapphire laser and a Nd:YAG laser

Ding Xin, Yao Jianquan, Yu Yizhong, Yu Xuanyi, Xu Jingjun, Zhang Guangyin
Chinese Journal of Lasers, 2001, B10(6): 407–410.

Abstract: In this paper the theoretical and experimental results of sum-frequency mixing of a Ti:sapphire laser and a 1.064 μm Nd:YAG laser are presented. By using two KTP crystals cut at $\theta=76°$ and $85°$ ($\varphi=90°$ in both crystals), respectively, the sum-frequency mixing tuning range from 459.3 to 509.6 nm in one Ti:sapphire laser setup is experimentally achieved. The maximum output energy was 14.6 mJ and the energy conversion efficiency was up to 15.2 %.
Key words: Ti:sapphire laser, Sum frequency mixing, Phase-matching condition, KTP crystal

100. A micro blue-violet laser by frequency doubling of semiconductor laser

Jin Jie, Jiao Qiang, Guo Shuguang, Fang Xiaojun, Yao Jianquan, Lu Fuyun, Zhang Guangyin
Chinese Journal of Lasers, , 2001, B10(2): 81–83.

Abstract: In this paper, a micro blue-violet laser by frequency doubling of a semiconductor laser with room temperature, the blue-violet laser output of 11.8 mW at 404 nm and the conversion efficiency of the second harmonic generation (SHG) of 0.60% were obtained with a 1.98 W, 808 nm semiconductor laser and a 4 mm crystal.
Key words: Blue-violet laser, Semiconductor laser, Frequency doubling, CMTC crystal

101. 基于混频效应的宽带激光谐波转换理论

王杰, 姚建铨, 于意仲, 王鹏, 张帆, 王涛

物理学报, 2001, 50(6): 1092-1096.

摘要：提出了一种新的宽带激光谐波转换理论，指出实际宽带激光谐波转换不能简单应用倍频理论，提出同时存在不同波长之间满足和频相位匹配情况。因而宽带谐波转换应该是一个倍频加和频的混合转换过程。在合理设计下，不同波长之间的和频效应可以在整个谐波转换中占主导地位，突破了传统允许波长的限制，这种理论和方法不仅适合二倍频，同时适合三倍频等高次谐波转换。

关键词：宽带激光, 谐波转换, 混频

102. 准相位匹配 PPLN、PPKTP、PPRTA 光学参量振荡器及其应用

姚建铨

中国计量学院学报, 2001, 12(2): 42-43.

摘要：非线性光学频率变换技术主要用来产生传统激光光源所不能产生的相干辐射，这些激光光源或者限于离散的谱线或者仅能相对较窄的光谱范围。而实际上许多应用所需要的波长是直接激光光源所产生不了的。基于周期性极化的铌酸锂 (PPLN) 的准相位匹配光学参量振荡器 (QPM-OPO) 由于其具有很大的非线性系数、低的损耗，以及可以运转在宽的范围和制作工艺多样性等特点，已经得到广泛的重视及应用。本文将简要介绍 PPLN (或 PPMgLN、PPKTP、PPRTA) 作为准相位匹配 OPO 的原理、主要的技术关键及可能的应用前景。

关键词：准相位匹配 (QPM), 光学参量振荡器 (OPO), 周期性极化铌酸锂 (PPLN)

103. 高功率准连续激光二极管抽运的 Q 开关内腔倍频固体激光器的研究

宁继平,詹仰钦,鲁笑春,姚建铨

光学学报, 2001, 21(12): 1516–1518.

摘要: 对高功率准连续激光二极管抽运的声光调 Q 内腔倍频 Nd:YAG 和 Nd:YVO$_4$ 激光器进行了实验研究,分别获得了输出功率为 3.5 W 和 3.2 W、输出波长为 532 nm 的激光脉冲,其重复率达 22 kHz。

关键词: 高功率激光二极管抽运,声光 Q 开关,内腔倍频

104. 非线性晶体三波互作用允许参量的修正计算

王杰,姚建铨,李喜福,于意仲,施翔春

光学学报, 2001, 21(2): 139–141.

摘要: 采用传统级数展开方法计算晶体非线性三波互作用的允许参量。利用定义通过计算机数值模拟进行了精确计算,修正了传统计算结果,发现晶体在产生二次谐波时,在某些角度存在很宽的允许波长峰,超过传统文献值约几十倍。研究表明,在产生二次谐波时,10 mm 长的晶体在某些波段获得 100 nm 以上的允许倍频带宽是可能的,并提出了移动这些宽带的方法。

关键词: 三波互作用,允许参量,超宽允许波长带

105. 拉曼激发激光诱导电子荧光流场测量系统中标记过程的研究

施翔春,王杰,肖绪辉,王晓勇,王鹏,姚建铨

光学学报, 2001, 21(2): 206-210.

摘要: 研制了一套拉曼激发激光诱导电子荧光空气流场测速系统, 利用 Nd:YAG 激光器的二次谐波及其抽运的氧气受激拉曼散射作为标记光源, 以 ArF 准分子激光作为荧光再现光源, 并用像增强 CCD 摄像机 (ICCD) 记录荧光图像, 成功地获得了纯氧及空气中的标记线的荧光图像, 并进行了氧气喷流速度测量的初步研究。

关键词: 拉曼激发激光诱导电子荧光, 荧光图像, 标记

106. 宽波段温度调谐 MgO:LiNbO$_3$ 光学参量振荡器

吕卫,于意仲,丁欣,姚建铨

光学学报, 2001, 21(4): 440-443.

摘要: 采用电光调 Q 脉冲 Nd:YAG 激光的二次谐波 (532 nm) 抽运温度调谐的 MgO:LiNbO$_3$ 晶体光学参量振荡器, 调谐范围达 800 nm~1 750 nm。在单谐振运转条件下, 抽运阈值为 21.5 mJ/pulse, 最大抽运能量为 58 mJ 时, 输出为 6.45 mJ, 在大信号情况下的能量转换效率达 11%, 输出线宽 1 nm 左右。

关键词: MgO:LiNbO$_3$ 晶体, 温度调谐, 光学参量振荡器, 宽波段

107. 可调谐钛宝石激光抽运的 KTP 单谐振光学参量振荡器的研究

丁欣,施翔春,于意仲,吕卫,姚建铨

光学学报, 2001, 21(4): 444-446.

摘要： 介绍了 700 nm~980 nm 的脉冲可调谐钛宝石激光抽运的 KTP 单谐振光学参量振荡器。通过改变抽运光波长作为光学参量振荡器输出参量光的调谐方式,确定了 KTP 晶体最佳切割角度(θ= 62.5°, φ= 0°),实验上获得了 1 251 nm~2 532 nm 的连续参量光输出,最大输出能量约为 27.2 mJ,最大转换效率为 35.7 %(1 311 nm 处)。

关键词： 光学参量振荡器,$KTiOPO_4$ 晶体,可调谐钛宝石激光器

108. LD 抽运 Nd:YAG 激光器声光调 Q 高效内腔谐波转换

王杰,姚建铨,王鹏,于意仲,施翔春

中国激光, 2001, 28(1): 4-6.

摘要： 报道了一台半导体激光器 (LD) 抽运的声光调 Q 高效内腔谐波转换 Nd: YAG 激光器。当注入抽运功率为 12 W 时,声光调 Q 的基频波 (1.064 μm) 输出平均功率为 2.6 W,采用内腔倍频技术,在简单腔情况下,二次谐波 (532 nm) 输出平均功率达到了 2.1 W,光 - 光转换效率分别达到 21.7% 和 17.5%。

关键词： LD 抽运,声光调 Q,内腔倍频,高转换效率

109. LD 抽运的内腔倍频 Nd:YAG 激光器输出倍频绿光达 40 W

王鹏, 姚建铨, 张帆, 陈进, 张百钢
中国激光, 2001, 28(1): 112.

摘要：采用美国激光二极管列阵（RE63-2C2-CA1-0021 型）抽运的 Nd:YAG 激光器, 工作物质尺寸为 Φ6.35 mm×146 mm, 5 付列阵对 YAG 棒横向抽运, 恒温控制系统可实现 0.2～0.5 ℃的监测及控制, 在激励电流为 21 A（抽运功率约 2 900 W）时, YAG 连续输出（1 064 nm）功率为 402 W。采用声光 Q 开关（27 MHz）KTP 晶体内腔倍频方案, 用两种腔型（直腔及 L 型腔）, 在激励电流为 13.7 A（抽运功率约 1 700 W）时, 532 nm 准连续输出为 41 W（f = 540 kHz, 脉宽为 70 250 ns）。对于 KTP 在高功率时的热效应问题, 可采用调整相位匹配角等进行补偿。为进一步提高绿光输出功率, 正准备采用 Z 型腔, 可望在激励电流为 2 021 A 时, 倍频绿光输出可达 80~100 W。

关键词：LD 抽运, 内腔倍频, 绿光激光器

110. 瑞利散射用于气体流场二维瞬态密度测量

王杰, 姚建铨, 于意仲, 陈进, 王鹏
光电子·激光, 2001, 12(1): 62–64.

摘要：研究了气体流场分子的瑞利散射光强与流场密度的关系。采用 ArF 准分子深紫外激光（波长 193 nm）, 利用带像增强的 ICCD 照相机拍摄了 O_2 喷流的瑞利散射图像, 得出喷流二维密度结构图像, 并通过分子标记示踪的方法测量了流场截面速度分布的不均匀性。

关键词：瑞利散射, 流场检测, 二维密度测量, 分子标记

111. 非线性光学频率变换及准相位匹配技术

姚建铨

人工晶体学报, 2002, 31(3): 201–207.

摘要：本文综述了非线性光学频率变换技术的发展及在激光与光电子领域中的地位与作用，概述了它们的基本原理、特点及关键技术，展示了其广阔的应用前景。

关键词：非线性光学频率变换技术，光学参量振荡器，准相位匹配，周期极化晶体，全固态激光器，可调谐固体激光器

112. 含预反馈的激光自混合干涉型位移测量结构

禹延光, 姚建铨, 叶会英

光学学报, 2002, 22(3): 308–312.

摘要：提出了含预反馈的激光自混合干涉型位移测量结构，建立了系统模型，经理论分析和实验验证，结果表明：该结构具有提高温度量程、提高测量信号信噪比以及获得粗糙表面的锯齿干涉信号等特点。

关键词：半导体激光，自混合干涉，预反馈，位移测量

113. 掺铜 KNSBN 光折变晶体的光致吸收特性

胡居广，林晓东，刘毅，阮双琛，姚建铨

光子学报，2002, 31(2): 169–172.

摘要： 掺铜 KNSBN（钾钠铌酸锶钡）光折变晶体具有光致吸收 (LIA) 特性。实验表明，泵浦光的光强越强，波长越短，LIA 系数就越大，探测光为 e 偏振态时的 LIA 系数是 o 光时的 1.2 倍。这种现象将被用来提高光场的空间或时间均匀性。

关键词： 掺铜 KNSBN 晶体，光致吸收，光强均匀性

114. 多模激光自混合干涉实验与理论分析

禹延光，叶会英，姚建铨

光电子·激光，2002, 13(11): 1190–1193.

摘要： 根据单模激光自混合干涉的基本理论，基于多模竞争，建立多模激光自混合干涉模型，分析多模激光的自混合干涉波形。理论分析与实验观察表明，多模自混合干涉信号是激光管腔长的谐波函数，利用光强波动的包络信号可进行大量程测量；激光模数增加有利于测量分辨率的提高；光强波动信号对应激光管等效腔长的整数倍时，可获得最大干涉信号调制深度。

关键词： 自混合干涉，多模管激光，位移测量

115. 准连续 660 nm Nd:YAG 内腔倍频激光器的研究

王涛,姚建铨,李喜福,王志勇,陈进,张百钢,王鹏

激光杂志, 2002, 23(2): 10–12.

摘要: 研制了一台 Nd:YAG 内腔倍频输出 660 nm 红光激光器。分析了腔长对激光功率的影响,采用新型径向调整式光学镜片调整架,平 - 平腔结构,腔长 390 mm,双灯泵浦,KTP 晶体内腔倍频,并设置声光 Q 开关,获得 660 nm 红光输出 2 W。

关键词: 红光 660 nm 激光器,KTP 晶体,Nd:YAG 内腔倍频

116. LD 抽运的高效率内腔倍频绿光激光器的研究

于意仲,陈进,张百钢,王鹏,张帆,王涛,朱孟,李喜福,姚建铨

中国激光, 2002, A29(增刊): 89–90.

摘要: 考虑了高功率 LD 抽运的倍频 YAG 激光器中 KTP 晶体的热效应,计算了在高功率抽运条件下的最佳相位匹配角应为 $\theta=90°$, $\varphi=23.27°+0.7°$。采用传统直腔和 L 形腔,在声光调 Q 及高功率 LD 抽运的情况下,获得了 40 W 的内腔倍频绿光输出。

关键词: 二极管抽运,声光调 Q,最佳耦合,内腔倍频

117. LD 抽运的全固态三倍频紫外激光器

陈进,姚建铨,王鹏,于意仲,朱孟,张百钢,王涛,王杰
中国激光,2002, A29(增刊): 86-88.

摘要: 报道了一台 LD 抽运的内腔三次谐波转换 Nd:YAG 全固态激光器,采用内腔倍频技术,当激光二极管注入抽运功率为 8 W 时,产生约 3 mW 连续运转的 355 nm 紫外激光,当采用声光调 Q 运转时,产生的三倍频紫外激光输出平均功率超过 50 mW。

关键词: 激光二极管固体激光器,紫外激光,内腔三倍频

118. 高功率全固态绿光激光器的设计与研究

张帆,陈进,于意仲,王鹏,陆颖,禹延光,王涛,张勇,姚建铨
中国激光,2002, A29(增刊): 91-92.

摘要: 报道了一种激光二极管 (LD) 侧面抽运 Nd:YAG 晶体,采用 KTP 晶体作内腔倍频的 Q 开关激光器。在重复频率为 13 kHz 时,绿光 (532 nm) 输出平均功率达到 40 W,电-光转换效率约为 3%。

关键词: 激光二极管抽运,Nd:YAG 激光,声光调 Q,内腔倍频

119. 红光 (660 nm) 准连续 Nd:YAG 内腔倍频激光器

王涛，姚建铨，李喜福，郭玲，陈进，于意仲，朱孟，王鹏

中国激光，2002, A29(增刊): 108-110.

摘要：一种准连续 Nd:YAG 内腔倍频输出 660 nm 红光激光器。采用平 - 平腔结构，双灯泵浦，KTP 晶体内腔倍频，并设置声光 Q 开关，获得 660 nm 红光输出 2 W。从激光光束的有效回射率的情况分析了腔长对激光功率的影响。

关键词：红光 660 nm 激光器，KTP 晶体，Nd:YAG 内腔倍频

120. 半导体激光自混合干涉系统的稳态分析

禹延光，叶会英，赵新，姚建铨

中国激光，2002, A29(增刊): 631-632.

摘要：通过求解激光自混合干涉位移测量系统模型的相位方程，确定其单值解和多值解时边界条件，由此确定自混合干涉系统稳态运行的参数选择规则。

关键词：半导体激光器，自混合干涉，稳态解

121. 光互连中两微腔激光器耦合机制研究

张勇, 张存善, 姚建铨, 王鹏, 陈进, 王涛, 张百钢

中国激光, 2002, A29(增刊): 406–408.

摘要： 报道用等效 F-P 腔的方法计算了多层 DBR 的反射率。然后对垂直腔面发射激光器 (VCSEL) 远场分布曲线进行插值拟合，得到其远场光强分布。最后，着重对光互连中的最小作用单元，即两 VCSEL 的相互耦合机制进行了研究。

关键词： 垂直腔面发射激光器, 光互连, 远场分布, 耦合机制

122. 激光动态识别 Line 元件三维形状及动态的研究

王涛, 李喜福, 郭玲, 姚建铨, 于意仲, 王鹏, 陈进, 张百钢

中国激光, 2002, A29(增刊): 608–610.

摘要： 一种利用激光扫描及动态定位技术实现对 Line 元件的三维形状、状态及动态进行在线实时动态识别及检测的新方法。提出了激光三维形状信息的采集方式，推导出有关数理模型，建立了激光动态识别 Line 元件三维形状及动态的系统。

关键词： 激光扫描, 动态识别, 三维形状, Line 元件

123.（1 064/1 319，532/659 nm）Nd:YAG 激光治疗仪的实验研究

宁喜发，姚建铨，王鹏，吴峰

光电子·激光，2002,13(12)：1311–1313.

摘要： 叙述了激光治疗仪双波长（1 064/1 319 nm）输出的最佳方案及其性能指标；介绍了双波长输出的 CW-Nd:YAG 激光器参数的确定及其倍频实验；提出了用 Ti:MgO:LiNbO₃ 晶体倍频的 4 波长 Nd:YAG 激光器 5 种输出的可行性方案。

关键词： 4 波长，CW-Nd:YAG 激光器，治疗仪，倍频

124. Study on high power laser diode pumped Nd:YAG laser and its frequency conversion

Jianquan Yao, Yizhong Yu, Fan Zhang, Jin Chen, Baigang Zhang, Degang Xu, Peng Wang

Proceedings of Society of Photo-Optical Instrumentation Engineers, 2002, 4914:258–262.

Abstract: A high power intracavity frequency doubled Nd:YAG laser with KTP crystal and A-O Q-switcher pumped by 1 600 Watt-808 nm laser diodes and its thermal effect are discussed. Also we proved that the title angle of KTP crystal can be to compensate for the phase mismatching and to solve the problem of the drop of green laser output power along with the increasing temperature of KTP crystal. Then based on optical parametric oscillator (KTP-OPO) pumped by 532 nm laser and their frequency doubling (with KTP and BBO) a Watt-level red and blue laser system which would be provided as RGB laser projection display are described.

Key words: High-power green laser, Nd:YAG, Intracavity frequency doubling, KTP

125. Different configurations of continuous-wave single resonant cavity and design of periodically poled LiNbO₃

Guiyan Zang, Jianquan Yao, Baigang Zhang, Yanguang Yu, Tao Wang, Peng Wang

Proceedings of Society of Photo-Optical Instrumentation Engineers, 2002, 4913: 267−271.

Abstract: In this letter we introduce three typical cw, singly resonant OPO configurations based on PPLN and the different advantages and disadvantages of them. And we also show three different kinds of PPLN's and describe the differences between them.

Key words: Optical parametric oscillator (OPO), Quasi phase matching (QPM), Periodically poled LiNbO₃ (PPLN)

126. All solid-state wide beam flat-top laser

Jin Chen, Jiaju Ma, Jianquan Yao, Meng Zhu, Yizhong Yu, Yong Zhang, Peng Wang, Pu Du

Proceedings of Society of Photo-Optical Instrumentation Engineers, 2002, 4914: 331−334.

Abstract: A solid-state laser source with flat-top wide beam profile was designed and made. The laser source was mainly made up of an Nd:YAG solid-state laser at 1 064 nm and a maximum 267 × collimating telescope. A 2 W laser diode (LD) pumped Nd:YAG laser with flat-concave cavity was adopted as the primary laser source at 1 064 nm. Then the output laser (1 064 nm) was injected into a collimating telescope. We designed three schemes according to spherical wave theory and chose the best one based on our experiments. We found out the spherical aberration (including axial and off-axial points) must be corrected. At last, an output laser beam with 80 mm diameter, 0.4 mrad divergence angle, > 50 mW output power, and ± 10% intensity distribution of cross section, was obtained.

Key words: All solid-state laser, Flat-top, Gaussian beam, Spherical aberration

127. High power high efficiency CW operation of a LD-pumped Nd:YAG rod laser

Degang Xu, Jianquan Yao, Baigang Zhang, Juguang Hu, Jin Chen, Peng Wang

Proceedings of Society of Photo-Optical Instrumentation Engineers, 2002, 4914: 165–167.

Abstract: We have describe a simple diode-side-pumped high-power and high efficiency Nd:YAG laser that produces 240 W of power continuous-wave(CW) with 33% optical-optical efficiency. We used the technology of diode laser close-couple to realize high pumping efficiency, We measured the thermal lensing focal length to design the laser resonator, which decrease the rod thermal effect. By changing the resonator parameters, we can adjust the fundamental mode size and the output beam quality.

Key words: Laser diode (LD), Thermal lensing, Side pumped, All-solid-state

128. A new method to measure the ultrashort thermal focal length of high-power solid-state laser

Baigang Zhang, Jianquan Yao, Degang Xu, Yizhong Yu, Tao Wang, Jin Chen, Peng Wang

Proceedings of Society of Photo-Optical Instrumentation Engineers, 2002, 4914: 460–463.

Abstract: Thermal lens effect is a very important factor in designing stable resonators for high-power solid-state laser pumped by laser diodes. In a symmetric plane-parallel resonator, if the thermal lens of the laser crystal is close to a thin lens, according to the theory of transfer matrix we can get that there is a critical hollow point where the resonator changes from stable region to unstable state, then back to stable region. So we can measure the ultrashort thermal focal length of high-power solid-state laser based on this fact. We used this method to measure the ultrashort thermal focal length of a 100 W magnitude Nd:YAG laser. The experimental results have shown this is a simple and effective technique.

Key words: Thermal lens effect, Ultrashort thermal focal length, Cavity stability conditions, High-power solid-state laser

129. The study on high average power all-solid-state green laser

Yizhong Yu, Baigang Zhang, Jin Chen, Fan Zhang, Jianquan Yao, Peng Wang

Proceedings of Society of Photo-Optical Instrumentation Engineers, 2002, 4914: 155–160.

Abstract: During design and experiments, multi-diode-laser-module was employed to pump a double AO switched Quasi-CW YAG laser with KTP or LBO crystal as frequency doubler. Thermal lensing effect in laser rod and frequency doubling crystal were both considered. Cavity design with ABCD law and crystal thermal compensation given in detail by numerical calculation. High average power output was achieved.

Key words: Green laser, Diode laser pumped, Cavity design, Thermal effects, Frequency doubling

130. The method research of used laser to processes the slot pipe applied to petroleum exploitation

Fan Zhang, Jianquan Yao, Ning Hou, Peng Wang

Proceedings of Society of Photo-Optical Instrumentation Engineers, 2002, 4915: 295–297.

Abstract: The action of slot pipe is much importance in petroleum exploitation. Despite the method to defend sand contain variety in course of artesian well and over well, it is still the most in common method to use slot pipe to over well and defend the sand. Use laser process slot pipe is function superiority and price cheap. Not only is this type of slot pipe relatively inexpensive but also durable, easily installed, and efficiently developed. So many oil-fields are positive to use it.

Key words: Slot pipe, Laser-cutting screen tube

131. Analysis of self-mixing interference signals in LD pumped solid-state laser using fast Fourier transform technique

Yong Zhang, Yanguang Yu, Jianquan Yao, Peng Wang, Jin Chen

Proceedings of Society of Photo-Optical Instrumentation Engineers, 2002, 4919: 488–492.

Abstract: Self-mixing interference (SMI) effects have been widely used in measuring the distance and displacement and velocity. Conventional methods are fringe counting and phase analysis. The former method has a precision of $\lambda/10$ in a multi-mode LD pumped solid-state laser (LDPSSL), while the latter has a precision of nearly $\lambda/50$ concerning the multiple reflections. Theoretical analysis and simulation calculations are presented.

Key words: Self-mixing interference effect, LD pumped solid-state laser, Fast Fourier transform technique, Distance, Displacement

132. Studies on the relation between temperature tuning value and grating period deviation of periodically poled crystal for SHG

Baigang Zhang, Jianquan Yao, Degang Xu, Yanguang Yu, Guiyan Zang, Peng Wang, Yong Zhang

Proceedings of Society of Photo-Optical Instrumentation Engineers, 2002, 4919: 97–101.

Abstract: It's unavoidable that there is a deviation (called "grating period deviation" in this paper) between the real QPM period of periodically poled crystal and the ideal value to cause phase mismatching. By tuning temperature, we can get phase matching again. We present the formula of the relation between temperature tuning value ΔT and grating period deviation $\Delta\Lambda$ of periodically poled crystal for SHG in order to get phase matching. Our numerical analyses show that ΔT of PPLN-SHG is approximately directly proportional to the $\Delta\Lambda$, and when the pumping wavelength or the quasi-phase matching order increase, the slope of $\Delta T \sim \Delta\Lambda$ curve will go down.

Key words: QPM, PPLN crystal, Grating period deviation, PPLN-SHG, Temperature tuning value

133. View for the development of theory on the self-mixing interference and the general model of the displacement measurement

Yanguang Yu, Jianquan Yao

Proceedings of Society of Photo-Optical Instrumentation Engineers, 2002, 4919: 235–241.

Abstract: The development of the theory on self-mixing interference is reviewed. A general model is deduced. It can be used to describe self-mixing interference phenomena at high feedback levels in a laser diode with single mode or multiple modes. It is helpful to design self-mixing interference systems. At the last part of the paper, some existing questions in this area are summarized and the future research trends are listed.

Key words: Laser diode, Self-mixing interference, Optic feedback, Displacement, General model, Multi-mode laser

134. Research on the laser disturbing and damage of TV tracking system

Niu Yanxiong, Yao Jianquan, Wang Yuefeng, Duan Xiaofeng, Zhang Chu, Sun Yujie

Proceedings of Society of Photo-Optical Instrumentation Engineers, 2002, 4914: 247–252.

Abstract: In the modern weapon systems, more and more the TV tracking systems have been widely used. It has more good characteristics than infrared tracking, laser tracking, radar tracking. When the high-power or high-energy laser irradiates the CCD video camera, it will be disturbed or damaged and lead to the TV tracking system damage, such as losing the tracking target. In the paper, the laser damage of the CCD has been studied theoretically and experimentally.

Key words: TV tracking system, Laser damage, CCD detecting system

135. 从 LASERS' 2000 看激光技术发展趋向

姚建铨

国际学术动态, 2002, 6: 15.

摘要: LASERS2000 国际会议于 2000 年 12 月 4—8 日在美国新墨西哥州 Albuquerque 举行,该会议是由光学量子电子学会举办的系列会议之一,原名激光科学会议,已是第 23 届,着重讨论激光场、激光科学、各类激光器件及激光应用。会议虽然规模不大,但水平很高,这次会议共录用 300 多篇论文,其中特邀报告 70 篇,一般报告加张贴论文 230 余篇,美国、日本、德国、法国、意大利、俄罗斯和中国的科学家出席了会议。本文根据该会议的信息,总结了激光技术发展的若干趋向。

关键字: 激光技术,激光器件,发展趋向

136. Analysis for the self-mixing interference effects in a laser diode at high optical feedback levels

Yanguang Yu, Huiying Ye, Jianquan Yao

Journal of Optics A: Pure and Applied Optics, 2003,5:117–122.

Abstract: Based on the study of the self-mixing interference effects at different optical feedback levels, a model is presented in terms of the threshold condition of a compound cavity laser. This model considers the optical feedback effect on the real and imaginary parts of the complex refractive index of the laser active medium and the asymmetric external cavity effect. Complex waveforms experimentally observed at high feedback levels can be interpreted well by this model. Theoretical and experimental results show a potential increase in the resolution of the displacement measurement at a high optical feedback level or misalignment condition.

Key words: External cavity semiconductor laser, Optical feedback, Displacement measurement, Self-mixing interference

137. Temperature tunable infrared optical parametric oscillator with periodically poled LiNbO$_3$

Zhang Baigang, Yao Jianquan, Zhang Hao, Zang Guiyan, Xu Degang, Wang Tao, Li Xuejin, Wang Peng

Chinese Physics Letters, 2003, 20(7): 1077−1080.

Abstract: We demonstrate a temperature tunable infrared optical parametric oscillator (OPO) based on periodically poled LiNbO$_3$ (PPLN) pumped by an acousto-optically Q-switched cw-diode-end-pumped Nd:YVO$_4$ laser with the signal output from 1.48 μm to 1.54 μm by tuning the work temperature from 60 ℃ to 250 ℃ and the maximum average signal power (1 500 nm) of 137 mW. We report the demonstration of the sum frequency of the pump and idler in a single-grating PPLN crystal. In addition, we present the theoretical analysis of the signal wavelength change rate with the changing temperature for a quasi-phase matching OPO, which is in good agreement with our experimental results.

Key words: Temperature Tunable, OPO, PPLN

138. High power diode single-end-pumped Nd:YVO$_4$ laser

Hongrui Zhang, Mingju Chao, Mingyi Gao, Liwen Zhang, Jianquan Yao

Optics & Laser Technology, 2003, 35(6): 445−449.

Abstract: A fiber-coupled diode-single-end-pumped Nd:YVO$_4$ laser with an Nd:YVO$_4$ crystal of 0.3 at% doping concentration and $3 \times 3 \times 10$ mm^3 dimensions was reported. 14.850 W of continuous-wave output power in an M^2 factor of 1.12 was obtained under pump power of 27.365 W, with an optical conversion efficiency of 60.49%, and a slope efficiency of 64.5%.

Key words: Nd:YVO$_4$ laser, Doping concentration, Crystal dimensions, Thermal effects

139. Study on 660 nm quasi-contionuous-wave intracavity frequency-doubled Nd:YAG laser

Tao Wang, Jianquan Yao, Baigang Zhang, Guiyan Zang, Peng Wang, Yizhong Yu

Chinese Optics Letters, 2003, 1(5): 289–291.

Abstract: A quasi-continuous-wave-intracavity frequency-doubled Nd:YAG laser which operates at 660 nm is studied. By using a flat-flat laser cavity, 2 Kr-lamps, KTP crystal and an acousto-optically Q-switch, 2-W output power at 660 nm is obtained. The relationship between laser cavity length and output power is analyzed.

Key words: Quasi-continuous-wave, Intracavity frequency-doubled, Nd:YAG laser

140. Angle-tuned signal-resonated optical parametric oscillator based on periodically poled lithium niobate

Baigang Zhang, Jianquan Yao, Hao Zhang, Degang Xu, Peng Wang, Xuejin Li, Xin Ding

Chinese Optics Letters, 2003, 1(6): 346–349.

Abstract: We demonstrate an angle tuned signal-resonated optical parametric oscillator (OPO) with periodically poled lithium niobate (PPLN) pumped by a diode pumped Nd:YVO$_4$ laser. 1 499.8-1 506.6 nm of signal wavelength is achieved at 140℃ by rotating a 29-μm period PPLN from 0 - 10.22° in the x-y plane while keeping the pump wave vertical to the resonator mirrors. Two pairs of the signal and idler waves of the same wavelengths can be achieved symmetrically for each pair of angles of rotation with same absolute value and opposite sign. Theoretical analyses on angle-tuned PPLN-OPO with pump wave vertical to the resonator mirrors are presented and in good agreement with our experimental results. It is also found that all interacting waves in the cavity (not inside the crystal) are always collinear for PPLN-OPO with the pump wave vertical to the resonator mirrors while phase matching is noncollinear within the crystal.

Key words: Optical parametric oscillator (OPO), PPLN, Angle-tuned

141. Study on CW Nd:YAG infrared laser at 1 319 nm

Tao Wang, Jianquan Yao, Guojun Yu, Peng Wang, Xifu Li, Yizhong Yu
Chinese Optics Letters, 2003, 1(11): 661−663.

Abstract: A continuous wave (CW) Nd:YAG infrared laser at 1 319 nm is reported in this paper. The energy level of 1 319-nm wave was analyzed. The repression of 1 064-nm lasing and enhancement of 1 319-nm output power were discussed. Mirror coating and cavity structure were studied and a maximum CW output power of 43 W at 1 319 nm was achieved in experiments.
Key words: Nd:YAG laser, 1 319 nm, Continuous wave

142. LD pumped high efficient intracavity frequency-doubled green laser

Chen Jin, Yao Jianquan, Yu Yizhong, Zhu Meng, Wang Peng, Wang Jie
Transactions of Tianjin University, 2003, 9(1): 1−6.

Abstract: A laser diode (LD)pumped Q-switched high efficient intracavity frequency-doubled Nd：YAG laser is reported here. The authors have designed an optical coupler and pointed out that the key to increasing harmonic conversion efficiency is to decrease the loss of fundamental wave. In the experiments, a fundamental mode output laser was acquired. When the pumping power was 12 W, 2.6 W average output power at 1 064 nm with AO Q-switch was obtained. 2.1 W average output power at 532 nm was obtained with intracavity frequency doubling, and the highest second harmonic conversion efficiency was 82%.
Key words: LD pumped solid-state laser, High efficient intracavity frequency-doubled, Green laser

143. Self-mixing interference effects in LD pumped multi-mode solid-state Laser

Zhang Yong, Yao Jianquan, Wang Peng, Yu Yanguang, Chen Jin
Transactions of Tianjin University, 2003, 9(4): 261–263.

Abstract: Based on the effective structure of the self-mixing interference effects, a general model for the self-mixing interference effects in the LD pumped solid-state laser has been established for the first time. The numerical simulation of the self-mixing interference signal has been done, the results show that when the external cavity length is integral times of 1/2, 1/3, 2/3, 1/4, 3/4 of the effective cavity length, the intensity of the selfmixing interference signals reach maximum in value. While that of single mode laser is integral times of half of the effective cavity length, the measuring precision of displacement of single mode laser is $\lambda/2$. A conclusion can be drawn from the above results that the measuring precision of displacement of multi-mode laser is higher than that of single mode laser.

Key words: General model, Self-mixing interference effects, LD pumped multi-mode solid-state laser, Numerical simulations

144. 激光自混合干涉位移测量系统的稳态解

禹延光,叶会英,姚建铨
光学学报,2003, 23(1):80–84.

摘要: 基于三镜腔等效自混合干涉位移测量系统,对自混合干涉系统的稳定性问题,建立系统模型,通过求解模型的相位方程,确定其单值解和多值解时边界条件,从理论上确定自混合干涉系统稳态运行的光反馈水平。该结论对系统设计具有指导意义。

关键词: 光学测量,半导体激光器,自混合干涉,稳态解

145. 104 W 全固态 532 nm Nd: YAG 激光器

徐德刚,姚建铨,周睿,郭丽,陈进,于意仲,丁欣,张百钢,温午麒,王鹏

中国激光,2003, 30(9): 864.

摘要: 我们采用单 Nd:YAG 棒平凹谐振腔设计及临界相位匹配的 KTP 晶体,利用内腔倍频技术获得了平均功率为 104 W 的绿光输出。

关键词: 高功率,全固态,Nd:YAG,绿光

146. 波长为 1 319 nm 的连续输出 Nd:YAG 激光器的研究

王涛,姚建铨,李喜福,郁道银,禹国俊,王鹏

中国激光,2003, 30(10): 881–884.

摘要: 介绍了一种波长为 1 319 nm 的连续 Nd:YAG 激光器,分析了波长 1 319 nm 激光的辐射跃迁能级,论述了抑制 1 064 nm 激光的生成从而提高 1 319 nm 激光输出等关键技术,研究了光学镜片的镀膜参数与腔型结构,实现 1 319 nm 激光连续输出最高功率 43 W。

关键词: 激光技术,1 319 nm 激光器,Nd:YAG 晶体,能级跃迁

147. 85 W 绿光激光器中的 KTP 晶体热效应研究

郭丽,姚建铨,禹国俊,周睿,温午麒,王鹏

激光杂志, 2003, 24(6): 20-22.

摘要: 本文强调了倍频晶体的热效应问题是影响高功率激光器输出功率的主要因素之一,利用热传导方程对 KTP 晶体热效应问题进行了理论分析,模拟出了在一定泵浦功率下 KTP 晶体内部温度分布情况。并提出角度调节及加强冷却的方法来解决晶体热效应问题,在实验中获得了 85 W 的高功率绿光输出。

关键词: 全固态绿光激光器,KTP 晶体,热效应,角度调节

148. 激光在人头部立体形状建模中的应用研究

王涛,姚建铨,郭玲,王鹏,李喜福,徐德刚

激光杂志, 2003, 24(3): 62-63.

摘要: 用激光对人头部进行立体扫描,CCD 摄像仪摄取扫描信息,移植航空摄影空间三维测量原理,基于人头部、脸部的特征测量部位,创建实现人头部外型的立体空间数学模型,提出了人头部形状的模型集成及特征识别模式原理,由此建立了人头部的激光扫描三维立体建模系统。

关键词: 人头部外型,激光扫描,特征识别

149. 送粉激光熔覆结合界面组织结构与特性

张三川, 姚建铨

激光杂志, 2003, 24(3): 58-59.

摘要：采用 OM、SEM、EDS 等先进表面检测技术, 研究了送粉激光熔覆 Ni60 自熔合金涂层与基体材料 (45 钢) 间的结合界面区的一种固溶体组织结构特征, 并分析了该组织形成的原因。

关键词：激光熔覆, 结合界面, 显微组织, 固溶体

150. 波长 1 319 nm 连续 Nd:YAG 激光器的研究

王涛, 姚建铨, 李喜福, 王鹏, 于意仲, 王志勇, 禹国俊, 郭丽

激光杂志, 2003, 24(4): 20-22.

摘要：使用连续 Nd:YAG 输出 1 319 nm 激光器, 分析了波长 1 319 nm 激光的辐射跃迁能级, 论述了抑制 1 064 nm 激光的生成从而提高 1 319 nm 激光输出等关键技术, 研究了光学镜片的镀膜参数与腔型结构, 实现 1 319 nm 激光连续输出最高功率 43 W。

关键词：1 319 nm 激光器, Nd:YAG 晶体, 能级跃迁

151. 天津大学精仪学院激光与光电子研究所研制出全固态绿光激光器已突破百瓦

王鹏, 姚建铨, 徐德刚, 周睿, 郭丽
激光杂志, 2003, 24(4): 3.

摘要: 高功率全固态 532 nm 激光器在工业加工、海底探测、分离铀同位素以及大功率三色显示等领域有着重要的应用, 国外在此项研究中绿光的最高输出功率早已超过百瓦, 而在国内报道绿光的最高输出功率仅是 68 W, 我们实验室利用美国 CEO 公司生产的半导体激光器组件、采取单声光调 Q 装置、KTP 晶体 II 类临界相位匹配、谐振腔为简单的平凹腔结构, 在泵浦电流 18.3 A、调制频率 20.7 KHz 时获得了 104 W 准连续绿光 (532 nm) 输出, 脉冲宽度低于 130 ns, 倍频效率达到了 50%, 使得我们国内在此项研究中达到了国际水平。

关键词: 全固态绿光激光器, 高功率, 百瓦

152. LD 纵向泵浦的 946 nm Nd:YAG 准三能级激光系统的研究

禹国俊, 姚建铨, 郭丽, 张百刚, 张浩, 王鹏
量子电子学报, 2003, 20(6): 675-679.

摘要: 对准三能级系统进行了研究, 分析了激光上、下能级不同的能级寿命以及上能级自发辐射对下能级粒子数所造成的影响, 修正了 Fan 等提出的纵向泵浦准三能级系统的理论模型。进行了数值模拟, 得出了系统参数对 LD 纵向泵浦的 946 nm Nd:YAG 激光器性能的影响规律。认为在一定掺杂浓度和输出损耗情况下, 存在最佳晶体长度使阈值最低, 及最佳模式匹配使斜效率最高。理论得到了实验验证。

关键词: 准三能级系统, 946 nm Nd:YAG 激光, 模式匹配

153. 激光微加工中运动控制系统的设计

杨雷, 姚建铨, 裴红星, 刘晓旻

天津工业大学学报, 2003, 22(5): 38–41.

摘要：设计了激光微加工系统中的运动控制系统，探讨了系统的组成、设计与实现。系统由目标值设定器、数字控制器、运动执行机构和位置检测用光栅尺构成。文中重点讨论了闭环运动控制系统中控制目标值的形成、执行机构 X-Y 平台的模型以及运动控制的算法等问题，控制精度达 2 μm。

关键词：激光微加工, 运动控制, X-Y 平台, 光栅尺

154. 片式电阻激光调阻机检测系统的设计

赵新, 姚建铨, 杨洗陈

天津工业大学学报, 2003, 22(5): 46–48.

摘要：在激光调阻机自动调阻过程中，需要对被调电阻进行实时在线精确测量，以控制调阻进程。传统的伏安法测量电阻，虽然可以进行精密测量，但不适合在线动态测量。本文将微处理机技术应用于有源电桥的测量，利用伏安法测量电阻的灵活性，配以高精度的精密运算放大器，构成高精密有源数字电桥，达到了工程应用要求。

关键词：激光调阻, 有源电桥, 在线测量

155. 送粉激光熔敷耐磨涂层制备工艺与组织结构

张三川,刘怀喜,姚建铨

天津工业大学学报, 2003, 22(5): 76-78.

摘要：采用 20%WC 作为镍基自熔合金的掺杂增强相，研究了该复合材料在送粉激光熔敷工艺条件下的熔敷层显微组织、显微硬度与熔敷工艺规范间关系，得出了在实验条件下的优化工艺参数：激光功率为 2.5 kW（光斑离焦量 60 mm）；扫描速度为 2.4 mm/s；送粉量为 3.0 g/min。

关键词：送粉激光熔敷,组织结构,显微硬度,耐磨涂层

156. 用偏振光学相干层析技术分析研究组织偏振特性

姜宇,姚建铨,王瑞康

激光与光电子学进展, 2003, 40(1): 36-39.

摘要：胶原质丰富而且结构排列整齐的生物组织受光照射时会显示双折射的特性。通过精确地控制入射光和反射光的偏振状态,偏振光学相干层析技术能够显示出组织的双折射性。我们利用斯托克斯参数和琼斯矢量,着重讨论点光源照射样品后其背景反射光的偏振特性,从而导出组织的偏振特性。

关键词：光学相干层析技术,胶原纤维组织,偏振,双折射

157. 天津大学激光与光电子研究所研制成功 104 W 全固态 532 nm 绿光激光器

周睿, 姚建铨, 徐德刚, 王鹏, 郭丽, 温午麒

激光与光电子学进展, 2003, 40(7): 64.

摘要： 高平均功率、高重复率全固态 532 nm 绿光激光器在可调谐激光器的抽运源、流场显示、海洋探测、光电对抗、污染检测、特别是受控热核聚变的驱动器,铀同位素分离的抽运源,以及大功率大能量激光加工及激光医疗设备,激光微加工,激光的军事应用(激光雷达,激光制导等)等科学和工业领域有着重要的应用。为了满足上述科学和工业领域的应用,全固态内腔倍频激光器是获得高效高功率稳定绿光光源的重要途径之一。天津大学激光与光电子研究所采用美国 CEO 公司的 1 600 w 半导体抽运组件,英国 NEOS 公司的声光调 Q 组件,利用 KTP 晶体内腔倍频平－凹谐振腔,在抽运电流为 184 A,声光重复频率为 20.7 kHz 时,获得了功率达 104 W 的高功率、高重复率 532 nm 绿光输出。在输出功率为 104 W 时,脉冲宽度小于 130 ns,倍频效率约为 50%,不稳定度小于 1%。天津大学激光与光电子研究所的这项成果填补国内高功率全固态绿光激光器方面的空白,达到了国际先进水平。

关键词： 高功率, Nd:YAG 激光器, KTP, 绿光

158. Low-threshold, high-efficiency, high-repetition-rate optical parametric generator based on periodically poled LiNbO₃

Zhang Baigang, Yao Jianquan, Ding Xin, Zhang Hao, Wang Peng, Xu Degang, Yu Guojun, Zhang Fan

Chinese Physics, 2004, 13(3): 364−368.

Abstract: We report a high-repetition-rate optical parametric generator (OPG) with a periodically poled lithium niobate(PPLN) crystal pumped by an acousto-optically Q-switched CW-diode-end-pumped Nd:YVO$_4$ laser. For the maximum 1 064 nm pump power of 970 mW, the maximum conversion efficiency is 32.9% under the conditions of 250 ℃, 1 064 nm pulse repetition rate of 22.6 kHz and pulse width of 12 ns, and the PPLN OPG threshold in the collinear case is less than 23.7 μJ. The output power increases with the increase of the crystal temperature. The 1 485-1 553 nm signal wave and 3 383-3 754 nm idler wave are obtained by changing the temperature and the angle of the PPLN crystal.

Key words: Quasi-phase-matching, Periodically poled lithium niobate, Optical parametric generator

159. Wavelength tunable optical parametric oscillator based on periodically poled lithium niobate

Yao Jianquan, Zhang Baigang, Lu Yang, Ding Xing, Xu Degang, Wang Peng
Journal of Synthetic Crystals, 2004, 33(4): 465–470.

Abstract: We report wavelength tunable quasi-phase-matching (QPM) optical parametric oscillators (OPOs) based on periodically poled lithium niobate (PPLN) pumped by an acousto-optically Q-switched CW-diode-end-pumped Nd:YVO$_4$ laser. Widely wavelength tunable outputs have been obtained by using temperature tuning, angle tuning and grating period tuning. The analysis of comparison about these tuning methods is given in this paper which is helpful to realize widely, rapidly and continuously tunable coherent sources.

Key words: Quasi-phase-matching (QPM), Temperature tuning, Angle tuning, Pump tuning, Grating period tuning

160. 在 PPLN-OPO 中伴生四波混频的研究

王涛,姚建铨,郁道银,李喜福,禹国俊,王鹏,张百钢
中国激光, 2004, 31(3): 284–288.

摘要: 提出了在周期极化 LiNbO$_3$ 光学参量振荡(PPLN-OPO)中伴有四波混频(FWM)的发生,论述了 PPLN-OPO 中的抽运光 λ_3(1 064 nm),信号光 λ_1(1 500 nm),闲频光 λ_2(3 660.55 nm),信号光 λ_1 的倍频光 λ_4(750 nm),满足 FWM 的频率匹配与准相位匹配的条件,初步建立了 PPLN-OPO-FWM 的过程理论体系,进行了光学系统装置的实验,理论分析与实验结论相吻合。

关键词: 非线性光学,周期极化 LiNbO$_3$,光学参量振荡,四波混频,准相位匹配

161. 绿光输出达 85 W 的全固态绿光激光器谐振腔研究

周睿, 姚建铨, 徐德刚, 郭丽, 温午祺, 王鹏, 于意仲

中国激光, 2004, 31(6): 641-645.

摘要：报道了对平均功率达 85 W 的高功率、高稳定性全固态绿光激光器谐振腔特性的研究，将高平均功率运转条件下的 KTP 倍频晶体的热透镜效应等效为一个薄透镜，利用 ABCD 传输矩阵，通过图解方法定性地讨论了 KTP 晶体的热透镜效应对谐振腔的稳定性和腔内激光模式的巨大影响，理论分析表明适当大小的倍频晶体热透镜焦距不但可以有效地补偿 Nd:YAG 棒的热透镜效应，而且对增大激光介质中的模体积和在倍频晶体处提高功率密度都有积极作用。实验中采用了 80 个 20 W 的高功率半导体激光器侧面抽运的单 Nd:YAG 棒、两个声光 Q 开关、高效平 - 凹谐振腔结构、对大尺寸 KTP 晶体进行角度偏离法补偿相位失配等技术，并通过对 KTP 晶体采取适当的冷却方式，最终实现了高功率内腔倍频激光器的高稳定性运转；在抽运功率约为 1 080 W 时，实现了重复频率为 20.4 kHz，脉冲宽度 230 ns，输出功率达 85 W 的高功率、高重复频率绿光（532 nm）输出，不稳定性仅为 ±1.03 %。

关键词：激光技术，全固态绿光激光器，谐振腔，平 - 凹腔，稳定性，KTP 晶体的热透镜效应

162. 8.1 W 全固态准连续红光 Nd:YAG 激光器

温午麒, 姚建铨, 丁欣, 周建勇, 李君, 魏权夫, 徐德刚, 周睿, 张强, 于意仲, 王鹏

中国激光, 2004, 31(11): 1281-1284.

摘要：报道了利用 II 类临界相位匹配的 KTP 晶体（相位匹配角选为 θ=59.9°, Φ=0°）对 Nd:YAG 在 1.3 μm 附近的振荡进行腔内倍频，产生高功率准连续红光激光的实验结果。激光器使用了一个连续运转的高功率激光二极管(LD)侧面抽运组件（组件内由 30 个 20W 的二极管阵列呈三角形阵列分布抽运一根 Nd: YAG 圆棒），使用声光调 Q 技术实现高重复频率输出，并选用了平 - 凹直腔的腔体结构。对该激光器的基频 (1.3 μm 波长) 调 Q 和倍频红光的功率输出特性及光谱特性进行了研究。在 LD 抽运功率 453 W 时产生了最大输出功率 8.1 W 的准连续红光激光，测量了此时的 M^2 值并给出了光强分布图。

关键词：激光技术，全固态激光器，红光激光，倍频，KTP 晶体

163. 高功率、高效率、全固态准连续钛宝石激光器

邹雷，丁欣，魏权夫，于意仲，温午麒，张百钢，徐德刚，陆颖，王鹏，姚建铨

中国激光, 2004, 31(12): 1409-1412.

摘要：以激光二极管阵列抽运 Nd:YAG 内腔倍频激光器作为抽运源,实现了全固态准连续钛宝石激光器的高功率、高效率运转。当 532 nm 的抽运光为 24 W 时,得到了 4.7 W 输出功率及 19.6% 的高转换效率。为了获得钛宝石激光器理想的宽带输出,分别使用了两组膜。第一组是 750~850 nm,第二是 850~950 nm,每膜片都有三种透过率,分别为 5%, 10%, 15%。由于钛宝石荧光谱线的中心波长在 795 nm 附近,使用第一组膜片在透过率为 10% 的情况下获得了最大 4.7 W 的输出功率。使用第二组膜片也获得 3 W 左右的输出功率,为将要进行的宽带调谐提供了必要的前提。

关键词：激光技术,钛宝石激光器,全固态,准连续,高功率,高效率

164. 104 W 内腔倍频全固态 Nd:YAG 绿光激光器

徐德刚，姚建铨，郭丽，周睿，张百钢，丁欣，温午麒，王鹏

光学学报, 2004, 24(7): 925-928.

摘要：报道了一台高功率内腔倍频全固态 Nd:YAG 绿光激光器,针对 KTP 晶体热效应和激光热稳定腔,采取了对 KTP 晶体进行低温冷却的优化措施,以便减少 KTP 晶体的热效应导致的相位失配,同时兼顾了 Nd:YAG 棒的热致双折射效应和 KTP 晶体热透镜效应,设计了热稳定谐振腔；实验中采用 80 个 20 W 激光二极管阵列侧面抽运 Nd:YAG 棒和 II 类相位匹配 KTP 晶体 (在 27℃时相位匹配角为 $\Phi=23.6°$; $\theta=90°$, 尺寸为 7 mm × 7 mm × 10 mm) 内腔倍频技术,谐振腔腔长为 530 mm, KTP 晶体的冷却温度为 4.3℃,抽运电流 18.3 A 时,实现平均功率达 104 W、脉冲宽度为 130 ns 的 532 nm 激光输出；其重复频率为 20.7 kHz。光 - 光转换效率为 10.2%。

关键词：激光器,Nd:YAG 激光器,532 nm 绿光,KTP 晶体,内腔倍频,低温冷却

165. 1 319 nm 与 660 nm 双波长 Nd:YAG 激光器的研究

王涛, 姚建铨, 郁道银, 李喜福, 禹国俊, 王鹏, 于意仲
天津大学学报, 2004, 37(5): 377–381.

摘要：为研究获得高功率红光的有效方法，研制了一种准连续输出 1 319 nm 与 660 nm 双波长 Nd:YAG 激光器。文中分析了波长 1 319 nm 激光的辐射跃迁能级，论述了抑制 1 064 nm 激光的生成从而提高 1 319 nm 激光输出等关键技术，研究了光学镜片的镀膜参数与腔型结构，实现 1 319 nm 激光连续输出最高功率 43 W，以 1 319 nm 激光为基频，置入 KTP 晶体内腔倍频，并设置声光 Q 开关，获得 660 nm 红光准连续输出 2 W，实现 1 319 nm 与 660 nm 双波长输出。

关键词：1 319 nm 激光器，Nd:YAG 晶体，红光 660 nm 激光器，KTP 晶体，内腔倍频，1 319 nm 与 660 nm 双波长输出

166. Self-mixing interferences effects in LD pumped multi-mode solid-state laser

Zhang Yong, Yao Jianquan, Wang Peng, Yu Yanguang, Chen Jin
Transactions of Tianjin University, 2004, 9(4): 261–263.

Abstract: Based on the effective structure of the self-mixing interference effects, a general model for the selfmixing interference effects in the LD pumped solid-state laser has been established for the first time. The numerical simulation of the self-mixing interference signal has been done, the results show that when the external cavity length is integral times of 1/2 , 1/3 , 2/3, 1/4, 3/4 of the effective cavity length, the intensity of the selfmixing interference signals reach maximum in value. While that of single mode laser is integral times of half of the effective cavity length, the measuring precision of displacement of single mode laser is $\lambda/2$. A conclusion can be drawn from the above results that the measuring precision of displacement of multi-mode laser is higher than that of single mode laser.

Key words: General model, Self-mixing interference effects, LD pumped multi-mode solid-state laser, Numerical simulations

167. 强激光对星载光电探测系统的干扰与破坏研究

牛燕雄,张鹏,姚建铨,段晓峰,汪岳峰,郭丽

光子学报, 2004, 33(7): 793-796.

摘要: 建立了激光辐照星载光电探测器的数学模型,对激光干扰和破坏星载光电系统的机理进行了研究,分析了激光光束与探测器光轴的偏角对干扰和破坏的影响。以 CO_2 激光辐照 HgCdTe 探测器实例,研究了强激光对探测单元的破坏,得到其损伤阈值曲线,并从理论上证明了强激光对星载光电探测系统干扰和破坏的可行性。

关键词: 激光干扰,激光破坏,星载光电探测,激光辐照

168. 利用 QPM-OPO 的调谐特性精确测定晶体的极化周期

张百钢,姚建铨,路洋,王鹏,丁欣,徐德刚,张浩

光电子·激光, 2004, 15(3): 337-340.

摘要: 利用准相位匹配(QPM)光学参量振荡器(OPO)的角度和温度调谐特性来精确测定晶体的极化周期。以 1 064.3 nm Nd:YVO$_4$ 全固态激光器泵浦的周期极化 LiNbO$_3$(PPLN)-OPO 为研究对象,测量、计算了 3 块 PPLN 的极化周期。通过理论温度调谐曲线与实验曲线的对比证明,此方法是一个精确测定 PPLN 极化周期的有效方法。

关键词: 准相位匹配(QPM),光学参量振荡器(OPO),温度调谐,角度调谐,周期极化铌酸锂晶体(PPLN)

169. LD 侧泵激光器抽运光和温度分布数值研究

蔡志强，姚建铨，温午麒，赵士勇，周建勇，李君，王鹏
光电子·激光，2004, 15(11): 1305–1310.

摘要： 基于光线追迹法和有限差分法，计算了不同参数下光 LD 侧面抽运激光棒时吸收的功率和温度分布。模拟计算结果表明不同的抽运头结构对抽运光功率和温度分布有重要影响。随着抽运方向的增多，抽运光功率和温度分布越来越对称。介绍了一些提高抽运光和温度分布均匀性的方法。研究了激光棒横截面内的温度与热传导系数的关系，该温度随热传导系数的增大非线性减小，在一定的抽运功率条件下，热传导系数存在着一个最佳值。

关键词： 光线追迹法，有限差分法，抽运光功率分布，温度分布

170. SESAM mode-locked YAG laser with LD sidepumping module

Yizhong Yu, Jianquan Yao, Zhigang Zhang, Wuqi Wen, Jianyong Zhou, Jun Li
Proceedings of Society of Photo-Optical Instrumentation Engineers, 2004, 5627:41–44.

Abstract: Pico-second solid-state lasers with multi-watt average power and multi-kilowatt peak power are required for numerous applications such as UV generation and pumping of optical parametric oscillators for RGB laser TV display. In the past few years laser diode bars as pump source with continue output power of hundreds watts or more have become commercially available. LD side pumping YAG laser module can be used to generate high reputation mode locked output pulse with SESAM (Semiconductor Saturable Absorber Mirror) as an end cavity mirror. In this paper, we demonstrate an all solid-state mode-locked YAG laser with three-mirror folded cavity with a SESAM mirror, 80 MHz pulse rate, about 10 ps pulse width, and 2 W 1 064 nm output power. Thermal effect of laser rod, and polarization of intra-cavity beam are considered in laser cavity design. The enhancement of laser performance and decreasing Q switching effect are discussed. Further investigation is greatly needed.

Key words: SESAM, Mode locked YAG laser, Folded cavity, Picosecond laser, LD pumped

171. Study on CW Nd:YAG laser at 1 319 nm and 660 nm

Tao Wang, Jianquan Yao, Peng Zhao, Binjing Cai, Peng Wang
Proceedings of Society of Photo-Optical Instrumentation Engineers, 2004, 5627: 121-127.

Abstract: A dual-wave Quasi-CW Nd:YAG laser has been demonstrated at the wavelengths of 1 319 nm and 660 nm. The radiation energy level of the 1 319 nm transition was analyzed. The critical technology of restraining resonance of radiation energy level of the 1 319 nm transition was analyzed. The critical technology of restraining resonance of 1 064 nm so as to improve that of 660 nm was discussed. Sophisticated optical coating and cavity structure was studied. And a maximum CW Output power of 43 W at 1 319 nm was acquired. Based 1 319 nm laser, a intracavity frequency-doubling laser of 660 nm was also demonstrated by using KTP crystal and a acousto-optically Q-switch. And a Quasi-CW red light output power of 2 W at 660 nm was acquired, accordingly, dual-wave output was realized.
Key words: 1 319 nm laser, Nd:YAG crystal, Transition level, Red-light 660 nm laser, KTP crystal, Intracavity frequency doubling

172. The influence of KTP's thermal effects on the stabilization of a high-power diode-side-pumped all-solid-state QCW intracavity-frequency-doubled Nd:YAG laser and the compensating methods

Zhou Rui, Xu Degang, Wen Wuqi, Ding Xin, Zhang Qiang, Wang Peng, Yao Jianquan
Proceedings of Society of Photo-Optical Instrumentation Engineers, 2004, 5627: 299-306.

Abstract: The stabilization and modes of a high-power intracavity frequency-doubled Nd:YAG laser are numerically analyzed, the great influence of frequency-doubler's thermal lensing on the stabilization and modes of this laser is demonstrated, and a compensating method is developed. A high-power QCW 532 nm nm green laser has been fabricated in the experiment, with a KTP crystal ($\theta=90°$, $\varphi=24.7°$, $6 \times 6 \times 9.2$ mm^3, cut for high-temperature (80 ℃) application) as frequency-doubler. With the KTP crystal warmed up to 48.8 ℃ and resonator parameters adjusted optimum according to the calculated thermal focal length of KTP crystal, a maximum 110 W green laser is generated at 10.6 kHz repetition rate, and it's pulse width is 142 ns, instability 2%, and optical-to-optical efficiency 11%.
Key words: All-solid-state, Green laser, QCW, KTP thermal effect, Resonator optimize

173. Phase mismatch compensation of second harmonic generation with controlling boundary temperature of type-II KTP crystal in high power green laser

Degang Xu, Jianquan Yao, Rui Zhou, Baigang Zhang, Peng Wang

Proceedings of Society of Photo-Optical Instrumentation Engineers, 2004, 5627: 307–311.

Abstract: We reported phase mismatch compensation of second harmonic generation with controlling boundary temperature of type-II KTP crystal in high power intracavity frequency-doubled Nd:AG laser. Thermal induced phase mismatching of the KTP crystal was analyzed theoretically by numerical computations of temperature derivative of refractive indices. The temperature gradient of the KTP crystal, phase matching angles change with difference boundary temperature of the KTP crystal, and tolerance temperature was analyzed. In the experiment, when two KTP crystals of difference type II phase matching condition (ϕ=23.6°, θ=90° at 27 ℃ temperature, ϕ=24.7°, θ=90° at 80 ℃ temperature) were applied to compensate the phase mismatching of the type-II KTP crystals. The maximum average 532 nm output power of 85 W and 110 W were generated when the boundary temperature of KTP were kept in 4 ℃ and 48.8 ℃ respectively. The corresponding conversion efficiency is 9.03% and 11%.

Key words: Phase mismatching, Second harmonic generation, Conversion efficiency, Temperature distribution, Boundary temperature, KTP, Temperature gradient

174. LD pumped Q-CW Nd:YAG/KTP red laser

Wuqi Wen, Xin Ding, Zhiqiang Cai, Jianing Zhou, Lei Zou, Meng Zhu, Yizhong Yu, Peng Wang, Jianquan Yao

Proceedings of Society of Photo-Optical Instrumentation Engineers, 2004, 5627: 449–452.

Abstract: High-power red laser are of great interest in the fields of medical application, laser display and also as a pumping source for tunable lasers such as Cr:LiSAF. This letter reports the generation of a 12 W Q-CW red laser beam by intracavity frequency-doubling of a Nd:YAG laser operating at 1.3 μm with a KTP crystal. A laser module that consisting of a Nd:YAG rod side-pumped by thirty 20 W LDs of a triangle radial pump geometry, a acousto-optic Q switch and a KTP crystal were used in the experiment. Because the efficiency of SHG is sensitive to the type II phase-matching angles of KTP, we calculate the value of phase-matching angles according to several of Sellemier equations of KTP, and modify the phase-matching angles to $\theta=59.9°$ and $\Phi=0°$ by experiment. The maximum average power of 12 W of red laser is achieved at 10 kHz when the pump power of LDs is about 470 W.

Key words: LD-pumped laser, Red laser, Nd:YAG, Frequency-doubling, KTP crystal

175. The study of stability on a laser-diode-pumped high-power high-repetition-rate intracavity frequency doubled 532 nm laser

Shiyong Zhao, Jianquan Yao, Degang Xu, Rui Zhou, Baigang Zhang, Jianing Zhou, Peng Wang

Proceedings of Society of Photo-Optical Instrumentation Engineers, 2004, 5627: 461–467.

Abstract: High power laser-diode-pumped 532 nm laser sources (including continuous wave and high repetition rate operation) are directly used for precise processing of metals and plastics. Furthermore, high power green laser will be used in some fields such as ocean exploration, laser probe and underwater communication. Recently, we reported a 110 W diode-side-pumped Nd:YAG intracavity frequency doubled high stability 532 nm laser. In the experiment, we found that the average output power of second harmonic fluctuated acutely with the variety of pumping current. Moreover, the length of arms between the mirrors were very sensitive to this cavity. We consider that one of the reason is the focus length of thermal lens of Nd:YAG rod alter with the variational pumping current, which makes the cavity be unstable. We consider the KTP crystal as a thin lens for its short length. As thermal lensing effect of the Nd:YAG rod is quite severe, so we consider it as thermal lensing medium. By ray matrix methods, we have obtained the stable regions and beam waist radii distribution in the flat-concave cavity. In our experiment, we used a pump head consisting of 80 diode bars with pentagon pump model and employed flat-concave cavity structure in order to achieve high stability output and increase output power. The total cavity length is 505 mm. By using an acousto-optic Q-switching with high diffraction loss and the KTP crystal which is type II phase matching, 110 W high stability 532 nm laser is achieved. The experimental result is in good agreement with the calculation.

Key words: All-solid-state green laser, Flat-concave resonator, Stability, Thermal focus lens, Thermal lensing effect

176. High power high efficiency all-solid-state quasi-continuous-wave tunable Ti:sapphire laser system

Lei Zou, Xin Ding, Wuqi Wen, Yue Zou, Hongmei Ma, Peng Wang, Jianquan Yao

Proceedings of Society of Photo-Optical Instrumentation Engineers, 2004, 5627: 468–472.

Abstract: This paper reports a high output power, high conversion efficiency. all-solid-state. quasi-continuous-wave Ti:sapphire laser system pumped by frequency-doubled Nd: YAG laser with DPL and by employing it as a pump source, we got the tunable broadband wavelength varying from 750 nm to 950 nm. Comparing with correlative research fields, two from 750 to 850 nm, the other is from 850 to 950 nm. Because the centric wavelength of the resonator laser is nearly ideal broadband output of Ti:sapphire laser, two sets of Ti: sapphire resonator mirrors were used respectively One is transmission rate of the output mirror is 10%. The power tops this field so far and the higher conversion efficiency is transmission rate of the output mirror is 10%. The power tops this field so far and the higher conversion efficiency is 22.2%. The second set of mirror we used can generate the output power of 3 W, which is high enough to achieve broadband tuning for the future. And then, by using a tunable and line width compressed implement-birefringent filter which was Brewster angle placed, we achieved continuous tuning from 750 nm to 950 nm with reasonably high power.

Key words: High power, All-solid-state, Quasi-continuous-wave, Ti:sapphire laser, broadband, Birefringent filter, Continuous tuning

177. New study on self-mixing interference effects in LD pumped laser

Zhang Yong, Yao Jianquan, Xu Kexin, Wang Peng

Proceedings of Society of Photo-Optical Instrumentation Engineers, 2004, 5628: 263−270.

Abstract: Self-mixing interference effects in the LD pumped green laser under the threshold current are observed. The behaviors of laser under the threshold current are more like LED. Based on this, we put forward for the first time the idea that replacing the light source with LED. If so, the lifetime and price of this kind of interferometer, and also the cubage of the interferometer is reduced. The accurate threshold current of lasers can also be measured using this equipment. This will be great contribution to self-mixing interference effects and its applications.

Key words: New study, LD pumped, Green laser, Self-mixing interference effects, Under the threshold current, LED, Light source, Interferometer, Accurate threshold current

178. Study on self-mixing interference effects in LD pumped laser with compensatory cavity

Zhang Yong, Yao Jianquan, Xu Kexin, Wang Peng

Proceedings of Society of Photo-Optical Instrumentation Engineers, 2004, 5628: 277−283.

Abstract: A new self-mixing interference structure with compensatory cavity is proposed. The model of this structure is also established. With compensatory cavity, the defects of self-mixing interference signals, such as phase and environmental disturbance, can be improved. In addition, the range of distance measuring can be extended.

Key words: Self-mixing interference, Compensatory cavity, LD pumped laser, Model, Phase and environmental disturbance, Range of distance measuring

179. Quantum mechanism method for self-mixing interference effects in LD pumped laser

Zhang Yong, Yao Jianquan, Xu Kexin, Wang Peng
Proceedings of Society of Photo-Optical Instrumentation Engineers, 2004, 5627: 284-291.

Abstract: Self-mixing interference effects in the LD pumped green laser are observed. Then small disturbance theory is used for the first time to analysis the quantum behaviors in the laser. The cavity system without feedback is considered to be the initial status function, the influence of feedback is considered to be the disturbance. The interaction Hamilton is deduced finally.

Key words: Quantum mechanism method, LD pumped, Green laser, Self-mixing interference, Small disturbance theory, Quantum behaviors, Cavity system, Initial status function, Influence of feedback, Interaction Hamilton

180. Diode end-pumped passively mode-locked Nd:YVO$_4$ picosecond laser with SESAM

Zhiqiang Cai, Wuqi Wen, Jianquan Yao, Yonggang Wang, Zhigang Zhang, Jianyong Zhou, Jianing Zhou, Peng Wang
Proceedings of Society of Photo-Optical Instrumentation Engineers, 2004, 5628: 311-317.

Abstract: We presented a diode end-pumped passively mode-locked Nd: YVO$_4$ picosecond laser with a semiconductor saturable absorber mirror homemade. Choosing a low-transmission output coupler and extending the cavity length suppressed the Q-switching mode-locked tendency in V-shaped cavity. We observed the pulse trains gone with a continuous background which was decreased with the decreasing of pulse repetition rate. The stable continuous mode-locked pulse was attained. The pulse repetition rate was 80.4 MHz and the corresponding spectrum width was measured to be 0.15 nm.

Key words: Passively mode-locked, SESAM, Diode-pumped laser

181. The study on laser scanning three-dimensional modeling of human head

Wang Tao, Yao Jianquan, Guo Ling, Cai Binjing, Lu yang, Wang peng, Liu Baoli
Proceedings of Society of Photo-Optical Instrumentation Engineers, 2004, 5630:74–76.

Abstract: A novel three-dimensional human head modeling method on laser technology, using the CCD camera to receive the scanning information, based on the theory of aerial photography surveying, is reported here. We introduce the mode integration of human head, its mathematical modeland the principle of human head feature identifying. The laser scanning human head three-dimensional modeling system is established.
Key words: Human head modeling, Laser scanning, Feature identifying

182. LD pumped Nd:YAG/KTP intracavity frequency-doubled 1.2 W CW red laser

Jianing Zhou, Wuqi Wen, Rui Zhou, Xin Ding, Zhiqiang Cai, Jianquan Yao
Proceedings of Society of Photo-Optical Instrumentation Engineers,2004, 5638: 582–588.

Abstract: This letter reports the CW red laser radiation at 659.5 nm by intracavity frequency doubling a side pumped Nd:YAG laser with a KTP crystal (type II phase match, θ=59. 8°, φ=0°). The thermal lens effect related to the laser crystal is analyzed and parameters of the three-mirror folded cavity are calculated. To enhancing the high gain at 1 319 nm of the operating wave of Nd:YAG, the reflectivity of the mirrors is designed, and the red laser radiation at 659.5 nm is achieved by intracavity frequency-doubling. The generation of 1.2 W of the CW red laser beam is obtained with 260 W pumped power.
Key words: Red laser, LD pumped laser, CW lasers, 659.5 nm, Intracavity frequency doubling

183. Research on the correlation of the signal and output in the stimulated Raman scattering

Wang Tao, Yao Jianquan, Liu Baoli, Cai Binjing, Wang Peng, Lu Yang
Proceedings of Society of Photo-Optical Instrumentation Engineers, 2004, 5646: 78–85.

Abstract: Potential effects were analyzed after incidence of signal wave firstly. Based on the electromagnetic theory and combined with the nonlinear effect of Stimulated Raman Scattering (SRS) of optical fiber, correlative equations were set up for the Raman amplifying Pump wave of optical fiber, Stokes wave and signal wave and the conditional expression which made system gain form positive feedback was given at the same time. It comes to the conclusion that under the condition that system gain positive feedback is satisfied by optical fiber parameters, the input signal wave is amplified when the SRS gain corresponds to unsaturated gain state; the higher is the SRS gain the more is amplification times; the correlation degree is relatively increased when the SRS gain is relatively low in some region.

Key words: Fiber Raman amplification, Correlation between pump wave and signal wave, Correlation function

184. Ultra-broad band wavelength tuning from 400 nm to 1 700 nm with near-NCPM OPO and frequency doubling

Yizhong Yu, Jianquan Yao, Xin Ding, Wei Lu
Proceedings of Society of Photo-Optical Instrumentation Engineers, 2004, 5646:179–182.

Abstract: Based on near-NCPM (Non Critical Phase Matching) design, type I temperature phase matching of MgO doped $LiNbO_3$ OPO (Optic Parametric Oscillation) with selected cutting angle at 82 degree. Replacing NCPM with near-NCPM to limit the operation temperature range to a relative low level, a temperature tuned OPO pumped by 532 nm is developed and 800~1 700 nm tunable signal lights are achieved. Output pulse energy is higher than 5mJ by a pumping laser of 58 mJ, 532 nm and 6 ns. With frequency doubling of OPO light through a 24.5 degree cut BBO crystal, 2 mJ, 400~800 nm output is obtained by a 19.8~29.2 degree rotating accordingly. The threshold of pump peak intensity for OPO is 57.3 MW/cm^2. On the position of two times of threshold energy, SRO conversion efficiency is over 11%.

Key words: OPO, NCPM, Temperature tuned, $LiNbO_3$, BBO

185. Total internal reflection noncollinear quasi-phase-matching PPLN OPO

Baigang Zhang, Jianquan Yao, Yang Lu, Xin Ding, Degang Xu, Xuejin Li, Peng Wang

Proceedings of Society of Photo-Optical Instrumentation Engineers, 2004, 5646: 206−211.

Abstract: The refractive index of the periodically poled lithium niobate (PPLN) is about twice as many as the refractive index of the air. When the rotation angle of the periodically poled crystal is bigger, the pump, signal and idler waves can form total internal reflection (TIR) inside the crystal, and the TIR noncollinear quasi-phase-matching (TIR-NQPM) optical parametric oscillator (OPO) can be realized if properly setting the angle of the OPO output coupler synchronously. In this paper, TIR-NQPM OPO based on PPLN pumped with a 1 064 nm all-solid-state laser is reported for the first time. In our experiment, the angle of rotation for PPLN is 5.57° and 1 504~1 542 nm of signal wavelength was achieved by tuning the work temperature from 140 ℃ to 250 ℃ Although adjustment of the cavity is difficult for this method, it broadens the rotation range of angle-tuned QPM OPO. As a result, by one time or more than one time total internal reflections inside the crystal, TIR-NQPM has less limitation of the transverse size of the periodically poled crystal, and the output wavelength range could be increased remarkably.

Key words: Quasi-phase-matching (QPM), PPLN, OPO, Angle-tuned, Total internal reflection (TIR), Noncollinear QPM(NQPM), All-solid-state

186. All-solid-state quasi-continuous-wave tunable Ti:Sapphire laser and pump-tuning PPLN-OPO

Xin Ding, Lei Zou, Wuqi Wen, Hongmei Ma, Yue Zou, Yizhong Yu, Peng Wang, Jianquan Yao

Proceedings of Society of Photo-Optical Instrumentation Engineers, 2004, 5646: 400−403.

Abstract: In this paper, high power and high efficiency operation of an all solid state, quasi-continuous-wave, Ti:sapphire laser is obtained with a laser diode pumped frequency doubled Nd:YAG laser. Discussing the output characters with three kinds of oscillate cavity, we acquire high output power, high light quality Ti: sapphire laser with flat to flat oscillate cavity. The maximum output power is 6.1 W at 797 nm with pump power of 27 W, conversion efficiency is 22.6%, FWHM is 38.4 ns and M^2 is 3.6. Tuning by birefringent filter, all solid state titanium doped sapphire laser's output wavelength is from 750~950 nm and average output power is 4.7 W. Pump-tuning method is very important and useful to OPO because it can relieve the OPO of any requirement for crystal translation or temperature variation. In this way, using PPLN crystal with poled period of 20.5 μm, 1 017~3 384 nm quasi-continuous-wave output is obtained pumped by Ti: sapphire laser which is tuned from 770 to 820 nm. The maximum output power is 1.9 W at 1 208 nm and conversion efficiency is 32%. OPO's average output power is 1.3 W in whole output wavelength.

Key words: All-solid-state, Quasi-continuous-wave, Tunable Ti:sapphire laser, Pump-tuning, PPLN-OPO

187. High conversion efficiency continuous wave quasi-phase-matched second harmonic generation in MgO doped stoichiometric lithium tantalite

Yang Lu, Baigang Zhang, Xin Ding, Peng Wang, Feng Ji, Tieli Zhang, Jianquan Yao

Proceedings of Society of Photo-Optical Instrumentation Engineers, 2004, 5646: 421−426.

Abstract: In this paper, for the first time we have demonstrated a continuous wave single-pass extracavity frequency doubling by use of MgO doped stoichiometric lithium tantalite (PP-MgO: SLT). The maximum output of the second harmonic green light is 905 mW with the pumping power of 7.69 W. The second harmonic conversion efficiency is 11.77%. We have theoretically calculated the tolerance of the temperature of SLT SHG which is about 1.5 ℃. The experimental tolerance of the temperature is consistent with the calculated value.

Key words: Quasi-phase matching (QPM), Continuous wave (CW), PP-MgO: SLT, Second harmonic generation (SHG)

188. Continuous-wave operation at 1 386 nm in a diode-end-pumped Nd:YVO$_4$ laser

Rui Zhou, Baigang Zhang, Xin Ding, Zhiqiang Cai, Wuqi Wen, Peng Wang, Jianquan Yao

Optics Express, 2005, 13(15): 5818−5824.

Abstract: We report a diode-end-pumped continuous wave (cw) Nd:YVO$_4$ laser operation at 1 386 nm. A maximum output power of 305 mW is achieved at an incident pump power of 4.24 W, achieving a slope efficiency of 13.9%. To the best of our knowledge, this is the first time that cw operation at this transition of Nd:YVO$_4$ crystal is reported. By using the experimentally measured threshold data, the stimulated-emission crosssection of this gain medium at 1 386 nm transition is determined to be 3×10^{19} cm^2. In addition, simultaneous cw operation at 1 342 nm and 1 386 nm is also observed.

Key words: Diode-end-pumped, CW operation, Nd: YVO$_4$ laser

189. 8.3 W diode-end-pumped continuous-wave Nd:YAG laser operating at 946-nm

Rui Zhou, Tieli Zhang, Enbang Li, Xin Ding, Zhiqiang Cai, Baigang Zhang, Wuqi Wen, Peng Wang, Jianquan Yao

Optics Express, 2005, 13(25): 10115–10119.

Abstract: This paper reports a diode-end-pumped continuous-wave (CW) Nd:YAG laser operating at 946-nm by utilizing the $^4F_{3/2}$-$^4I_{9/2}$ transition. We demonstrated that at an incident pump power of 27.7 W, an output power of 8.3-W could be achieved with a slope efficiency of 33.5%. To the best of our knowledge, this is the highest CW output power at 946 nm generated by LD end-pumped Nd:YAG lasers. By using intracavity frequency doubling with an LBO crystal, we further obtained a 473-nm blue laser with an output power of 1.2 W, achieving an optical-to-optical conversion efficiency of 7.1% at a pump power of 16.9 W. The short-term power instability of the blue laser was less than 1 %.

Key words: Diode-end-pumped, CW opreating, Nd: YAG

190. Multi-point laser Doppler velocimeter

E.B. Li, J. Xi, J.F. Chicharo, J.Q. Yao, D.Y. Yu

Optics Communications, 2005, 245: 309–313.

Abstract: This paper describes a novel multi-point laser Doppler velocimeter. By employing two acousto-optic modulators (AOMs) for both beam splitting and frequency shifting, we have demonstrated that multiple measuring volumes can be formed by intersecting the multiple diffraction orders generated by the AOMs. Each measuring volume has its frequency shift, which acts as a carrier frequency to that channel. Since the signals from different channels are multiplexed in the frequency domain, only a single photo-detector and a single-channel signal processor are necessary for multi-point measurements. The system has been demonstrated by making multi-point velocity measurements on a rotating disk.

Key words: Laser Doppler velocimeters, Acousto-optical devices

191. 110 W high stability green laser using type II phase matching KTiOPO$_4$ (KTP) crystal with boundary temperature control

Degang Xu, Jianquan Yao, Baigang Zhang, Rui Zhou, Enbang Li, Shiyong Zhao, Xin Ding, Wuqi Wen, Yanxiong Niu, Juguang Hu, Peng Wang

Optics Communications, 2005, 245: 341−347.

Abstract: We have developed a diode-pumped high power and high stability green laser. By controlling and stabilizing the boundary temperature of type II phase matching KTP crystal, 110 W high stability green laser output have been achieved. Temperature distribution inside the KTP crystal has been analyzed by solving the thermal conductivity equation. From the temperature distribution inside the KTP crystal, we have calculated the optimal phase matching angles and temperature bandwidth of the type II KTP crystal as a function of temperature. The second harmonic conversion efficiency as a function of temperature has also been calculated. In the experiment, the type II phase matched KTP crystal (optimum phase matching angles are ϕ = 24.68°, θ = 90° under the condition of phase matching temperature 353 K) was used in the intrcavity frequency-doubling resonator. An average output power of 110 W at 532 nm has been achieved with values of 11% and 2% for the optical-to-optical conversion efficiency and the instability, respectively. The optimal boundary temperature of the KTP crystal has been found to be 321.8 K. The experiment results are in good agreement with the theoretical calculation.

Key words: Frequency-doubling, Boundary temperature control, KTP crystal, High stability, Green laser

192. Broadening angular acceptance bandwidth of second-harmonic generation by using nearly ideal temperature quasi-phase-matching

Zhang Baigang, Yao Jianquan, Lu Yang, Xu Degang, Li Xuejin, Ji Feng, Zhang Tieli, Xu Kexin

Optics Communications, 2005, 254:344–352.

Abstract: We propose the method for broadening the angular acceptance bandwidth of quasi-phase-matched (QPM) frequency doubling by using nearly ideal temperature QPM (NIT-QPM). By NIT-QPM we mean that the crystal temperature is not controlled at the temperature, which satisfies the collinear phase matching condition and is termed ideal temperature, but near it. The scalar analysis on periodically poled MgO doped of stoichiometric lithium tantalate(PP-MgO: SLT) second-harmonic generation (SHG) shows that, for single-pass extracavity frequency doubling, the angular acceptance bandwidth of NIT-QPM with a lower temperature would be greater than that of ideal temperature QPM (IT-QPM) which is usually used. In the experiment, the angular acceptance bandwidth of PP-MgO: SLT SHG is broadened from 4.6° to 6.4° when the heating oven temperature falls from the ideal temperature to a lower temperature.

Key words: PP-MgO: SLT, Nearly ideal temperature quasi-phase-matching, Angular acceptance bandwidth, Harmonic generation

193. High-power continuous-wave Nd:YAG laser at 946 nm laser and intracavity frequency-doubling with a compact three-element cavity

Rui Zhou, Zhiqiang Cai, Wuqi Wen, Xin Ding, Peng Wang, Jianquan Yao

Optics Communications, 2005, 255: 304−308.

Abstract: Efficient operation of $^4F_{3/2}$–$^4I_{9/2}$ transition in diode-end-pumped Nd:YAG lasers at 946 nm is reported. At incident pump power of 23.9 W, up to 6.6 W continuous-wave (cw) output power has been obtained with a conventional non-composite Nd:YAG rod, leading to 27.1% optical-to-optical efficiency and 29.8% slop efficiency. To our knowledge, this is the highest value of LD pumped Nd:YAG 946 nm laser with a conversional Nd:YAG rod as gain medium. By intracavity frequency-doubling with an LBO crystal in a compact three-element resonator, cw output power of 1.05 W in the blue spectral range at 473 nm is achieved with 8% optical-to-optical conversion efficiency versus incident pump power of 13.1 W. The instability of the blue laser is less than 1% during half an hour.

Key words: Nd: YAG laser, Diode-end pumped, 946 nm, Blue laser, Intracavity frequency-doubling

194. High-efficiency single-pass cw quasi-phase-matched frequency doubling based on PP-MgO:SLT

Zhang Baigang, Yao Jianquan, Lu Yang, Xu Degang, Ding Xin, Wang Peng, Zhang Tieli, Ji Feng

Chinese Physics, 2005, 14(2): 353−358.

Abstract: We demonstrate a single-pass cw quasi-phase-matched (QPM) frequency doubling based on periodically poled MgO-doped stoichiometric lithium tantalate (PP-MgO:SLT) with crystal length of 20 mm. For the maximum 1 064 nm fundamental power of 7.69 W, the maximum conversion efficiency and the maximum output second-harmonic power are 11.8% and 905 mW, respectively. In the experiment we found that the optimum temperature of the oven for PP-MgO:SLT should be adjusted with the change of the fundamental power and the focal length. In addition, angular acceptance bandwidth and temperature acceptance bandwidth are studied. Their experimental results are in good agreement with the theoretical calculations.

Key words: Quasi-phase-matching, Periodically poled MgO-doped stoichiometric lithium tantalate (PP-MgO: SLT), Second-harmonic generation, Angular acceptance bandwidth

195. High power diode-end-pumped Nd:YAG 946-nm laser and its efficient frequency doubling

Zhou Rui, Zhao Shiyong, Cai Zhiqiang, Zhang Qiang, Wen Wuqi, Ding Xin, Wang Peng, Ding Lili, Yao Jianquan

Chinese Physics Letters, 2005, 22(6): 1413−1415.

Abstract: We report a high power operation of the $^4F_{3/2} \rightarrow {}^4I_{9/2}$ transition in diode-end-pumped laser at 946 nm. The maximum output of 5.1 W is obtained with a short linear plano-concave cavity, and the slope efficiency is 24.5% at incident pump power of 23.3 W. To our knowledge, this is the highest value of the LD-pumped Nd:YAG 946 nm lasers that employ the conversional Nd:YAG rod as the gain medium. By intracavity frequency doubling with an LBO crystal, up to 982 mW cw output power in the blue spectral range at 473 nm is achieved at an incident pump power of 10.9 W with a compact three-element cavity, leading to optical-to-optical conversion efficiency of 9%. The conversion efficiency should be increased to 15.1%, if the rather low absorption coefficient of this Nd:YAG is considered.

Key words: High power, Diode-end-pumped, Nd:YAG laser, Frequency doubling

196. High-average-power nanosecond quasi-phase-matched single–pass optical parametric generator in periodically poled lithium niobate

Zhang Baigang, Yao Jianquan, Lu Yang, Xu Degang, Ji Feng, Zhang Tieli, Zhao Xin, Wang Peng, Xu Kexin

Chinese Physics Letters, 2005, 22(7): 1691−1693.

Abstract: A pulsed nanosecond optical parametric generator (OPG) in periodically poled lithium niobate (PPLN) crystal is presented. The pump laser is an acousto-optically Q-switched Nd:YVO$_4$ laser with the maximum average power of 6.58 W. When the repetition rate is 50 kHz and the pulse width of the pump source is 80 ns, the maximum average total output power of the single-pass PPLN OPG is about 1.9 W, which includes 1.322 W of 1.536 μm signal radiation. The length of the PPLN crystal is only 38.7 mm (at room temperature) with a grating period of 28.93 μm (at room temperature). The 1.502-1.536 μm signal radiation and 3.652-3.465 μm idler radiation are obtained by adjusting the PPLN crystal temperature from 155 ℃ to 250 ℃.

Key words: High-average-power, Quasi-phase-matched, OPO, PPLN

197. Experimental 511 W composite Nd: YAG ceramic laser

Li Haifeng, Xu Degang, Yang Yang, Wang Yuye, Zhou Rui, Zhang Tieli, Zhao Xin, Wang Peng, Yao Jianquan

Chinese Physics Letters, 2005, 22(10): 2565–2567.

Abstract: We demonstrate a 511 W laser diode pumped composite Nd: YAG ceramic laser. The optical pumping system is consisted of five laser diode stacked arrays arranged in a pentagonal shape around the ceramic rod whose size is $\Phi 6.35 \times 144$ mm. When the pumping power is 1 600 W, the cw laser output up to 511 W at 1 064 nm can be obtained with a linear plano–plano cavity, and the optical-to-optical efficiency is 31.9%. To our knowledge, this is the highest value of laser output by using a newly invented composite Nd: YAG ceramic rod as the gain medium.
Key words: LD-pumped, Nd: YAG ceramic laser

198. Influence of the KTP crystal boundary temperature on conversion efficiency in high power green laser

Degang Xu, Jianquan Yao, Baigang Zhang, Shiyong Zhao, Rui Zhou, Xin Ding, Wuqi Wen, Peng Wang

Chinese Optics Letters, 2005, 3(2): 85–88.

Abstract: The influence of the $KTiOPO_4$ (KTP) crystal boundary temperature on conversion efficiency in high power green laser has been studied theoretically and experimentally. Temperature distribution inside the KTP crystal has been analyzed by solving the thermal conductivity equation. From the temperature distribution inside the KTP crystal, we have calculated the optimal phase-matching angles of the type-II KTP crystal as a function of temperature. The second-harmonic conversion efficiency as a function of temperature has also been calculated. In the experiment, two KTP crystals with different phase-matching angles were used in the intrcavity-frequency-doubled resonator. When the boundary temperature of KTP-A ($\Phi= 23.6°$, $\theta= 90°$ under the condition of 27 °C temperature) was setting at 4 °C, a maximum green light power of 104 W was generated at repetition rate of 20.7 kHz and pulse width of 132 nm with pumping current of laser diode of 18.3 A, leading to 10.2% optical-to-optical conversion efficiency. When KTP-B crystal ($\Phi= 24.68°$, $\theta= 90°$ under the condition of 80 °C temperature) was employed, an average output power of 110 W at 532 nm has been achieved with values of 11.5% and 2% for the optical-to-optical efficiency and the instability, respectively. The optimal boundary temperature of this KTP crystal has been found to be 48.8 °C.
Key words: KTP boundary temperature, Conversion efficiency, High power green laser

199. High power all-solid-state quasi-continuous-wave tunable Ti: sapphire laser system

Lei Zou, Xin Ding, Yue Zou, Hongmei Ma, Wuqi Wen, Peng Wang, Jianquan Yao

Chinese Optics Letters, 2005, 3(4): 208–209.

Abstract: This paper reports a high power, all-solid-state, quasi-continuous-wave tunable Ti: sapphire laser system pumped by laser diode (LD) pumped frequency-doubled Nd: YAG laser. The maximum tuned output power of 4.2 W (797 nm) and tuned average power of 3.7 W were achieved when fixing the Ti: sapphire broadband output power at 5.0 W and applying 750-850 nm broadband coated mirror.

Key words: High power, All-solid-state, Quasi-continuous-wave tunable, Ti: sapphire laser

200. Diode end-pumped 1 123-nm Nd: YAG laser with 2.6-W output power

Zhiqiang Cai, Meng Chen, Zhigang Zhang, Rui Zhou, Wuqi Wen, Xin Ding, Jianquan Yao

Chinese Optics Letters, 2005, 3(5): 281–282.

Abstract: We present a compact and high output power diode end-pumped Nd: YAG laser which operates at the wavelength of 1 123 nm. Continuous wave (CW) laser output of 2.6 W was achieved at the incident pump power of 15.9 W, indicating an overall optical-optical conversion efficiency of 16.4%, and the slope efficiency was 18%.

Key words: Diode end-pumped, Nd: YAG laser, Output power

201. 5.3-W Nd:YVO$_4$ passively mode-locked laser by a novel semiconductor saturable absorber mirror

Zhiqiang Cai, Wuqi Wen, Yonggang Wang, Zhigang Zhang, Xiaoyu Ma, Xin Ding, Jianquan Yao

Chinese Optics Letters, 2005, 3(6): 342−344.

Abstract: We report a diode end-pumped continuous wave (CW) passively mode-locked Nd:YVO$_4$ laser with a home-made semiconductor saturable absorber mirror (SESAM). The maximum average output power is 5.3 W at the incident pump power of 17 W, which corresponds to an optical-optical conversion efficiency of 31.2% and slope efficiency of 34.7%. The corresponding optical spectrum has a 0.2-nm full width at half maximum (FWHM), and the pulse repetition rate is 83 MHz.

Key words: Passively mode-locked, Laser, Semiconductor saturable absorber mirror

202. Efficient stable simultaneous CW dual-wavelength diode-end-pumped Nd: YAG laser operating at 1.319 and 1.338 μm

Rui Zhou, Wuqi Wen, Zhiqiang Cai, Xin Ding, Peng Wang, Jianquan Yao

Chinese Optics Letters, 2005, 3(10): 597−599.

Abstract: An efficient, stable diode-end-pumped simultaneous continuous-wave (CW) dual-wavelength laser operating at 1.319 and 1.338 μm in a Nd:YAG crystal has been demonstrated. A total output power of 6.3 W is achieved at an absorbed pump power of 15 W, with a slope efficiency of 43.5%. The instability of output power is less than 1%. With a type II critical phase-matched KTP crystal inserted into the cavity as frequency doubler, a maximum outputpower of 200 mW in red region is acquired. In addition, a six-wavelength laser operation at 1.319 μm, 1.338 μm, 1.356 μm, 659.5 nm, 669 nm, and 678 nm is observed.

Key words: Dual-wavelength, Diode-end-pumped, Nd: YAG laser

203. High power output quasi-continuous-wave nanosecond optical parametric generator based on periodically poled lithium niobate

Lu Yang, Zhang Baigang, Xu Degang, Ding Xin, Wang Peng, Zhang Tieli, Ji Feng, Yao Jianquan

Optoelectronics Letters, 2005, 1(1): 0010–0012.

Abstract: We report on a high power output quasi-continuous-wave nanosecond optical parametric generator (OPG) of congruent periodically poled lithium niobate (PPLN) pumped by a 1 064 nm acousto-optically Q-switched Nd: YVO$_4$ laser (duration: 70 ns, repetition rate: 45 kHz, spatial beam quality M^2 < 1.3). The OPG consists of a 38.7 mm long PPLN crystal with a domain period of 28.93 μm. With 5.43 W of average pump power the maximum average output power is 991 mW at 1 517.1 nm signal wave of the PPLN OPG.

Key words: Quasi-continuous-wave, OPG, PPLN

204. 脉冲激光对类金刚石(DLC)薄膜的热冲击效应研究

牛燕雄, 黄峰, , 段晓峰, 江岳峰, 张鹏, 何琛娟, 禹晔, 姚建铨

物理学报, 2005, 54(10): 4816–4821.

摘要：强激光辐照红外热像系统时，可造成系统的干扰和破坏，激光的波长不同，对系统的破坏效果也不同。为了保护红外系统窗口以及提高窗口的透过率，红外窗口广泛沉淀类金刚石(DLC)薄膜。当入射的激光波长位于红外系统响应波段外时，激光对系统的破坏首先是激光对 DLC 薄膜的破坏。以波长为 1.06 μm 的激光为例，研究了脉冲激光对 DLC 薄膜的损伤机理，建立了 DLC 薄膜的热冲击效应模型，并通过求解热传导和应力平衡方程，得出了薄膜的温度场和应力场分布。理论分析表明，热应力破坏在脉冲强激光对 DLC 膜的损伤机理中占主导地位。当辐照能量密度为 E_0=100 mJ·cm^{-2} 时，在薄膜表面距光斑中心约 40 μm 区域内的压应力明显超出其断裂强度，将造成膜层的剥离、脱落。理论分析与实验结果基本相符，表明建立热冲击效应模型的正确性。

关键词：激光辐照, 类金刚石(DLC)薄膜, 热冲击效应

205. LD 泵浦准连续 Nd:YAG/KTP 12 W 红光激光器

温午麒, 姚建铨, 王涛, 周佳凝, 蔡志强, 朱孟, 丁欣, 周睿, 张强, 于意仲, 王鹏

光电子·激光, 2005, 16(3): 271-273.

摘要：报道了使用国产大功率全固态 Nd：YAG 泵浦组件产生 1.3 μm 附近波长的激光振荡，利用 II 类临界相位匹配的 KTP 晶体腔内倍频产生高功率的红光激光输出。泵浦组件内包含 30 个 20 W 的 808 nm 二极管阵列，呈三角型阵列分布连续抽运 $\Phi 5~\text{mm} \times 125~\text{mm}$ 的 Nd：YAG 圆棒。为产生高功率的倍频输出，激光器采用 V 型折叠腔结构，并使用 1 个声光 Q 开关。在泵浦功率大约 470 W 时，产生了 12 W 的准连续高功率红光激光。

关键词：全固态激光器, 红光激光, 倍频, Nd:YAG, KTP 晶体

206. 高功率全固态准连续钛宝石激光器的实验研究

邹雷, 丁欣, 温午麒, 于意仲, 王鹏, 姚建铨

光电子·激光, 2005, 16(6): 650-652.

摘要：以激光二极管阵列 (LDA) 抽运 Nd:YAG 内腔倍频激光器作为抽运源，实现了全固态准连续钛宝石激光器的高功率、高效率运转。实验中，选择 750~850 nm 宽带膜片组，钛宝石输出镜的透过率分别为 T = 5%、10% 和 15%。当 532 nm 的抽运光为 27 W 时，得到了 6 W 输出功率及 22.2% 的转换效率。

关键词：全固态, 准连续, 激光二极管抽运 (DPL), 钛宝石激光器, 高功率, 高效率

207. LD 抽运 Nd:YAG/KTP 腔内倍频连续波 1.2 W 红光激光器

周佳凝,温午麒,周睿,丁欣,赵士勇,牛燕雄,王鹏,姚建铨

光电子·激光, 2005, 16(7): 767–770.

摘要：报道了用Ⅱ类相位匹配 KTP(相位匹配角选为 θ=59.9°,Φ=0°) 对激光二极管 (LD) 侧向抽运的 Nd:YAG 腔内倍频的红光激光器。通过分析大功率抽运 Nd:YAG 棒热透镜效应的影响,优化设计了三镜折叠腔参数。采用镜片镀膜的方法使 Nd:YAG 工作在 1 319 nm 波长,经腔内倍频获得单一波长 659.5 nm 的红光激光。在抽运电流 13 A 和输出镜曲率半径为 200 mm 时,达到 1.2 W 的红光连续波输出。

关键词：红光激光器,激光二极管 (LD) 抽运,连续波,659.5 nm,腔内倍频,望远镜谐振腔

208. 大功率连续锁模皮秒激光器单端输出超过 5 W

蔡志强,温午麒,王勇刚,马晓宇,张志刚,丁欣,王鹏,姚建铨

中国激光, 2005, 32(5): 693.

摘要：近年来,国际上采用半导体可饱和吸收镜(SESAM)进行被动锁模以获得大功率皮秒脉冲激光的技术已经日趋成熟。平均输出功率在几 W 到 10 W 的大功率皮秒脉冲激光在激光显示技术和自由电子激光器等方面有潜在的应用价值。我们利用半导体可饱和吸收镜,采用 6 镜折叠腔结构,实现了单端连续锁模(CWML)输出,平均输出功率高于 5 W。

关键词：大功率,锁模,皮秒激光器

209. 半导体可饱和吸收镜连续被动锁模端面抽运 Nd:YVO$_4$ 激光器

蔡志强, 温午麒, 姚建铨, 王勇刚, 张志刚, 丁欣, 周建勇, 周佳凝, 李君, 王鹏
中国激光, 2005, 32(6): 734–738.

摘要： 报道了利用国产半导体可饱和吸收镜 (SESAM) 实现端面抽运 Nd:YVO$_4$ 激光器连续锁模 (CWML) 运转的实验结果。在 V 型腔结构的实验中观察到调 Q 锁模 (QML) 波形不稳并有较强的直流成份, 通过选择小透过率输出镜和增加腔长, 解决了这两个问题。在腔长从 0.66 m 增至 1.27 m 和 1.86 m 过程中, 直流分量从 52% 降到 13.6%、0.3%, 最终获得了重复频率约 80 MHz 的稳定连续锁模脉冲输出, 光谱宽度为 0.15 nm。在考虑晶体热效应基础上, 利用 ABCD 矩阵方法设计了四镜 Z 型腔, 获得了 1 W 左右的连续锁模输出, 重复频率约 150 MHz。分析和比较了 V 型腔和 Z 型腔的优缺点。

关键词： 激光技术, Nd:YVO$_4$ 激光器, 半导体可饱和吸收镜, 连续锁模, 端面抽运

210. 准相位匹配周期极化高掺镁铌酸锂 532 nm 倍频准连续输出研究

陈云琳, 罗勇锋, 袁建伟, 郭娟, 张万林, 陈绍林, 黄自恒, 张光寅, 张百钢, 路洋, 丁欣, 姚建铨
光学学报, 2005, 25(1): 63–66.

摘要： 对周期性极化高掺镁铌酸锂倍频过程进行了准相位匹配倍频理论研究。在室温下通过外加电场极化法, 用较低的极化开关电场 ~ 5.5 kV/mm, 在厚为 1 mm、长为 10 mm、宽为 10 mm 的掺镁铌酸锂基片上成功地制备了周期为 5.8 ~ 7.3 μm (间隔 0.3 μm) 的一阶准相位匹配倍频周期性极化光学微结构。将温度控制在 70 ℃ 左右, 以波长为 1.064 μm 的 Nd:YAG 激光为基频光源, 对所研制的光学微结构样品进行倍频通光实验验证。当入射基频光为 920 mW 时, 可以获得约 15 mW 的 532 nm 准连续倍频蓝光输出, 其归一化转换效率高达 1.77 %/W。

关键词： 非线性光学, 倍频, 准相位匹配, 周期性极化掺镁铌酸锂

211. 高功率绿光激光器热稳定性的研究

赵士勇,周睿,徐德刚,温午麒,王鹏,姚建铨

激光杂志, 2005, 26(4): 11–13.

摘要：本文报道了对平均功率达 110 W 的高功率,高重复频率全固态绿光激光器热稳定性的研究。针对实验中出现的 532 nm 绿光输出平均功率不稳定的情况,分析了泵浦电流,谐振腔长度等因素的变化对平 - 凹腔的稳区范围和腔内激光模式特性的影响。理论分析中分别把 KTP 晶体和 Nd:YAG 棒看做薄透镜和热透镜,通过计算谐振腔传输矩阵的方法,得到了稳区范围和腔内光束分布情况。实验中采用的泵浦头由 80 个 20 W 的高效率半导体激光器组成,按照五角形等间距侧面泵浦 Nd:YAG 棒,采用高效率的声光 Q 开关,倍频晶体为 II 类相位匹配的 KTP 晶体 ($\varphi=24.7°$, $\theta=90°$),最终得到了平均功率为 110 W 的 532 nm 激光输出,实验结果与理论计算符合得很好。

关键词：全固态绿光激光器,平 - 凹腔,热稳定性,热透镜介质,KTP 晶体的热透镜效应

212. 外差椭偏测量技术及其混频误差分析

邓元龙,姚建铨,阮双琛,孙秀泉,王鹏

激光与红外, 2005, 35(6): 438–440.

摘要：结合激光外差干涉术和反射式椭偏测量技术,设计了一种抗干扰能力强,快速,高精度测量纳米厚度薄膜光学参数的方法。着重分析并计算了非线性混频误差对测量精度的影响,其中塞曼激光和波片产生的光束椭偏化是关键因素。定义了评价因子以比较非线性混频误差的相对大小,这对外差椭偏纳米薄膜测量系统的设计有指导意义。

关键词：椭偏测量术,外差干涉,薄膜,混频误差,评价因子

213. 纳米厚度薄膜外差椭偏测量技术研究

邓元龙,姚建铨,阮双琛,孙秀泉,王鹏

光学技术, 2005, 31(3): 391-393.

摘要：结合激光外差干涉法和透射式椭偏测量原理,研究了一种快速、高精度测量纳米厚度薄膜光学参数的方法。计算并分析了复灵敏度因子随薄膜参数和入射角度的变化规律、椭偏参数的选择及容许测量误差。两个声光调制器产生 20 kHz 的拍频,采用简单的直接比相方法即可获得优于 0.1°的相位分辨率,而且测量系统中没有使用任何波片和运动部件,抗干扰能力强且测量过程完全自动化,适用于工业现场在线连续测量。实验数据和理论分析表明,此方法可以达到亚纳米级测量精度。

关键词：光学测量,薄膜,外差干涉法,椭偏测量术,声光调制器,纳米精度

214. 透射式外差椭偏测量及非线性误差分析

邓元龙,姚建铨,阮双琛,孙秀泉,吴玉斌

测试技术学报, 2005, 19(3): 245-248.

摘要：结合激光外差干涉法和透射式椭偏测量原理,研究了一种快速、高精度测量纳米厚度薄膜光学参数的方法。给出了光学系统设计和理论分析,使用两个声光调制器产生 20 kHz 的差频,直接比较平行分量和垂直分量外差信号的幅值和相位,得到所需要的椭偏参数。光束偏振态的椭圆化及偏振分光不完全所引起的非线性误差是影响纳米薄膜测量精度的主要因素,推导出椭偏参数非线性误差的近似解析表达式,计算结果表明由此导致的膜厚测量误差可达几个 nm,相对而言,激光器和反射镜等器件产生的光束椭偏化是其主要原因。

关键词：椭偏测量术,外差干涉,非线性误差,薄膜测量,声光调制器

215. A laser-diode-pumped 7.36 W continuous-wave Nd:YVO$_4$ laser at 1 342 nm

Zhang Yuping, Zheng Yi, Zhang Huiyun, Yao Jianquan
Chinese Physics Letters, 2006, 23(2): 363−365.

Abstract: An efficient and high-power diode-laser single-end-pumped Nd:YVO$_4$ laser with cw emission at 1 342 nm is presented. With a crystal single-end-pumped by a bre-coupled diode laser, an output power of 7.36 W is obtained from the laser cavity of concave-convex, corresponding to an optical-to-optical conversion efficiency of 32.8%. The laser is operated in TEM$_{00}$ mode with small rms noise amplitude of 0.3%. This represents, to the best of our knowledge, the highest power obtained from a diode-laser single-end-pumped Nd:YVO$_4$ cw laser at 1 342 nm so far.

Key words: Laser-diode-pumped, CW, Nd:YVO$_4$ laser

216. A low-pump-threshold high-repetition-rate intracavity optical parametric generator based on periodically poled lithium niobate

Ji Feng, Zang Baigang, Li Enbang, Zhou Rui, Zhang Tieli, Lu Yang, Zhao Pu, Wang Peng, Yao Jianquan
Chinese Physics Letters, 2006, 23(8): 2113−2116.

Abstract: A low-pump-threshold high-repetition-rate intracavity optical parametric generator (IOPG) by using a periodically poled lithium niobate (PPLN) is reported. The PPLN, which is 18.7 mm long and has a grating period of 28.93 μm at room temperature, is inserted in a diode-end-pumped Nd:YVO$_4$ laser with an acousto-optic Q switch. The parametric generation threshold is 1.3 W (diode laser power) at a Q-switch repetition rate of 19 kHz. At an incident diode pump power of 5 W, an average signal output power of 280 mW has been achieved. The signal pulse duration is approximately 85 ns. By changing the crystal temperature from 120℃ to 250 ℃, the signal wavelength can be tuned from 1.493 μm to 1.538 μm.

Key words: Low-pump-threshold, High-repetition-rate, OPG, PPLN

217. The study on the high-power LD-pumped Nd:YAG cw dual-wavelength laser

Zhang Qiang, Yao Jianquan, Wen Wuqi, Liu Huan, Gan Yu
Proceedings of Society of Photo-Optical Instrumentation Engineers, 2006, 6344: 63442C-1–63442C-6.

Abstract: Though controlling the coating reflectivity and intracavity diffraction loss, in the four cavity mirror configuration, a dual-wavelength continuous wave (cw) diode-side-pumped Nd:YAG laser that generates simultaneously at wavelengths of 1 064 nm and 1 319 nm is demonstrated. The relationship between power ratio of the two wavelengths and cavity length is studied. The cw dual-wavelength output power reaches 85 W when the average pump power is more than 500 W. The output power of respective wavelength exceeds 40 W, which is the best record as we know.
Key words: Laser, High power, CW dual-wavelength, Four cavity mirror

218. High-power continuous-wave diode-end-pumped intracavity frequency doubled Nd:YVO$_4$ laser at 671 nm with a compact three-element cavity

Zhou Rui, Ding Xin, Wen Wuqi, Cai Zhiqiang, Wang Peng, Yao Jianquan
Chinese Physics Letters, 2006, 23(4): 849–851.

Abstract: We report a high-power high-efficient continuous-wave (cw) diode-end-pumped Nd:YVO$_4$ 1 342-nm laser with a short plane-parallel cavity and an efficient cw intracavity frequency-doubled red laser at 671 nm with a compact three-element cavity. At incident pump power of 20.6 W, a maximum output power of 7 W at 1 342 nm is obtained with a slope efficiency of 37.3%. By inserting a type-I critical phase-matched LBO crystal as intracavity frequency-doubler, a cw red output as much as 2.85-W is achieved with an incident pump power of 16.9 W, inducing an optical-to-optical conversion efficiency of 16.9%. To the best of our knowledge, this is the highest output of diode-pumped solid-state Nd:YVO$_4$ red laser. During half an hour, the red output is very stable, and the instability of output power is less than 1%.
Key words: Intracavity frequency doubled, Nd:YVO$_4$ laser at 671 nm, Compact three-element cavity

219. A tunable optical parametric generator by using a quasi-phase-matched crystal with different wedge angles

Tieli Zhang, Baigang Zhang, Haifeng Li, Feng Ji, Yang Lu, Xin Zhao, Peng Wang, Jianquan Yao

Chinese Optics Letters, 2006, 4(4): 234−236.

Abstract: We report a tunable quasi-phase-matched optical parametric generator (OPG) with different wedge angles, pumped by a commercially available Q-switched diode-pumped Nd:YVO$_4$ laser with a repetition of 50 kHz. The nonlinear crystal is a periodically poled MgO-doped LiNbO$_3$ (PPMgOLN) with a period of 30 μm. A congruent bulk LiNbO$_3$ (LN) with three different wedge angles of 0°, 4°, and 9° is placed in front of PPMgOLN. Rapid tuning has been achieved by simply moving the LN crystal along its lateral direction and over 60-mW average signal output power was obtained in the whole wavelength tuning range of 1 539-1 570 nm.

Key words: Tunable OPG, Quasi-phase-matched crystal

220. All-solid-state quasi-continuous-wave high power dispersion cavity tunable Ti:sapphire laser

Xin Ding, Yue Zou, Lei Zou, Hongmei Ma, Wuqi Wen, Yizhong Yu, Ying Lu, Peng Wang, Jianquan Yao

Chinese Optics Letters, 2006, 4(2): 96−98.

Abstract: An all-solid-state quasi-continuous-wave dispersion cavity tunable Ti:sapphire laser pumped by a laser diode pumped frequency-doubled Nd:YAG laser is reported. Using a dense flint glass prism as the dispersion element, a tuning range from 730 to 880 nm with the linewidth of 3 nm and the pulse width of 17.2 ns was obtained. The maximum output power of this laser system was 5.6 W at 786.3 nm corresponding to an optical-to-optical conversion efficiency of 25.5% under the pump power of 22 W.

Key words: All-solid-state, Quasi-continuous-wave, Tunable Ti:sapphire laser

221. Precise calculation of the KTP crystal used as both an intracavity electro-optic Q-switch and a second harmonic generator

Yuye Wang, Jianquan Yao, Degang Xu, Pu Zhao, Peng Wang
Chinese Optics Letters, 2006, 4(7): 419–421.

Abstract: A method of precisely calculating the external applied voltage and the optimum type-II phase matching angles for KTP crystal, which is used as both an intracavity electro-optic (EO) Q-switch and a frequency doubler, is presented. The effective EO coefficient along the phase-matching direction is defined to calculate the half-wave voltage and the quarter-wave voltage, and the precise calculation for the phase matching angles in the condition of KTP crystal optimum second harmonic phase matching is theoretically realized.
Key words: KTP crystal, Electro-optic Q-switch, SHG

222. Efficient and high-power laser-diode single-end-pumped Nd:YVO$_4$ continuous wave laser at 1 342 nm

Zhang Yuping, Zheng Yi, Zhang Huiyun, Wang Peng, Yao Jianquan
Chinese Physics, 2006, 15(9): 2018–2021.

Abstract: A compact, efficient and high-power laser diode (LD) single-end-pumped Nd:YVO$_4$ laser with continuous-wave emission at 1 342 nm is reported. With a single crystal single-end-pumped by fibre-coupled LD array, an output power of 7.36 W is obtained from the laser cavity of concave-convex shape, corresponding to an optical-to-optical efficiency of 32.8%. The laser is operated in TEM$_{00}$ mode with small rms amplitude noise of 0.3%. The influences of the Nd concentration, transmissivity of the output mirror and the cavity length on the output power have been studied experimentally.
Key words: Laser diode, Single-end-pumped Nd:YVO$_4$ at 1 342 nm, Continuous wave, Efficiency

223. Dual-signal-wavelength optical parametric generator based on ppr-PP-MgO:LN

Feng Ji, Baigang Zhang, Yang Lu, Tieli Zhang, Pu Zhao, Peng Wang, Jianquan Yao
Conference digest of the 2006 Joint 31ST International Conference on Infrared and Millimeter Waves and 14TH International Conference on Terahertz Electronics, Shanghai, 2006:106.

Abstract: A pulsed nanosecond optical parametric generator(OPG) in a periodically phased-reversed MgO-doped periodically poled lithium niobate(ppr-PP-MgO:LN) was demonstrated. Two pairs of signal waves and idler waves were obtained with a maximum output power of 135 mW repeating at a rate of 25 kHz.

Key words: Dual-signal-wavelength, OPG, ppr-PP-MgO:LN

224. Theoretical analysis of broadband source by using retracing behavior of collinear quasi-phase-matching optical parametric generator

Y. Lu, B. G. Zhang, D. G. Xu, X. Ding, X. Zhao, T. L. Zhang, F. Ji, P. Wang, J. Q. Yao
Conference digest of the 2006 Joint 31ST International Conference on Infrared and Millimeter Waves and 14TH International Conference on Terahertz Electronics, 2006:111-111.

Abstract: According to the concept of phase matching bandwidth the band width of broadband source by using retracing behavior of collinear quasi-phase-matching (CQPM) optical parametric generator (OPG) in periodically poled lithium niobate (PPLN) is presented firstly. Comparing the various pump wavelengths we obtain the pumping condition and parameters of PPLN to realize the broadband source near 1 550 nm in the CQPM-OPG. In theory we present the optimum pump wavelength and the maximum ideal bandwidth range to be 940.75 nm and 1 475~1 681 nm respectively.

Key words: Broadband source, Collinear quasi-phase-matching, OPG

225. All-solid-state high power dual-wavelength Ti: sapphire laser

Lei Zou, Xin Ding, Hongmei Ma, Yue Zou, Wuqi Wen, Peng Wang, Jianquan Yao
Journal of Optoelectronics and Advanced Materials. 2006, 8(2):843–846.

Abstract: A diode-pumped Q-cw (quasi-continuous wave) with simultaneous dual-wavelength laser operation at 744.8 nm and 860.9 nm in a single Ti: sapphire crystal is demonstrated. The birefringent filter is employed as the tuning apparatus for its low loss. A total output power of 4.8 W at the two wavelengths was achieved at the incident pump power of 23 W. To our knowledge, this is the highest value of the all-solid-state quasi-continuous dual-wavelength lasers which employ the Ti: sapphire as the gain medium.
Key words: All-solid-state, Dual-wavelength, Ti: sapphire, High power laser

226. Theoretical and experimental research on the birefringent filter applied for the high power tunable Ti: sapphire laser

Lei Zou, Xin Ding, Yue Zou, Hongmei Ma, Wuqi Wen, Peng Wang, Jianquan Yao
Journal of Optoelectronics and Advanced Materials. 2006, 8(4): 1584–1588.

Abstract: Investigating from the classical theory of the birefringent filter (BF), we calculate out the appropriate parameters that fit for the high power tunable Ti: sapphire laser experiment greatly. Then from the experiment we validate its effect and achieve continuous tuning from 750 nm to 920 nm with the maximum output power of 5.8 W and average power of 3.83 W that tops this field so far.
Key words: Birefringent filter, All-solid-state, Ti: sapphire laser, High-power laser, Continuously tunable

227. Continuous-wave, 15.2 W diode-end-pumped Nd:YAG laser operating at 946 nm

Rui Zhou, Enbang Li, Haifeng Li, Peng Wang, Jianquan Yao
Optics Letters, 2006, 31(12): 1869–1871.

Abstract: A high-power continuous-wave (cw) Nd:YAG laser operating at 946 nm by utilizing a quasi-three-level transition is reported. The laser consists of a composite Nd:YAG rod end pumped by a fiber-coupled diode laser and a simple plane-concave cavity. At an incident pump power of 40.2 W, a maximum cw output of 15.2 W at 946 nm is obtained, achieving a slope efficiency of 45%. To the best of our knowledge, this is the highest output at 946 nm ever generated by diode-pumped Nd:YAG lasers. In addition, at an incident pump power of 15.2 W, a 1.25 W blue output at 473 nm is achieved with a simple compact three-element cavity and a type-I lithium triborate (LiB_3O_5) crystal as a frequency doubler.
Key words: CW, Diode-end-pumped, Nd:YAG laser

228. Theoretical study of the electro-optic effect of aperiodically poled lithium niobate in a Q-switched dual-wavelength laser

Feng Ji, Baigang Zhang, Enbang Li, Haifeng Li, Rui Zhou, Tieli Zhang, Peng Wang, Jianquan Yao
Optics Communications, 2006, 262: 234–237.

Abstract: The electro-optic effect of aperiodically poled lithium niobate (APLN) has been theoretically investigated and proposed to use as a Q-switch in a simultaneous dual-wavelength laser. Our analysis shows that the polarization planes of the z-polarized (or y-polarized) dualwavelength beams can be simultaneously rotated by 90° through a well-constructed APLN with an external electric field applied along the y-axis, which enables Q-switch function in a dual-wavelength laser cavity. Using a $Nd:YVO_4$ laser operating at 1.064 3 μm and 1.341 9 μm as an example, we present a design method of APLN by using the so-called simulated annealing algorithm. The influence of the domain errors in fabricating an APLN device is also studied. The results show that the device is not susceptible to the fabrication errors.
Key words: Dual-wavelength laser, Aperiodically poled lithium niobate, Electro-optic effect, Q-switch

229. Design of a broadband source by using the retracing behavior of a collinear quasi-phase-matching optical parametric generator

Yang Lu, Baigang Zhang, Enbang Li, Jianquan Yao
Optics Express, 2006, 14(25): 12316–12326.

Abstract: In this paper, we introduce a concept of phase matching bandwidth of broadband sources by using the retracing behavior of collinear quasi-phase-matching (CQPM) optical parametric generation (OPG) in periodically poled lithium niobate (PPLN). By comparing various pump wavelengths we derive the pumping condition and parameters of PPLN to realize a broadband source near 1 550 nm in the CQPM-OPG. We predict the optimum pump wavelength and the maximum ideal bandwidth range to be 940.75 nm and 1 475-1 681 nm respectively. Experimentally we have demonstrated a 946 nm Nd:YAG laser which serves as the pumping source of CQPM-OPG.
Key words: Broadband source, CQPM-OPG, PPLN, Nd: YAG

230. Experimental study on a high conversion efficiency, low threshold, high-repetition-rate periodically poled lithium niobate optical parametric generator

Pu Zhao, Baigang Zhang, Enbang Li, Rui Zhou, Degang Xu, Yang Lu, Tieli Zhang, Feng Ji, Xueyu Zhu, Peng Wang, Jianquan Yao
Optics Express, 2006, 14(16): 7224–7229.

Abstract: A high-conversion-efficiency, low-threshold, quasi-continuouswave optical parametric generator (OPG) based on a periodically poled lithium niobate (PPLN) crystal is presented. Pumped by an acoustooptically Q-switched 1 064 nm Nd:YAG laser with a power output of 848 mW, the OPG generated an output power of 452 mW for the signal and the idle waves, achieving an internal conversion efficiency of 62.7% and a slope efficiency of 75.6%. To the best of our knowledge, this is the highest efficiency ever reported for single-pass, quasi-continuous-wave OPGs by using periodically poled crystals.
Key words: High conversion efficiency, Low threshold, High repetition rate, PPLN, OPG

231. LD 端泵浦 1 319 nm 3.11 W-CW Nd:YAG 激光器

赵欣,姚建铨,张百钢,路洋,张铁犁,纪峰,李海峰

光电子·激光, 2006, 17(6):673-677.

摘要：报道了 LD 端面泵浦 Nd:YAG 晶体实现 1 319 nm 单波长连续输出的研究。通过对输出镜镀选择性介质膜，抑制了 1 064 nm 和 1 338 nm 光波在腔内的振荡。在功率为 26.03 W 的 808 nm LD 泵浦下，得到 1 319 nm 单波长最大输出 3.11 W。对 1 319 nm 单波长运转条件进行了理论分析，得到 1 319 nm 单波长运转区域图，描述了 1 319 nm 单波长运转时输出镜对 1 319 nm 和 1 338 nm 光波透过率应满足的关系，为实现 1 319 nm 单波长运转提供了理论指导。实验结果与理论分析结果相互吻合。

关键词：端面泵浦, 1 319 nm, 1 338 nm, 单波长运转, Nd:YAG

232. 基于复合晶体的全固态 1 064 nm 激光器的实验研究

杨扬,徐德刚,姚建铨,李海峰,王与烨,王鹏

光电子·激光, 2006, 17(2): 171-174.

摘要：研究了最大输出功率超过 500 W 的全固态连续高功率 1 064 nm 激光器。为了改善热效应，选用了扩散键合方法制成的复合 Nd:YAG，有效地降低了高功率激光器中的激光晶体热透镜问题。在实验中，1 064 nm 输出镜的透过率分别为 T=10%、30 %、35 % 和 40 %，得到了 512 W 1 064 nm 输出，光-光转换效率达到 32 %。

关键词：全固态, 扩散耦合, 复合 Nd:YAG, 高功率, 1 064 nm

233. 激光二极管端面抽运 Nd:YVO$_4$ 实现 1 386 nm 连续波激光输出

李海峰,周睿,赵璞,姚建铨,王鹏

光学学报, 2006, 26(7):1069-1072.

摘要: 在 Nd:YVO$_4$ 晶体的 $^4F_{3/2}$-$^4I_{13/2}$ 跃迁带内,除了 1 342 nm 激光辐射之外,其它的跃迁谱线由于小的受激发射截面和强的寄生振荡,很难形成激光振荡。通过调整谐振腔损耗,获得了光纤耦合激光二极管端面抽运 1 386 nm Nd:YVO$_4$ 激光器激光连续输出。在抽运功率达到 4.24 W 时,得到了 305 mW 的 1 386 nm 激光连续输出,最高输出功率下的斜效率为 13.9 %。实验中还观察到了 1 342 nm 和 1 386 nm 的双波长运转。根据抽运阈值能量和实验数据,计算得到了 Nd:YVO$_4$ 晶体中 1 386 nm 激光辐射处的受激发射截面大约为 $(3 \pm 1) \times 10^{-19}$ cm^2。

关键词: 激光器,激光二极管,1 386 nm,端面抽运,Nd:YVO$_4$

234. 高稳定 LD 泵浦腔内倍频 Nd:YVO$_4$/KTP 连续绿光激光器

张玉萍,郑义,张会云,王鹏,姚建铨

光子光报, 2006, 35(7): 970-973.

摘要: 设计出一种能够较好地补偿激光晶体热效应的激光谐振腔,实现了高稳定 LD 单端泵浦 KTP 腔内倍频 Nd:YVO$_4$ 连续绿光激光器。当晶体吸收的泵浦功率为 24.56 W 时,532 nm 激光功率达到 5.3 W,光-光转换效率达到 21.6 %,激光模式为 TEM$_{00}$ 模。在输出功率 5 W 左右时,激光器 1 h 功率不稳定度优于 0.6 %。

关键词: LD 端面泵浦,连续,Nd:YVO$_4$,KTP,532 nm

235. 激光在机械制造技术中的应用

王涛, 古桂茹, 李玉翔, 姚建铨, 刘宝利, 蔡彬晶

激光杂志, 2006, 27(3): 75-76.

摘要：激光制造技术迅猛发展, 正逐渐取代传统的机械制造。本文介绍了激光在机械制造技术中的几个典型应用, 激光三维建模技术有效地解决了无人自动化生产线上元件三维信息的获取问题, 激光热处理技术解决了其它表面处理方法无法解决或不好解决的材料强化问题, 另外, 激光在智能识别、快速成型、焊接、熔覆涂层、微加工中也得到了广泛的应用。随着激光技术的进一步发展和市场的不断扩大, 激光技术必将在更宽更广的领域中得到更充分的应用。

关键词：激光应用, 三维建模, 激光热处理, 智能识别, 快速成型, 激光焊接, 激光熔覆涂层, 微加工

236. 抽运光角度调谐准相位匹配光学参量振荡器的研究

张百钢, 姚建铨, 路洋, 纪峰, 张铁犁, 徐德刚, 王鹏, 徐可欣

物理学报, 2006, 55(3): 1231-1236.

摘要：针对利用周期极化晶体实现的抽运光角度调谐准相位匹配 (QPM) 光学参量振荡器 (OPO) 进行了系统的理论分析, 给出了描述 QPM OPO 中抽运光旋转角与三波波长关系的精确公式和近轴公式。研究发现, 对信号光单谐振的情况而言, 抽运光与空闲光沿晶体 x 轴的同侧出射, 而对空闲光单谐振而言, 抽运光与信号光沿晶体 x 轴的同侧出射。另外, 信号光单谐振下信号光与空闲光间的夹角要大于空闲光单谐振下两者间的夹角。更重要的是, 信号光单谐振时的波长调谐速度也较空闲光单谐振时的大。

关键词：准相位匹配, 空闲光单谐振光学参量振荡器, 抽运光角度调谐, 调谐速度

237. 高功率激光二极管抽运 Nd:YAG 连续双波长激光器

张强, 姚建铨, 温午麒, 刘欢, 丁欣, 周睿
中国激光, 2006, 33(5): 577–581.

摘要：通过双波长理论计算确定了双波长运转时腔镜介质膜在不同波长的最佳透射率以及激光腔内不同波长的衍射损耗，最终利用四腔镜双谐振腔结构实现了激光二极管 (LD) 侧面抽运 Nd:YAG 激光器在 1 064 nm 和 1 319 nm 的双波长同时连续运转，并分析了激光腔长与双波长激光输出功率比值之间的关系以及抑制 1 338 nm 等其他波长运转的方法。在抽运功率为 500 W 时，实现了平均功率超过 45 W 的连续激光输出，1 064 nm 和 1 319 nm 单一波长连续输出功率均超过 20 W。两波长输出的光束质量因子 M^2 分别为 32 和 39。输出功率不稳定性均小于 5 %。
关键词：激光器, 高功率, 连续双波长, 四腔镜

238. The study on the high-power LD-pumped Nd:YAG cw dual-wavelength laser

Zhang Qiang, Yao Jianquan, Wen Wuqi, Liu Huan, Gan Yu
Proceedings of Society of Photo-Optical Instrumentation Engineers, 2006, 6344: 63442C-1–63442C-16.

Abstract: Though controlling the coating reflectivity and intracavity diffraction loss, in the four cavity mirror configuration, a dual-wavelength continuous wave (cw) diode-side-pumped Nd:YAG laser that generates simultaneously at wavelengths of 1 064 nm and 1 319 nm is demonstrated. The relationship between power ratio of the two wavelengths and cavity length is studied. The cw dual-wavelength output power reaches 85 W when the average pump power is more than 500 W. The output power of respective wavelength exceeds 40 W, which is the best record as we know.
Key words: Laser, High power, CW dual-wavelength, Four cavity mirror

239. High-power continuous-wave diode-end-pumped intracavity frequency doubled Nd:YVO$_4$ laser at 671 nm with a compact three-element cavity

Zhou Rui, Ding Xin, Wen Wuqi, Cai Zhiqiang, Wang Peng, Yao Jianquan
Chinese Physics Letters, 2006, 23(4): 849–851.

Abstract: We report a high-power high-efficient continuous-wave (cw) diode-end-pumped Nd:YVO$_4$ 1 342-nm laser with a short plane-parallel cavity and an efficient cw intracavity frequency-doubled red laser at 671 nm with a compact three-element cavity. At incident pump power of 20.6 W, a maximum output power of 7 W at 1 342 nm is obtained with a slope efficiency of 37.3%. By inserting a type-I critical phase-matched LBO crystal as intracavity frequency-doubler, a cw red output as much as 2.85-W is achieved with an incident pump power of 16.9 W, inducing an optical-to-optical conversion efficiency of 16.9%. To the best of our knowledge, this is the highest output of diode-pumped solid-state Nd:YVO$_4$ red laser. During half an hour, the red output is very stable, and the instability of output power is less than 1%.

Key words: Intracavity frequency doubled, CW, High power

240. 104 W high stability green laser generation by using diode laser pumped intracavity frequency-doubling Q-switched composite ceramic Nd:YAG laser

Degang Xu, Yuye Wang, Haifeng Li, Jianquan Yao, Yuen H. Tsang
Optics Express, 2007, 15(7): 3991–3997.

Abstract: By use of CW diode laser stacked arrays, side-pumping Q-switched composite ceramic Nd:YAG rod laser based on a type II KTP crystal intracavity frequency-doubled, a high power high stability green laser has been demonstrated. Average output power of 104 W is obtained at a repetition rate of 10.6 kHz with a diode-to-green optical conversion efficiency of 10.9%. For the average output power of about 100 W, the measured pulse width is 132 ns with power fluctuation of less than 0.2%. The experimental results show that the green laser system using this novel ceramic Nd:YAG offers better laser performance and output stability than the traditional single Nd:YAG crystal green laser system with the same operating conditions and experimental configuration.

Key words: Laser and laser optics, Lasers, Diode pumped, Frequency conversion, Solid-state

241. Widely tunable, high-repetition-rate, dual signal-wave optical parametric oscillator by using two periodically poled crystals

Tieli Zhang, Jianquan Yao, Xueyu Zhu, Baigang Zhang, Enbang Li, Pu Zhao, Haifeng Li, Feng Ji, Peng Wang

Optics Communications, 2007, 272:111−115.

Abstract: This paper reports on a high-repetition-rate dual signal-wave (DSW) optical parametric oscillator (OPO) operating at the 1.5 μm band with tunable wavelength intervals from 2.5 nm to 69.1 nm. Two periodically poled crystals, a periodically poled lithium niobate (PPLN) with multiple gratings and a single grating MgO-doped PPLN (PPMgOLN), are cascaded in the same OPO cavity to generate dual signal-waves by using quasi-phase-matched (QPM) technique. The pump source was a Q-switched diode-pumped Nd:YVO$_4$ laser operating at 50 kHz. At an incident pump power of 3 W, an average output power of 169.6 mW at 1 489.2 nm and 1 558.3 nm has been achieved.

Key words: Dual signal-wave, Quasi-phase-matched, Optical parametric oscillator, Periodically poled lithium niobate

242. Tunable quasi-phase-matched optical parametric generator by translating a segmented crystal

Zhang Baigang, Yao Jianquan, Zhang Tieli, Ji Feng, Lu Yang, Zhao Pu

Optics Communications, 2007, 270: 368−372.

Abstract: A simple way to realize a fast tunable quasi-phase-matched optical parametric generation without crystal rotation, temperature changing or multi-grating periodically poled crystal translation is presented. The key is to translate a segmented crystal which is stacked by a lithium niobate crystal with three teeth and a single grating periodically poled crystal. The signal wavelength is tunable near 1.55 μm with the signal power of ⩾90 mW.

Key words: Quasi-phase-matching, Segmented crysta, Optical parametric generator

243. Tunable dual-signal PPLN optical parametric generator by using an acousto-optic beam splitter

Pu Zhao, Baigang Zhang, Enbang Li, Denggang Xue, Yang Lu, Tieli Zhang, Feng Ji, Yueye Wang, Peng Wang, Jianquan Yao

Journal of Optics: Pure and Applied Optics, 2007, 9(3): 235–238.

Abstract: A tunable dual-signal optical parametric generator (OPG) based on multi-grating periodically poled lithium niobate (PPLN) is presented. An acousto-optic device (AOD) was used as a beam splitter outside the cavity of a high-repetition-rate, Q-switched 1 064 nm Nd:YAG laser. A synchronous and widely tunable dual-signal OPG was obtained by coupling the split beams into two parts with different poled periods in the same PPLN crystal and by tuning the temperature of the crystal. A total conversion efficiency of 27.9% was achieved when the system operated at 1 687 nm and 1 864 nm.

Key words: Diode-pumped lasers, Frequency conversion, Optical parametric oscillators, Acousto-devices

244. High-repetition-rate dual-signal intracavity optical parametric generator based on periodically-phase-reversal PPMgLN

Feng Ji, Jianquan Yao, Fanghua Zheng, Enbang Li, Tieli Zhang, Pu Zhao, Peng Wang, Baigang Zhang

Journal of Optics: Pure and Applied Optics, 2007, 9(10): 797–801.

Abstract: A high-repetition-rate dual-signal intracavity optical parametric generator (IOPG) inside a diode-end-pumped acousto-optically (AO) Q-switched Nd:YVO$_4$ laser is presented. The nonlinear material is a periodically-phase-reversal periodically poled MgO-doped lithium niobate (ppr-PPMgLN). At an incident diode pump power of 6.1 W and a Q-switch repetition rate of 20 kHz, an average dual-signal output power of 0.44 W is achieved. The dual-signal wavelengths of 1 477 and 1 491 nm are obtained at a crystal temperature of 40 ℃. The measurements of the beam quality factor of 1.4 and the pulse duration of 77 ns show good spatial and temporal overlaps for the dual-signal radiation.

Key words: High repetition rate, Intracavity optical parametric generator, Dual signal, ppr-PPMgLN

245. An all-solid-state high power quasi-continuous-wave tunable dual-wavelength Ti:sapphire laser system using birefringence filter

Ding Xin, Ma Hongmei, Zou Lei, Zou Yue, Wen Wuqi, Wang Peng, Yao Jianquan
Chinese Physics, 2007, 16(7): 1991–1995.

Abstract: This paper describes a tunable dual-wavelength Ti:sapphire laser system with quasi-continuous-wave and high-power outputs. In the design of the laser, it adopts a frequency-doubled Nd:YAG laser as the pumping source, and the birefringence filter as the tuning element. Tunable dual-wavelength outputs with one wavelength range from 700 nm to 756.5 nm, another from 830 nm to 900 nm have been demonstrated. With a pump power of 23 W at 532 nm, a repetition rate of 7 kHz and a pulse width of 47.6 ns, an output power of 5.1 W at 744.8 nm and 860.9 nm with a pulse width of 13.2 ns and a line width of 3 nm has been obtained, it indicates an optical-to-optical conversion efficiency of 22.2%.

Key words: All-solid-state, Dual-wavelength, Ti:sapphire laser, Birefringence filter

246. A bidirectional diode-pumped, passively mode-locked Nd:YVO$_4$ ring laser with a low-temperature-grown semiconductor saturable absorber mirror

Cai Zhiqiang, Yao Jianquan, Wang Peng, Wang Yonggang, Zhang Zhigang
Chinese Physics Letters, 2007, 24(5): 1270–1272.

Abstract: We report the operation of a bidirectional picosecond pulsed ring Nd:YVO$_4$ laser based on a low-temperature-grown semiconductor saturable absorber mirror. Except for the laser crystal, the six-mirror ring laser cavity has no intra-cavity elements such as focusing lens or mirror. The bidirectional mode locked pluses are obtained at the repetition rate of 117.5 MHz, pulse duration of 81 ps, power of 2×200 mW.

Key words: Passively mode-locked, Bidirectional Diode-pumped, Ring laser

247. An all-solid-state tunable dual-wavelength Ti: sapphire laser with quasi-continuous-wave outputs

Ding Xin, Pang M Ing, Yu Xuanyi, W Ang Xiaoheng, Zhang Shaoming, Zhang Heng, Wang Rui, Wen Wuqi, Wang Peng, Yao Jianquan
Chinese Physics Letter, 2007, 24(7): 1938–1940.

Abstract: A high power dual-wavelength Ti:sapphire laser system with wide turning range and high efficiency is described, which consists of two prism-dispersed resonators pumped by an all-solid-state frequency-doubled Nd:YAG laser. Tunable dual-wavelength outputs, with one wavelength range from 750 nm to 795 nm and the other from 800 nm to 850 nm, have been demonstrated. With a pump power of 23 W at 532 nm, a repetition rate of 6.5 kHz and a pulse width of 67.6 ns, the maximum dual-wavelength output power of 5.6 W at 785.3 nm and 812.1 nm, with a pulse width of 17.2 ns and a line width of 2 nm, has been achieved, leading to an optical-to-optical conversion efficiency of 24.4%.

Key words: Ti: sapphire laser, All-solid-state, Tunable dual-wavelength, Quasi-continuous-wave

248. High-power diode-side-pumped intracavity-frequency-doubled continuous wave 532 nm laser

Zhang Yuping, Zhang Huiyun, Zhong Kai, Li Xifu, Wang Peng, Yao Jianquan
Chinese Physics Letter, 2007, 24(8): 2242–2244.

Abstract: An efficient and high-power diode-side-pumped cw 532 nm green laser based on a V-shaped cavity geometry, and capable of generating 22.7 W green radiation with optical conversion efficiency of 8.31%, has been demonstrated. The laser is operated with rms noise amplitude of less than 1% and with M^2-parameter of about 6.45 at the top of the output power. This laser has the potential for scaling to much higher output power.

Key words: Diode-Side-Pumped, Intracavity-Frequency-Doubled, Continuous Wave

249. Efficient nanosecond dual-signal optical parametric generator with a periodically phase reversed PPMgLN

Ji Feng, Li Xifu, Zhang Baigang, Zhang Teili, Wang Peng, Xu Denggang, Yao Jianquan

Chinese Physics Letters, 2007, 24(11): 3157–3159.

Abstract: We report an efficient nanosecond optical parametric generator (OPG) with a periodically-phase-reversed periodically poled MgO:LiNbO$_3$(ppr-PPMgLN), which produces two pairs of signal and idler waves. The OPG is pumped by a 1.064 μm Q-switched Nd:YVO$_4$ laser. When the repetition rate is set at 10 kHz, the maximum average total output power of 570 mW is achieved, including 410 mW of dual-signal radiations and 160 mW of dual-idler radiations. The total conversion efficiency is 32.5%. The tunable dual-signal wavelengths in the range of 1.474–1.518 μm and 1.490–1.539 μm and the dual-idler of 3.826–3.558 μm and 3.726–3.451 μm are obtained by changing the crystal temperature from 30 ℃ to 200 ℃

Key words: OPG, Nanosecond Dual-Signal, PPMGLN

250. Analysis of frequency mixing error on heterodyne interferometric ellipsometry

Yuanlong Deng, Xuejin Li, Yubin Wu, Juguang Hu, Jianquan Yao

Measurement Science & Technology, 2007, 18(11): 3339–3343.

Abstract: A heterodyne interferometric ellipsometer, with no moving parts and a transverse Zeeman laser, is demonstrated. The modified Mach–Zehnder interferometer characterized as a separate frequency and common-path configuration is designed and theoretically analyzed. The experimental data show a fluctuation mainly resulting from the frequency mixing error which is caused by the imperfection of polarizing beam splitters (PBS), the elliptical polarization and non-orthogonality of light beams. The producing mechanism of the frequency mixing error and its influence on measurement are analyzed with the Jones matrix method; the calculation indicates that it results in an error up to several nanometres in the thickness measurement of thin films. The non-orthogonality has no contribution to the phase difference error when it is relatively small; the elliptical polarization and the imperfection of PBS have a major effect on the error.

Key words: Ellipsometry, Heterodyne interferometer, Films measurement, Frequency mixing error

251. High-peak-power, high-repetition-rate intracavity optical parametric oscillator at 1.57 μm

Yueye Wang, Denggang Xu, Yizhong Yu, Wuqi Wen, Jingping Xiong, Peng Wang, Jianquan Yao

Chinese Optics Letters, 2007, 5(2): 93−95.

Abstract: We report a high-peak-power, high-repetition-rate diode-side-pumped Nd:YAG Q-switched intracavity optical parametric oscillator (IOPO) at 1.57 μm with a type-II non-critically phase-matched x-cut KTP crystal. The average power of 1.15 W at 1.57 μm is obtained at 4.3-kHz repetition rate. The peak power of the pulses amounts to 33.4 kW with 8-ns duration. The average conversion efficiency from Q-switched 1.064-μm-wavelength input power to OPO signal output power is up to 10.5%.

Key words: Parametric oscillators and amplifiers, Lasers, Diode-pumped, Nonlinear optical materials, Lasers, Q-switched

252. 端面抽运全固态皮秒被动锁模激光器

蔡志强，王鹏，温午麒，丁欣，姚建铨

中国激光，2007, 34(7): 901−907.

摘要： 利用国产半导体可饱和吸收镜 (SESAM)，设计了不同的腔型结构，实现了平均功率 5 W 单路输出，5 W 双路输出，输出透过率可调节的半导体可饱和吸收镜线型腔连续锁模 (CWML) 激光器，双向输出六镜环行腔连续锁模激光器等。将半导体可饱和吸收镜放在腔内的特殊位置，利用一种新的技术方法实现了锁模激光器的频率翻倍。利用由非线性晶体 KTP 和双色镜构成的非线性镜 (NLM)，实现了端面抽运 Nd:YVO$_4$ 激光器的 4 W 锁模输出。

关键词： 激光器，皮秒被动锁模，半导体可饱和吸收镜，非线性镜

253. 端面抽运高功率连续单频 1 064 nm Nd:YVO₄ 环形腔激光器

张铁犁,姚建铨,王鹏,朱雪玉,蔡志强,张百钢
中国激光, 2007, 34(9): 1194–1197.

摘要：采用 808 nm 光纤耦合输出激光二极管 (FCLD) 单端端面抽运 Nd:YVO₄ 晶体,采用四镜折叠环行腔,在腔内插入法拉第旋光器和半波片实现激光的单向运转以抑制空间烧孔效应,并在腔内加入标准具,最终实现连续单频 1 064 nm 激光输出。在 24.6 W 抽运功率时,最高输出功率达到 9 W, 光 - 光转换效率为 36.6%, M^2 因子约为 1.14, 频率漂移约 200 MHz。

关键词：激光器, 单频, 环行腔, Nd:YVO₄ 晶体

254. 高稳定 LD 端面泵浦腔内倍频 Nd:YVO₄/LBO 连续红光激光器

张会云,郑义,张玉萍,王鹏,姚建铨
光子学报, 2007, 36(5): 769–772.

摘要：设计出一种能够较好地补偿激光晶体热效应的激光谐振腔,实现了高稳定 LD 单端泵浦 LBO 腔内倍频 Nd:YVO₄ 连续红光激光器。当晶体吸收的泵浦功率为 24.56 W 时,671 nm 激光功率达到 1.203 W, 光 - 光转换效率 4.9%, 激光模式为 TEM_{00} 模。在输出功率为 1.08 W 时, 激光器 1 h 功率不稳定度为 0.52%。

关键词：LD 单端泵浦, Nd:YVO4, LBO, 腔内倍频, 671 nm

255. New Progress on all solid state laser technology in China

YAO Jianquan

OPTICS&OPTOELECTRONIC TECHNOLOGY, 2007, 5(1): 1–10.

Abstract: Because of the advantages of high conversion efficiency, good beam quality, small size and light weight, DPL becomes the hotspot and priority of development of laser technology. It may be the main body of laser in the future and replace gas laser and liquid laser. It is a great revolution of laser technology. The developed countries vie in developing DPL. China has achieved great success in this field, but there is a wide gap between the developed countries and us. We should attach great importance to it.

Key words:DPL, Laser technology

256. 15.2 W LD 端面抽运 946 nm 连续波 Nd:YAG 激光器

周睿，李恩邦，李海峰，王鹏，姚建铨

激光与光电子学进展, 2007, 44(2): 11.

摘要：利用离子扩散键合 Nd:YAG 棒作为激光增益介质，成功实现了 15.2 W 高功率连续运转 946 nm 激光输出，光 - 光转换效率为 38%，斜率效率为 45%。另外，通过使用 I 类临界相位匹配的 LBO 作为内腔倍频晶体和简单的紧凑线性平 - 凹直腔实现了 1.25 W、473 nm 蓝光的输出。

关键词：Nd:YAG，蓝光，倍频

257. 激光电视的研究进展及趋势分析

李玉翔,姚建铨,蔡彬晶,王 涛,王 鹏,古贵茹,刘宝利

激光杂志, 2007, 28(1): 1–2.

摘要： 在显示技术日益重要的今天,激光电视以其他技术无法比拟的优势受到各国的关注。本文介绍了激光电视的基本概念和特点,从激光电视机的发展历史入手,主要介绍了代表世界先进水平的美、日、德三国的激光电视的发展现状,国内激光电视及相关技术的发展情况。总结出我国激光电视今后的发展目标———产业化。

关键词： 激光电视,激光显示

258. 氧化锆掺杂激光熔覆涂层成形、结构与增韧机制

张三川,姚建铨

激光杂志, 2007, 28(2): 73–74.

摘要： 本文利用氧化锆陶瓷在不同凝固成形条件下具有相结构变化的特点,将其作为激光熔覆涂层的增韧相。激光熔覆试验结果表明含氧化锆增韧激光熔覆涂层成形关键在于控制熔池熔体的流动性,低的激光线功率密度有助于分层现象的消除；扫描电镜和能谱分析表明氧化锆陶瓷在熔覆层中没有显著的富集,且点状弥散分布较均匀,同时 XRD 图谱证明激光熔覆层中氧化锆为单斜相结构,达到了利用氧化锆相变消除残余热应力裂纹的目的,从而可以解决激光熔覆裂纹产生的关键问题。

关键词： 激光熔覆,增韧机制,成形,相结构,氧化锆

259. 1 064 nm 泵浦温度调谐 PPLN 光学参量振荡器

徐龙浩,刁述妍,姚建铨,郑义,刘强,杨剑

激光与红外, 2007, 37(8): 728–730.

摘要： 对脉冲泵浦的温度可调准相位匹配 (QPM) 光学参量振荡器 (OPO) 进行了研究。LD 泵浦的声光调 Q Nd:YAG 激光器输出的 1 064 nm 脉冲激光做泵浦源，极化周期为 30.7 μm 的单周期 PPLN 做光学参量振荡器的参量晶体,通过控制参量晶体的温度可以得到信号光的波长调谐输出。在 LD 电流 17 A 得到信号光的平均功率最大为 230 mW,转化效率达 13%。

关键词： 光学参量振荡 (OPO),温度调谐,准相位匹配 (QPM)

260. 高功率全固态内腔倍频调 Q 激光器中相位失配的理论研究

熊景平,姚建铨,徐德刚,王与烨

科学技术与工程, 2007, 7(2): 182–185.

摘要： 给出了在考虑相位失配情况下高功率全固态内腔倍频调 Q 激光器的速率方程解的情况。以 Nd:YAG 声光调 Q 内腔倍频激光器为例,建立并数值求解相位失配条件下的速率方程组,从理论上分析了相位失配和泵浦功率等参数对内腔倍频调 Q 激光器输出功率、脉冲宽度等的影响。

关键词： 速率方程,相位失配,声光调 Q,内腔倍频

261. 基于周期极化铌酸锂的可调谐内腔光参量产生的实验研究

朱雪玉, 张铁犁, 张百钢, 丁欣, 王鹏, 姚建铨

科学技术与工程, 2007, 7(7): 1425−1427.

摘要：实现了基于周期极化铌酸锂的人眼安全波段全固态可调谐高重复频率内腔光学参量产生过程，得到了 1.6 W 的二极管泵浦阈值，最大输出信号光功率为 330 mW，并在 (1.497 ~ 1.517)μm 范围波长可调。

关键词：内腔光学参量产生, 人眼安全, 准相位匹配, 周期极化铌酸锂

262. 基于掺氧化镁周期性极化铌酸锂的光学参量产生器

赫文哲, 李喜福, 耿优福, 谭晓玲, 姚建铨

科学技术与工程, 2007, 7(15): 3645−3648.

摘要：报道了基于掺氧化镁的周期性极化铌酸锂 (PPMgLN) 光学参量产生器的研究。采用 1 064 nm 声光调 Q 准连续 Nd:YAG 激光器作为泵浦源，重复频率为 4.3 kHz。光学参量晶体 PPMgLN 的周期为 30.7 μm。采用温度调谐，调谐的范围为 30℃ ~ 180℃，输出信号光的调谐范围为 1 561 nm ~ 1 672 nm，最高输出功率为 180 mW，相应的 1 064 nm 光的泵浦功率为 1.017 W，光光转换效率为 17.7%。

关键词：相位匹配 (QPM), PPMgLN, 光学参量产生 (OPG), 温度调谐

263. PPLN 准相位匹配内腔光学参量振荡器的实验研究

赵伟,纪峰,张百钢,张铁犁,赵璞,王鹏,姚建铨

科学技术与工程, 2007, 7(16): 4009–4012.

摘要：将准相位匹配光学参量振荡器 (QPM-OPO) 置于激光二极管 (LD) 端面泵浦的声光调 Q 1 064 nm-Nd:YVO$_4$ 激光器谐振腔之内,获得了信号光单谐振的内腔光参量输出。在声光 Q 开关重复频率为 25 kHz 的条件下,内腔参量振荡的阈值仅为 0.9 W (LD 功率)。在 6 W LD 泵浦功率下,获得了 350 mW 的信号光输出。通过在 120 ℃~250 ℃范围内改变 PPLN 晶体的温度,信号光输出实现了在 (1 493~1 538)nm 波段的调谐。

关键词：准相位匹配,内腔光学参量振荡器,周期极化铌酸锂,温度调谐

264. 三镜折叠腔内腔倍频高功率绿光激光器

王占鳌,于意仲,王鹏,姚建铨

科学技术与工程, 2007, 7(18): 4459–4461.

摘要：随着 LD 侧面泵浦技术的提高,高功率连续绿光激光器得到了更广泛的应用。通过推导出三镜折叠腔的传输矩阵,讨论了三镜折叠腔的稳定性与各参数之间的关系。经过参数优化和实验调整,在驱动电流为 20 A 时,获得功率高达 7 W 的连续绿光输出。

关键词：三镜折叠腔,KTP 晶体,绿光激光器,稳定性

265. High-efficiency direct-pumped Nd : YVO$_4$ laser operating at 1.34 μm

X. Ding, H. Zhang, R. Wang, W. Q. Wen, P. Wang, J. Q. Yao, X. Y. Yu

Optics Express, 2008, 16(15): 11247–11252.

Abstract: We report a high-efficiency Nd:YVO$_4$ laser operating at 1 342 nm pumped by an all-solid-state Q-switched Ti:Sapphire laser at 879 nm. A plano-concave cavity was optimized to obtain high efficiency and good beam quality. Output power for two Nd:YVO$_4$ crystals with 1.0- and 3.0- at.% Nd^{3+} doping under 879-nm pumping was measured respectively. Comparative result obtained by traditional pumping at 808 nm into the highly absorbing $^4F_{5/2}$ level were presented, showing that the slope efficiency of the 1.0-at.% Nd:YVO4 laser under 879-nm pumping was 10.5% higher than that of 808-nm pumping. In a 4-mm-thick, 1.0-at.% Nd:YVO$_4$ crystal, a high slope efficiency of 64% was achieved under 879-nm pumping, with an optical-to-optical conversion efficiency of 41.3%.

Key words: Lasers, Solid-state, Pumping, Neodymium, Q-switched

266. Generation of 3.5 W high efficiency blue-violet laser by intracavity frequency-doubling of an all-solid-state tunable Ti:sapphire laser

X. Ding, R. Wang, H. Zhang, W. Q. Wen, L. Huang, P. Wang, J. Q. Yao, X.Y.Yu, Z. Li

Optics Express, 2008, 16(7): 4582–4587.

Abstract: In this paper, we report a high power, high efficiency blue-violet laser obtained by intracavity frequency-doubling of an all-solid-state Q-switched tunable Ti:sapphire laser, which was pumped by a 532 nm intracavity frequency-doubled Nd:YAG laser. A β-BaB$_2$O$_4$ (BBO) crystal was used for frequency-doubling of the Ti:sapphire laser and a V-shape folded three-mirror cavity was optimized to obtain high power high efficiency second harmonic generation (SHG). At an incident pump power of 22 W, the tunable output from 355 nm to 475 nm was achieved, involving the maximum average output of 3.5 W at 400 nm with an optical conversion efficiency of 16% from the 532 nm pump laser to the blue-violet output. The beam quality factor M^2 was measured to be M_x^2=2.15, M_y^2=2.38 for characterizing the tunable blue laser.

Key words: Lasers, Diode-pumped, Solid-state, Titanium, Tunable, Frequency doubled

267. 26.3-W continuous-wave, diode-double-end-pumped all-solid-state Nd:GdVO$_4$ laser operating at 1.34 μm

Rui Zhou, Shuangchen Ruan, Chenlin Du, Jianquan Yao
Optics Communications, 2008, 281(13): 3510–3513.

Abstract: A high-power continuous-wave (CW) all-solid-state Nd:GdVO$_4$ laser operating at 1.34 μm is reported here. The laser consists of a low doped level Nd:GdVO$_4$ crystal double-end-pumped by two high-power fiber-coupled diode lasers and a simple plane-parallel cavity. At an incident pump power of 88.8 W, a maximum CW output of 26.3 W at 1.34 μm is obtained with a slope efficiency of 33.7%. To the best of our knowledge, this is the highest output at 1.34 μm ever generated by diode-end-pumped all-solid-state lasers.

Key words: Nd:GdVO$_4$ laser, Diode-double-end-pumped, Continuous-wave, 1.34 μm

268. 1-W, high-repetition-rate room temperature operation of mid-infrared optical parametric oscillator based on periodically poled Mgo-doped LiNbO$_3$

Zhang Tieli, Zhang Baigang, Xu Degang, Wang Peng, Ji Feng, Yao Jianquan
Chinese Physics B, 2008, 17(2): 633–636.

Abstract: In this paper a high-repetition-rate mid-infrared (mid-IR) optical parametric oscillator based on periodically poled MgO-doped LiNbO$_3$ (PPMgLN) at room temperature was demonstrated. The maximum average mid-IR output power at 3.63 μm was 1.02 W with the repetition rate of 60 kHz and corresponding efficiency from the pump to the idler was 26.7%. The temperature tuning and the period tuning characteristics were also discussed.

Key words: Optical parametric oscillator, Mid-infrared, Periodically poled MgO-doped LiNbO$_3$, Room temperature

269. Compact efficient optical parametric generator internal to a Q-switched Nd:YVO₄ laser with periodically poled MgO:LiNbO₃

Ji Feng, Yao Jianquan, Zhang Baigang, Zhang Tieli, Xu Degang, Wang Peng
Chinese Physics B, 2008, 17(4): 1286–1290.

Abstract: This paper demonstrates a compact efficient optical parametric generator internal to a Q-switched diode-endpumped Nd:YVO₄ laser with periodically poled MgO:LiNbO₃(PPMgLN). With the Q-switch set at a repetition rate of 25 kHz and the PPMgLN crystal operated at room temperature (25 ℃), the intracavity optical parametric generator threshold was reached as a diode pump power of 0.9 W. A maximum signal output power of 0.34 W with a pulse width of 25 ns and a beam quality factor of 1.4 was obtained at an incident diode power of 3.4 W, leading to a conversion efficiency of 10% with a slope efficiency of 14.4%. By varying the crystal temperature from 25 to 200 ℃, the output signal wavelengths were tuned in range of 1 506–1 565 nm. Over a 30-minutes interval, the instability of the signal power was measured to be less than 1%. In addition, the threshold pump intensity for the intracavity optical parametric generator is theoretically investigated, and the obtained result is in good agreement with the experimental results.

Key words: Quasi-phase-matching, Intracavity optical parametric generator, Periodically poled MgO-doped lithium niobate (PPMgLN)

270. Continuous-wave mid-infrared intracavity singly resonant optical parametric oscillator based on periodically poled lithium niobate

Ding Xin, Zhang Shaomin, Ma Hongmei, Pang Ming, Yao Jianquan, Li Zhuo
Chinese Physics B, 2008, 17(1): 211–216.

Abstract: This paper reports a continuous-wave (CW) mid-infrared intracavity singly resonant optical parametric oscillator based on periodically poled lithium niobate (PPLN) pumped by a diode-end-pumped CW Nd:YVO₄ laser. Considering the thermal lens effects, it adopted an optical ballast lens and the near-concentric cavity for better operation. At the PPLN's grating period of 28.5 μm and the temperature of 140 ℃, the maximum idler output power of 155 mW at 3.86 μm has been achieved when the 808 nm pump power is 8.5 W, leading to an optical-to-optical conversion efficiency of 1.82%.

Key words: Continuous-wave, Mid-infrared, Intracavity optical parametric oscillator, Periodically poled lithium niobate crystal

271. Performace of gain-switched all-solid-state quasi-continuous-wave tunable Ti:sapphire laser system

Ding Xin, Zhang Heng, Wang Rui, Yu Xuanyi, Wen Wuqi, Zhang Baigang, Wang Peng, Yao Jianquan
Chinese Physics B, 2008, 17(10): 3759–3764.

Abstract: We have made a gain-switched all-solid-state quasi-continuous-wave (QCW) tunable Ti:sapphire laser system, which is pumped by a 532 nm intracavity frequency-doubled Nd:YAG laser. Based on the theory of gain-switching and the study on the influencing factors of the output pulse width, an effective method for obtaining high power and narrow pulse width output is proposed. Through deliberately designing the pump source and the resonator of the Ti:sapphire laser, when the repetition rate is 6 kHz and the length of the cavity is 220 mm, at an incident pump power of 22 W, the tunable Ti:sapphire laser from 700 to 950 nm can be achieved. It has a maximum average output power of 5.6 W at 800 nm and the pulse width of 13.2 ns, giving an optical conversion efficiency of 25.5% from the 532 nm pump laser to the Ti:sapphire laser.

Key words: Gain-switched, All-solid-state, Quasi-continuous-wave, Narrow pulse width

272. Numerical modelling of QCW-pumped passively q-switched Nd:YAG lasers with Cr^{4+}:YAG as saturable absorber

Wang Yuye, Xu Degang, Xiong Jingping, Wang Zhuo, Wang Peng, Yao Jianquan
Chinese Physics Letters, 2008, 25(8): 2880–2883.

Abstract: Passively Q-switched quasi-continuous-wave (QCW) diode-pumped Nd:YAG laser with Cr^{4+}:YAG as saturable absorber is numerically investigated by solving the coupled rate equations. The threshold pump rate for passively Q-switched QCW-pumped laser is derived. The effects of the pump rate and pump-pulse duration on the laser operation characteristics are studied theoretically. The pump power range can be estimated according to the number of output pulses. The numerical simulation results are in good agreement with the experimental results.

Key words: QCW, Nd:YAG lasers, Passively Q-switched

273. Passively Q-switched quasi-continuous-wave diode-pumped intracavity optical parametric oscillator at 1.57 μm

Wang Yuye, Xu Degang, Wen Wuqi, Wang Peng, Yao Jianquan
Chinese Physics Letters, 2008, 25(11): 4009−4012.

Abstract: We report on a passively Q-switched quasi-cw diode-pumped Nd:YAG including an intracavity optical parametric oscillator. The dynamics of this system is described by solving the coupled equations. The effect of the initial transmission of Cr^{4+}:YAG saturable absorber on the signal wave operation is studied. Under optimum conditions, we achieve 2.3 mJ energy at 1.57 μm wavelength for 40 Hz repetition rate. The peak power of the pulses amounts to 0.88 MW with the pulse width of 2.6 ns. When the Fresnel reflection losses of the filters are taken into account, the pulse energy would be higher than 2.3 mJ. To the best of our knowledge, this is the highest pulse energy and peak power for such a type of single resonant quasi-cw diode pumped Nd:YAG/Cr^{4+}:YAG IOPO laser.

Key words: Optical parametric oscillator, Passively Q-Switched, Quasi-Continuous-Wave

274. Threshold characteristic of intracavity optical parametric oscillator pumped by all-solid-state *Q*-switched laser

Yuye Wang, Degang Xu, Yizhong Yu, Wuqi Wen, Xifu Li, Jianquan Yao
Chineses Optics letters, 2008, 6(3): 207−210.

Abstract: We derive the threshold pump intensity for a singly resonant intracavity optical parametric oscillator (IOPO) based on a temporal coupled field model. Particular attention is paid to the dependence of the intracavity singly resonant OPO (SRO) threshold intensity on the signal wave output coupling. Meanwhile, a Nd:YAG laser pumped $KTiOPO_4$ (KTP) IOPO for eye-safe laser output is studied experimentally. The experiment is performed with four signal wave output reflectivities of 60%, 70%, 80%, and 90%, respectively. The measured values are in good agreement with the theoretical results. With an output coupler reflectivity of 80%, a peak power of 70 kW at 1 572 nm has been obtained at a repetition rate of 3.5 kHz. The pulse width is 4.9 ns. Such investigation is helpful to identifying suitable operational regime of low pump intensity.

Key words: Q-switched laser, Intracavity optical parametric oscillator

275. LD 端面抽运 Nd:YAG 1 319 nm/1 338 nm 双波长激光器研究

刘欢,姚建铨,郑芳华,路洋,王鹏
物理学报, 2008, 57(1): 230-237.

摘要：从 LD 端面抽运固体激光器的激光阈值公式出发,建立了双波长激光同时振荡的阈值条件,理论计算了腔镜对于两个波长的透过率关系,实现了 LD 端面抽运 Nd:YAG 1 319 nm/1 338 nm 双波长激光连续和准连续输出。双波长激光连续输出功率可达 6 W,斜效率为 30%;准连续输出功率在重复频率 50 kHz 时可达 4.75 w,斜效率为 24.73%,脉冲宽度为 55.05 ns;腔内插入布儒斯特片,在重复频率为 50 kHz 时,双波长激光准连续线偏振输出功率可达 2.22 W,不稳定性小于 0.52%,M^2 因子仅为 1.16。这两条非常接近的谱线为进一步通过非线性光学差频方法获得高相干性太赫兹波提供了实验基础。

关键词：端泵 Nd:YAG 激光器,319 nm/1 338 nm 双波长,声光调 Q,太赫兹波

276. 高效高功率侧面抽运腔内倍频连续绿光激光器

张会云,张玉萍,钟凯,王鹏,李喜福,姚建铨
中国激光, 2008, 35(1): 3-5.

摘要：激光二极管 (LD) 侧面抽运的内腔倍频激光器技术是实现高功率、高稳定且低成本连续绿光激光器的有效方法。为满足激光彩色显示、激光加工、数据存储、医疗卫生和科研等领域对连续绿光激光器的需求,研制了一台高效、高功率侧面抽运腔内倍频 Nd:YAG/KTP 连续绿光激光器。采用优化的平 - 凹 - 平三镜折叠腔结构,Ⅱ类相位匹配 KTP 晶体内腔倍频,当 808 nm 激光二极管抽运功率约为 180 W 时,得到最高 18.7 W 的连续绿光激光输出,对应的光 - 光转换效率为 10.4%。在输出功率 15.4 W 时测量激光功率稳定性,其功率不稳定度小于 0.5%。输出光束平滑,远场为类高斯分布,用刀口法测量了激光器不同输出功率时的光束质量,光束传输因子 M^2 小于 7。

关键词：激光器,侧面抽运,连续波,腔内倍频

277. 全固态激光电视的研究

王与烨, 于意仲, 徐德刚, 王鹏, 姚建铨

光电子·激光, 2008, 19(6): 739–742.

摘要：阐述了转鼓振镜扫描激光电视系统的设计，系统采用瓦级以上的红、绿、蓝全固态激光器作为光源，合理计算三基色的功率配比，采用多面转鼓和振镜分别实现行、场扫描。设计中采用图像存储、转鼓镜面位置检测的方法，即利用检测到的转鼓位置信号来控制存储的图像信息的读取，从而进一步控制调制器和振镜，很好的实现了同步扫描。该方法降低了光学扫描系统的制作难度，避免了转鼓转速起伏、镜面加工时的角度分割误差对扫描造成的不利影响。制成的激光电视样机扫描面积为 5 m^2，显示的彩色视频图像亮度高、色彩鲜明。

关键词：激光电视, 扫描, 转鼓, 全固态激光器

278. 准连续 LD 侧面泵浦的 Nd:YAG 双制式脉冲激光器

徐德刚, 姚建铨, 王与烨, 张铁犁, 王鹏

光电子·激光, 2008, 19(10): 1297–1300.

摘要：报道了采用 5 个峰值功率为 100 W 的准连续激光器二极管侧面泵浦 Nd:YAG 棒，通过泵浦电流触发铌酸锂晶体加压式 Q 开关，实现宽脉冲和窄脉冲交替输出的双制式脉冲激光。当泵浦能量为 239 mJ、重复频率为 40 Hz、泵浦脉宽 500 μs 时，获得长脉冲、窄脉冲交替输出的双制式脉冲激光，其中长脉冲、窄脉冲输出能量分别为 33.5 mJ 和 18.3 mJ、脉冲宽度分别为 480 μs 和 10 ns，光 - 光转换效率分别为 14% 和 7.7%。

关键词：准连续激光器二极管, 侧面泵浦, 电光 Q 开关, 双侧式, 交替输出

279. 电光调 Q 1 064 nm/532 nm 脉冲激光器

李继武，李忠洋，钟凯，徐德刚，李喜福，姚建铨

应用激光, 2008, 28(3): 230–233.

摘要：报道了一种用于产生 THz 波的电光调 Q 脉冲激光器，重复频率为 1 Hz、10 Hz，输出 1 064 nm、532 nm 可方便转换。当腔长 310 mm，输入能量为 16 J 时，输出的 1 064 nm 调 Q 脉冲能量 210 mJ，电光转换效率 1.3%，平均脉宽 13 ns，532 nm 脉冲能量 110 mJ，倍频效率 50%，脉宽 9~12 ns。并对实验结果进行了分析和讨论。

关键词：电光调 Q，脉冲激光器，THz 波

280. Ar^{3+} 激光器泵浦的连续波 KTP 光学参量振荡器的研究

刘强，丁春峰，邴丕彬，郑义，姚建铨

量子光学学报, 2008, 14(2): 218–222.

摘要：对 Ar^{3+} 激光器泵浦的连续波 KTP 光参量振荡器进行了理论设计。通过对 KTP-OPO 相位匹配、有效非线性系数以及振荡阈值的数值计算，从而确定了晶体的切割角为 $\theta=90°$，$\phi=40°$。同时决定采用双振荡的运转方式。本文对连续波 KTP 光参量振荡器的初步设计和优化有一定的参考价值。

关键词：非线性光学，连续波光参量振荡器 (CW-OPO)，KTF 晶体，相位匹配，非线性系数，振荡阈值

281. 双棒 Nd:YAG 激光器双折射补偿未对准灵敏度的研究

李忠洋, 张会云, 闫昕, 李继武, 邴丕彬, 姚建铨

激光与红外, 2008, 38(5): 437–440.

摘要：在双棒 Nd:YAG 激光器中，利用一块 90° 石英旋转片来改变激光在两棒间的偏振态，利用一个望远镜系统在激光棒的两个主面间构成像传递光学系统，从而达到热致双折射的补偿。在这个补偿系统中，可能会产生激光棒和透镜横向或纵向未校准以及激光棒倾斜而导致的未校准等情况，我们对于激光腔内每一种未校准而产生的退偏损耗进行了理论分析。

关键词：应力双折射，双折射补偿，热应力，Nd:YAG

282. 514.5 nm 泵浦 KTP 晶体的连续光学参量振荡器的研究

邴丕彬, 丁春峰, 闫昕, 樊心民, 姚建铨, 郑义

激光杂志, 2008, 29(2): 6–7.

摘要：对 Ar^+ 激光器产生的 514.5 nm 的连续光泵浦 KTP 晶体的光学参量振荡器 (CW-KTP-OPO) 做了理论上的分析，分别讨论了 KTP 晶体在 II 类相位匹配时 x-z 平面和 x-y 平面内的匹配情况，数值模拟了其角度调谐曲线，并且计算了两种匹配方式下的有效非线性系数及其阈值功率，证实了其实验的可行性，并最终确定了晶体的切割角为 $\theta=-90°$ $\varphi=50.60°$，可以有效输出 1.029 μm 左右的的参量光。

关键词：连续波光学参量振荡器 (CW-OPO)，KTP 晶体，Ar^+ 激光器，II 类相位匹配

283. Ar^+ 激光器泵浦的连续光学参量振荡器的理论设计

邴丕彬,丁春峰,姚建铨,郑义,李忠洋,王冠军

激光与红外, 2008, 38(2): 132–134.

摘要:对 Ar^+ 激光器产生的 514.5 nm 的连续光泵浦 KTP 晶体的光学参量振荡器 (CW - KTP-OPO) 做了理论分析,讨论了 KTP 晶体采用 II 类临界相位匹配 (CPM) 时的角度调谐曲线、有效非线性系数以及走离角,并针对所设计的条件计算了 KTP 晶体的泵浦阈值。合理设计腔内各个器件的参数,最终确定晶体切割角度为 $\theta=67°$, $\phi=0°$,此时可以产生中心波长为 1.55 μm 的信号光。

关键词:连续波光学参量振荡器 (CW-OPO),KTP 晶体,Ar^+ 激光器,临界相位匹配

284. Efficient electro-optic Q-switched eye-safe optical parametric oscillator based on $KTiAsO_4$

K.Zhong, Y.Y.Wang, D.G.Xu, Y.F.Geng, J.L.Wang, P.Wang, J.Q.Yao

Applied Physics B Lasers and Optics, 2009, 97: 61–66.

Abstract: An efficient high-energy, high-peak-power eye safe optical parametric oscillator (OPO) at 1 536 nm based on a noncritically phase-matched KTA crystal pumped by a quasi-cw diode side-pumped electro-optic Q-switched Nd:YAG laser is presented. The maximum output energy is 74.9 mJ, corresponding to an optical-to-optical conversion efficiency of 12.14% and an electrical-to-optical conversion efficiency of 4.73%. The pulse width (FWHM) is about 3.5 ns with a peak power of 21.4 MW. Moreover, the maximum electrical-to-optical conversion efficiency and optical-to-optical conversion efficiency are up to 4.94% and 15.6%, respectively.

Key words: Electro-optic Q-switched, OPO, $KTiAsO_4$

285. High-energy pulsed laser of twin wavelengths from KTP intracavity optical parametric oscillator

Y.Y. Wang, D.G. Xu, K. Zhong, P. Wang, J.Q. Yao

Applied Physics B Lasers and Optics, 2009, 97: 439–443.

Abstract: We have demonstrated an efficient, high-energy singly resonant pulsed KTP-based intracavity optical parametric oscillator pumped by a quasi-continuous wave (QCW) diode-pumped electro-optical Q-switched Nd:YAG laser. 31.5-mJ energy at 1 572-nm wavelength and 50.4-mJ pulse energy at 1 064 nm were obtained at 10 Hz simultaneously. The total conversion efficiency with respect to diode pump energy was 14%.

Key words: Optical parametric oscillator, High-energy pulsed laser

286. Optimum design for 160-Gb/s all-optical time-domain demultiplexing based on cascaded second-order nonlinearities of SHG and DFG

Jing Shen, Song Yu, Wanyi Gu, Jianquan Yao

IEEE Journal of Quantum Electronics, 2009, 45(6): 694–699.

Abstract: All-optical time-domain demultiplexing based on cascaded second-order nonlinearities of second-harmonic generation and difference-frequency generation in quasi-phase-matched periodically poled lithium niobate waveguides is discussed in order to maximize conversion efficiency and control crosstalk under the given level. The three parameters, including the waveguide length, clock pulse width, and clock offset, are investigated carefully and comprehensively. The concepts of maximum waveguide length, maximum clock pulse width, and optimal clock offset are defined, and their theoretical expressions are derived for the demultiplexing from 160 to 10 Gb/s. The results calculated from the theoretical expressions are well-consistent with those from the numerical simulation. For two kinds of practical situations, the waveguide length or the clock offset being given, an optimum scheme is proposed to determine the other two parameters readily. The maximum conversion efficiency can therefore be achieved while the quality of the converted wave is sufficient for the practical situations.

Key words: All-optical time-domain multiplexing (OTDM), Cascaded second-order nonlinearities, Conversion efficiency, Crosstalk, Difference-frequency generation (DFG), Periodically poled lithium niobate (PPLN), Quasi-phase-matching (QPM), Second-harmonic generation (SHG)

287. A high-power continuous-wave diffraction-limited Nd:GdVO$_4$ laser operating at 1.34 μm

Rui Zhou, Shuangchen Ruan, Chenlin Du, Jianquan Yao

Optics & Laser Technology, 2009, 41: 651–653.

Abstract: High-power and high beam quality continuous-wave (CW) Nd:GdVO$_4$ lasers operating at 1.34 mm were experimentally demonstrated. The lasers consisted of either one or two crystals, which were both end-pumped by high-power fiber-coupled diode lasers. With one crystal, the maximum CW output power generated was 8.4 W. When two crystals were used, a maximum output power of 15.7 W was achieved with the incident pump power of 76.2 W, showing a slope efficiency of 26.2% and an optical-to-optical efficiency of 20.6%. The beam divergence at an output power of 15 W was measured to be about two times that of the diffraction limit.

Key words: Diode-end-pumped, Nd:GdVO$_4$ crystal, 1.34 μm

288. Efficient dual signal output from a Nd:YVO$_4$ laser-pumped PPMgLN OPO

F. JI, R. S. LU, B. S. LI, B. G. ZHANG, J. Q. YAO

Modern Physics Letters B, 2009, 23(18):2251–2259 .

Abstract: We report an efficient dual signal wave intracavity optical parametric oscillator (IOPO) inside a diode-end-pumped acousto-optically Q-switched Nd:YVO$_4$ laser. The nonlinear crystal in the IOPO is periodically poled MgO-doped lithium niobate (PPMgLN) with a phase-reversed grating. In the experiment, dual signal waves at 1 475.1 nm and 1 489.8 nm are achieved at a crystal temperature of 40 ℃. At an incident diode pump power of 6.1 W and a Q-switch repetition rate of 19 kHz, an average dual signal output power of 0.66 W is achieved, leading to a conversion efficiency of 10.8%. During half an hour, the instability of the dual signal waves is less than 1.5% (rms).

Key words: Quasi-phase-matching, IOPO, Dual signal wave, PPMgLN

289. High efficiency continuous-wave tunable signal output of an intracavity singly resonant optical parametric oscillator based on periodically poled lithium niobate

Ding Xin, Sheng Quan, Chen Na, Yu Xuanyi, Wang Rui, Zhang Heng, Wen Wuqi, Wang Peng, Yao Jianquan
Chinese Physics B, 2009, 18(10): 4314–4318.

Abstract: In this paper we report on a continuous-wave (CW) intracavity singly resonant optical parametric oscillator (ICSRO) based on periodically poled $LiNbO_3$ (PPLN) pumped by a diode-end-pumped CW $Nd:YVO_4$ laser. Considering the thermal lens effects and difraction loss, an optical ballast lens and a near-concentric cavity are adopted for better operation. Through varying the grating period and the temperature, the tunable signal output from 1 406 nm to 1 513 nm is obtained. At a PPLN grating period of 29 μm and a temperature of 413 K, a maximum signal output power of 820 mW at 1 500 nm is achieved when the 808 nm pump power is 10.9 W, leading to an optical-to-optical conversion efficiency of 7.51%.

Key words: Continuous-wave, Intracavity optical parametric oscillator, Periodically poled lithium niobate crystal, Tunable eye-safe laser

290. High power continuous-wave diode-end-pumped 1.34-μm $Nd:GdVO_4$ laser

Zhou Rui, Ruan Shuangchen, Du Chenlin, Yao Jianquan
Chinese Physics Letters, 2009, 25(12): 4273–4275.

Abstract: A high power cw all-solid-state 1.34-μm $Nd:GdVO_4$ laser is experimentally demonstrated. With a diode-double end-pumped configuration and a simple plane-parallel cavity, a maximum output power of 27.9 W is obtained at incident pump power of 96 W, introducing a slope efficiency of 35.4%. To the best of our knowledge, this is the highest output power of diode-end-pumped 1.3-μm laser. With the experimental data, the thermal-stress resistance figure of merit of $Nd:GdVO_4$ crystal with 0.3 at% Nd^{3+} doped level is calculated to be larger than 9.94 W/cm.

Key words: $Nd:GdVO_4$ laser, Diode-end-pumped, High power

291. Green-beam generation of 40 W by use of an intracavity frequency-doubled diode side-pumped Nd:YAG laser

Zhang Yuping, Zhang Huiyun, Geng Youfu, Tan Xiaoling, Zhong Kai, Li Xifu, Wang Peng, Yao Jianquan

Chinese Physics Letters, 2009, 26(7): 074217.

Abstract: A cw diode side-pumped Nd:YAG laser is frequency doubled to 532 nm with an intracavity KTP crystal in a V-shaped arrangement, achieving an output power of 40 W corresponding to an optical–optical conversion efficiency of 9.7%. The instabilities and the M^2-parameters of the laser are measured at different output powers after the beam is filtered.

Key words: Intracavity frequency-doubled, Green-beam generation, Nd:YAG Laser

292. All-solid-state Nd:YAG laser operating at 1 064 nm and 1 319 nm under 885 nm thermally boosted pumping

Ding Xin, Chen Na, Sheng Quan, Yu Xuanyi, Xu Xiaoyan, Wen Wuqi, Zhou Rui, Wang Peng, Yao Jianquan

Chinese Physics Letters, 2009, 26(9): 094207-1–094207-4.

Abstract: We report a high-efficiency Nd:YAG laser operating at 1 064 nm and 1 319 nm, respectively, thermally boosted pumped by an all-solid-state Q-switched Ti:sapphire laser at 885 nm. The maximum outputs of 825.4 mW and 459.4 mW, at 1 064 nm and 1 319 nm respectively, are obtained in a 8-mm-thick 1.1 at.% Nd:YAG crystal with 2.1 W of incident pump power at 885 nm, leading to a high slope efficiency with respect to the absorbed pump power of 68.5% and 42.0%. Comparative results obtained by the traditional pumping at 808 nm are presented, showing that the slope efficiency and the threshold with respect to the absorbed pump power at 1 064 nm under the 885 nm pumping are 12.2% higher and 7.3% lower than those of 808 nm pumping. At 1 319 nm, the slope efficiency and the threshold with respect to the absorbed pump power under 885 nm pumping are 9.9% higher and 3.5% lower than those of 808 nm pumping. The heat generation operating at 1 064 nm and 1 319 nm is reduced by 19.8% and 11.1%, respectively.

Key words: All-Solid-State, Nd:YAG

293. Low threshold and high conversion efficiency nanosecond mid-infrared KTA OPO

Zhong Kai, Lijiansong, Cui Haixia, Xu Denggang, Wang Yuye, Zhou Rui, Wang Jingli, Wang Peng, Yao Jianquan

Chinese Physics Letters, 2009, 26(12): 124213.

Abstract: Based on a Type Ⅱ non-critically phase-matched KTA crystal, a low-threshold and high conversion efficiency midinfrared optical parametric oscillator (OPO) pumped by a diode-end-pumped Nd:YVO$_4$ laser is demonstrated. The OPO threshold is only 0.825 W. The maximum output power of 435 mW at 3.47 μm is achieved with the repetition rate of 30 kHz, corresponding to an optical-to-optical conversion efficiency of 4.4%. The photon conversion efficiency is as high as about 64%. The pulse width is 3.5 ns with a peak power of 4 kW for the maximum output power.

Key words: OPO, Low Threshold, Mid-Infrared

294. 高效高稳定高光束质量声光调 Q 绿光激光器的研究

张玉萍，张会云，钟凯，王鹏，李喜福，姚建铨

物理学报，2009, 58(5):3193-3197.

摘要：通过优化双棒串接直腔结构设计，利用大功率 LD 侧面抽运，声光 Q 开关，Ⅱ类相位匹配 S-KTP 内腔倍频获得高效大功率绿色激光输出。当抽运电流为 45 A，重复频率为 15 kHz 时，激光平均功率为 132 W，光 - 光转换效率为 13.2%，脉宽约为 120 ns。在输出 130 W 时，测得 1 h 功率不稳定度小于 0.5%，光束质量因子 M^2 为 6.7。对高功率抽运情况下激光介质的热透镜效应以及谐振腔稳定运转工作区域也进行了理论分析和实验研究。

关键词：绿光激光器，腔内倍频，声光调 Q，侧面抽运

295. 36 W 侧面抽运腔内倍频 Nd:YAG/KTP 连续绿光激光器

张玉萍, 张会云, 何志红, 王 鹏, 李喜福, 姚建铨

物理学报, 2009, 58(7): 4647–4651.

摘要：通过优化平 - 凹 - 平三镜折叠腔结构设计, 利用大功率半导体激光器侧面抽运、Ⅱ类相位匹配 KTP 晶体腔内倍频, 获得高效高功率连续绿色激光输出。当抽运电流约为 36 A 时, 得到最高 36.6 W 的连续绿光激光输出, 对应的光 - 光转换效率为 8.71%。在输出功率 33 W 时测量激光功率稳定性, 其功率不稳定度为 0.27%。用刀口法测量了激光器高输出功率时的光束质量, 光束质量因子小于 8。对高功率抽运情况下三镜折叠腔的像散补偿、失调灵敏度和基模在腔内分布情况做了数值模拟。

关键词：侧面抽运, 腔内倍频, 连续波

296. 准连续 TEM_{00} 模被动调 Q Nd:YAG 激光器的研究

王与烨, 徐德刚, 王鹏, 姚建铨

光电子·激光, 2009, 20(2):143–147.

摘要：采用峰值功率为 500 W 准连续激光二极管 (LD) 阵列侧面泵浦 Nd：YAG 晶体, Cr^{4+}：YAG 可饱和吸收体作为被动调 Q 元件, 在重复频率 40 Hz 时实现了单脉冲能量为 8.24 mJ、脉冲宽度约 7 ns 的 1 064 nm TEM_{00} 模调 Q 激光脉冲输出, 脉冲峰值功率达到 1.2 MW, 从而为主振荡功率放大 (MOPA) 系统提供高峰值功率、高光束质量的种子源。对泵浦功率和泵浦脉冲宽度对输出脉冲特性的影响进行了实验研究。

关键词：准连续泵浦, Cr^{4+}：YAG 晶体, 被动调 Q

297. 双棒串接 Nd:YAG 激光器的稳区分析和实验研究

张玉萍,李喜福,张会云,王鹏,何志红,姚建铨

光子学报, 2009, 38(6):1317–1321.

摘要：用传输矩阵法理论分析了在含热致双折射补偿和不含热致双折射补偿两种情况下，γ 偏振和 ∅ 偏振对双棒串接激光器谐振腔稳区的影响。选用低掺杂浓度的 Nd:YAG 棒，实验上用含 90° 石英旋光片的双棒对称平行平面短腔获得了最佳实验结果。1 064 nm 激光最高输出功率达 482.3 W，对应光-光转换效率为 40.2%。

关键词：双棒串接,Nd:YAG 激光器,传输矩阵法

298. 大功率衍射极限准连续 1.34 μm 激光器的研究

周睿,阮双琛,杜晨林,姚建铨

深圳大学学报理工版, 2009, 26(1):12–15.

摘要：采用两台大功率光纤输出半导体激光器端面泵浦两块 Nd:GdVO$_4$ 晶体，以声光 Q 开关作为腔内调制元件，用对称结构双晶体串接平行平面谐振腔。在注入泵浦功率为 66 W，重复频率为 100 kHz 时，获得 10 W 的大功率准连续 1.34 μm 激光输出，斜率效率为 18.3%，脉冲宽度为 96 ns，激光输出光束发散角约为衍射极限的 2 倍。

关键词：大功率激光器,Nd:GdVO$_4$ 晶体,1.34 μm 激光,全固态,声光调 Q

299. Image processing in spectral domain optical coherence tomography with phase shifting interferometry

Gong Qiang, Wang Ruikang, Ma Zhenhe, Zhang Fan, Yao Jianquan
Nanotechnology and Precision Engineering, 2009, 7(2):168–172.

Abstract: To decrease the noise in spectral domain optical coherence tomography(SDOCT), phase shifting interferome try was applied to eliminating the autocorrelation noise and virtual image in SDOCT. And the main factor affecting the quality of noise elimination is miscalibration error. Results of the slide imaging experiment show that five-phase shifting method is the most insensitive to PZT phase-shifting miscalibration error; it controls the autocorrelation term near zero and virtual image under 0.25 in intensity, with 10% miscalibration error. Moreover, in the experiment of in vivo rabbit cornea imaging, over three phase (including three-phase) shifting methods can eliminate the virtual image and autocorrelation noise, simultaneously avoiding the overlapping between the real and virtual images, and the setting of optical path length (OPL) matching point in experimental process.

Key words: Optical coherence tomography(OCT), Phase shifting interferometry, Spectral domain, PZT

300. High-pulse-energy high-efficiency mid-infrared generation based on KTA optical parametric oscillator

K. Zhong, J. Q. Yao, D. G. Xu, J. L. Wang, J. S. Li, P. Wang
Applied Physics B-Lasers and Optics, 2010, 100: 749–753.

Abstract: A high-energy, high-efficiency mid-infrared KTA OPO at 3.47 μm intracavity pumped by a Nd:YAG laser is presented. The maximum output energy is 31 mJ at the repetition rate of 10 Hz with a V-shaped cavity, corresponding to the absolute optical-to-optical conversion efficiency of 4.76% from the diode and the photon conversion efficiency of 87% from the fundamental to mid-infrared energy. The pulse width is 5.8 ns at the maximum output energy and the peak power reaches higher than 5 MW. The line width of the mid-infrared wave is about 1.1 nm (or 0.9 cm^{-1} in wave number). The output energy demonstrates good stability. To our knowledge, these are the highest pulse energy and conversion efficiency of mid-infrared OPOs using bulk nonlinear crystals in the 3–5 μm range.

Key words: :OPO, Mid-infrared generation, High-pulse-energy, KTA

301. High efficiency generation of 355 nm radiation by extra-cavity frequency conversion

Bin Li, Jianquan Yao, Xin Ding, Quan Sheng, Peng Wang

Optics Communications, 2010, 283: 3497–3499.

Abstract: High efficiency extra-cavity third harmonic generation (THG) of 355 nm has been developed. A laser diode (LD) end-pumped, acoustic-optical Q-switched Nd:YAG laser was used as the fundamental wave source. With an input pump power of 25 W, average power of 6.75 W at 1 064 nm was generated with the repetition rate 12 kHz and pulse duration 10 ns. Using the extra-cavity frequency conversion of three critical phase match (CPM) LiB_3O_5 (LBO) crystals, 3.2 W third harmonic radiation at 355 nm was obtained. The optical-to-optical (1 064 nm to 355 nm) conversion efficiency was up to 47.4%.

Key words: Extra-cavity frequency conversion, 355 nm radiation

302. High-power continuous wave green beam generation by use of simple linear cavity with side-pumped module

Y. P. Zhang, X. D. Zhao, H. Y. Zhang, L. H. Meng, L. Li, X. F. Li, P. Wang, J. Q. Yao

Optics Communications, 2010, 283: 5161–5164.

Abstract: High-power continuous wave green radiation has been generated by means of type-II phase-matched frequency doubling in a KTP crystal located in a simple linear cavity incorporating a diode side-pumped Nd:YAG laser module. The cavity was designed to make the fundamental beam radius at the KTP crystal smaller than that at the gain medium, as is required for obtaining large mode volume in Nd:YAG crystal and realizing efficient CW intracavity frequency doubling. Output power of 51.2 W is obtained in the experiment with a diode-to-green optical conversion efficiency of 10.3%. The M^2-parameters of the laser are measured at different output powers. For the output power of about 47 W, the power fluctuation is measured less than 1%. The experimental results show that the continuous wave green laser system using this simple linear cavity offers good laser performance and output stability.

Key words: Diode-side-pumped, Continuous-wave, Frequency doubled

303. Using relaxation time approximation for collisions in laser-cluster interaction simulation

Miao Wang, Yinfei Lu, Rao Fu, Guizhong Zhang, Degang Xu, Jianquan Yao
Optics Communications, 2010, 283: 993–997.

Abstract: Numerical experiments were performed for cluster dynamics on a one-dimensional cluster model consisting of hydrogen atoms exposed to intense laser pulses. The algorithm deployed the relativistic equations for solving motions of both protons and electrons, respectively. In contrary to traditionally and extensively used method which treats collisions within a jumbo cluster in a statistical way, we introduced a phenomenological relaxation time parameter to cope collisions among particles in a cluster, thereby profoundly reducing computational workload while still attaining the essential physics. Positive ion's maximum kinetic energy and maximum kinetic energy saturation with laser intensity were explored, which were found to be in fairly good accordance with experimental observations, corroborating the effectiveness of our method and inferring that this treatment is useful in numerical experiments on light-cluster interaction.

Key words: Atomic hydrogen cluster, One-dimensional model, Relaxation time, Parallel computing

304. Multi-wavelength generation based on cascaded Raman scattering and self-frequency doubling in KTA

K. Zhong, J. S. Li, D. G. Xu, X Ding, R. Zhou, W. Q. Wen, Z. Y. Li, X. Y. Xu, P. Wang, J. Q. Yao
Laser Physics, 2010, 20(4): 750–755.

Abstract: A multi-wavelength laser is developed based on cascaded stimulated Raman scattering (SRS) and self-frequency-doubling in an x-cut KTA crystal pumped by an A-O Q-switched Nd:YAG laser. The generation of 1 178 nm from cascaded SRS of 234 and 671 cm^{-1} Raman modes is observed. The six wavelengths, including the fundamental 1 064 nm, four Stokes waves at 1 091, 1 120, 1 146, 1 178 nm, and the second harmonic generation (SHG) of 1 146 nm, are tens to hundreds of millwatts for each at 10 kHz, corresponding to a total conversion efficiency of 8.72%.

Key words: Multi-wavelength generation, Cascaded Raman scattering, Self-frequency doubling

305. Compact and tunable mid-infrared source based on a 2 μm dual-wavelength KTiOPO₄ optical parametric oscillator

Geng Youfu, Tan Xiaoling, Li Xuejin, Yao Jianquan
Chinese Physics B, 2010, 19(11): 114209.

Abstract: Using a double resonant KTiOPO₄ (KTP) intracavity optical parametric oscillator operating at degenerated point of 2 μm, we demonstrate a unique mid-infrared source based on difference frequency generation in GaSe crystal. The output tuning range is 8.42–19.52 μm, and a peak power of 834 W for type-I phase matching scheme and 730 W for type-II phase matching scheme are achieved. Experimental results show that this oscillator is a good alternative to the generator of a compact and tabletop mid-infrared radiation with a widely tunable range.

Key words: Mid-infrared radiation, Difference frequency generation, Optical parametric oscillator, Phase matching scheme

306. High-average-power, high-repetition-rate dual signal optical parametric oscillator based on PPMgLN

Feng Ji, Rongsheng Lu, Baosheng Li, Baigang Zhang, Jianquan Yao
Chinese Optics Letters, 2010, 8(5): 505–507.

Abstract: A high-average-power, high-repetition-rate dual signal optical parametric oscillator based on periodically poled MgO-doped lithium niobate (PPMgLN) with a phase-reversed grating is reported. The pump laser is an acousto-optically Q-switched Nd:YVO₄ laser with a maximum average power of 7.6 W. When the repetition rate is 50 kHz and the pulse width of the pump source is 23 ns, the maximum average dual signal output power is about 1.9 W, leading to a conversion efficiency of 25%. Over a 30-min interval, the instability of the signal power measured is less than 0.5%.

Key words: Optical parametric oscillator, High-average-power, PPMgLN

307. 基于常规组合光谱仪下的 LIBS 快速铅检测研究

姜琛昱，张贵忠，傅饶，陆茵菲，汪淼，姚建铨

光谱学与光谱分析，2010, 30(6): 1652-1656.

摘要：激光诱导击穿光谱技术(LIBS)是一种实时在线的无损检测技术，可以满足工农业生产以及生物医学领域检测仪器高速化和自动化的要求。采用波长为 1 064 nm 的调 Q 脉冲为激发光源，常规组合光谱仪中的 CCD 为探测器，以纯铅、玻璃、焊锡等为样品，在建立的 LIBS 实验装置上对样品的铅含量进行定性定量检测。通过对不同铅含量样品的检测，初步确定了在自行搭建的全国产化系统下，铅的检测极限为 0.007 4%，定量检测误差多在 4% 左右，2~3 min 即可完成定性定量分析。实验结果表明，采用光栅光谱仪和非增强型 CCD 完全可以实现 LIBS 的快速铅检测，其检测极限和检测精度基本达到了商用标准，并大大降低了实验装置成本，对 LIBS 在国内同行的相关实用化研究具有一定的借鉴作用。

关键字：激光诱导击穿光谱，铅，快速检测，常规光谱仪，CCD

308. 角度调谐双信号光运转准相位匹配光学参量的产生

纪峰，卢荣胜，李保生，张百钢，姚建铨

中国激光，2010, 37(2): 358-361.

摘要：利用声光调 Q Nd:YVO$_4$ 激光器输出的 1 064 nm 激光抽运周期极化掺氧化镁 LiNbO$_3$(PPMgLN) 晶体，实现了角度调谐的双信号光运转准相位匹配(QPM) 光参量输出。PPMgLN 晶体长 50 mm，具有周期相位反转结构，极化周期 29.6 μm，相位反转周期 6.808 mm，利用外加电场极化法制作，极化沿晶体的 z 向进行。在晶体外部旋转角度为 0°~8° 范围内，实现了双信号光波在 1 474.8~1 479 nm 和 1 489.5~1 495.2 nm 波段的快速调谐输出。对角度调谐准相位匹配进行了相应的理论分析，理论结果与实验值较为一致。

关键字：非线性光学，光学参量产生，准相位匹配，双信号光运转，角度调谐

309. 径向、切向热透镜效应对高亮度激光器输出特性的影响

丁欣，李雪，盛泉，李斌，周睿，温午麒，王鹏，姚建铨，宋峰，禹宣伊，张光寅

中国激光，2010, 37(11): 2780-2783.

摘要：在所有激光二极管 (LD) 侧面抽运的固体激光器的输出功率曲线中都存在一个平缓区域。通过对径向和切向热透镜焦距的理论计算，发现平缓区域出现在径向稳区边缘，在平缓区之后输出功率将继续升高，并在切向稳区边缘达到最大值，此时 M^2 理论值最小。得出平缓区是由径向、切向稳区的共同作用导致的，并且在切向稳区边缘可以实现高亮度激光器的结论。基于以上理论分析，设计了一个具有简单平-平腔结构、LD 侧面抽运的 Nd:YAG 高亮度激光器。实验中采用短腔型，并使其工作在稳功率点，当腔长为 200 mm，入射的抽运光功率为 220 W 时，获得了输出功率为 50 W，M^2 理论值为 2 的高亮度 1 064 nm 连续激光输出。

关键字：激光器，激光二极管侧面抽运，热透镜，径向稳区，切向稳区

310. 连续波可调谐内腔光学参量振荡器及橙红光源

盛泉，丁欣，陈娜，李雪，禹宣伊，温午麒，王鹏，姚建铨

中国激光，2010, 37(11): 2821-2824.

摘要：报道了利用半导体激光器 (LD) 端面抽运的钒酸钇 (Nd:YVO$_4$) 激光器作为抽运源，多周期周期极化铌酸锂 (PPLN) 为非线性晶体的连续波内腔光学参量振荡器 (OPO) 及基于此的连续波可调谐橙红光光源。为实现 OPO 的连续波运转，采用了内腔抽运方式，并对谐振腔进行了合理设计。实验得到调谐范围 1 406~1 513 nm 的信号光及 3.66~4.1 μm 的中红外闲频光连续波输出，在 10.9 W 的 LD 功率下，最大输出功率分别为输出波长 1 500 nm 处的 820 mW 和 3.86 μm 处的 195 mW，相对 LD 功率的转换效率分别为 7.5% 和 1.8%。利用 BaB$_2$O$_4$(BBO) 晶体对 OPO 的 1 064 nm 抽运光和 1.4~1.5 μm 信号光进行内腔和频，获得了调谐范围 606~624 nm 的橙红波段连续波输出，最大输出功率为 624 nm 处的 120 mW，转换效率为 1.1%。

关键字：非线性光学，光学参量振荡器，连续波，可调谐，中红外，橙红光

311. 正向抽运脉冲染料激光放大时间特性研究

谈小虎，张志忠，张云兴，张晓卫，张贵忠，姚建铨

中国激光, 2010, 37(9): 2346–2350.

摘要： 主振荡-功率放大是获得高功率脉冲染料激光的有效模式，染料激光脉冲在放大过程中，与抽运激光脉冲的时序匹配是获得高效转换的基本条件之一。采用速率方程对高重复频率、调 Q 和倍频 Nd:YAG 激光正向抽运脉冲放大过程进行了数值模拟，研究了实际应用中，抽运激光脉宽大于染料激光脉宽条件下的放大过程的时间特性。结果显示，当最佳时序匹配时，抽运激光脉冲峰值相对信号激光延迟具有纳秒量级的滞后，抽运激光脉冲越宽，滞后越多，同时，染料分子的激发-受激辐射-再激发的动态过程造成放大后的染料激光脉冲被展宽，分析显示采用固定抽运激光脉冲峰值位置的方法进行时序控制是可行的。

关键字： 激光器, 时间特性, 速率方程, 染料激光放大

312. Theoretical study of tunable mid-infrared radiation based on sum-frequency generation in nonresonant fresnel phase matching ZnSe and GaP crystal

Zhuo Wang, Yuye Wang, Jianquan Yao

Chinese Journal of Lasers, 2010, 37(5): 1192–1197.

Abstract: Based on the theory of nonresonant Fresnel phase matching(FPM),tuning characteristics of second-harmonic generation(SHG) wave,the full-width at half-maximum(FWHM) acceptance bandwidth for total internal reflection angle,and the effective nonlinear coefficient of sum-frequency generation in ZnSe and GaP crystals are calculated under different polarization configurations.In theoretical calculation,the parameters,which have been reported to generate the tunable infrared waves by SHG in experiments,are adopted.The optimal quasi-phase-match conditions in ZnSe and GaP are summarized by comparing the simulated results,respectively.The calculated results show a sound theoretical basis for using FPM method to generate tunable mid-infrared wave in isotropic crystals.

Key words: Nonlinear optics, Nonresonant Fresnel phase match, Second harmonic generation, ZnSe crystal, Gap crystal

313. LDA 侧面泵浦 1 319 nm / 1 338 nm 双波长激光器的研究

温午麒, 张安, 丁欣, 陆颖, 周睿, 康建翊, 王鹏, 姚建铨

光电子·激光, 2010, 21(6): 799-802.

摘要：使用最大泵浦功率为 600 W 的 LDA 侧面泵浦组件，并采用布儒斯特偏振片以及同时优化全反镜和输出镜透过率两种方法，获得了功率基本相等的 1 319 nm/1 338 nm 双波长激光输出。在连续输出时，获得 30 W 的输出功率，利用声-光 Q 开关调 Q 输出，在重复频率为 4 kHz 时，单脉冲能量为 6 mJ，脉冲宽度为 237 ns，峰值功率为 25 kW。

关键词：全固态激光器，双波长，1 319 nm，1 338 nm，Nd: YAG

314. 高效率紧凑紫外 355 nm 激光器

李斌, 崔海霞, 丁欣, 徐德刚, 姚建铨, 张帆

光电子·激光, 2010, 21(9): 1283-1286.

摘要：报道一种激光二极管(LD)端面抽运的 Nd:YAG 激光晶体腔外三倍频 355 nm 紫外激光器，实验中采用声光调 Q 技术，选用结构紧凑的平平腔结构，在腔外对 1 064 nm 基波采用了 I 类相位匹配 $Li_3B_3O_5$(LBO) 晶体二倍频，II 类相位匹配 LBO 晶体实现了三倍频，获得了较好的光束质量的准连续 355 nm 紫外激光输出，在激光二极管泵浦功率为 28 W 时，声光 Q 开关调制频率为 10 kHz 时，获得了 8.1 W 的红外 1 064 nm 红外激光，紫外单脉冲能量 165 μJ，脉宽 6 ns，峰值功率 27.5 kW，808 nm 到 355 nm 的光-光转换率为 5.89%，整个系统长度控制在 150 mm 以内，该激光器结构紧凑，适合产品化。

关键词：LD 端面抽运，Nd: YAG，声光 Q 开关，LBO，紫外激光

315. 高峰值 266 nm 紫外激光器

李斌,崔海霞,姚建铨,王鹏

强激光与粒子束, 2010, 22(9): 1991–1994.

摘要：报道了一种激光二极管 (LD) 端面泵浦的 Nd:YAG 声光 Q 开关高峰值功率 266 nm 紫外激光器。该激光器采用紧凑的平-平腔结构,LBO 和 BBO 分别作为其二倍频和四倍频晶体。分别利用高偏振比 LD 阵列 (40:1),低偏振比 LD 阵列 (5:1) 及低偏振 LD 阵列腔内放置布氏片结构进行了实验。当注入功率为 25 W,调制频率为 10 kHz 时,以上结构分别得到功率 0.85、0.61 和 0.72 W 的 266 nm 紫外光输出。其中,采用高偏振比 LD 阵列的输出功率最高,单脉冲能量为 85 μJ,脉宽为 5 ns,峰值功率高达 17 kW,泵浦光到紫外光的光-光转换率达 3.4%。

关键词：激光技术,激光二极管,Nd:YAG,紫外激光器,偏振比

316. Tunable and coherent nanosecond 7.2-12.2 μm mid-infrared generation based on difference frequency mixing in ZnGeP$_2$ crystal

Zhong Kai, Li Jiansong, Xu Degang, Wang Jingli, Wang Zhuo, Wang Peng, Yao Jianquan

Optoelectronics Letters, 2010, 6(3): 0179–0182.

Abstract: A coherent mid-infrared laser source, which can be tuned from 7.2 μm to 12.2 μm based on the type-Ⅰ phase-matched difference frequency generation(DFG) in an uncoated ZnGeP$_2$(ZGP) crystal, is reported. The two pump waves are from a type-Ⅱ phase-matched dual-wavelength KTP optical parametric oscillator(OPO) of which the signal and idler waves are tuned during 1.85-1.96 μm (extraordinary wave) and 2.5-2.33 μm (ordinary wave), respectively. The maximum energy of the generated mid-infrared laser is 10 μJ at 9.22 μm, corresponding to the peak power of 2.2 kW.

Key words: Mid-infrared generation, Difference frequency mixing

317. 基于 GaSe 差频产生 8~19 μm 可调中红外辐射

李建松，姚建铨，徐小燕，钟凯，徐德刚，王鹏

光子学报，2010, 39(8): 1491–1495.

摘要：利用 Nd:YAG 激光器抽运内腔 KTP 光学参量振荡器，得到了输出为 2 μm 附近的双波长，并在满足 I 类相位匹配条件下，在 GaSe 中共线差频产生 8 ～ 19 μm 连续宽调谐调的中红外辐射。在 8.76 μm 处脉冲最大能量达 33.66 μJ，峰值功率 7.4 kW，光子转换效率达 4.43% 左右。分析了影响能量转化效率的原因，提出了补偿走离的措施。

关键词：中红外辐射，差频，相位匹配

318. 165 W high stability green laser based on composite ceramic Nd:YAG crystal

Degang Xu, Da Lv, Yuye Wang, Kai Zhong, Yingjin Lv, Jianquan Yao

Proceedings of Society of Photo-Optical Instrumentation Engineers - Photonics Asia, 2010, 7843: 78430.

Abstract: A high stability high average-power green laser was reported with composite ceramic Nd:YAG as gain material and KTP crystal as frequency doubler. Average output power of 165 W is obtained at a repetition rate of 25 kHz with a diodeto- green optical conversion of 14.68% and measured pulse width of 162 ns. For the average output power of about 160 W, the power fluctuation is less than 0.6%. The experimental results show that the green laser system using this novel ceramic Nd:YAG crystal offers better laser performance and output stability.

Key words: High power, High stability, Ceramic green laser, Intracavity frequency-doubled

319. Generation of 1 178 nm based on cascaded stimulated Raman scattering in KTA crystal

Kai Zhong, Jianquan Yao, Degang Xu, Peng Wang

Proceedings of Society of Photo-Optical Instrumentation Engineers - Photonics Asia, 2010, 7846: 784605.

Abstract: We report the generation of 1 178 nm based on cascaded Raman scattering in KTA crystal intracavity pumped by a AO Q-switched Nd:YAG laser. The output power at 1 178 nm is around 80 mW when the diode pump power was 7.6 W at 808 nm. At the same time, the low-order Stokes waves at 1 091 nm, 1 120 nm, 1 146 nm and visible yellow laser at 573 nm (the second harmonic wave of 1 146 nm) are also detected. The total Stokes output power was 240 mW and the yellow laser was 115 mW. The power at 1 178 nm can be increased with output mirrors that are more suitable. The spectra of the generated wavelengths were experimentally analyzed and they accords well with theoretical results.

Key words: KTA, Raman lasers, Stimulated Raman scattering, All-solid-state lasers, Yellow lasers

320. High power widely tunable all-solid-state pulsed titanium-doped sapphire laser

Quan Sheng, Xin Ding, Xue Li, Na Chen, Bin Li, Xuanyi Yu, Wuqi Wen, Rui Zhou, Jianquan Yao

Proceedings of Society of Photo-Optical Instrumentation Engineers - Photonics Asia, 2010, 7843: 78431H.

Abstract: We report a pulsed, widely tunable Ti:sapphire laser pumped by an all-solid-state Q-switched intra-cavity frequency-doubled Nd:YAG laser with repetition rate of 7 KHz. Using two dense flint glass prisms as dispersion elements, the output wavelength could be continuously tuned over 675-970 nm, with spectral line-width of 2 nm. Gain-switching characteristics of Ti: sapphire laser shortened the pulse width to 17.6 ns. Well mode matching between pump and laser beam in the sapphire crystal and thermal design of the cavity ensured stable, efficient laser operation. The maximum output power was 6.2 W at 780 nm when the 532 nm pump power was 22 W; corresponding conversion efficiency was 28.2%.

Key words: Ti:sapphire, Tunable, Q-switched, All-solid-state

321. Generation of tunable coherent nanosecond 8-12 μm mid-infrared pulses based on difference frequency generation in GaSe and ZnGeP₂

Kai Zhong, Jianquan Yao, Jiansong Li, Degang Xu, Peng Wang

Proceedings of Society of Photo-Optical Instrumentation Engineers - Photonics Asia, 2010, 7846: 78460S.

Abstract: Based on the theoretical analysis on the phase-matching relations, effective nonlinear coefficients, walk-off and acceptance angles, the generation of tunable and coherent nanosecond mid-infrared radiation covering the 8-12 μm range is realized by use of difference frequency generation (DFG) in a GaSe and a ZnGeP₂ (ZGP) crystal. Using an 8-mm-long GaSe crystal, we achieve the mid-infrared generation that is continuously tunable from 8.28 μm to 18.365 μm. The maximum pulse energy is 31 μJ at 8.76 μm, corresponding to the conversion efficiency of 0.9% and the maximum midinfrared peak power of about 7 kW. In the case of using an 8-mm-long ZGP crystal, the tuning range is from 7.2 μm to 12.2 μm. The maximum pulse energy is about 10 μJ at 9.22 μm, corresponding to the conversion efficiency of 0.45% and the peak power of 2.2 kW.

Key words: Mid-infrared, 8-12 μm, Difference frequency generation, Nanosecond, GaSe, ZGP

322. A low-threshold efficient KTA OPO by a fiber-coupled diode-end-pumped Nd:YVO₄ laser

Zhong Kai, Yao Jianquan, Xu Degang, Wang Yuye, Wang Peng

Optoelectronics Letters, 2010, 6(6): 0412–0416.

Abstract: A low-threshold efficient high-repetition-rate eye-safe optical parametric oscillator (OPO) is presented. The OPO is based on an x-cut non-critically phase-matched (NCPM) KTA intra-cavity pumped by an acousto-optically (A-O) Q-switched Nd:YVO₄ laser. At 10 kHz, the lowest threshold of 0.75 W and the signal power of 0.6 W are got, corresponding to thesingle pulse energy of 60 μJ and the peak power of 20 kW. Tuning the frequency, the maximum output power at 1 536 nmis 1.03 W at 30 kHz with an optical-to-optical conversion efficiency of 12.26%. The fluctuation of the output power is 2.1% during 2 h operation.

Key words: OPO, Nd:YVO₄ laser, Low-threshold

323. Efficient diode-end-pumped dual-wavelength Nd, Gd:YSGG laser

Kai Zhong, Jianquan Yao, Chongling Sun, Chengguo Zhang, Yueyang Miao, Ran Wang, Degang Xu, Fan Zhang, Qingli Zhang, Dunlu Sun, Shaotang Yin

Optics Letters, 2011, 36(19): 3813–3815.

Abstract: We demonstrate a dual-wavelength laser based on a new laser material—Nd, Gd:YSGG, or Nd:GYSGG for short—for the first time to our knowledge. Besides its attractive properties such as antiradiation, high segregation coefficient, etc., this kind of laser crystal also shows excellent laser performance. For continuous-wave operation, the maximum output power is 10.1 W with the absorbed power of 18.45 W at 808 nm, corresponding to the slope efficiency of nearly 60%. The maximum single pulse energy and peak power reach 277 μJ and 4.6 kW (60 ns) when the absorbed pump power is 11.4 W for acousto-optic Q-switched operation.

Key words: dual-wavelength laser, Nd,Gd:YSGG, diode-end-pumped

324. High efficiency 1 342 nm Nd:YVO$_4$ laser in-band pumped at 914 nm

Xin Ding, Sujia Yin, Chunpeng Shi, Xue Li, Bin Li, Quan Sheng, Xuanyi Yu, Wuqi Wen, Jianquan Yao

Optics Express, 2011, 19(15):14315–14320.

Abstract: A high-efficiency 1 342 nm Nd:YVO$_4$ laser in-band pumped at 914 nm is demonstrated for the first time. Using an all-solid-state Nd:YVO$_4$ laser operating at 914 nm as pump source, 0.86 W output was obtained with 1.82 W absorbed pump power. Corresponding slope efficiency of 65.4% was the highest of Nd:YVO$_4$ lasers operating at 1 342 nm to the best of our knowledge. Effects of crystal's doping concentration and temperature on laser power and conversion efficiency were also investigated.

Key words: Lasers, Neodymium, Solid-state, Pumping

325. Effects of crystal parameters on power and efficiency of Nd:YVO$_4$ laser in-band pumped by 914 nm laser

X. Ding, C. P. Shi, S. J. Yin, X. Li, B. Li, Q. Sheng, X. Y. Yu, W. Q. Wen, J. Q. Yao

Journal of Physics D: Applied Physics, 2011, 44: 395101–395105.

Abstract: To overcome the main drawback of poor pump absorption of Nd:YVO$_4$ lasers in-band pumped at 914 nm, three possible methods of higher temperature, longer crystal length and higher doping concentration were experimentally studied. A Nd:YVO$_4$ laser operating at 914 nm was used as high beam quality, narrow line-width pump source to make the measurement of pump absorption accurate. Four different Nd:YVO$_4$ crystals were used as gain mediums to investigate the effects of crystal parameters. We found that heating the crystal effects differently on output power and overall efficiency when crystal doping concentration changed. To improve the pump absorption and meanwhile achieve high efficiency, longer crystal with proper doping concentration and pump feedback should be adopted instead of highly doped crystal.

Key words: Nd:YVO$_4$ laser, Crystal parameters

326. A continuous-wave tunable orange-red source based on sum-frequency generation in an intra-cavity periodically poled LiNbO$_3$ singly resonant optical parametric oscillator cavity

Quan Sheng, Xin Ding, Sujia Yin, Chunpeng Shi, Xue Li, Bin Li, Xuanyi Yu, Wuqi Wen, Jianquan Yao

Journal of Optics, 2011, 13: 095201.

Abstract: This work demonstrates a tunable continuous-wave (CW) all-solid-state orange-red source based on sum-frequency generation (SFG), using a tunable 1.4–1.5 μm signal and a 1 064 nm pump of an intra-cavity periodically poled LiNbO$_3$ optical parametric oscillator in a BBO (β-BaB$_2$O$_4$) crystal. Elaborate intra-cavity design ensured efficient CW singly resonant optical parametric oscillator (SRO) operation. Intra-cavity SFG made full use of the high circulating pump and signal power of the SRO under an external laser diode (LD) pump power that was ten times higher than the SRO threshold. The output wavelength tuning range was 605–623 nm. The maximum output power was 370 mW at 622.8 nm with 17.1 W LD pump power, corresponding to an overall conversion efficiency of 2.2%.

Key words: Tunable orange-red, Continuous-wave, Intra-cavity optical parametric oscillator, Intra-cavity sum-frequency generation

327. Comparison of eye-safe KTA OPOs pumped by Nd:YVO₄ and Nd:YLF lasers

K. Zhong, J. Q. Yao, Y. Y. Wang, D. G. Xu, P. Wang

Optics & Laser Techonology, 2011, 43: 636–641.

Abstract: An eye-safe KTA OPO pumped by a Nd:YLF laser is demonstrated and a comparison with that pumped by a Nd:YVO₄ laser is performed. Although the slope efficiency of the continuous-wave free-running Nd:YLF laser is lower than that of the Nd:YVO₄ laser, the performance of KTA OPOs pumped by the Q-switched Nd:YLF laser is better, especially at lower repetition rates. The slope efficiency of KTA OPO pumped by a Nd:YLF laser is 14.6% at 30 kHz and 11.04% at 10 kHz. The better energy storage ability of Nd:YLF makes it an excellent laser medium in IOPOs.

Key words: Eye-safe, Optical parametric oscillator, KTA

328. High power widely tunable narrow linewidth all-solid-state pulsed titanium-doped sapphire laser

Ding Xin, Li Xue, Sheng Quan, Shi Chunpeng, Yin Sujia, Li Bin, Yu Xuanyi, Wen Wuqi, Yao Jianquan

Chinese Physics Letters, 2011, 28(9): 094205.

Abstract: We report a widely tunable, narrow linewidth, pulsed Ti:sapphire laser pumped by an all-solid-state Q-switched intra-cavity frequency-doubled Nd:YAG laser. By using four dense flint glass prisms as intra-cavity dispersive elements, the output wavelength can be continuously tuned over 675–970 nm and the spectral linewidth is shortened to 0.5 nm. The maximum output power of 6.65 W at 780 nm is obtained under 23.4 W pump power with repetition rate of 5.5 kHz; corresponding to an conversion efciency of 28.4%. Due to the gain-switching characteristics of the Ti:sapphire laser, the output pulse duration is as short as 17.6 ns.

Key words: Ti:sapphire laser, Narrow Linewidth, All-Solid-State

329. 飞秒抽运随机激光输出波形的可控性研究

吕健滔，王可嘉，刘劲松，姚建铨，朱启华，张清泉
物理学报，2011, 60(7): 074203.

摘要：本文基于随机激光的时域理论,研究了飞秒脉冲抽运下二维随机激光的辐射特性,并着重讨论了抽运脉冲的峰值强度、脉宽和脉冲波形对辐射光时域波形的影响。结果表明,辐射光的时域波形强烈依赖于抽运光脉冲的参数,通过调整抽运方式可以控制辐射光的输出波形。数值模拟结果为研究随机激光输出波形的可控性技术提供了理论依据。

关键词：激光物理,随机激光器,飞秒抽运,脉冲波形

330. 激光二极管抽运共轴双晶体黄光激光器

李斌，姚建铨，丁欣，王鹏，张帆
物理学报，2011, 60(2): 024208.

摘要：提出一种全新的单抽运源、共轴双晶体实现黄光激光的方法,并对其进行了理论分析和实验研究。利用该方法 在抽运功率为 1.5 W 时,获得了 54 mW 的黄光激光输出,激光不稳定度为 5%,光-光转换率为 3.6%,而利用对 1 064 nm 损耗的方法在相同条件下只获得 15 mW 的黄光输出。该方法结构简单、灵活多样,可以应用到很多弱增益谱线与强增益谱线和频的结构当中,实现诸如 491、488、593.5、555 和 500.8 nm 的激光输出。

关键词：黄光激光器,单抽运源,双晶体,和频

331. 铅的单脉冲 LIBS 定量光谱检测与比较

陆茵菲,金鑫,张贵忠,姚建铨

光谱学与光谱分析, 2011, 31(5): 1332–1335.

摘要： 单脉冲激光诱导击穿光谱技术 (single-shot, LIBS) 是一种在国际上被广泛使用的物质及元素检测技术,具有快速、准确、无需样品制备等诸多优点。为了满足单脉冲 LIBS 的实验要求,实验用激光器的选取是非常重要的,它直接关系到诱导激光的强度以及脉冲宽度的大小,而这些又对实验结果产生深刻的影响。因此,选取正确的激光器,是单脉冲 LIBS 成功的关键。正是基于这个目的,比较了不同波长、不同激光能量的激光脉冲作用下,纯铅发射光谱的定量变化,对于单脉冲 LIBS 中激光器的选取有着很好的借鉴意义。

关键词： 单脉冲激光诱导击穿光谱,铅,定量检测

332. 常温条件下 KTP 晶体应用于 1 319 nm 激光三倍频相位匹配角的测量

温午麒,康建翊,瞿宁,丁欣,陆颖,周睿,盛泉,姚建铨

中国激光, 2011, 38(8): 0802007.

摘要： 针对常温工作条件下利用磷酸氧钛钾（KTP）晶体对钇铝石榴石（Nd:YAG）晶体 1 319 nm 激光三倍频产生 440 nm 蓝色激光的实验,对三倍频 KTP 晶体的相位匹配角进行了理论计算和实验研究。通过多组色散方程得到 KTP 晶体的相位匹配角,并计算出相应的有效非线性系数。选取一组结果（$\theta = 84.6°$,$\varphi = 0°$）对 KTP 晶体切割,利用一台 1 319 nm 激光器,将晶体放入腔中,采用旋转晶体偏角和调节温度的方法寻找出三倍频 KTP 晶体最佳匹配角度（$\theta = 85.04°$,$\varphi = 0°$）。该晶体经过重新切割,440 nm 蓝色激光输出的光束强度有了明显的提高,最佳工作温度为 18℃。

关键词： 非线性光学,三倍频相位匹配角,KTP 晶体,440 nm 蓝色激光

333. 高效腔外频率变换紫外激光器

李斌, 姚建铨, 丁欣, 张帆, 王鹏

强激光与粒子束, 2011, 23(2): 290-292.

摘要：为了得到高效的腔外频率变换 355 nm 紫外激光输出，提出了一种利用 3 块 LBO 作为非线性频率变换晶体的新方案。采用 LD 端面泵浦 Nd：YAG 声光 Q 开关激光器作为基波源，当入射泵浦功率为 25 W、调制频率 12 kHz 时，获得了 6.20 W 的 1 064 nm 激光输出，经过非线性频率变换后，获得了 2.7 W 的紫外 355 nm 激光输出，光-光转换率 43.4%。

关键词：激光技术, 紫外激光, LBO 晶体, 腔外频率变换

334. 激光二极管侧面泵浦高功率 266 nm 紫外激光

李斌, 姚建铨, 丁欣, 张帆, 王鹏

强激光与粒子束, 2011, 23(8): 2065-2068.

摘要：报道了一种半导体激光列阵侧面泵浦 Nd：YAG 四倍频 266 nm 全固态紫外激光器，采用 Z 型腔结构，I 类临界相位匹配 LBO 和 BBO 晶体分别作为二倍频晶体和四倍频晶体。在调制频率为 5 kHz 时，最终获得了 2.1 W 的 266 nm 紫外激光输出，单脉冲能量 420 μJ，绿光到紫外激光的转换率为 13.13%，在相同的泵浦功率下利用 V 型腔结构仅获得 305 mW 的 266 nm 紫外激光输出。

关键词：激光二极管, 侧面泵浦, 四倍频, Z 型腔, I 类临界相位匹配, BBO 晶体, 紫外激光

335. 强光电离团簇并行数值模拟

傅饶, 汪淼, 陆茵菲, 张贵忠, 向望华, 徐德刚, 姚建铨
强激光与粒子束, 2011, 23(2): 527-530.

摘要：为研究强激光电离氢原子团簇，在理论上采用1维氢原子团簇的经典粒子动力学模型，结合粒子对(PP)算法及粒子模拟(PIC)方法，采用自行搭建的9节点并行集群系统，利用消息传递接口(MPI)与OpenMP混合编程模型进行了并行数值模拟计算，获得了较为理想的计算加速比。并且引入了弛豫时间参数，有效地处理了粒子间的碰撞过程，在极大简化计算量的同时，保留了物理本质。所得模拟结果与已有的实验结果符合较好，表明该并行计算模型是稳定、可行的。

关键词：激光与团簇互作用, 强光电离, 并行计算

336. 基于序列脉冲激光法高速空气流场速度的研究

崔海霞, 姚建铨, 郑义, 王杰, 肖绪辉, 万春明
光电子·激光, 2011, 22(1): 103-109.

摘要：针对高速空气流场中气体瞬时速度的非接触测量，提出采用序列脉冲作为光源的拉曼激发-激光诱导电子荧光(YRELIEF)法，通过对其测量原理的分析、激光器件及测量系统的选择、主要部件的设计及最终的实验研究，得到系统的最佳参数及风洞流场的标记图像。研究结果表明，本文提出的方法是风洞高速不稳定流场速度非接触测量的最佳方案之一，在未来流场检测技术中具有应用前景。

关键词：拉曼激发-激光诱导电子荧光(RELIEF)法, 序列脉冲激光, 高速空气流场, ICCD

337. 利用通用多媒体技术的激光光斑测量方法的研究

康建翊,温午麒,梅达华,李洋,丁欣,陆颖,周建勇,姚建铨

激光杂志, 2011, 32(4): 11–13.

摘要: 本文对利用当前通用的计算机多媒体技术测量激光光斑进行了理论分析、软件设计和实验研究。使用流行的多媒体图像输入硬件——USB 接口数码摄像头进行图像采集,由于摄像头直接输出数字化的图像及视频流,从而省掉专用的数据采集卡。利用 VC 编程,调用摄像头进行图像采集,将获取的 JPG 图像和 AVI 视频流解码,并在输出设备上绘制三维强度曲面图。软件设计中调用了多媒体领域的一些公开的函数库或软件如 Direct Show 和 Open GL,从而使软件实现高的运行速度和操作系统对程序的支持。实验证明该方案能实现激光光斑测量的功能,并在 640 × 480 分辨率条件下获得每秒约 3 次的测量速度。

关键词: 激光光斑,计算机多媒体技术,数码摄像头,Open GL,Direct Show

338. Continuous-wave mid-infrared intra-cavity singly resonant PPLN-OPO under 880 nm in-band pumping

Quan Sheng, Xin Ding, Chunpeng Shi, Sujia Yin, Bin Li, Ce Shang, Xuanyi Yu, Wuqi Wen, Jianquan Yao

Optics Express, 2012, 20(7): 8041–8046.

Abstract: We report herein a continuous-wave mid-infrared intra-cavity singly resonant optical parametric oscillator (ICSRO) which is the first example of ICSRO that utilize in-band pumped Nd-doped vanadate laser as pump source. A 1 064 nm Nd:YVO$_4$ laser in-band pumped by 880 nm LD and a periodically poled lithium niobate (PPLN) crystal are employed as the parent pump laser and the nonlinear medium, respectively. The idler output wavelength tuning range is 3.66-4.22 μm. A maximum output power of 1.54 W at 3.66 μm is obtained at absorbed pump power of 21.9 W, with corresponding optical efficiency being 7.0%. The control experiment of ICSRO under 808 nm traditional pumping is also carried out. The results show that in-band pumped ICSRO has better performance in terms of threshold, power scaling, efficiency and power stability than ICSRO traditionally pumped at 808 nm.

Key words: PPLN-OPO, mid-infrared intra-cavity, CW

339. Continuous-wave intra-cavity singly resonant optical parametric oscillator with resonant wave output coupling

Quan Sheng, Xin Ding, Ce Shang, Bin Li, Chen Fan, Haiyong Zhang, Xuanyi Yu, Wuqi Wen, Yila Ma, Jianquan Yao

Optics Express, 2011, 20(25): 27953–27958.

Abstract: We report herein the enhancement in both power and efficiency performance of a continuous-wave intra-cavity singly resonant optical parametric oscillator (ICSRO) by introducing finite resonant wave output coupling. While coupling out the resonant wave to useful output, the output coupling increases the SRO threshold properly thus suppresses the back conversion under high pump power. Therefore, the down-conversion efficiency is maintained under high pump without having to raise the threshold by defocusing. With a T = 9.6% signal wave output coupler used, the SRO threshold is 2.46 W and the down-conversion efficiency is 72.9% under the maximum pump power of 21.4 W. 1.43 W idler power at 3.66 μm and 5.03 W signal power at 1.5 μm are obtained, corresponding to a total extraction efficiency of 30.2%. The resonant wave out coupling significantly levels up the upper limit for the power range where the ICSRO exhibits high efficiency, without impeding its advantage of low threshold.

Key words: Parametric oscillators and amplifiers, Infrared and far-infrared lasers, Nonlinear optics, Devices

340. Continuous-wave Nd:GYSGG laser around 1.3 μm

K. Zhong, C.L. Sun, J.Q. Yao, D.G. Xu, Y.Q. Pei, Q.L. Zhang, J.Q. Luo, D.L. Sun, S.T. Yin

Laser Physics Letters, 2011, 9(7): 491−495.

Abstract: The operating characteristics of a Nd:GYSGG laser in the 1.3 μm wavelength band based on the transition $^4F_{3/2}$-$^4I_{13/2}$ is demonstrated. Using two different output couplers, single wavelengths operation at 1 336 nm and dual-wavelength operation at 1 321/1 336 nm are obtained. The 1 329 nm laser spectrum is also found when the pump level is relatively weak. The stimulated emission cross section of the 1 336 nm laser line is estimated to be 1.12 time of that of the 1 321 nm laser line. The output power of 1.98 W at 1 336 nm is achieved with the pump power of 13.5 W, corresponding to the conversion efficiency of 14.7%. This is the first report on a 1.3-μm Nd:GYSGG laser to our knowledge.
Key words: Nd:GYSGG, Diode-end-pumped, Continuous wave, 1.3 μm

341. Stimulated emission cross section of the $^4F_{3/2} \rightarrow ^4I_{11/2}$ transition of Nd:GYSGG

C.L. Sun, K. Zhong, C.G. Zhang, J.Q. Yao, D.G. Xu, F. Zhang, Y.Q. Pei, Q.L. Zhang, J.Q. Luo, D.L. Sun, S.T. Yin

Laser Physics Letters, 2011, 9(6): 410−414.

Abstract: The stimulated emission cross section of a 1.1-at.% doped Nd:$Gd_{3x}Y_{3(1-x)}Sc_2Ga_{3(1+\delta)}O_{12}$ ($x = 0 - 1$, $\delta = -0.2 - 0.2$) (Nd:GYSGG) crystal at 1.06 μm ($^4F_{3/2}$-$^4I_{11/2}$ transition) is measured at room temperature, using both the laser efficiency comparison method with an Nd:$Y_3Al_5O_{12}$ (Nd:YAG) laser and the threshold formula method, with their results of 1.59×10^{-19} and 1.50×10^{-19} cm^2, respectively. The measured results accord well with each other and they are of great use in designing a Nd:GYSGG laser system.
Key words: Nd:GYSGG, Stimulated emission cross section, Laser efficiency comparison method, Threshold formula method

342. High-power high-stability Q-switched green laser with intracavity frequency doubling of diode-pumped composite ceramic Nd:YAG laser

Wang Yuye, Xu Degang, Liu Changming, Wang Weipeng, Yao Jianquan
Chinese Physics B, 2011, 21(9): 094212.

Abstract: We successfully obtain high-average-power high-stability Q-switched green laser based on diode-side-pumped composite ceramic Nd:YAG in a straight plano-concave cavity. The temperature distribution in composite ceramic Nd:YAG crystal is numerically analyzed and compared with that of conventional Nd:YAG crystal. By use of a composite ceramic Nd:YAG rod and a type-II high gray track resistance KTP (HGTR-KTP) crystal, a green laser with an average output power of 165 W is obtained at a repetition rate of 25 kHz, with a diode-to-green optical conversion of 14.68%, and a pulse width of 162 ns. To the best of our knowledge, both the output power and optical-to-optical efficiency are the highest values for green laser systems with intracavity frequency doubling of this novel composite ceramic Nd:YAG laser to date. The power fluctuation at around 160 W is lower than 0.3% in 2.5 hours.

Key words: High-average-power green laser, High stability, Composite ceramic Nd:YAG rod

343. Efficient continuous-wave eye-safe region signal output from intra-cavity singly resonant optical parametric oscillator

Li Bin, Ding Xin, Sheng Quan, Yin Sujia, Shi Chunpeng, Li Xue, Yu Xuanyi, Wen Wuqi, Yao Jianquan
Chinese Physics B, 2011, 21(1): 014207.

Abstract: We report an efficient continuous-wave (CW) tunable intra-cavity singly resonant optical parametric oscillator based on the multi-period periodically poled lithium niobate and using a laser diode (LD) end-pumped CW 1 064 nm Nd:YVO$_4$ laser as the pump source. A highly efficiency CW operation is realized through a careful cavity design for mode matching and thermal stability. The signal tuning range is 1 401–1 500 nm obtained by varying the domain period. The maximum output power of 2.2 W at 1 500 nm is obtained with a 17.1 W 808 nm LD power and the corresponding conversion efficiency is 12.9%.

Key words: Continuous-wave, Intracavity optical parametric oscillator, Periodically poled lithium niobate, Tunable eye-safe laser

344. 双电子 CO_2 模型及其高次谐波模拟

付宁,于川,代春燕,张贵忠,姚建铨

中国激光, 2011, 39(10): 1017001.

摘要：根据三原子 CO_2 分子的电子结构,建立了双电子一维模型,采用对称分裂算符法求解含时薛定谔方程,计算了 CO_2 分子模型体系在 800 nm 波长脉冲激光激发下的高次谐波(HHG)谱。结果发现,该模型计算量适中,可很好地描述 CO_2 分子的 HHG 谱性质,用于激发 HHG 的激光脉冲宽度对 HHG 谱性质影响很大,多正电中心导致 HHG 谱呈现干涉相消现象。

关键词：光电子学,高次谐波,含时薛定谔方程,干涉相消

345. H_2^+ 和 H_3^+ 系统的电子频谱与高次谐波谱

代春燕,于川,张贵忠,向望华,姚建铨

强激光与粒子束, 2011, 24(6): 1311-1314.

摘要：针对单电子 H_2^+ 和双电子 H_3^+ 系统,采用对称分裂算符法求解波函数的时间演化。计算了电子的含时频谱和体系的高次谐波谱,结果表明:电子频谱和高次谐波谱呈互补规律,即电子频谱峰值出现在基波的偶数倍频率位置,高次谐波则出现在奇数位置。对高次谐波谱随软化库仑参数的变化规律进行了数值研究,发现只在软化库仑参数的一定范围内才产生高次谐波,对该现象作了解释。

关键词：高次谐波,H_3^+ 系统,库仑势参数,电子频谱

346. Dynamic Stark effect on XUV-laser-generated photoelectron spectra: Numerical experiment on atomic hydrogen

Chuan Yu, Ning Fu, Guizhong Zhang, Jianquan Yao
Physical Review A, 2013, 87(4): 043405.

Abstract: We present the results of our numerical simulation on the photoelectron spectra of hydrogen atoms ionized by the single-photon process of intense XUV laser pulses. The dynamic interference structure in the photoelectron spectra is investigated with respect to the dynamic Stark effect, XUV laser parameters, and properties of the ionization continuum states. The research outcome is that plane waves can be readily deployed to study photoelectron spectroscopy, which is crucial to the wave-function ansatz for generic systems of atoms and molecules. The dynamic Stark effect is prominently demonstrated above certain carrier frequencies of the XUV laser pulse, which mandates a photon energy threshold for the dynamic Stark effect to be discerned in the photoelectron spectra. A one-dimensional model for the hydrogen atom is feasible, which is of fundamental significance in modeling atoms and molecules with reduced dimensionality models. We also point out that detailed characterization of the interference fringes in the photoelectron spectra will facilitate more quantitative investigations of the Stark effect contained in the photoelectron spectra created by single XUV photons.

Key words: Hotoelectron spectra, Stark effect, Atomic hydrogen

347. Dynamic Stark effect and interference photoelectron spectra of H_2^+

Chuan Yu, Ning Fu, Tian Hu, Guizhong Zhang, Jianquan Yao

Physical Review A, 2013, 88: 043408.

Abstract: For atomic hydrogen, an interference structure has been theoretically found to appear in the photoelectron spectra induced by strong extreme ultraviolet (XUV) laser pulses [Demekhin and Cederbaum, Phys. Rev. Lett. 108, 253001 (2012)]. For molecular hydrogen ion H_2^+, our present work provides the results of the numerical simulation on its photoelectron spectra in an XUV photoionization process. The interference feature in the photoelectron spectra is investigated with respect to the dynamic Stark effect and the XUV laser parameters. The research outcome is that our numerical findings corroborate the prediction for the emergence of the modulation structures in the photoelectron spectra, similar to those already found for a hydrogen atom. For an XUV laser pulse with photon energy well above the ionization threshold of H_2^+, the dynamic Stark effect is prominently demonstrated for certain values of laser parameters such as intensity and pulse duration. The deployed onedimensional model and the plane-wave description for the continuum states for the molecular hydrogen ion are of crucial significance in modeling the molecular system. The numerical computation is made feasible by using a hybrid algorithm. We advocate an ensemble physical picture to decipher the physical mechanism for the formation of the interference photoelectron spectra by taking into account the nuclear degree of freedom. It is anticipated that more quantitative investigations on the Stark effect concealed in the photoelectron spectra created by single XUV photons are needed plus experimental verifications.

Key words: Dynamic Stark effect, Interference photoelectron

348. Continuous-wave Nd:GYSGG laser properties in 1.3 and 1.4 μm regions based on $^4F_{3/2}$ to $^4I_{13/2}$ transition

K Zhong, W Z Xu, C. L. Sun, J Q Yao, D G Xu, X L Cao, Q L Zhang, J Q Luo, D L Sun, S T Yin

Journal of Physics D: Applied Physics, 2013, 46:315106.

Abstract: Continuous-wave laser properties of Nd : GYSGG crystal based on the transition from $^4F_{3/2}$ to $^4I_{13/2}$ are investigated. With different output couplers, single or multiple wavelength lasers operating at 1 321, 1 336, 1 404 and 1 424 nm are obtained, which provide new options in these wavelength bands. In the 1.3 μm region, the maximum output power is 2.5 W, corresponding to the conversion efficiency of 18.5%. In the 1.4 μm region, the maximum output power at 1 424 nm is 707 mW, corresponding to the conversion efficiency of 6.8%. The output power and conversion efficiency can be improved with better anti-reflection coatings for the laser crystal.

Key words: Nd:GYSGG laser, Continuous-wave

349. Time-dependent Born–Oppenheimer approximation approach for Schrödinger equation: Application to H_2^+

Chuan Yu, Ning Fu, Chunyan Dai, Hua Wang, Guizhong Zhang, Jianquan Yao

Optics Communications, 2013, 300:199–203.

Abstract: We deploy a time-dependent Born–Oppenheimer approximation approach for numerically solving the time-dependent Schrödinger equation (TDSE) by reducing the wavefunction dimensionality to nuclear and electronic degrees of freedom, and apply it to a one-dimensional model of H_2^+. Based upon our three distinct error evaluation schemes, we quantitatively compare the wavefunctions and HHG spectra which are computed by the present approximation method, with those obtained by fully solving the TDSE. The similarities of both the wavefunctions and the HHG spectra justify that our approach is feasibly precise for medium laser intensity. It is also anticipated that this approximation can be adopted for other polyatomic molecules with a dimensionality reduction and computational simplification in calculating the time-dependent wavefunctions.

Key words: Born–Oppenheimer approximation, Time-dependent Schrödinger equation, 1D H_2^+ model, HHG

350. Diode-pumped continuous-wave quasi-three-level Nd:GYSGG laser at 937 nm

Chongling Sun, Kai Zhong, Jianquan Yao, Degang Xu, Xiaolong Cao, Qingli Zhang, Jianqiao Luo, Dunlu Sun, Shaotang Yin
Optics Communications, 2013, 294: 229−232.

Abstract: A continuous-wave (cw) 937 nm Nd:GYSGG laser based on the quasi-three-level transition $^4F_{3/2}$ to $^4I_{9/2}$ is reported. The output power of 0.80 W at 937 nm is achieved with the pump power of 8.2 W, corresponding to the conversion efficiency of 9.8%. The variation of output power and pump threshold with increased working temperature is also studied, which demonstrates the deterioration operation because of the serious reabsorption in a quasi-three-level Nd^{3+} doped laser.
Key words: Nd:GYSGG, Diode-pumped, Quasi-three-level, Reabsorption

351. The double-ended 750 nm and 532 nm laser output from PPLN-FWM

Wang Tao, Li Yuxiang, Yao Jianquan, Guo Ling, Wang Zhuo, Han Shasha, Zhang Cuiying, Zhong Kai
Chinese Physics Letters, 2013, 30(6): 064203.

Abstract: We investigate 750 nm and 532 nm dual-wavelength laser for applications in the internet of things. A kind of optical maser is developed, in which the semiconductor module outputs the 808 nm pump light and then it goes into a double-clad Nd^{3+}:YAG monocrystal optical fber through the intermediate coupler and forms a 1 064 nm laser. The laser outputs come from both left and right terminals. In the right branch, the laser goes into the right cycle polarization Li_nNbO_3 (PPLN) crystal through the right coupler, produces the optical parametric oscillation and forms the signal light λ_1 (1 500 nm), the idle frequency light λ_2 (3 660.55 nm), and the secondharmonic of the signal light λ_3 (750 nm). These three kinds of light and the pump light λ_4 together form the frequency matching and the quasi-phase matching, then the four-wave mixing occurs to create the high-gain light at wavelength 750 nm. Meanwhile, in the left branch, the laser goes into the left PPLN crystal through the left coupler, engenders frequency doubling and forms the light at wavelength 532 nm. That is to say, the optical maser provides 750 nm and 532 nm dual-wavelength laser outputting from two terminals, which is workable.
Key words: FWM, Dual-wavelength laser

352. 双电子 H_3^+ 体系的高次谐波谱的数值模拟

代春燕,于川,张贵忠,姚建铨

光谱学与光谱分析, 2013, 33(2): 297–299.

摘要：提出了双电子 H_3^+ 体系的一维计算模型。采用对称分裂算符法计算波函数的时间演化,数值模拟了在 800 nm 波长激光脉冲作用下体系的高次谐波(HHG)谱性质。通过哈密顿算符中的电子相互作用项研究了电子间互作用对 HHG 谱的作用,结果发现电子间的相互作用对 HHG 谱性质的影响不大,但是会对截止频率有一定影响;还模拟了体系在不同核间距下的 HHG 谱的截止频率,发现较大的核间距的截止频率大于半经典三步模型给出的截止频率,并对其进行了初步分析。

关键词：高次谐波,H_3^+ 系统,三步模型,波包能量

353. 914 nm 基态高斯塔克能级共振抽运的高效率 Nd:YVO$_4$ 激光器

丁欣,张海永,盛泉,李斌,尚策,范琛,禹宣伊,温午麒,姚建铨

中国激光, 2013, 40(7): 0702008.

摘要：利用全固态 Nd:YVO$_4$ 激光器的 914 nm 输出作为抽运源,对 Nd:YVO$_4$ 晶体进行基态高斯塔克能级共振抽运,获得了高效率的 1 064 nm 激光输出。为解决基态高斯塔克能级共振抽运吸收差、光-光转换效率低的问题,详细分析了掺杂浓度、温度以及长度等晶体参数对抽运光吸收和激光转换效率的影响。在此基础上,使用长度为 20 mm 的 Nd:YVO$_4$ 晶体获得了相对入射抽运光 56.9% 的高光-光转换效率。914 nm 基态高斯塔克能级共振抽运 Nd:YVO$_4$ 激光器的光-光转换效率达到了可与传统抽运相比拟的实用化水平。

关键词：激光器,共振抽运,Nd:YVO$_4$ 激光器,基态高斯塔克能级

354. 880 nm 共振抽运连续波内腔单谐振光学参量振荡器及其逆转换

丁欣，尚策，盛泉，李斌，范琛，张海永，禹宣伊，温午麒，姚建铨
中国激光，2013, 40(6): 0602008.

摘要： 报道了 880 nm 激光二极管 (LD) 共振抽运的连续波 (CW) Nd:YVO$_4$-PPLN 内腔单谐振光学参量振荡器 (ICSRO)。在 21.9 W 抽运功率下，获得了 1.54 W 的 3.66 μm CW 中红外闲频光输出，光-光转换效率为 7.0%；与 808 nm 传统抽运相比，共振抽运 ICSRO 在振荡阈值、输出功率、转换效率和功率稳定性等方面都显示出明显优势。针对高抽运功率下逆转换过程影响单谐振光学参量振荡器 (SRO) 转换效率的问题，研究了振荡信号光的耦合输出透射率对 SRO 阈值和下转换效率的影响。通过提高振荡光输出镜透射率优化 SRO 阈值，可在高抽运功率下保持下转换效率的同时获得高效的信号光输出；21.4 W 抽运功率下同时获得 1.54 W 闲频光和 5.03 W 信号光输出，总提取效率为 30.2%。

关键词： 非线性光学，内腔光学参量振荡器，共振抽运，连续波，中红外，逆转换

355. 提高全固态内腔倍频绿光激光器功率稳定性的一种新方法

姚建铨，张卓，徐德刚，张昊，李佳起，王与烨，钟凯
光电子·激光，2013, 24(4): 629-634.

摘要： 报道了一种基于温控倍频 KTP 晶体的低噪声全固态内腔倍频绿光激光器。实验中，将 KTP 晶体作为 1/4 波片，当注入电功率为 300 W 时，控制 KTP 至 19.8℃获得了 2.5 W 连续绿光，其功率不稳定度小于 0.9% (10 min)，相比最差温度点稳定性提高约 70%。理论上，从基频纵模偏振态出发，利用 Jones 矩阵分析了倍频 KTP 晶体相位延迟对谐振腔纵模振荡的影响，通过调节 KTP 温度改变相位延迟，有效遏制了基波偏振模耦合从而得到稳定的绿光输出。理论分析与实验结果相吻合，这为低噪声全固态内腔倍频绿光激光器提供了一种新的技术途径。

关键词： 绿光问题，Jones 矩阵，偏振态，低噪声，温度控制

356. High-power picosecond 355 nm laser based on $La_2CaB_{10}O_{19}$ crystal

Kai Li, Ling Zhang, Degang Xu, Guochun Zhang, Haijuan Yu, Yuye Wang, Faxian Shan, Lirong Wang, Chao Yan, Yicheng Wu, Xuechun Lin, Jianquan Yao

Optics Letters, 2014, 39(11): 3305–3307.

Abstract: Third harmonic generation experiments were performed on a type-I phase-matching $La_2CaB_{10}O_{19}$ crystal cut at $\theta=49.4°$ and $\varphi=0.0°$ with dimensions of 4.0 mm × 4.0 mm × 17.6 mm. A 1 064 nm laser with a maximum average power of 35.2 W was employed as the fundamental light source, which has a pulse width of 10 picoseconds and a pulse repetition rate of 80 MHz. A type-I noncritical phase-matching LBO crystal was used to generate 532 nm lasers. By investigating a series of focusing lens combinations, a picosecond 355 nm laser of 5.3 W was obtained, which is the highest power of picosecond 355 nm laser based on a $La_2CaB_{10}O_{19}$ crystal so far. The total conversion efficiency from 1 064 nm to 355 nm was up to 15.1%.

Key words: Nonlinear optical materials, Lasers, Ultraviolet, Harmonic generation and mixing

357. 5.2-W high-repetition-rate eye-safe laser at 1 525 nm generated by Nd:YVO$_4$–YVO$_4$ stimulated Raman conversion

Xin Ding, Chen Fan, Quan Sheng, Bin Li, Xuanyi Yu, Guizhong Zhang, Bing Sun, Liang Wu, Haiyong Zhang, Jian Liu, Pengbo Jiang, Wei Zhang, Cen Zhao, Jianquan Yao

Optics Express, 2014, 22, (23): 29121–29126.

Abstract: We report herein an efficient eye-safe Raman laser, which is based upon Nd:YVO$_4$-YVO$_4$ and in-band pumped by a wavelength-locked laser diode array at 878.6 nm. By virtue of mitigated thermal load and improved pump absorption, a maximum average output power of 5.2 W at 1 525 nm is obtained under the incident pump power of 30.6 W with the pulse repetition frequency of 140 kHz, corresponding to an optical efficiency of 17.0%.

Key words: Lasers, Raman, Pumping, Q-switched

358. A 7.81 W 355 nm ultraviolet picosecond laser using $La_2CaB_{10}O_{19}$ as a nonlinear optical crystal

Ling Zhang, Kai Li, Degang Xu, Haijuan Yu, Guochun Zhang, Yuye Wang, Lirong Wang, Faxian Shan, Chao Yan, Yingying Yang, Baohua Wang, Nan Wang, Xuechun Lin, Yicheng Wu, Jianquan Yao

Optics Express, 2014, 22(14): 17187–17192.

Abstract: We demonstrate high-power 355 nm ultraviolet (UV) picosecond (ps) laser using a type I phase-matching nonlinear optical crystal of $La_2CaB_{10}O_{19}$ (LCB), which possesses the characteristic of non-hygroscopicity. The high-power third harmonic generation was successfully achieved from two types of 1 064 nm ps fundamental lasers. The maximum output power of 7.81 W of 355 nm UV laser was obtained with a pump of 35.2 W 1 064 nm ps laser (80 MHz repetition rate, 10 ps pulse width) with optical conversion efficiency of 22.2%. The experimental results show that the LCB crystal is a promising candidate for generating high-power UV laser.

Key words: Ultrafast lasers, Mode-locked lasers, Nonlinear optical materials, Harmonic generation and mixing

359. 12.45 W wavelength-locked 878.6 nm laser diode in-band pumped multisegmented Nd:YVO₄ laser operating at 1 342 nm

Bin Li, Xin Ding, Bing Sun, Quan Sheng, Jian Liu, Zhang Wei, Pengbo Jiang, Chen Fan, Haiyong Zhang, Jianquan Yao

Applied Optics, 2014, 53(29): 6778–6781.

Abstract: A multisegmented Nd:YVO₄ laser operating at 1 342 nm that is in-band pumped by a wavelength-locked 878.6 nm laser diode is reported here. We achieve an output power of 12.45 W at 1 342 nm for an absorbed pump power of 35.6 W, corresponding to an optical-to-optical efficiency of 34.9% and a slope efficiency of 36.1%. To the best of our knowledge, it is the highest optical-to-optical efficiency of Nd:YVO₄ lasers operating at 1 342 nm with an output power more than 10 W.

Key words: Lasers, Diode-pumped, Neodymium, Pumping

360. Efficient Nd:YVO$_4$ self-Raman laser in-band pumped by wavelength-locked laser diode at 878.7 nm

Quan Sheng, Xin Ding, Bin Li, Xuanyi Yu, Chen Fan, Haiyong Zhang, Jian Liu, Pengbo Jiang, Wei Zhang, Wuqi Wen, Bing Sun, Jianquan Yao

Journal of Optics, 2014, 16: 105206.

Abstract: A frequency-doubled Nd:YVO$_4$ self-Raman laser in-band pumped by an 878.7 nm wavelength-locked laser diode (LD) is demonstrated, with the purpose of improving the pump absorption of in-band pumping and thus enhancing optical efficiency. 2.64 W average output power at 588 nm is obtained under incident LD power of 9.7 W, corresponding to an optical efficiency of 27.2%. The results of control experiments show that in-band pumping leads to higher conversion efficiency of a self-Raman laser than does traditional pumping. Moreover, the application of a wavelength-locked LD further improves optical efficiency significantly.

Key words: Self-Raman laser, In-band pumping, Frequency-doubled

361. A novel CW yellow light generated by a diode-end-pumped intra-cavity frequency mixed Nd:YVO$_4$ laser

Bin Li, Jianquan Yao, Xin Ding, Quan Sheng, Sujia Yin, Chunpeng Shi, Xue Li, Xuanyi Yu, Bing Sun.

Optics & Laser Technology, 2014, 56: 99–101.

Abstract: We report an efficient continuous-wave (CW) yellow beam based on a Nd:YVO$_4$ laser with a dual-wavelength wave plate, which is a full-wave plate at 1 342 nm and a quarter-wave plate at 1 064 nm. By inserting the dual-wavelength wave plate into the laser cavity and tuning the fast axis angle, based on the polarized property emission of Nd:YVO$_4$, two simultaneous wavelengths of 1 064 nm and 1 342 nm were achieved. A LiB$_3$O$_5$ (LBO) crystal that is cut for critical type I phase matching at room temperature is used for summing frequency generation (SFG) of the laser. At an incident pump power of 5 W, a maximum CW output of 224 mW was obtained for a yellow laser with a wavelength of 593.5 nm and an optical-to-optical efficiency of 4.48%. To the best of our knowledge, this is the first time that a dual-wavelength wave plate has been used to realize a Nd:YVO$_4$ yellow laser output.

Key words: Yellow laser, Dual-wavelength, Summing frequency generation

362. 28.2 W 波长锁定 878.6 nm 激光二极管共振抽运双晶体 1 064 nm 激光器

李斌,丁欣,孙冰,盛泉,姜鹏波,张巍,刘简,范琛,张海永,姚建铨

物理学报, 2014, 63(21): 214206.

摘要：报道了一种由波长锁定 878.6 nm 半导体激光器共振抽运两块不同掺杂浓度 $Nd:YVO_4$ 晶体串接的 1 064 nm 激光器,并与使用单块的低掺杂浓度晶体和高掺杂浓度晶体情况进行比较,实验表明利用波长锁定 878.6 nm 半导体激光器共振抽运双晶体串接的方式,有利于降低晶体的热效应,提高光光转换效率。当抽运功率为 40 W 时,获得了 28.2 W 的 1 064 nm 激光输出,光 - 光转换率为 70.5%,斜率效率为 70.6%,相对吸收光的光光转换率 76%,斜率效率为 76.4%,同时该激光器在 10 ℃-40 ℃ 的温度变化范围内具有极好的温度稳定性。

关键词：锁波长, 878.6 nm 抽运, $Nd:YVO_4$ 晶体, 1 064 nm

363. 波长锁定 878.6 nm 激光二极管抽运 $Nd:YVO_4$ 1 064 nm 激光器

李斌,丁欣,张巍,盛泉,范琛,张海永,姜鹏波,刘简,姚建铨,孙冰,王鹏

中国激光, 2014, 41: 0502010.

摘要：报道了一种由波长锁定 878.6 nm 半导体激光器抽运 $Nd:YVO_4$ 晶体的 1 064 nm 激光器,当晶体吸收 7.41 W 的抽运功率时获得了 5.75 W 的 1 064 nm 激光输出,相对于吸收功率的斜率效率为 80.2%,光 - 光转换率为 77.6%,并且对波长锁定 878.6 nm,非波长锁定的 808 nm, 878.6 nm 抽运的激光器的温度特性进行了研究,结果表明利用波长锁定 878.6 nm 作为抽运源的激光器在 10 ℃~40 ℃ 的温度变化范围内具有很好的输出稳定性。

关键词：激光器, 波长锁定, 878.6 nm 抽运, $Nd:YVO_4$ 晶体, 1 064 nm

364. Efficient eye-safe Nd:YVO$_4$ selfRaman laser in-band pumped at 914 nm

Xin Ding, Jian Liu, Quan Sheng, Bin Li, Xuanyi Yu, Liang Wu, Guizhong Zhang, Bing Sun, Pengbo Jiang, Wei Zhang, Cen Zhao, Jianquan Yao

IEEE Photonics Journal, 2015, 7(6): 1503807.

Abstract: An efficient Q-switched eye-safe Nd:YVO$_4$ self-Raman laser, which we believe to be the first one in-band pumped at 914 nm, is presented, and 2.58 W of 1 525-nm Stokes output was generated under the absorbed pump power of 10.2 W with a high repetition rate of 100 kHz, corresponding to a conversion efficiency of 25.3%. The optical efficiency with respect to incident pump also reached 16.4%. Influences of crystal temperature and dopant concentration on the performance of the eye-safe self-Raman laser were experimentally investigated.

Key words: Eye-safe, In-band pumping, Stimulated Raman scattering, Vanadate laser

365. A non-critically phase matched KTA optical parametric oscillator intracavity pumped by an actively Q-switched Nd:GYSGG laser with dual signal wavelengths

Kai Zhong, Shibei Guo, Maorong Wang, Jialin Mei, Degang Xu, Jianquan Yao

Optics Communications, 2015, 344: 17–20.

Abstract: A non-critically phase matched eye-safe KTA optical parametric oscillator intracavity pumped by a dualwavelength acousto-optically Q-switched Nd:GYSGG laser is demonstrated. Simultaneous dual signal wavelength at 1 525.1 nm/1 531.2 nm can be realized using only one laser crystal and one nonlinear crystal. When the absorbed diode pump power at 808 nm is 7.48 W, the maximum output power, single pulse energy and peak power are 296 mW, 2.96 μJ and 6.4 kW, respectively. As the signal wavelengths exactly locates at the absorption band of C_2H_2, such an Nd:GYSGG/KTA eye-safe laser has good application prospects in differential absorption lidar (DIAL) for C_2H_2 detection and difference frequency generation for terahertz waves at 0.77 THz.

Key words: Optical parametric oscillator, Non-critically phase matched, Dual wavelength, Intracavity pumped

366. High-order Stokes generation in a KTP Raman laser pumped by a passively Q-switched ND:YLF laser

Maorong Wang, Kai Zhong, Jialin Mei, Shibei Guo, Degang Xu, Jianquan Yao
Optics Communications, 2015, 356: 411–415.

Abstract: High-order Stokes wave was observed in an x-cut KTP crystal based on stimulated Raman scattering (SRS) pumped by a passively Q-switched Nd:YLF laser with a Cr^{4+}:YAG saturable absorber. Output spectra including the fundamental wave at 1 047 nm and six Stokes wavelengths at 1 077 nm, 1 110 nm, 1 130 nm, 1 143 nm, 1 164 nm, 1 180 nm based on two Raman frequency shift at 267.4 cm^{-1} × 1 cm^{-1} and 693.0 cm^{-1} were obtained simultaneously. We also detected green light generation with output power of 12 mW from self frequency mixing in the KTP crystal. The maximum total output power reached 452 mW at the repetition frequency of 8.1 kHz, corresponding to the optical-to-optical conversion efficiency of 4.61% and pump-to-Raman conversion efficiency of 3.6%.

Key words: Raman laser, KTP, Nd:YLF, Passively Q-switched

367. 飞秒激光等离子体在高超声速飞行器减阻中的应用

付宁，徐德刚，张贵忠，姚建铨
中国激光，2015, 42(2): 0202003.

摘要： 报道了利用飞秒激光产生的等离子体冲击波对高超声速飞行的钝体飞行器进行减阻的研究。通过模拟计算了距离地球表面 30 km、来流马赫数为 5 的大气环境中，飞秒激光能量注入后产生的等离子体冲击波与钝体飞行器头部正激波相互耦合的演化过程，分析了飞秒激光等离子体减阻的机理。通过求解 Navier-Stokes 方程，计算了飞秒激光能量对飞行器减阻效果的影响。结果发现，利用飞秒激光产生的等离子体冲击波比纳秒激光等离子体冲击波对飞行器的减阻效果更明显。当飞秒激光能量为 0.06 mJ 时，能使飞行器所受的阻力减小 98%，飞秒激光能量越高，减阻比越高，低阻力持续的时间越长，减阻效果越好。采用 3 个飞秒激光能量点源沉积的方式能够更好地实现飞行器的减阻，提高了最佳减阻比，节省了激光能量。

关键词： 光电子学，飞秒激光等离子体，Navier-Stokes 方程，减阻，高超声速飞行器

368. 激光二极管抽运 Nd:YAG/Nd:YVO$_4$ 共轴双晶体 Cr:YAG 被动调 Q 激光器

李斌,丁欣,孙冰,盛泉,姜鹏波,张巍,刘简,范琛,张海永,姚建铨
中国激光, 2015, 42(4): 0402003.

摘要:报道了一种由激光二极管抽运的 Nd:YAG/Nd:YVO$_4$ 共轴双晶体的 Cr:YAG 被动调 Q 激光器,利用这种方式相比于传统的 Nd:YAG/Cr:YAG 激光器提高了输出激光的偏振比,在非线性频率变换过程中得到了更高的转换效率,当抽运功率为 10 W 时获得了 2.8 W 的被动调 Q 1 064 nm 激光输出,偏振比大于 80∶1,激光重复频率为 15 kHz,脉冲宽度为 7 ns,采用 LBO 作为非线性频率变换晶体,最终获得了 223 mW 的 355 nm 紫外激光输出。
关键词:激光器,Nd:YAG/Nd:YVO$_4$,偏振比,Cr:YAG,355 nm

369. 矢量多高斯 - 谢尔模型光束在大气湍流中上行链路中的传输特性

刘侠,吴国华,曹丁象,邓剑钦,肖青,姚建铨
激光与光电子学进展, 2015, 52(2): 91-96.

摘要:为了有效地提高星地激光通信系统的通信性能,从推广的惠更斯 - 菲涅耳衍射原理出发,系统研究了矢量多高斯 - 谢尔模型光束在上行链路中的传输特性。研究结果表明,改变光束的横向相干长度能有效地改变其远场光强分布和大气湍流引起的光束展宽。另外,大气湍流会改变矢量多高斯光束的偏振度,但是在远场其偏振度会趋于一个固定值。该研究结果对星地激光通信等有着潜在的应用价值。
关键词:光通信,星地激光通信,光束展宽,光束漂移

370. Random lasing in a colloidal quantum dot-doped disordered polymer

Mingxuan Cao, Yating Zhang, Xiaoxian Song, Yongli Che, Haiting Zhang, Haitao Dai, Guizhong Zhang, Jianquan Yao
Optics Express, 2016, 24(9): 9325−9331.

Abstract: We report random lasing in colloidal quantum dots (CQDs) doped disordered polymer. The CdSe/ZnS core-shell CQDs are dispersed in hybrid polymer including two types of monomers with different rates of polymerization. After UV curing, spatially localized random resonators are formed owing to long range refractive-index fluctuations in inhomogeneous polymer with gain. Upon the optical excitation, random lasing action is triggered above the threshold of 7 mJ/cm^2. Through the investigation on the spectral characteristics of random laser, the wavelengths of random lasers strongly depend on pump position, which confirms that random laser modes originate from spatially localized resnonators. According to power Fourier transform of emission spectrum, the average size of equivalent micro resonators is attributed to be 50 μm. The proposed method provides a facile route to develop random lasers based on CQDs, showing potential applications on random fiber laser and laser displays.
Key words: Nanomaterials, Laser materials

371. Simultaneous dual-wavelength eye-safe KTP OPO intracavity pumped by a Nd:GYSGG laser

Maorong Wang, Kai Zhong, Jialin Mei, Shibei Guo, Degang Xu, Jianquan Yao
Journal of Physics D: Applied Physics, 2016, 49(6): 065101.

Abstract: A simultaneous dual-wavelength intracavity pumped non-critical eye-safe optical parametric oscillator (OPO) is realized using a Nd:GYSGG laser crystal and a KTP nonlinear crystal. A folded cavity is used for thermal stability and mode matching, which greatly improves the output characteristics versus a linear cavity. The maximum output power of the 1 562.1 nm/1 567.4 nm dual-wavelength eye-safe laser is 750 mW at 10 kHz, corresponding to the optical-to-optical conversion efficiency, single-pulse-energy and peak power of 5.8%, 75 μJ and 22.7 kW. Such a dual-wavelength OPO provides a good laser source for remote sensing for CO and CO_2 gases or difference frequency generation for terahertz wave at the important 0.65 THz band.
Key words: Optical parametric oscillator, Eye-safe, Dual-wavelength, Nd:GYSGG

372. Efficient 914-nm Nd:YVO$_4$ laser under double-end polarized pumping

Pengbo Jiang, Xin Ding, Quan Sheng, Bin Li, Xuanyi Yu, Guizhong Zhang, Bing Sun, Liang Wu, Jian Liu, Wei Zhang, Cen Zhao, Jianquan Yao

Applied Optics, 2016, 55(5): 1072–1075.

Abstract: We report herein the enhancement of output power and optical efficiency of a quasi-three-level Nd:YVO$_4$ laser through a double-end polarized pumping scheme, which improves the usually insufficient pump absorption of the short laser gain medium with low doping concentration, and meanwhile alleviates the influence of thermal effect. 17.7 W laser output at 914 nm is obtained under the launched 808-nm pump of 53.0 W, corresponding to an optical efficiency of 33.4%.
Key words: Lasers, Diode-pumped, Pumping, Thermal effects

373. High efficiency actively Q-switched Nd:YVO$_4$ self-Raman laser under 880 nm in-band pumping

Ding Xin, Zhang Wei, Liu Junjie, Sheng Quan, Li Bin, Liu Jian, Jiang Pengbo, Sun Bing, Zhao Cen, Yao Jianquan

Infrared and Laser Engineering, 2016, 45(1): 0105002.

Abstract: An efficient acousto-optic Q-switched Nd:YVO$_4$ self-Raman laser in-band pumped at 880 nm was demonstrated. Using two 10-mm-long Nd:YVO$_4$ crystals as gain medium, 6.11 W of average output power at 1 176 nm Stokes wavelength was obtained under the incident pump power of 26.8 W and a high pulse repetition rate of 190 kHz, corresponding optical efficiency was 22.8%. The influence of Raman-gain-length on conversion efficiency was investigated in the experiment and the dips on Stokes output power was also discussed. Control experiment of the self-Raman laser under 808 nm traditional pumping shows that in-band pumping help improve the conversion efficiency and maximum output power greatly.
Key words: Self-Raman laser, In-band pumping, Nd:YVO$_4$ laser

374. Dynamic interference photoelectron spectra in double ionization: numerical simulation of 1D helium

Fu Sun, Dong Wei, Guizhong Zhang, Xin Ding, Jianquan Yao

Chinese Physics Letters, 2016, 33(12): 123202.

Abstract: We report our numerical simulation on the dynamic interference photoelectron spectra for a one-dimensional (1D) He model exposed to intense ultrashort extreme ultraviolet (XUV) laser pulses. The results demonstrate an unambiguous interference feature in the photoelectron spectra, and the interference is unveiled to originate from the dynamic Stark effect. The interference photoelectron spectra are prompted for intense sub-femtosecond XUV laser pulses in double ionization. The stationary phase picture is corroborated qualitatively in the two-electron system. The ability of probing the dynamic Stark effect by the photoelectron spectra in a pragmatic experiment of single-photon double ionization of He may shed light on further investigation on multi-electron atoms and molecules.

Key words: 1D Helium, Photoelectron Spectra, Double Ionization

375. 轨道电子的三重动量相关：氖原子的从头计算

胡沺，丁欣，王与烨，张贵忠，姚建铨

激光与光电子学进展，2016, 6: 062601.

摘要： 通过求解密度泛函理论中的含时科恩-沈（TDKS）方程，对 Ne 原子光电离过程进行了数值模拟，发现了在高强度极紫外（XUV）激光脉冲作用下的三重动量相关（TMC）现象。计算结果显示了不同轨道电子具有不同电离特性，发现对于高强度 XUV 激光脉冲，Ne 原子 p 轨道电子的电离主要发生在沿着轨道纵向的方向上。通过计算各轨道电子的动量分量，发现轨道电子的平行动量相互关联，垂直动量也相互关联，但平行动量和垂直动量之间并不关联。这些相互关联的关系可以由轨道形状、轨道朝向和激光偏振来解释。模拟结果显示了内层轨道电子也可以发生显著电离现象。

关键词： 物理光学，光致电离，三重动量相关，含时密度泛函，极紫外激光

376. Optical parametric oscillation in a random polycrystalline medium

Qitian Ru, Nathaniel Lee, Xuan Chen, Kai Zhong, Georgiy Tsoy, Mike Mirov, Sergey Vasilyev, Sergey B. Mirov, Konstantin L. Vodopyanov

Optica, 2017, 4(6): 617−618.

Abstract: We demonstrate an optical parametric oscillator (OPO) based on random phase matching in polycrystalline ZnSe. The OPO was pumped by Cr:ZnS laser pulses (2.35 μm, 62 fs, 79 MHz), had a pump threshold of 90 mW, and produced an ultrabroadband spectrum spanning 3–7.5 μm.
Key words: Parametric processes, Nonlinear optics, Parametric processes

377. 9.80-W and 0.54-mJ actively Q-switched Nd:YAG/Nd:YVO₄ hybrid gain intracavity raman laser at 1 176 nm

Pengbo Jiang, Xin Ding, Bin Li, Jian Liu, Xuanyi Yu, Guizhong Zhang, Bing Sun, Cen Zhao, Liang Wu, Jianquan Yao

Optics Express, 2017, 25(4): 3387−3393.

Abstract: Here we propose an efficient diode-end-pumped actively Q-switched 1 176-nm Nd:YAG/Nd:YVO₄ hybrid gain intracavity Raman laser. By virtue of the construction of a coaxial double crystal, the laser not only can operate efficiently at low pulse repetition frequencies (PRFs), thereby realizing relatively high-energy and high-peak-power pulsed output, but also is capable of generating a high average output power at high PRFs. A maximum pulse energy of 0.54 mJ for the 1 176-nm Stokes light is achieved at the PRF of 10 kHz, and the maximum average output power up to 9.80 W is obtained at the PRF of 100 kHz, while the incident pump power is 42.0 W.
Key words: Lasers, Raman, Solid-state, Diode-pumped

378. 16.7 W 885 nm diode-side-pumped actively Q-switched Nd:YAG/YVO$_4$ intracavity raman laser at 1 176 nm

Pengbo Jiang, Guizhong Zhang, Jian Liu, Xin Ding, Quan Sheng, Xuanyi Yu, Bing Sun, Rui Shi, Liang Wu, Rui Wang, Jianquan Yao

Journal of Physics D: Applied Physics, 2017, 50: 465303.

Abstract: We proposed and experimentally demonstrated the generation of high-power 1 176 nm Stokes wave by frequency shifting of a 885 nm diode-side-pumped Nd:YAG laser using a YVO$_4$ crystal in a Z-shaped cavity configuration. Employing the 885 nm diode-side-pumped scheme and the Z-shaped cavity, for the first time to our knowledge, we realized the thermal management effectively, achieving excellent 1 176 nm Stokes wave consequently. With an incident pump power of ~190.0 W, a maximum average output power of 16.7 W was obtained at the pulse repetition frequency of 10 kHz. The pulse duration and spectrum linewidth of the Stokes wave at the maximum output power were 20.3 ns and ~0.08 nm, respectively.

Key words: Raman laser, Diode-side-pumped, Q-switched, Lasers resonators

379. High power, widely tunable dual-wavelength 2 μm laser based on intracavity KTP optical parametric oscillator

Dexian Yan, Yuye Wang, Degang Xu, Wei Shi, Kai Zhong, Pengxiang Liu, Chao Yan, Jialin Mei, Jia Shi, Jianquan Yao

Journal of Physics D: Applied Physics, 2017, 50: 035104.

Abstract: We presented a high power, widely tunable narrowband 2 μm dual-wavelength source employing intracavity optical parametric oscillator with potassium titanium oxide phosphate (KTP) crystal. Two identical KTP crystals were oriented oppositely in the OPO cavity to compensate the walk-off effect. The output average power of dual-wavelength 2 μm laser was up to 18.18 W at 10 kHz with the peak power of 165 kW. The two wavelengths can be tuned in the range of 2 070.7 nm to 2 191.1 nm for ordinary light while in the range of 2 190.7 nm to 2 065.9 nm for extraordinary light with the full width at half maximum (FWHM) about 0.8 nm. The pulse width of the tunable laser was as narrow as 11 ns. The beam quality factor M^2 was less than 4 during wavelength tuning.

Key words: Intracavity optical parametric oscillator, Widely tunable, Dual-wavelength, 2 μm laser, High power

380. Thermal management of Nd:YVO$_4$ laser by 808-/880-nm dual-wavelength pumping

Quan Sheng, Lu Liu, Xin Ding, Pengbo Jiang, Jian Liu, Xuanyi Yu, Liang Wu, Guizhong Zhang, Cen Zhao, Bing Sun, Jianquan Yao

IEEE Photonics Journal, 2017, 9(2): 1501307.

Abstract: Here, we present a thermal load model of a Nd:YVO$_4$ laser under dual-wavelength pumping. Pumping the gain medium with both traditional and in-band pump light in adjustable proportion, the design takes full advantage of high efficiency and low heat of the two pump schemes, and the thermal load can be controlled actively, thus being capable of achieving optimal laser performance allowed under certain restrictions. The range of optimal proportion under different restrictions is discussed. Experimental results of a Nd:YVO$_4$ laser under 808-nm and 880-nm dual-wavelength pumping validated the theoretical analysis.

Key words: Pumping, Thermal effects, Solid-state, Diode-pumped

381. Compact and flexible dual-wavelength laser generation in coaxial diode-end-pumped configuration

Yang Liu, Kai Zhong, Jialin Mei, Maorong Wang, Shibei Guo, Chu Liu, Degang Xu, Wei Shi, Jianquan Yao

IEEE Photonics Journal, 2017, 9(1): 1500210.

Abstract: A novel coaxial diode-end-pumping configuration with combined two laser crystals was proposed for simultaneous, compact, and flexible dual-wavelength laser generation. Theoretical simulations showed that by balancing the gain in both laser crystals by varying the pump focusing depth or pump wavelength, the power ratio for each wavelength could be tuned for either continuous wave or Q-switched operation, and the time interval between two pulses at different wavelengths in Q-switched mode was also tunable. Experimental verifications were performed, demonstrating coincident conclusions. As there was no gain competition between two wavelengths, the output characteristics were much more stable than dual-wavelength generation in a single-laser crystal. It is believed that this is a feasible and promising method for generating dual-wavelength laser for applications of spectroscopy, precision measurement, nonlinear optical frequency conversion to terahertz wave, and so on.

Key words: Dual-wavelength lasers, Solid-state lasers, Diode end pumping

382. Characterizing carrier envelope phase of an isolated attosecond pulse with annular photoionization momentum spectra

Meng Li, Guizhong Zhang, Tianqi Zhao, Xin Ding, Jianquan Yao
Chinese Optics Letters, 2017,15(11): 110201.

Abstract: The carrier envelope phase (CEP) has a direct impact on the physical properties of an isolated attosecond pulse (IAP) and many strong field processes, but it is difficult to measure in reality. Aiming at obtaining more accurate and complete characterization of CEP, we numerically investigate the annular photoelectron momentum spectra of the hydrogen atom ionized by overlapped fields of an IAP and an infrared (IR) pulse. By defining an overlapping parameter, the momentum patterns are classified and optimized for unambiguously measuring the rotation angle of a momentum pattern versus the CEP value. A series of simulations verify its robustness.
Key words: Single-shot measurement, Laser-pulses electron localization

383. Electron vortices in photoionization by a pair of elliptically polarized attosecond pulses

Meng Li, Guizhong Zhang, Tianqi Zhao, Xin Ding, Jianquan Yao
Chinese Optics Letters, 2017, 15(12): 120202.

Abstract: The photoionization by two elliptically polarized, time delayed attosecond pulses is investigated to display a momentum distribution having the helical vortex or ring structures. The results are obtained by the strong field approximation method and analyzed by the pulse decomposition. The ellipticities and time delay of the two attosecond pulses are found to determine the rotational symmetry and the number of vortex arms. For observing these vortex patterns, the energy bandwidth and temporal duration of the attosecond pulses are ideal.
Key words: Electron vortices, Attosecond pulses, Photoionization

384. High-pulse-energy mid-infrared optical parametric oscillator based on BaGa$_4$Se$_7$ crystal pumped at 1.064 μm

Wentao Xu, Yuye Wang, Degang Xu, Chao Li, Jiyong Yao, Chao Yan, Yixin He, Meitong Nie, Yicheng Wu, Jianquan Yao

Applied Physics B-Laser and Optics, 2017, 123: 80.

Abstract: We have demonstrated a high-pulse-energy nanosecond mid-infrared optical parametric oscillator (OPO) based on the nonlinear crystal BaGa$_4$Se$_7$ pumped by 1.064 μm Nd:YAG laser. The experimental OPO threshold of 7.97 MW/cm^2 was in good agreement with the theoretical calculation of 8.05 MW/cm^2. The maximum pulse energy of the idler wavelength was 2.56 mJ at 4.11 μm wavelength when the pump energy was 61.6 mJ, corresponding to an optical-to-optical conversion efficiency of 4.16%. The idler wavelength can be continuously tuned in the range from 3.12 to 5.16 μm.

Key words: Nonlinear crystals, 1 064 nm, Generation, Laser

385. Ultrastable, high efficiency picosecond green light generation using K$_3$B$_6$O$_{10}$Br series nonlinear optical crystals

Z Y Hou, M J Xia, L R Wang, B Xu, D X Yan, L P Meng, L J Liu, D G Xu, L Zhang, X Y Wang, R K Li, C T Chen

Laser Physics, 2017, 27: 095401.

Abstract: Two perovskite-structure K$_3$B$_6$O$_{10}$Br$_{1-x}$Cl$_x$ (x = 0 and 0.5) series nonlinear optical crystals were thoroughly investigated for their picosecond 532 nm laser pulses abilities and high power outputs were achieved via second harmonic generation (SHG) technique for the first time. SHG conversion efficiency of 57.3% with a 13.2 mm length K$_3$B$_6$O$_{10}$Br (KBB) crystal was achieved using a laser source of pulse repetition rate of 10 Hz and pulse width of 25 ps, which is the highest conversion efficiency of ps visible laser based on KBB crystal. And by employing an 80 MHz, 10 ps fundamental laser beam, maximum power outputs of 12 W with K$_3$B$_6$O$_{10}$Br$_{0.5}$Cl$_{0.5}$ (KBBC) and 11.86 W with KBB crystals were successfully demonstrated. Furthermore, the standard deviation jitters of the average power outputs are less than 0.6% and 1.17% by KBB and KBBC, respectively, showing ultrastable power stabilities favorable for practical applications. In addition, the other optical parameters including acceptance angle and temperature bandwidth were also investigated.

Key words: Second-harmonic generation, Nonlinear optics, Materials, Laser, Picosecond solid-state

386. Power-ratio tunable dual-band Nd:GYSGG laser at 0.94 μm and 1.06 μm

Kai Zhong, Chu Liu, Yang Liu, Jialin Mei, Jie Shi, Degang Xu, Jianquan Yao
Laser Physics, 2017, 27(12): 125804.

Abstract: A dual-band diode pumped solid-state laser at 0.94 μm and 1.06 μm based on a Nd:$Gd_{3x}Y_{3(1-x)}Sc_2Ga_{3(1+\delta)}O_{12}$ (x = 0-1, δ = -0.2-0.2) crystal was demonstrated, in which a novel power-ration tuning method with temperature controlling was used. With the incident pump power of 3.14 W, the power proportion of the 1.06 μm component could be tuned from 12.4% to 100% when the outside cooling temperature was changed from 2 ℃ to 20 ℃. As far as we know, it was the first report of temperature tuning for the power ratio in a multiwavelength laser. Analysis showed that reducing the actual operating temperature was critical to the performance of the power-ratio tunable dual-band laser. Several schemes were proposed for optimization and a power ratio tuning range covering 0-100% was expected. This method could be extended to all neodymium ion doped solid-state lasers.
Key words: Nd:GYSGG, Diode-end-pumped, Power-ratio tunable, Dual-wavelength

387. High efficiency actively Q-switched Nd:YVO₄ self-Raman laser under 914 nm in-band pumping

Ding Xin, Zhao Cen, Jiang Pengbo, Sheng Quan, Li Bin, Liu Jian, Sun Bing, Yao Jianquan
Infrared & Laser Engineering, 2017, 46(10): 1005001.

Abstract: An high efficient actively Q-switched Nd:YVO₄ self-Raman laser based on in-band pumped at 914 nm was reported. Particles from thermally excited Stark leveled in the ground-state were pumped to the upper lasing level directly, the Stokes factor loss was reduced to a minimum and the quantum efficiency loss was eliminated. This alleviated the thermal effect of the laser and consequently realized better performance. The effect of pump absorption on conversion efficiency was investigated experimentally in detail with the in-band pumping. Using two Nd:YVO₄ crystals with different doping concentration,1.51 W (1.0-at.%, 20℃) and 2.11 W (2.0-at.%, 20 ℃) output power were obtained respectively, corresponding to the conversion efficiency of 42.7%(1.0-at.%) and 39.0%(2.0-at.%) and the optical efficiency of 28.5%(1.0-at.%) and 35.2%(2.0-at.%).
Key words: Self-Raman, In-band pumping, Nd:YVO₄ laser

388. Generation of high-output yellow light by intracavity doubling Nd:YAG-Nd:YVO$_4$ hybrid gain Raman laser

Pengbo Jiang, Xin Ding, Quan Sheng, Jian Liu, Xuanyi Yu, Guizhong Zhang, Jianquan Yao

Proceedings of Society of Photo-Optical Instrumentation Engineers, 2017, 10457: 104571T.

Abstract: A promising method for generation of relatively high-peak-power and high-energy yellow light based on doubling actively Q-switched Nd:YAG-Nd:YVO$_4$ hybrid gain intracavity Raman laser at 1 176 nm has been proposed and experimentally demonstrated for the first time. Under the incident pump power of 42.0 W, pulse energy of 0.37 mJ and peak power of 75 kW at 1 176 nm were generated in our experiment.

Key words: Raman laser, Yellow light, Hybrid gain

389. Efficient yellow-light generation based on a Q-switched frequencydoubled self-Raman laser

Jian Liu, Xin Ding, Quan Sheng, Pengbo Jiang, Xuanyi Yu, Bing Sun, Rui Shi, Guizhong Zhang, Jianquan Yao

Proceedings of Society of Photo-Optical Instrumentation Engineers, 2017, 10457: 104570W.

Abstract: Efficient 588-nm yellow laser with highpulse energy and peak power is generated based on an intracavity frequencydoubled Q-switched Nd:YVO$_4$ self-Raman laser. The cavity is elaborately designed to ensure high conversion efficiencies in both Raman and second harmonic generation processes, and meanwhile mitigate the thermal load deposited in gain medium so as to help improve power scalability. At the pulse repetition frequency of 60 kHz, 6.33 W of 588-nm yellow output is obtained under 30.3-W incident pump. The pulse energy of 167 μJ and peak power of 33.4 kW are also obtained at 30 kHz.

Key words: Yellow output, Self-Raman laser, Q-switching

390. High-performance second-Stokes generation of a Nd:YVO$_4$/YVO$_4$ Raman laser based on a folded coupled cavity

Jian Liu, Xin Ding, Pengbo Jiang, Quan Sheng, Xuanyi Yu, Bing Sun, Rui Shi, Jingbo Wang, Yuntao Bai, Lei Zhao, Guizhong Zhang, Liang Wu, Jianquan Yao

Optics express, 2018, 26(8): 10171−10178.

Abstract: We report an actively Q-switched Nd:YVO$_4$/YVO$_4$ intracavity Raman laser at second-Stokes wavelength of 1 313.6 nm, which is capable of operating efficiently under pulse repetition frequency higher than 80 kHz. A folded coupled cavity is adopted to optimize the fundamental and the Stokes resonators individually and make full use of the high pump intensity on the Raman crystal. With relatively high output coupling of 82% at 1 313 nm, the average output power of 5.16 W at 1 313 nm is achieved under the incident pump power of 36.7 W. The cascaded Raman emission at both the first- and second-Stokes wavelength of 1 176 and 1 313 nm is investigated to discuss the optimization of the second-Stokes generation.

Key words Lasers, Raman, Q-switched, Diode-pumped

391. Dual-signal-resonant optical parametric oscillator intracavity driven by a coaxially end-pumped laser with compound gain media

Yang Liu, Kai Zhong, Jie Shi, Hongzhan Qiao, Xin Ding, Degang Xu, Jianquan Yao
Optics express, 2018, 26(16): 20768–20776.

Abstract: A flexible method of generating stable dual-wavelength laser pulses with tunable power ratio and pulse interval is proposed, through integrating a coaxially end-pumped laser with compound gain media and an intracavity pumped optical parametric oscillator (IOPO). A theoretical model was built by a set of time-domain coupling wave equations containing both the generation of two fundamental waves from a shared pump source and the conversion to signal waves through the parametric process. Simulations showed that by simply varying the pump focal position or pump wavelength, the gains in two laser crystals could be changed, leading to simultaneous change in average power ratio and time interval between two wavelengths. Experimental verifications were performed with combined laser crystals (Nd:YAG and a-cut Nd:YLF) and a nonlinear crystal (KTA), which enabled dual-wavelength signal output in the 1.5–1.6 μm eye-safe region and demonstrated coincident conclusions with theoretical results. As there was no gain competition between two fundamental waves, stable signal output was obtained. Moreover, various wavelength pairs in any wavelength ranges are possible by using different laser crystals and nonlinear crystals. It is believed that this is a promising method for generating simultaneous dual-wavelength laser pulses for applications in lidar, remote sensing, nonlinear frequency conversion, etc.

Key words: Coaxially end-pumped laser, Optical parametric oscillator, Compound gain media

392. Wavelength tuning and power enhancement of an intracavity Nd:GdVO$_4$-BaWO$_4$ Raman laser using an etalon

Quan Sheng, Andrew Lee, David Spence, Helen Pask
Optics express, 2018, 26(24): 32145–32155.

Abstract: We report the wavelength tuning, linewidth narrowing and power enhancement of a continuous-wave intracavity Raman laser by incorporating solid etalons in the high-Q fundamental resonator. With a-cut Nd:GdVO$_4$ and a-cut BaWO$_4$ serving as the laser and Raman crystals respectively, tilting of a 50 μm-thick etalon in the high-Q fundamental cavity enabled the fundamental to be tuned from 1 061.00 nm to 1 065.20 nm. This gave rise to Stokes output which was tunable from 1 176.46 nm to 1 181.63 nm whilst the narrowed fundamental linewidth resulted in higher effective Raman gain and as a consequence enhanced output power, as well as the narrow-linewidth Stokes output. Frequency-doubling of the Stokes field resulted in yellow output tunable from 588.23 nm to 590.81 nm, which covers the guide star wavelength of 589.16 nm.
Key words: Raman laser, Wavelength tuning

393. Dynamic Stark induced vortex momentum of hydrogen in circular fields

Meng Li, Guizhong Zhang, Xuelian Kong, Tianqing Wang, Xin Ding, Jianquan Yao
Optics express, 2018, 26(2): 878–886.

Abstract: In this paper, we report our numerical simulation on the symmetry distortion and mechanism of the vortex-shaped momentum distribution of hydrogen atom by taking into account of the dynamic Stark effect. By deploying the strong field approximation (SFA) theory, we performed extensive simulation on the momentum pattern of hydrogen ionized by two time-delayed oppositely circularly polarized attosecond pulses. We deciphered that this distortion is originated from the temporal characteristics of the dynamic Stark phase which is nonlinear in time.
Key words: Atomic and molecular physics, Strong field laser physics

394. Symmetric electron vortices of hydrogen ionized by orthogonal elliptical Fields.

Meng Li, Guizhong Zhang, Xin Ding, Jianquan Yao.
IEEE Photonics Journal, 2018, 10(4): 1–9.

Abstract: In this paper, we report our numerical investigation on the photoionization of hydrogen by two orthogonal elliptically polarized and time-delayed attosecond pulses; we found that the momentum distributions exhibit a helical structure. The spectral patterns were obtained by making the use of the strong field approximation theory, and analyzed in terms of the irregularity of the overlapped fields. Some morphological characteristics of the generated vortex arms were found to be sensitive to the ellipticity and time delay of the two attosecond pulses. These phenomena could be used to investigate the intrinsic features of the photoionizaiton and complex polarized ultrashort pulses.
Key words: Helical structure, Orthogonal elliptically polarized, Strong field approximation theory

395. Odd-fold-symmetric spiral momentum distributions and their Stark distortions in hydrogen

Xuelian Kong, Guizhong Zhang, Meng Li, Tianqing Wang, Xin Ding, Jianquan Yao
Journal of the Optical Society of America B-optical Physics, 2018, 35(9): 2163–2168.

Abstract: Odd-fold-symmetric spiral photoelectron momentum distributions in hydrogen atoms ionized by two counterrotating circularly polarized and time-delayed laser pulses with different carrier frequencies are simulated by utilizing the strong field approximation theory. The odd-fold symmetry emerges when the total number of absorbed photons is odd. We also find that the symmetric patterns of the momentum distributions become distorted when the optical Stark effect is considered. We attribute this distortion to the nonlinear Stark phase. We anticipate that our result will be useful in studies on the atomic Stark effect.
Key words: Atomic and molecular physics, Strong field laser physics, Photoionization

396. Stable, high power, high efficiency picosecond ultraviolet generation at 355 nm in $K_3B_6O_{10}Br$ crystal

Z.Y. Hou, L.R. Wang, M.J. Xia, D.X. Yan, Q.L. Zhang, L. Zhang, L.J. Liu, D.G. Xu, D.X. Zhang, X.Y. Wang, R.K. Li, C.T. Chen

Optics Communications, 2018, 416: 71–76.

Abstract: We demonstrate a high efficiency and high power picosecond ultraviolet source at 355 nm with stable output by sum frequency generation from a Nd:YAG laser using a type-I critically phase matched $K_3B_6O_{10}Br$ crystal as nonlinear optical material. Conversion efficiency as high as 30.8% was achieved using a 25 ps laser at 1 064 nm operated at 10 Hz. Similar work is done by using a 35 W 10 ps laser at 1 064 nm as the pump source with a repetition rate of 80 MHz, and the highest average output power obtained was up to 5.3 W. In addition, the power stability of the 355 nm output power measurement shows that the standard deviation fluctuations of the average power are ±0.69% and ±0.91% at 3.0 W and 3.5 W, respectively.

Key words: Harmonic generation and mixing, Nonlinear optics, Materials, Laser, Solid-state

397. Compact, efficient and widely tunable 2-μm high-repetition-rate optical parametric oscillators

Jialin Mei, Kai Zhong, Yang Liu, Jie Shi, Hongzhan Qiao, Degang Xu, Wei Shi, Jianquan Yao

Optics Communications, 2018, 426: 119–125.

Abstract: Compact efficient high-repetition-rate tunable optical parametric oscillators (OPOs) in the 2-μm region were investigated, based on nonlinear crystals of KTP and periodically polarized $LiNbO_3$ (PPLN) intracavity pumped by a diode-end-pumped acousto-optical (AO) Q-switched $Nd:YVO_4$ laser. The cavity was designed and optimized from the theoretically analysis on thermal stability and mode matching. Using walk-off compensated twin KTP crystal pairs as the nonlinear media, the maximum output power of 4.15 W was obtained at 25 kHz and the tuning range spanned from 1 923.22 nm to 2 378.63 nm, pumped at the diode power of 18.41 W at 808 nm. The optical–optical conversion efficiency of 22.5% and the photon conversion efficiency of around 60% are both among the highest in intracavity pumped OPOs. To overcome the power degradation during angle tuning, a type-II quasi-phase matched (QPM) PPLN crystal with temperature tuning was also studied. With a 55-mm-long 14.1-μm-grating period PPLN crystal, the maximum output power was 3.41 W corresponding to the optical–optical conversion efficiency of 18.5%. Both the KTP and PPLN OPOs demonstrated good efficiency as well as temporal, spectrum and beam characteristics, providing excellent 2-μm laser source for applications of lidar, medical diagnosis, nonlinear frequency conversion to the long-wave infrared and terahertz range, etc.

Key words: 2-μm laser, Optical parametric oscillator, KTP, Periodically polarized $LiNbO_3$ (PPLN)

398. 10.3-W actively Q-switched Nd: YVO$_4$/YVO$_4$ folded coupled-cavity Raman laser at 1 176 nm

Jian Liu, Xin Ding, Pengbo Jiang, Quan Sheng, Xuanyi Yu, Bing Sun, Jingbo Wang, Rui Shi, Lei Zhao, Yuntao Bai, Guizhong Zhang, Liang Wu, Jianquan Yao

Applied optics, 2018, 57(12): 3154−3158.

Abstract: We report herein an efficient actively Q-switched Nd:YVO$_4$/YVO$_4$ intracavity Raman laser operating at 1 176 nm. Factors such as resonator geometry and pumping scheme are optimized to strengthen the power scalability and the conversion efficiency of the intracavity Raman laser. With a folded coupled cavity adopted to make full use of the high pump intensity on the Raman crystal, the first-order Stokes output of 10.32 W at 1 176 nm is achieved under the incident pump power of 39 W and pulse repetition frequency of 160 kHz. The corresponding optical efficiency reaches 26.4%, and even higher efficiency of 27.8% is obtained at lower incident pump of 34.4 W.

Key words: Lasers, Diode-pumped, Raman, Solid-state

399. Dual-wavelength eye-safe optical parametric oscillator intracavity driven by a coaxially end pumped laser

Yang Liu, Kai Zhong, Jie Shi, Hongzhan Qiao, Xin Ding, Degang Xu, Jianquan Yao

Quantum and Nonlinear Optics V. International Society for Optics and Photonics, 2018, 10825: 1082516.

Abstract: A compact and flexible dual-wavelength eye-safe intracavity optical parametric oscillator (IOPO) configuration driven by a coaxially end pumped laser was proposed. Two fundamental waves were provided by a coaxially end pumped Q-switched dual-wavelength laser with combined two laser crystals, and the OPO cavity was placed inside the laser cavity for efficient conversion. Theoretical simulations showed that the power ratio for each signal wave, as well as the time interval between two pulses at different wavelengths, were both tunable by tuning the pump focusing depth or pump wavelength. Experimental results were performed with combined laser crystals (Nd:YAG and a-cut Nd:YLF) and a nonlinear crystal (KTA), demonstrating coincident conclusions. The maximum OPO output power was 724 mW (388 mW at 1 506 nm and 336 mW at 1 535 nm) with the LD pump power of 10 W at 6 kHz, corresponding to the optical-optical conversion efficiency of 7.24%. As there was no gain competition between two fundamental waves, stable signal output could be obtained. Moreover, various wavelength pairs can be generated by using different laser crystal combinations. It is believed that this is a promising method for simultaneously generating dual-wavelength eye-safe lasers pulses.

Key words: Dual-wavelength, Eye-safe, Coaxial pumping, Optical parametric oscillator (OPO)

400. Compact low-energy high-repetition rate laser-induced breakdown spectroscopy for lead element detection

Feng Jiachen, Xu Degang, Wang Yuye, Zhang Guizhong, Li Jining, Yan Dexian, He Yixin, Yao Jianquan

Acta Optica Sinica, 2018, 47(1): 0114002.

Abstract: A compact laser-induced breakdown spectroscopy using a low-energy, high-repetition rate Nd: YAG DPSS laser as excitation source has been developed. The influences of laser repetition rate and laser power on LIBS signal were analyzed. When the laser operated at 4 kHz, pulse energy of 745 μJ and pulse width of 17 ns, the seven emission line of Lead (Pb) was obtained. Compared with the traditional single pulse LIBS, the number of emission lines is increase, also the signal-to-noise ratio is improved by almost 73%. Based on simplified setup and low energy, high-repetition rate laser-induced breakdown spectroscopy will be applied in more emerging fields.

Key words: Spectroscopy, Laser-induced breakdown spectroscopy, Plasma, Heavy metal, Spectral intensity

401. 基于相机式激光光束参数测量精度的影响因素分析

肖青, 刘侠, 邓剑钦, 姚建铨, 张大鹏, 王兴龙

激光与光电子学进展, 2018, 55(7): 302–307.

摘要: 针对相机式光束质量分析仪测量光束的结果不一致问题,分析了相关参数对测量误差的影响。基于高斯光束模型,并添加与实际对应的噪声,根据 ISO 11146 标准计算了光斑直径。结合实验测试时的实际操作过程,提炼出光强饱和度、积分面积比、信噪比和光斑尺寸 4 个关键参数。不仅阐明了相机参数对测量误差影响的关键趋势,还分析了各个参数之间的关联影响,最终对如何选择相机的像元、像素、动态范围、暗噪声,以及如何设置积分时间、计算区域等提出了更明确的要求。

关键词: 激光光学, 激光光束, 光束参数, 精度分析, 相机

402. 航天器用内螺纹的气体湍流激光多普勒式检测

王涛, 昝占华, 张翠亭, 姚建铨, 马俊杰, 蔡军, 李玉翔

红外与激光工程, 2018, 47(4): 0406006.

摘要：为了解决航天器操纵旋转套筒内螺纹的自动检测问题，特别是在内螺纹件表面有油污与锈蚀时，单纯地利用光学图像检测的方法很难更好地解决问题，利用平流气体吹向内螺纹，从而形成了映射着内螺纹形貌的湍流输出，然后利用多普勒激光测量内螺纹的湍流特性。研究了内螺纹气体湍流的拉伸与压缩过程的数据特性，利用箱线图的方法处理内螺纹湍流数据，建立了内螺纹湍流箱线图的标准样板库，将检测数据与标准样板对比检测，结果表明：当异常值阈值设为7%时，检测准确率为99.3%。气体湍流激光多普勒式检测内螺纹方法可行，克服了表面油污等难点，完全满足检测的要求。

关键词：内螺纹检测, 湍流边界层, 激光多普勒技术, 流场测量, 数据箱线图

403. 基于面阵探测器的激光光束参数测量精度的影响因素分析

肖青, 刘侠, 邓剑钦, 姚建铨, 张大鹏, 王兴龙

激光与光电子学进展, 2018, 55(7): 071401.

摘要：本文主要是针对相机式的光束质量分析仪对光束进行测量时会出现结果不一致的情况，来分析相关参数对其测量误差的影响。我们基于高斯光斑模型，同时给光斑添加与实际对应的噪声，根据 ISO 11146 标准里的计算方法来计算光斑直径。结合测试时的实际操作过程，并进行实验分析，提炼出光强饱和度、积分面积比、信噪比和光斑尺寸这四个关键参数。同以往的类似分析研究相比，本文不仅发现了相机参数影响的关键趋势，还发现了各个参数之间的关联影响，最终对如何选择相机的像元、像素、动态范围、暗噪声，以及该如何设置积分时间、计算区域等参数提出了更明确的要求。

关键词：激光光束, 光束参数, 测量, 精度分析, 面阵探测器

404. 基于小波阈值法的激光雷达回波信号去噪研究

王涛,沈永辉,姚建铨

激光技术, 2019, 43(1): 63–68.

摘要: 为了对激光雷达探测回波信号去噪进行研究,基于小波阈值法分析了激光雷达发射和探测回波信号,创建了基于 MATLAB 仿真平台的模型,研究了阈值法中选择各参量的策略,结合激光雷达信号和噪声特性,选取了去噪处理中的最佳参量,并进行了 4 种阈值策略的仿真实验,实现了对探测回波信号中目标信号的有效提取和去噪。结果表明,对于低信噪比信号,同一基函数分解层越高去噪效果越好,在分解层数 j 为 4 ~ 5 时去噪效果最好;对于高信噪比信号,同一基函数分解层越低去噪效果越好,在 j = 3 时去噪效果最好,并且信号 db9 基函数去噪效果好于 db2 基函数去噪效果。此研究获得了较为理想的去噪结果。

关键词: 信号处理,阈值去噪,MATLAB 仿真,最佳参量

405. 四波长激光准六自由度激光增材制造异形永磁件

王涛,朱惠芳,张翠亭,姚建铨,马俊杰,何晓阳,吕雪亮,牛世兴

激光与光电子学进展, 2018, 55(1): 134–140.

摘要: 提出了四波长激光多管喷粉准六自由度数控运动的激光增材新结构方法,分析了整体系统结构的线性空间与李子群函数,建立了系统的线性空间数学模型。采用多管喷粉机构喷涂定量 Nd-Fe-B 稀土永磁体成分,并利用闭环反馈控制四个激光束的功率与角度,进行了选择性与指向性激光辐照,实现了异形永磁件的激光增材成型,成型精度达 20 μm,最小宽度低于 80 μm,磁能积达 474 kJ·m^3。

关键词: 激光技术,激光增材制造,异形永磁件,准六自由度成型,钕铁硼永磁体材料,四波长激光,多管喷粉

406. Dual-wavelength intracavity Raman laser driven by a coaxially pumped dual-crystal fundamental laser

Yang Liu, Quan Sheng, Kai Zhong, Wei Shi, Xinding, Hongzhan Qiao, Kefei Liu, Hanchao Ma, Ran Li, Degang Xu, Jianquan Yao

Optics express, 2019, 27(20): 27797–27806.

Abstract: An actively Q-switched dual-wavelength intracavity Raman laser based on a coaxially pumped dual-crystal (Nd:YAG and *b*-cut Nd:YAP) fundamental configuration was theoretically and experimentally investigated. Stable dual-wavelength Stokes output at 1 176 nm and 1 195 nm was subsequently obtained by the Raman conversion from an *a*-cut YVO_4 crystal. A total average power of 1.8 W was produced at 10 kHz pulse repetition frequency under an incident diode pump power of 15.8 W, in which the power levels at the two Stokes wavelengths were nearly equivalent. The power proportion and the interval between the dual-wavelength Stokes pulses could be manipulated actively by changing the pump focal position or pump wavelength.

Key words: Coaxially pumped, Intracavity Raman laser, Dual-wavelength

407. Widely tunable eye-safe optical parametric oscillator with noncollinear phase-matching in a ring cavity

Kai Zhong, Jialin Mei, Yang Liu, Hongzhan Qiao, Kefei Liu, Degang Xu, Jianquan Yao

Optics express, 2019, 27(8): 10449–10455.

Abstract: An effective method for wavelength tuning in an optical parametric oscillator (OPO) was proposed using noncollinear phase-matching (PM) in a ring cavity. This method was particularly useful for noncritically phase-matched (NCPM) KTP/KTA OPOs where changing the crystal orientation or working temperature is ineffective. A wide tuning range in the eye-safe band from 1 572.9 nm to 1 684.2 nm was realized pumped by an Nd:YAG laser at 1.06 μm in an NCPM KTP OPO while the internal noncollinear angle was tuned from 0 to 3.1° r the external angle from 0° to 5.8°, with slight variation of the deflection angle of one cavity mirror. The good beam quality and high spectrum intensity of the narrow-linewidth Nd:YAG laser resulted in 33.3% conversion efficiency for the collinear case and above 11% throughout the tuning range. Such OPOs have many potential applications where tunable eye-safe lasers are required, and the proposed wavelength tuning method can be extended to all kinds of OPOs.

Key words: Optical parametric oscillator, Ring cavity, Noncollinear phase-matching

408. High energy and tunable mid-infrared source based on BaGa$_4$Se$_7$ crystal by single-pass difference-frequency generation

Yixin He, Yangwu Guo, Degang Xu, Yuye Wang, Xianli Zhu, Jiyong Yao, Chao Yan, Longhuang Tang, Jining Li, Kai Zhong, Chang Liu, Xiaoli Fan, Yicheng Wu, Jianquan Yao

Optics express, 2019, 27(6): 9241–9249.

Abstract: A high-energy and tunable mid-infrared source based on BaGa$_4$Se$_7$ crystal was demonstrated by single-pass difference-frequency generation (DFG). Orthogonally polarized wave at 1 064 nm (λ_1) and tunable idler wave (λ_2) generated by KTP-OPO, which could be tuned in the wavelength range of 1 360–1 600 nm, were used as the DFG dual-wavelength pump. The pump parameters including total pump energy and energy ratio were studied. Maximum pulse energy of 5.72 mJ at 3.58 μm was obtained at the dual-wavelength pump energy of 58.4 mJ/pulse. The wavelength tuning range was 3.36–4.27 μm with a flat tunability. Moreover, a saturation phenomenon of DFG output was observed and experimentally inferred to be related to the input energy of λ_2 in the BaGa$_4$Se$_7$ crystal.
Key words: Difference-frequency generation, Mid-infrared source

409. A single-frequency intracavity Raman laser

Quan Sheng, Ran Li, Andrew J. Lee, David J. Spence, Helen M. Pask
Optics express, 2019, 27(6): 8540–8551.

Abstract: A continuous-wave (CW) single-longitudinal-mode (SLM) intracavity Raman laser is demonstrated for the first time, by virtue of the spatial hole-burning free nature of stimulated Raman scattering (SRS) gain. By using a single etalon in the Nd:GdVO$_4$ fundamental laser cavity, the spectral linewidth of the multimode fundamental field is suppressed below the Raman linewidth of Raman crystal BaWO$_4$; hence power in all the longitudinal modes of fundamental field can be extracted by one single Stokes mode. Therefore, the hole-burning free SRS gain exhibits a spectral cleanup effect whereby a stable SLM Stokes field is derived from the multimode fundamental field within a simple standingwave cavity arrangement. The low-threshold SLM Raman laser delivered 3.42 W SLM Stokes and 1.53 W SLM one-way yellow harmonic at the guide-star wavelength of 589.16 nm. The results here provide a new approach to SLM laser operation with good simplicity and power dynamic range. Further engineering for power scaling and better stability is also discussed.
Key words: Raman laser, Single-frequency, Intracavity

410. Efficient and tunable 1.6-μm MgO:PPLN optical parametric oscillator pumped by Nd: YVO$_4$/YVO$_4$ Raman laser

Lei Zhao, Xin Ding, Jian Liu, Guizhong Zhang, Xuanyi Yu, Yang Liu, Bing Sun, Jingbo Wang, Yuntao Bai, Guoxin Jiang, Peng Lei, Tengteng Li, Liang Wu, Jianquan Yao

IEEE Photonics Journal, 2019, 12(1): 1–7.

Abstract: A highly-efficient tunable 1.6 μm MgO:PPLN-OPO, which we believe to be the first one pumped by a diode-end-pumped Raman laser, is demonstrated. A good beam-quality of actively Q-switched Nd:YVO$_4$/YVO$_4$ Raman laser at 1 176 nm and a good mode matching of the whole system resulted in a high-performance OPO. The highest Raman-to-signal conversion efficiency was 49.5% at 1 638.8 nm when the incident Raman power was 2 W. The maximum output power of 3.2 W at 1 663.5 nm with the pulse-width of 2.85 ns was obtained pumped by 7.57-W Raman laser. Meanwhile, the linewidth was less than 0.3 nm and a high beam quality with M^2 factor of 1.92 was obtained.

Key words: Beam quality, Raman pumping, 1.6-μm laser, Optical parametric oscillator

411. Intracavity-pumped, mid-infrared tandem optical parametric oscillator based on BaGa$_4$Se$_7$ crystal

Yixin He, Degang Xu, Jiyong Yao, Yuye Wang, Yangwu Guo, Xianli Zhu, Chao Yan, Longhuang Tang, Jining Li, Kai Zhong, Yicheng Wu, Jianquan Yao

IEEE Photonics Journal, 2019, 11(6): 1–9.

Abstract: We have demonstrated a tunable mid-infrared tandem optical parametric oscillator (OPO) based on BaGa$_4$Se$_7$ (BGSe) crystal, which was intracavity-pumped by the tunable signal wave at ~1 μm generated from a KTiOPO$_4$-OPO (KTP-OPO) in doubly-resonant oscillator (DRO) configuration. Wavelength tuning of the mid-infrared tandem OPO was realized by tuning the pump wavelength at the fixed phase matching angle to reduce the effects of beam cutting and Fresnel reflection loss. Together with pump-tuning method, angle-tuning method was supplied to extend tuning range. A tuning range of 4.1–4.5 μm was obtained with just 5 individual phase matching angles of the BGSe crystal. The maximum output energy of 1.92 mJ/pulse was obtained at 4.26 μm with an estimated slope efficiency of 9.5%.

Key words: Nonlinear crystals (BaGa$_4$Se$_7$ crystal), Optical parametric oscillator, Tunable mid-infrared source

412. AC Stark effect on vortex spectra generated by circularly polarized pulses

Meng Li, Guizhong Zhang, Xin Ding, Jianquan Yao

IEEE Photonics Journal, 2019, 11(3): 1–11.

Abstract: In this paper, we report the results of numerical simulation on vortex-shaped photoelectron momentum spectra of the hydrogen atom irradiated by a pair of time-delayed circularly polarized ultrashort pulses. Spectral alterations including broadening, splitting, and fusion are observed with stronger pulse intensity deploying the quantum wave-packet theory. These alterations of the dynamic interference structure are further investigated to stem from the ac Stark effect, by making the local approximation for an analytical ansatz. In addition, for evaluating the ac Stark effect on the hydrogen atom quantitatively, we propose a nonlinear-curve-fitting algorithm of the ground-state population for extracting the complex Stark coefficient, which we define. We find that the ac Stark effect can be well characterized by the complex coefficient, as supported by the overall agreement between the momentum spectra obtained from the wave-packet theory and the local approximation ansatz. The present research may shed further light on Stark effect in laser atom interactions.

Key words: Circularly polarized pulses, AC Stark effect

413. Laser performance of neodymium-and erbium-doped GYSGG crystals

Kai Zhong

Crystals, 2019, 9(220): 344−359.

Abstract: Garnet crystals possess many properties that are desirable in laser host materials, e.g., they are suitable for diode laser (LD) pumping, stable, hard, optically isotropic, and have good thermal conductivity, permitting laser operation at high average power levels. Recently, a new garnet material, GYSGG, was developed by replacing some of the yttrium ions (Y^{3+}) with gadolinium ions (Gd^{3+}) in YSGG, demonstrating great potential as a laser host material. GYSGG crystals doped with trivalent neodymium ion (Nd^{3+}) and erbium ions (Er^{3+}) were successfully grown for laser generation in the near- and mid-infrared range, with some of the laser performances reaching the level of mature laser gain media. This paper gives an overview of the achievements made in Nd^{3+}-and Er^{3+}-doped GYSGG lasers at different wavelength ranges. Additionally, full descriptions on Q-switching, mode-locking and wavelength-selecting methods for Nd:GYSGG, and the mechanisms of power scaling by co-doping sensitizers and deactivators in Er:GYSGG, are given. It is expected that this review will help researchers from related areas to quickly gain an understanding of these laser materials and promotes their commercialization and applications.

Key words: Nd:GYSGG, Er:GYSGG, Garnet laser crystal, Solid-state laser, Diode pumping, Laser performance, Radiation resistant

414. Evolution of the Raman beam quality in a folded-coupled-cavity Nd: YVO$_4$/YVO$_4$ Raman laser

Jinbao Wang, Xin Ding, Jian Liu, Guizhong Zhang, Xuanyi Yu, Bing Sun, Yang Liu, Tengteng Li, Yuntao Bai, Lei Zhao, Guoxin Jiang, Peng Lei, Liang Wu, Jianquan Yao

Applied Optics, 2019, 58(32): 8785−8790.

Abstract: An end-pumped actively Q-switched Nd:YVO$_4$/YVO$_4$ Raman laser with a folded coupled cavity is demonstrated to study the evolution of Raman beam quality. The theoretical mechanism of the beam cleanup effect of stimulated Raman scattering is analyzed. The beam quality (M^2) of the Raman beam and the fundamental beams before and after the Raman conversion are measured experimentally. The results show that with the incident pump power increasing, the M^2 of the fundamental beam increases from1.85 to 3.08, while the M^2 of the Raman beam increases from 1.21 to 1.69. The beam quality of the Raman laser and its degradation are better than that of the fundamental laser.
Key words: Raman laser, Folded-coupled-cavity

415. Carrier envelope phase description for an isolated attosecond pulse by momentum vortices

Meng Li, Guizhong Zhang, Xin Ding, Jianquan Yao
Chinese Physics Letters, 2019, 36(6): 063201−1−063201−5.

Abstract: As a crucial parameter for a few-cycle laser pulse, the carrier envelope phase (CEP) substantially determines the laser waveform. We propose a method to directly describe the CEP of an isolated attosecond pulse (IAP) by the vortex-shaped momentum pattern, which is generated from the tunneling ionization of a hydrogen atom by a pair of time-delayed, oppositely and circularly polarized IAP-IR pulses. Superior to the angular streaking method that characterizes the CEP in terms of only one streak, our method describes the CEP of an IAP by the features of multiple streaks in the vortex pattern. The proposed method may open the possibility of capturing sub-cycle extreme ultraviolet dynamics.
Key words: Momentum vortices, Isolated attosecond pulse

416. 由动态 Stark 效应诱导的氢原子涡旋动量分布

孔雪莲, 张贵忠, 汪天庆, 丁欣, 姚建铨
光学学报, 2019, 39(6): 0602001-1-0602001-8.

摘要: 采用强场近似方法,对处于两个具有时间延迟的圆偏振激光场中氢原子的涡旋状光电子动量分布进行了数值模拟。在两个延时激光脉冲的作用下,电子吸收光子后克服电离阈值,从基态经由两个不同的通道跃迁到连续态,产生的电子波包之间会发生干涉。模拟结果表明,所产生的光电子动量涡旋的旋向与两脉冲的偏振方向有关,涡旋臂的数目与激光载波频率有关。动态 Stark 效应是一种典型的强场现象。若在电离发生的同时考虑动态 Stark 效应,将会观察到动量涡旋的扭曲。对顺时针的动量涡旋及其扭曲进行分析,发现扭曲现象是由动态 Stark 效应引入的附加相位的时间非线性特性引起的。

关键词: 原子与分子物理学, 原子光电离, 强场近似, 氢原子, 超短脉冲, 数值模拟

417. 基于 Nd: YAG/LBO 倍频蓝光的全固态激光综合实验系统

钟凯, 温午麒, 徐德刚, 姚建铨
实验技术与管理, 2019, 36(2): 896–901.

摘要: 基于 Nd: YAG/LBO 内腔倍频开发了一套多功能全固态激光综合实验系统。创新性地引入准三能级 Nd: YAG 激光器、非线性光学倍频技术及独立温控技术,扩展了全固态激光器实验教学的内容,能够全面锻炼学生在全固态激光及非线性光学领域的实践能力,尤其是对激光二极管(LD)泵浦源、Nd:YAG 晶体能级结构及倍频的相位匹配等概念与技术的理解。该综合实验系统在经济性、适用性、功能性及科学前沿性等方面与现有装置相比均具有明显优势,相关实验受到了学生的欢迎。

关键词: 实验系统, 全固态激光, 非线性光学, 准三能级, 激光二极管, 倍频

418. 基于 TDLAS 技术的人体呼气末二氧化碳在线检测

王鑫，荆聪蕊，侯凯旋，张建涛，娄存广，姚建铨，刘秀玲

中国激光, 2020, 47(3): 902-913.

摘 要：呼出气体分析是一种测量呼出气体成分和含量的新技术，它具有很大的研究前景，在诸如幽门螺旋菌，哮喘等疾病无创检测和分析中的应用越来越广泛。本文基于低成本微型垂直腔面发射激光二极管（VSCEL）搭建了可调谐激光吸收光谱（TDLAS）气体分析系统，实现了人体呼出 CO_2 的在线测量。该系统主要是由激光管、驱动控制电路、光电探测器、放大电路、数据采集卡、控制软件、锁相放大器及赫里奥特气体池构成。该检测池体积为 400 mL，有效光程 20 m，激光光源中心波长 1 579.57 nm。采用波长调制吸收光谱技术中的二次谐波幅值反演计算人体呼出 CO_2 的浓度。该系统可以实现精准、无损、高效地在线测量人体呼出 CO_2 气体，波动范围小于 ±0.056%，灵敏度为 0.14%。这一研究为近红外 TDLAS 技术研究人体呼出气体与相关疾病标志物的无创检测提供了新的研究思路。

关键词：可调谐半导体吸收光谱，无损检测，呼出气体，二氧化碳

第二章 光电子及光纤技术

1. 掺铒光纤激光放大的研究

宁继平,何志宏,刘宏伟,傅靖,姚建铨,董孝义
光学学报, 1992, 12(8): 678–683.

摘要:本文研究了描述掺铒光纤激光放大的速率方程解,分析了光纤中几种场分布和铒离子横向分布时的解析解,并进行了数值计算。得到了任意信号强度(包括大信号)时的增益以及影响信号增益的主要参数,利用 532 nm 激光泵浦掺铒光纤,得到信号的最大增益为 29.5 dB。

关键词:光纤,掺铒光纤,激光放大,信号增益

2. 四能级系统光纤激光放大器的理论分析

宁继平,刘宏伟,姚建铨,郑义
中国激光, 1993, 20(6): 469–473.

摘要:本文从理论上分析了属于四能级系统的光纤激光放大器的增益特性,得到了信号增益与光纤长度、掺杂离子浓度、泵浦光和信号光功率等参数之间的关系式,并进行了数值计算。对四能级系统的激光放大器的设计有一定的指导作用。

关键词:掺钕光纤,四能级系统,光纤激光放大器,信号增益

3. 光学广义载波条纹图的计算机辅助分析

彭翔,高志,朱绍明,姚建铨

中国激光, 1995, 22(7): 541–545.

摘要: 从理论上和实验上研究了光学广义载波条纹图形成的机理及其数字解调技术,同时还给出了光学调制广义载波条纹的实验技术。利用计算机仿真的广义载波条纹图以及实验获得广义载波条纹图的数字解调结果与理论分析的结果一致。

关键词: 光学条纹,广义载波,数字解调

4. 1 477 nm LD 泵浦掺铒光纤放大器的研究

宁继平,孙小卫,汤声书,许晟,姚建铨

中国激光, 1995, 22(7): 485–489.

摘要: 报道了采用 1 477 nm 激光二极管(LD)泵浦的掺铒光纤放大器的实验结果。研究了放大器的增益和时域特性。对 1 520 nm 的信号光,获得了 23 dB 的增益,泵浦效率为 2.28 dB/mW。低频脉冲信号经过放大器后未发生波形畸变。

关键词: 掺铒光纤,光放大器,时域特性,激光二极管

5. 电极处势垒对新结构器件电子输运的影响

陈立春,王向军,徐叙瑢,姚建铨

物理学报, 1996, 45(4): 709–713.

摘要：在新结构薄膜电致发光器件中,电极处的势垒的高度决定电子的注入数量。在电极界面处插入不同的薄膜材料,可以改变势垒的高度,并对电子注入数量和器件的发光亮度产生影响。通过拟合计算得到 ZnO/SiO, ITO/SiO 的界面势垒高度分别为 0.51 和 1.87 eV。

关键词：电致发光器件,电极,势垒,电子输运

6. Multipoint velocity interferometer system for any reflector

Zeren Li, Ruchao Ma, Guanghua Chen, Jun Liu, Jianquan Yao

Review of Scientific Instruments, 1999, 70(10): 3872–3876.

Abstract: A multipoint velocity interferometer system for any reflector (VISAR) has been developed which can simultaneously measure velocity versus time histories of two to six points on a target or different objects during dynamic compression. A single-frequency laser beam is divided into two to six individual beams that are transmitted into an experimental device by incident fibers to illuminate measured points. Diffusely reflected laser beams from different measured points are separately collected by fiber detectors and guided by signal fibers into a common "push and pull" interfering cavity with the same delay etalon to interfere. This not only simplifies the system structure and the experimental operation, but also eliminates the system error among measured points and makes the system almost as small in volume as that of a single-point VISAR. The multipoint VISAR possesses all the advantages of the traditional single-point VISAR as well as the temporal and the space resolution ability. We have used it to monitor velocity histories of several points on a target in a few explosion experiments, and good experimental results were obtained.

Key words: Multipoint velocity interferometer system, VISAR, Single-frequency laser

7. Optimal conditions of coupling between the propagating mode in a tapered fiber and the given WG mode in a high-Q microsphere

Ying Lu, Jiyou Wang, Xiaoxuan Xu, Shihong Pan, Cunzhou Zhang
Optik, 2001, 112(3): 109–113.

Abstract: We have investigated the coupling between the propagating mode in a tapered fiber and the given WG mode in a high-Q microsphere by a simple model. Approximate analytic expressions for coupling are derived. The optimal conditions for coupling are discussed.

Key words: Microsphere, Tapered fiber, Coupling efficiency, Optimal coupling

8. An investigation of a tapered fiber-microsphere coupling system with gain and evanescent-field sensing device

Ying Lu, Jianquan Yao, Peng Wang, Cunzhou Zhang
Optik, 2001, 112(10): 475–478.

Abstract: We consider a coupling system composed of a tapered fiber and a microsphere with gain. Power amplification and strong sensitivity to the absorbent sample molecule, on the surface of the sphere are found when gain is larger than intrinsic losses in this system. The intrinsic losses being in some extent compensated by gain make this system possess more advantage in integrated optics and photonics devices application.

Key words: Microsphere, Tapered fiber, Coupling, Gain

9. 基于微球与锥形光纤耦合系统的窄带信道下载滤波器

陆颖，张帆，王鹏，姚建铨

激光与光电子学进展, 2001, (9): 27.

摘要： 系统地研究了以微球-锥形光纤耦合系统为基础的信道下载滤波器的特性及最佳化条件，并与共振器分别为环形和盘形的情况进行了比较。根据耦合区的场传输方程及微球内的共振回廊模特性，得出透过信号、下载信号的输出功率及自由光谱区的解析表达式，通过数值计算，分析了系统的参数对滤波特性的影响，进而进行优化设计。

关键词： 光纤耦合系统，滤波器，微球，锥形

10. Novel alternating current electroluminescent devices with an asymmetric structure based on a polymer heterojunction

Tan Haishu, Yao Jianquan, Wang Xin, Wang Peng, Xie Hongquan

Chinese Physics Letters, 2002, 19(9): 1359–1361.

Abstract: Novel alternating current electroluminescent devices with an asymmetric structure are successfully fabricated by using a hole-type polymer, poly(2,5-bis (dodecyloxy)-phenylenevinylene) (PDDOPV), and an electron-type polymer, poly(phenyl quinoxaline) (PPQ). The performance of the polymer devices with heterojunctions under ac operation is insensitive to the thickness of the two polymer layers, compared to that under dc operation. This new advantage means easy and cheap production facility on a large scale in the near future. Different emission spectra are obtained when our ac devices are operated in an ac mode, forward or reverse bias. The emission spectrum at reverse bias includes two parts: one from PDDOPV and the other from PPQ.

Key words: Alternating current electroluminescent devices, Hetero junctions, PDDOPV

11. 基于微球-锥形光纤耦合系统的窄带信道下载滤波器

陆颖,张帆,王鹏,赵新,姚建铨

中国激光, 2002, A29, Suppl: 379-381.

摘要: 根据耦合区的场传输方程及微球内的共振回廊模特性,得出微球与锥形光纤耦合系统的透过信号、下载信号的输出功率及自由光谱区的表达式,通过数值计算,分析了系统的参数对滤波特性的影响,进而进行优化设计,其结果对信道下载滤波器的设计有一定的指导意义。

关键词: 微球,锥形光纤,耦合系统,滤波器

12. 大气污染监测中的 DOAS 技术

朱孟,于意仲,陈进,姚建铨

中国激光, 2002, A29, Suppl: 561-564.

摘要: 从传统大气污染监测手段的局限性入手,简要介绍先进的光谱大气监测技术,并对紫外及可见光的差分吸收光谱技术(DOAS)进行详细的理论分析,建立该系统的结构框架,并设计 DDAS 系统,给出实验数据,准确地反映出真实污染。

关键词: DOAS,大气监测,开路监测

13. Analysis of speckle in optical coherence tomography

Yuan An, Jianquan Yao, Ruikang Wang

Proceedings of SPIE, 2002, 4916: 245–250.

Abstract: An analysis has been developed based on former theory analysis and a new deduction about the forming of speckles is given. The popular theories believe that the speckle of OCT image comes from the multiple scattering effect, which distorts the wave-front of light then the speckle induced. By the principle of scattering, least-scattered light still maintains coherent, whereas multiply scattered light will lose coherence for the reason of depolarization, phase shift and dispersion. So mostly the least scattered light contributes the amplitude of the detected coherence envelope. The speckle comes from the superposition of several coherence envelopes.

Key words: Optical coherence tomography, Multiple scattering, Refractive index, Coherence length, Bio-tissue

14. Formulation of beam propagating through the organized tissues with polarization sensitive OCT

Yu Jiang, Jianquan Yao, Ruikang K Wang, Peng Wang

Proceedings of SPIE, 2002, 4916: 293–298.

Abstract: It is known that the collagen-rich and well-organized biological tissues have birefringent characteristics when presented to the light. Determination of collagen fiber organization in tissue is of paramount importance in clinical diagnosis. With the precise control of the polarization state of incident and reflected light, polarization sensitive optical coherence tomography (PSOCT) could be used to visualize the tissue birefringence. To understand this better, the mathematical treatment relying on rigorous polarization optics would be needed. This paper is primarily for this purpose. The emphasis is placed on the discussions of light reflected from within a sample using a point source beam and analytical derivations of the polarization properties of tissue based on quasi-Stokes parameters (I, Q, U, V) and Jones matrix formalism.

Key words: OCT, Collagen fiber organization, Polarization, Birefringence

15. Tissue clearing of bio-tissues for optical coherence tomography

Ruikang K Wang

Proceedings of SPIE, 2002, 4916: 241−244.

Abstract: The high scattering nature of non-transparent human tissue limits the imaging depth of optical coherence tomography (OCT) to 1-2 millimeters. By using the longer wavelength of the light source, the penetration depth is improved; the imaging contrast is however decreased largely due to the reduced backscattering in microscopic scale and the reduced refractive heterogeneity in macroscopic scale. For more effective diagnosis using OCT, a concurrent improvement of penetration depth and imaging contrast are often needed. We report in this paper that the OCT imaging depth and contrast can be enhanced concurrently by the use of osmotic agents. We demonstrate experimentally, by examples, that the topical applications of glycerol and propylene glycol, two common biocompatible and osmotically active solutions, onto the tissue surfaces could significantly improve the OCT imaging contrast and depth capability. The biotissues demonstrated include the rat skin, human oesophageal and gastric tissues.

Key words: Bio-tissues, OCT

16. Analyses on propagation and imaging properties of GRIN lens

Li Xuejin, Yao Jianquan, Zhang Baigang

Proceedings of SPIE, 2002, 4919: 155−160.

Abstract: GRIN lenses, graded-index lenses, are widely used to construct micro-optic devices such as optical coupler, connector, beam splitters, optical attenuator, beam expanders, optical fiber switch and WDM, etc. Grin lens should be designed according to different uses to meet some special requirements. It is essential to realize the properties of propagation and imaging. Propagation properties and image properties of two kinds of Grin lens, respectively parabolic-index dielectric cylinder GRIN lens and tapered cylinder GRIN lens, are described in this paper. The propagation properties are studied using a ray-optics approach. We denote the passage of a ray through these GRIN lens can be described by simple 2×2 matrices. The parameters of this two kind GRIN lenses, such as focal length and magnifying power are given in this paper meanwhile. Comparison between cylinder GRIN lenses and tapered cylinder GRIN lenses is done in applications for coupling of light sources to fibers, beam expanding/refocusing in micro-optics devices, and splicing two fibers of different core sizes.

Key words: Fiber devices, Imaging, Tapered GRIN lens

17. Research on parameters option of a pressure fiber sensor

Li Xuejin, Zhang Baigang, Yao Jianquan, Wang Peng
Proceedings of SPIE, 2002, 4919: 149–154.

Abstract: A pressure sensor based on fiber-optical sensor(FOS) is discussed about the design parameters option in this paper. An elastic spherical diaphragm is used as a reflective target of displacement fiber sensor in this paper. The spherical diaphragm coils when force is performed upon, light intensity of receiving fiber changes depending on the deformation of it. The relationship between sensitivities of the sensor and diameters of the elastic diaphragm, diameters of fibers and the distance from the terminal of fiber to the initial diaphragm before force performed are calculated based on light distributing theory. From the consequence of calculations, we give out the optimum condition for this kind fiber sensor design. The receiving intensity of light have a good linear relationship to the fore is presented meanwhile.
Key words: Fiber sensor, Pressure sensor, Deformation

18. The amplified model of erbium-doped fiber amplifier (EDFA) with cross-phaser modulation (XPM)

Huang Jing, Yao Jianquan, Liang Ruisheng
Optics Communications, 2003, 220: 433–438.

Abstract: Based on the rate equations and the power propagation equations, the amplified model is obtained which considers the cross-phase modulation (XPM)-induced intensity fluctuation between pump power and signal power in erbium-doped fiber amplifier (EDFA). Simulation results show: a higher nonlinear coefficient and a stronger pump power cause a stronger XPM in EDFA. Before the optimal length of EDF, the intensity caused by XPM increases with the increasing length, but after the optimal length, the XPM presents saturation. A bigger nonlinear coefficient γ corresponds to a smaller optimal length and gain, and makes gain saturation more easy. XPM effect will decrease EDFA's amplified efficiency.
Key words: XPM, EDFA, Small-signal gain

19. 紫外差分光学吸收法测量污染气体的实验研究

于意仲,周小玉,张帆,李喜福,马家驹,姚建铨,王鹏

环境科学学报, 2003, 23(5): 630-634.

摘要: 采用差分光学吸收光谱法,对紫外波段的 SO_2、NO、NO_2 污染气体吸收光谱进行了测量。得到了紫外波段的差分吸收截面。提出的分波段计算方法对混合气体测量的误差达到15%以内,测量精度与所选测量波段有关。

关键词: 差分光学吸收光谱,最小二乘法拟合,大气污染监测

20. 新型单片式开关电源的电磁干扰及其抑制

马家驹,于意仲,陈进,姚建铨

河北工业大学学报, 2003, 32(2): 67-70.

摘要: 分析了 TOPSwitch 单片开关电源存在的电磁干扰源,并根据其差模干扰和共模干扰的特点分别建立电路模型,最后提出不同的抑制方法。

关键词: TOPSwitch,电磁兼容,共模干扰,差模干扰,开关电源

21. 长光程差分吸收光谱法

吴桢，虞启琏，张帆，姚建铨

环境科学与技术，2003, 26(1): 48–49.

摘要： 长光程差分吸收光谱法是一种实时、在线监测方法，可同时对多种气体进行连续监测。使用这种方法获得的数据经计算机处理后，可分析出大气中各种污染物的含量。介绍了这种方法的基本原理、仪器结构及它的应用。

关键词： 长光程，差分吸收光谱法，光谱仪

22. Power distribution in Yb^{3+}-doped double-cladding fiber laser

Qiang Zhang, Jianquan Yao, Peng Wang, Jianing Zhou, Yuanqin Xia, Baigang Zhang

Chinese Optics Letters, 2004, 2(8): 468–470.

Abstract: The distribution of pump light and signal light in Yb^{3+}-doped double-cladding fiber laser is analyzed based on a rate equation model. Numerical simulation results are obtained. The numerical solution of the rate equation is shown to be in excellent agreement with the experimental data.

Key words: Power distribution, Yb^{3+}-doped, Double-cladding, Fiber laser

23. Absorption and emission of ErNbO$_4$ powder

Delong Zhang, Yufang Wang, E.y.b. Pun, Yizhong Yu, Caihe Chen, Jianquan Yao
Optical Materials, 2004, 25(4): 379–392.

Abstract: Visible and near infrared absorption and emission (488 nm excitation) characteristics of ErNbO$_4$ powder, which were prepared by calcining the Er$_2$O$_3$ (50 mol%) and Nb$_2$O$_5$ (50 mol%) powder mixture at 1 100 and 1 600 ℃ for different durations, have been investigated at room temperature. The absorption and emission characteristics of these calcined ErNbO$_4$ powder were summarized and discussed in comparison with those of Er$_2$O$_3$. Weak emission of Er$_2$O$_3$ relative to the calcined ErNbO$_4$ is mainly conducted with absorption difference at the excitation wavelength 488 nm. The obvious spectral changes from Er$_2$O$_3$ to calcined ErNbO$_4$ samples are related to an elevated-temperature-assisted phase transformation according to the solid-state chemical reaction equation: Er$_2$O$_3$ + Nb$_2$O$_5$ ⇌ 2ErNbO$_4$, which results in the changes of the ion environment of Er^{3+} and hence changes of the Stark levels of Er^{3+}. The further spectral change as the strengthened calcination results from the improvement of ErNbO$_4$ purity in the calcined mixture. The borders between two green transitions and between two near infrared transitions in the emission spectra of both calcined samples and Er$_2$O$_3$ were tentatively identified by referencing earlier reported emission spectra of the precipitated Z-cut VTE Er(2.0 mol%):LiNbO$_3$ crystal and the match relation between absorption and emission spectra of the ErNbO$_4$ powder. A comparison was performed on the spectra of calcined ErNbO$_4$ powder and those of VTE Er(2.0 mol%):LiNbO$_3$ crystals. The results allow to preliminarily deem the contribution of ErNbO$_4$ precipitates, generated inside these crystals by the VTE procedure, to the spectra of these crystals.

Key words: ErNbO$_4$, Absorption, Emission

24. 一种光纤压力传感器的设计理论分析

李学金，张百钢，姚建铨，胡巨广
传感技术学报, 2004, 17(1): 133–135.

摘要：设计了一种用于流体压力的反射式光纤压力传感器。利用高反射率的弹性膜片作为光纤位移传感器的反射镜，由光纤光强分布理论出发，通过理论计算，分别给出了膜片直径、光纤芯径、光纤端面到膜片的初始距离与灵敏度的关系，为传感器设计最佳参数选择提供了参考。文中指出，传感器中光纤可以选用 200 μm 左右大芯径光纤，膜片直径可以选 2~3 mm，光纤端面与膜片初始距离可选 2 mm 左右。在此情况下，传感器具有较高灵敏度和较好的线性。

关键词：光纤位移传感器，光强分布，灵敏度，变形

25. Image distortion of optical coherence tomography

An Yuan, Yao Jianquan
Transactions of Tianjin University, 2004, 10(1): 81–84.

Abstract: A kind of image distortion in Optical Coherence Tomography (OCT)resulted from average refractive index changes between structures of bio-tissue is discussed for the first time. Analysis is given on following situations:1) Exact refraction index changes between microstructures; 2) The gradient of average refractive index change between different tissue layers is parallel to the probe beam ;3) The gradient of average refractive index change is vertical to the probe beam .The results show that the image distortion of situation 1) is usually negligible; in situation 2) there is a spread or shrink effect without relative location error; however, in situation 3) there is a significant image error inducing relative location displacement between different structures. Preliminary design to eliminate the distortion is presented, the method of which mainly based on the image classification and pixel array re-arrangement.

Key words: OCT, Image distortion, Refractive index, Bio-tissue

26. A monte carlo model of light propagation in nontransparent tissue

Yao Jianquan, Zhu Shuiquan, Hu Haifeng, Wang Ruikang
Transactions of Tianjin University, 2004, 10(3): 209–213.

Abstract: To sharpen the imaging of structures, it is vital to develop a convenient and efficient quantitative algorithm of the optical coherence tomography (OCT) sampling. In this paper a new Monte Carlo model is set up and how light propagates in bio-tissue is analyzed in virtue of mathematics and physics equations. The relations, in which light intensity of Class 1 and Class 2 light with different wavelengths changes with their permeation depth, and in which Class 1 light intensity (signal light intensity)changes with the probing depth, and in which angularly resolved diffuse reflectance and diffuse transmittance change with the exiting angle, are studied. The results show that Monte Carlo simulation results are consistent with the theory data.

Key words: OCT, Mathematics and physics equations, Bio-tissue, Monte Carlo model, Photon cluster, Weight

27. 2.2 W 掺 Yb^{3+} 双包层光子晶体光纤激光器

阮双琛,杨冰,朱春艳,林浩佳,姚建铨
光子学报, 2004, 33(1): 15–16.

摘要: 采用多模大功率 972 nm 半导体激光器泵浦 20 m 掺 Yb 双包层光子晶体光纤,详细研究了输出功率与泵浦功率的关系,获得了 1.09 μm,功率为 2.2 W 的激光输出。

关键词: 掺 Yb^{3+} 光子晶体光纤, 双包层

28. 15 W 光子晶体光纤激光器的研究

阮双琛，杜晨林，杨冰，朱春艳，姚建铨，林浩佳

光子学报，2004, 33(10): 1156–1158.

摘要：采用多模大功率 980 nm 半导体激光器泵浦 20 m 掺 Yb 双包层光子晶体光纤，获得了 1.09 μm，功率为 15 W 的激光输出。详细研究了输出功率与泵浦功率的关系。

关键词：掺 Yb^{3+} 光子晶体光纤，双包层

29. Supercontinuum generation at 1.6 μm region using a polarization-maintaining photonic crystal fiber

Yu Yongqin, Ruan Shuangchen, Chen Chao, Du Chenlin, Yao Jianquan

Acta Photonica Sinica, 2004, 33(11): 1301–1303.

Abstract: The generation of supercontinuum in the 1.6 μm region was reported in a polarization-maintaining photonic crystal fiber. The optical pulses produced by optical parameter amplifier (OPA) with the central wavelength of 1.593 8 μm, the repetition rate of 250 kHz, the pulse duration of 250 fs were coupled into a 0.2 m-long, polarization-maintaining photonic crystal fiber. The broadened spectrum with the bandwidth of 45.8 nm (1.589 2 ~ 1.635 0 μm) in the 1.6 μm region was obtained.

Key words: Photonic crystal fiber, Supercontinuum, Optical parameter amplifier

30. Theoretical study of double microcavity resonators system with absorption or gain

Ying Lu, Jianquan Yao, Xifu Li, Peng Wang, Yizhong Li, Xin Ding, Wuqi Wen, Meng Zhu

Proceedings of SPIE, 2004, 5451: 458-464.

Abstract: We present a theoretical study of a double microcavity resonators system with absorption or gain. The output intensity and the power intensity inside the microresonator are derived. Some interesting features and their possible application are discussed. The results may be useful for modulator, amplifier, laser and sensor.

Key words: Microcavity resonator, Coupling coefficient, Absorption, Gain

31. Conceptual design of LD side pumped high power disk fiber laser

Qiang Zhang, Jianquan Yao, Wuqi Wen, Jianing Zhou, Rui Zhou

Proceedings of SPIE, 2004, 5627: 481-487.

Abstract: Disk fiber laser is a novel fiber laser. The pumping scheme of such type of fiber laser is side pumping by LD arrays and the pumping scale is large in comparison with the core pumping manner in a typical clad pumping scheme. More pumping power could be coupled into the disk and higher output power could be achieved. To optimize the system of disk fiber laser, it is necessary to analyze the parameter of each part of it. In this paper, the configuration factors that influence the pumping efficiency of disk fiber laser were analyzed and propagation of the rays in disk fiber laser was simulated using the method of BPM. In the process of simulation, the fiber was treated as cylindrical lens. The optimal position of pump resource is obtained with a fixed size of the fiber cross section.

Key words: Disk fiber laser, Simulation, BPM, LD, Side pump

32. High-speed spectral domain optical coherence tomography for imaging of biological tissues

Zhenhe Ma, Ruikang K Wang, Fan Zhang, Jianquan Yao
Proceedings of SPIE, 2004, 5630: 286–294.

Abstract: Optical coherence tomography (OCT) is a new modality used to image biological tissues that weakly scatter and absorb light. It was demonstrated that this technique provides image with micrometer resolution in a noncontact and noninvasive way. Traditional OCT is time domain OCT (TDOCT). In this method the length of the reference arm is rapidly scanned over a distance corresponding to the imaging depth range. The mechanism of scanning largely limits the acquisition speed and makes real-time imaging impossible. In recent year a new model OCT based on Fourier domain interferometry is emerged, we called it spectral OCT (SOCT) or Fourier domain OCT (FDOCT). SOCT can avoid scanning of the reference, thus can reach very high acquisition speed.

Key words: Spectral OCT, Time domain OCT, Optical frequency domain imaging (OFDI)

33. Photoacoustic tomography imaging of biological tissues

Yixiong Su, Ruikang K Wang, Kexin Xu, Fan Zhang, Jianquan Yao
Proceedings of SPIE, 2004, 5630: 582–586.

Abstract: Non-invasive laser-induced photoacoustic tomography is attracting more and more attentions in the biomedical optical imaging field. This imaging modality takes the advantages in that the tomography image has the optical contrast similar to the optical techniques while enjoying the high spatial resolution comparable to the ultrasound. Currently, its biomedical applications are mainly focused on breast cancer diagnosis and small animal imaging. In this paper, we report in detail a photoacoustic tomography experiment system constructed in our laboratory. In our system, a Q-switched Nd: YAG pulse laser operated at 532 nm with a 10 ns pulse width is employed to generate photoacoustic signal. A tissue mimicking phantom was built to test the system. When imaged, the phantom and detectors were immersed in a water tank to facilitate the acoustic detection. Based on filtered back-projection process of photoacoustic imaging, the two-dimension distribution of optical absorption in tissue phantom was reconstructed.

Key words: Photoacoustic tomography, Optical absorption, Back projection, Imaging reconstruction

34. Fourier domain optical coherence tomography for imaging of biological tissues

Zhenhe Ma, Shuquan Zhu, Ruikang K Wang, Fan Zhang, Jianquan Yao

Proceedings of SPIE, 2004, 5630: 844–850.

Abstract: Optical coherence tomography (OCT) is a new imaging modality that is being actively used in a variety of medical applications. Currently, most of the OCT systems operate in the time domain, which requires scanning the optical path length in the reference arm in order to obtain the in-depth profile, i.e. A scan. This however limits the system scanning speed. To avoid the axial scanning and therefore improve the system scanning speed, a novel OCT system is recently proposed by a number of groups that operates in the frequency domain, i.e. the spectral OCT. In this paper, we report the spectral OCT system being constructed at Tianjin University. The system has a dynamic range at 78 dB and is capable of scanning speed at 12 seconds per image, largely limited by the bottleneck of data transferring from the CCD camera currently employed to the computer. The SOCT imaging results obtained from the animal tissues (cornea from an intact porcine eye) in vitro will be presented.

Key words: Time domain OCT, Spectral domain OCT, CCD camera, Biological tissues

35. Multi-channel FBG sensing system using a dense wavelength demultiplexing module

Enbang Li, Jiangtao Xi, Yanguang Yu, Joe Chicharo, Jianquan Yao
Proceedings of SPIE, 2004, 5634: 211−218.

Abstract: Fiber Bragg grating (FBG) sensing is gaining attention in both scientific research areas and engineering applications thanks to its distinguishing advantages including wavelength multiplexing capability, miniature size, high sensitivity, immunity from electro-magnetic interference, etc. FBG sensing is based on the detection of the shifted Bragg wavelength of the light reflected by a fiber grating which is sensitive to various physical parameters such as strain and temperature. One of the challenging tasks in FBG sensing is to determine the Bragg wavelength shift, which can be done by using an optical spectrum analyzer (OSA). An OSA is suitable for laboratory tests, but not an ideal solution for field applications in term of cost and convenience. Different wavelength demodulation methods have been developed for FBG sensing purpose. One of them is employing a bulk linear edge filter to convert the wavelength shifts to intensity variations. This method offers several obvious advantages including low cost and ability for dynamic measurements. However, most of the edge-filter based FBG sensing systems are designed for single-channel measurement. In this study, we propose and develop a multi-channel FBG sensing system based on the edge-filtering technique. In order to demodulate multi-channel signal from FBG sensors, we propose to use a dense wavelength division multiplexing (DWDM) module. The light signals coming from wavelength-multiplexed FBG sensors are demultiplexed into individual channels and demodulated by the pass-band edges. In the present study, a four-channel FBG sensing system has been demonstrated.

Key words: Fiber Bragg grating, Fiber sensing, Wavelength division multiplexing

36. Mobile on-line DOAS trace-gases monitoring system with fiber spectroscopy

Meng Zhu, Wuqi Wen, Xin Ding, Peng Wang, Jianquan Yao
Proceedings of SPIE, 2004, 5634: 675–679.

Abstract: The article mainly focused on mobile on-line air quality monitoring system. By deeply analyzed DOAS theory, we designed this new air quality monitoring system. It is mobile and may monitoring many pollution sources on-line everyday. The Differential Optical Absorption Spectroscopy (DOAS), based on the work by U. Platt and co-workers, is becoming increasingly popular for environment monitoring. DOAS may measure many trace gases like NO_2, O_3, BrO, NO_3, and SO_2. It is designed for the measurement of primary and secondary urban air pollutants with high precision and little cross interference. In the DOAS technique, the spectrum of an artificial light source within a given bandwidth is measured after passing through the open atmosphere for between 100 m and 10 km. After removing the emission spectrum of the light source, the remaining differential absorption features are compared with the absorption cross sections of relevant trace gases. This allows both the qualitative and quantitative determination of their concentration in the light path. After deeply research, we design this new system. It uses fiber spectroscopy, and it is mobile. User may monitor many pollution sources in a car. This makes auto-monitoring more easily. Our DOAS system has these merits: New Light Emission-Receiver Unit which united emission、receiver and collimation lenses. New Background Elimination Fiber. And dynamic-feedback self-adapting program. In our experiment by this design idea, we get accurate data.

Key words: Trace-gas monitoring, DOAS, On-line, Fiber spectroscopy, Sulphur dioxide, UV spectrum

37. Optical bistability and differential amplification in coupled nonlinear microcavity resonators

Ying Lu, Jianquan Yao, Yizhong Yu, Peng Wang

Proceedings of SPIE, 2004, 5646: 404–410.

Abstract: We present a theoretical study of optical bistability and differential amplification, arising as a consequence of the nonlinear Kerr effect in mutually coupled nonlinear microcavity resonators. The dependence of the bistability and differential amplification on various parameters of the coupling range and the resonators is investigated. Multiple closely spaced bistability of the internal intensity in the microresonator is found with appropriate parameters. The results may be useful for a complex logic, memory or switch operation system.

Key words: Optical bistability, Differential amplification, Microcavity resonator, Kerr effect

38. Optical phase shifting with acousto-optic devices

Enbang Li, Jianquan Yao, Daoyin Yu, Jiangtao Xi, Joe Chicharo

Optics Letters, 2005, 30(2): 189–191.

Abstract: A novel optical phase-shifting method based on a well-known acousto-optic interaction is proposed. By using a pair of acousto-optic modulators (AOMs) and properly aligning them, we construct an optical phase shifter that can directly control the phase of a collimated beam. The proposed phase shifter is insensitive to the polarization of the incident beam when polarization-insensitive AOMs are used, and no calibration is necessary. The proposed approach is confirmed with experimental results.

Key words: Phase shift, Phase modulation, Interferometry, Acousto-optical devices

39. Tunable asymmetrical Fano resonance and bistability in a microcavity-resonator-coupled Mach-Zehnder interferometer

Ying Lu, Jianquan Yao, Xifu Li, Peng Wang

Optics Letters, 2005, 30(22): 3069–3071.

Abstract: We propose a simple microresonator scheme for a Mach–Zehnder interferometer in which a microresonator is side coupled to one arm and a phase shifter is introduced into the other arm, to produce an asymmetric Fano-resonance line shape. In this system, a phase shifter is used to control the variation of the asymmetric line shape, with another reverse resonance next to a resonance minimum over a very narrow frequency range, which results from the interference between a direct channel and a high-Q resonance indirect channel. We also theoretically investigate the novel bistability characteristic based on these shapes.

Key words: Optical tuning, Optical bistability, Microcavities, Optical resonators, Mach-Zehnder interferometers, Micro-optics, Q-factor, Optical phase shifters, Nonlinear optics, Resonators

40. Multi-frequency and multiple phase-shift sinusoidal fringe projection for 3D profilometry

E. B. Li, X. Peng, J. Xi, J. F. Chicharo, J. Q. Yao, D. W. Zhang

Optics Express, 2005, 13(5): 1561–1569.

Abstract: In this paper, we report on a laser fringe projection set-up, which can generate fringe patterns with multiple frequencies and phase shifts. Stationary fringe patterns with sinusoidal intensity distributions are produced by the interference of two laser beams, which are frequency modulated by a pair of acousto-optic modulators (AOMs). The AOMs are driven by two RF signals with the same frequency but a phase delay between them. By changing the RF frequency and the phase delay, the fringe spatial frequency and phase shift can be electronically controlled, which allows high-speed switching from one frequency or phase to another thus makes a dynamic 3D profiling possible.

Key words: Acousto-optical devices, Interferometry, Three-dimensional image acquisition

41. Crosstalk characteristics of resonant dispersion EDFAs in WDM systems

Huang Jing, Yao Jianquan
Optics Communications, 2005, 251: 132–138.

Abstract: The nonlinear Schrödinger equation is solved using the small signal analysis approach. In the resonant dispersion erbium-doped fiber amplifiers (EDFAs) of WDM systems, the crosstalk between signals and induced by signals are discussed. The resonant dispersion of EDFAs has a strong effect on the cross-phase modulation instability gain frequency, and causes severe crosstalk when the multiple signal channels are taken into account. In a WDM system, for most channels where the resonant dispersion occurs, the crosstalk powers induced by other channels' XPM are stronger than the intensity modulation induced by XPM between the reference and other channels. The factors affecting crosstalk include signal power, nonlinear coefficient and channel separation. Pump power distribution can dominantly diminish crosstalk. To keep the system s performance, the proper dispersion compensation is a better method to minimize the crosstalk.

Key words: Crosstalk, Resonant dispersion, Small signal analysis, Erbium-doped fiber amplifer

42. A photoacoustic tomography system for imaging of biological tissues

Yixiong Su, Fan Zhang, Kexin Xu, Jianquan Yao, Ruikang K Wang
Journal of Physics D: Applied Physics, 2005, 38: 2640–2644.

Abstract: Non-invasive laser-induced photoacoustic tomography (PAT) is a promising imaging modality in the biomedical optical imaging field. This technology, based on the intrinsic optical properties of tissue and ultrasonic detection, overcomes the resolution disadvantage of pure-optical imaging caused by strong light scattering and the contrast and speckle disadvantages of pure ultrasonic imaging. Here, we report a PAT experimental system constructed in our laboratory. In our system, a Q-switched Nd:YAG pulse laser operated at 532 nm with a 8 ns pulse width is used to generate a photoacoustic signal. By using this system, the two-dimensional distribution of optical absorption in the tissue-mimicking phantom is reconstructed and has an excellent agreement with the original ones. The spatial resolution of the imaging system approaches 100 μm through about 4 cm of highly scattering medium.

Key words: Potoacoustic tomography, Imaging, Biological tissues

43. Analysis of cross-phase modulation in WDM systems

Huang Jing, Yao Jianquan

Journal of Modern Optics, 2005, 52(13): 1819–1825.

Abstract: A general theory on cross-phase modulation (XPM) intensity fluctuation has been derived. It is adapted to all kinds of modulation formats (continue wave, not-return-zero and return-zero pulse waves). In single-mode fibres (SMFs) and nonzero dispersion-shifted fibres (NZDSFs), this analytical solution for XPM is validated for the not-return-zero (NRZ) pulse modulation format. The relationship between the XPM intensity fluctuation and dispersion is discussed for different signal powers, channel numbers, wavelength separations, fibre lengths and nonlinear coefficients. The conclusions drawn can be used to explain some contradictory experimental results.

Key words: Cross-phase modulation, WDM systems, Single-mode fibres, Not-return-zero modulation

44. Small-signal analysis of cross-phase modulation instability in lossy fibres

Huang Jing, Yao Jianquan

Journal of Modern Optics, 2005, 52(14): 1947–1955.

Abstract: Based on the small signal analysis theory and split-step Fourier method, the complex nonlinear Schrodinger equation (NLSE) with fibre loss can be solved. This procedure is also adapted to the NLSE with the high order dispersion terms (β_3, β_4). Because the fibre loss is taken into account, experiment demonstrates that the cross-phase modulation instability gain spectrum is more similar to represent the actual systems. The fibre loss decreases the modulation instability (MI) gain and has an effect on SNR which is caused by the cross-phase modulation instability. The MI intensity fluctuation caused by the nonlinear and dispersion effects is directly derived. As a result, the initial stage of MI can be described, and the whole characteristic of MI can also be discussed in this way.

Key words: Cross-phase modulation, Lossy fibres, Modulation instability, Nonlinear Schrodinger equation

45. Supercontinuum generation using a polarization-maintaining photonic crystal fibre by a regeneratively amplified Ti:sapphire laser

Yu Yongqin, Ruan Shuangchen, Du Chenlin, Yao Jianquan
Chinese Physics Letters, 2005, 22(2): 384−387.

Abstract: Supercontinuum with an ultra-broad bandwidth in the range from 380 nm to 1 750 nm was generated by injecting 250 kHz 200 fs optical pulses produced by a regeneratively amplified Ti:sapphire laser into a 2.5-m-long polarization-maintaining photonic crystal fibre (PCF). It is indicated that the mechanism for the supercontinuum generation in the anomalous dispersion region of the PCF are directly related to the Raman effect, the fission of higher-order solitons, nonsolitonic radiation, and the coinstantaneous effect of four-wave mixing. The frequency components beyond 1.4 μm were also observed. It is interpreted that the energy of solitons is shifted beyond the OH absorption with a higher input power.

Key words: Supercontinuum Generation, Photonic Crystal Fibre, Polarization-Maintaining, Regeneratively Amplified

46. Spectral optical coherence tomography using two-phase shifting method

Ma Zhenhe, Ruikang K. Wang, Zhang Fan, Yao Jianquan
Chinese Physics Letters, 2005, 22(8): 1909−1912.

Abstract: A two-phase shifting method is introduced to eliminate the strong autocorrelation noise inherent in spectral optical coherence tomography and to mitigate the unwanted auto- and cross-coherent terms introduced by the reflections from various optical interfaces present in the system. Furthermore, this method is also able to amplify the desired signal by a factor of 2. The feasibility of such a method is demonstrated using a mirror-like object. An intact porcine cornea tissue in vitro is also used to show the potential of this method for biological imaging.

Key words: Spectral optical coherence tomography, Quantum theory

47. Spectral broadening in the 1.3 μm region using a 1.8-m-long photonic crystal fiber by femtosecond pulses from an optical parametric amplifier

Yu Yongqin, Ruan Shuangchen, Du Chenlin, Yao Jianquan
Acta Photonica Sinica, 2005, 34(4): 481–484.

Abstract: Supercontinuum with a spectral bandwidth of 700 nm (1.09 μm-1.79 μm) was achieved in a 1.8-m-long photonic crystal fiber with an average core radius of 2.0 μm pumped by optical femtosecond pulses at the wavelength of 1.275 9 μm, with the average power of 30 mW, the duration of 250 fs and the repetition rate of 250 kHz from an optical parametric amplifier. It was interpreted that the spectral broadening was due to the fission of higher-order solitons and four-wave mixing. The concave profile at the wavelength of about 1.4 μm was resulted from the OH absorption. The broadened spectra were applicable to a multi-channel optical source with ultra-short pulse width for WDM communication and photonic network systems.

Key words: Photonic crystal fiber, Supercontinuum, Optical parameter amplifier

48. Analysis and simulation of XPM intensity modulation

Jing Huang, Jianquan Yao
Chinese Optics Letters, 2005, 3(3): 129–131.

Abstract: Based on the split-step Fourier method and small signal analysis, an improved analytical solution which describes the cross-phase modulation (XPM) intensity is derived. It can suppress the spurious XPM intensity modulation efficiently in the whole transmission fiber. Thus it is more coincidence with the practical result. Furthermore, it is convenient, because it is independent of channel separation and the dispersion and nonlinear effects interact through the XPM intensity. A criterion of select the step size is described as the derived XPM intensity modulation being taken into account. It is non-uniform distribution and is the function of average signal power <P(z)> (or z). Compared with the conventional split-step method, the simulation accuracy is improved when the step size is determined by the improved XPM intensity.

Key words: Cross-phase modulation, Intensity modulation

49. A method of simulating intensity modulation-direct detection WDM systems

Huang Jing, Yao Jianquan, Li Enbang
Optoelectronics Letters, 2005, 1(1): 0057–0060.

Abstract: In the simulation of Intensity Modulation-Direct Detection WDM Systems, when the dispersion and nonlinear effects play equally important roles, the intensity fluctuation caused by cross-phase modulation may be overestimated as a result of the improper step size. Therefore, the step size in numerical simulation should be selected to suppress false XPM intensity modulation (keep it much less than signal power). According to this criterion, the step size is variable along the fiber. For a WDM system, the step size depends on the channel separation. Different type of transmission fiber has different step size. In the split-step Fourier method, this criterion can reduce simulation time, and when the step size is bigger than 100 meters, the simulation accuracy can also be improved.

Key words: Optical fiber communication, WDM, Dispersion, Nonlinear effects, Intensity Modulation, Direct detection

50. 时域光声技术及其在生物组织检测中的应用

苏翼雄，王瑞康，徐可欣，张帆，姚建铨
光谱学与光谱分析，2005, 25(8): 1176–1179.

摘要：当强度调制的光束照射于吸收物质，周期性热流使周围的介质热胀冷缩而激发声波，这种将光能转化为声能的现象称为光声效应。基于光声效应的时域光声谱技术将光学和声学有机地结合，为生物组织的无损检测技术提供了新的检测手段。该技术能够实现类似光学技术的高对比度和近似于声学技术的高精度和穿透深度，在生物医学检测中具有广阔的应用前景。文章介绍时域光声谱技术的原理及其在生物组织成分检测和层析成像检测中的应用。

关键词：光声效应，时域光声谱技术，成分检测，层析成像技术，生物医学

51. Bandgap extension of disordered 1D ternary photonic crystals

Zhang Yuping, Yao Jianquan, Zhang Huiyun, Zheng Yi, Wang Peng
Acta Photonica Sinica, 2005, 34(7): 1094–1098.

Abstract: Bandgap properties of disordered one-dimensional (1D) ternary photonic crystals are investigated by optical transfer matrix method for the first time. The results show that disordered structure provides strikingly extended bandgap compared with the corresponding periodic structure. The more ingredient of disordered dielectric multilayers adopted in the calculation, the wider stop band will be obtained. The influence of degree of disorder D and contrast of high and low refractive indices to the photonic bandgap are also calculated and discussed.
Key words: Photonic crystal, Degree of disorder, Band gap extension

52. 泵浦波长对光子晶体光纤产生超连续谱的影响

于永芹，阮双琛，曾剑春，姚建铨
光子学报，2005, 34(9): 1293–1296.

摘要： 采用钛宝石光参量放大器作为泵浦源，利用其输出波长可调谐性，研究了不同泵浦波长对光子晶体光纤中产生超连续谱的影响，结果表明光子晶体光纤中零色散点处的群时延和 1.4 μm 处的 OH 根离子的吸收对超连续谱的平坦度影响很大，并且泵浦波长离光纤的零色散点越远，产生的超连续谱平坦度越差，甚至在可见光区产生的各个频率峰还只是分离的，没有形成超连续谱。当泵浦波长为 1.2 μm 时，获得了带宽为 300 nm～1 350 nm 的超连续谱，谱宽超过了两个倍频程。
关键词： 光子晶体光纤，光参量放大器，超连续谱

53. 使用混合高渗制剂提高 OCT 的探测深度和清晰度

安源, 姚建铨

天津大学学报, 2005, 38(10): 927–930.

摘要：用高渗制剂可提高光学相干层析成像 (OCT) 的探测深度。然而单一高渗制剂由于个体敏感性差异而效果不够稳定，为此提出了使用混合高渗制剂涂敷在被测表面以增透并提高其稳定性。使用自行建立的 OCT 系统对涂敷 100% 丙三醇和饱和葡萄糖混合制剂、100% 丙三醇和 50% 丙三醇前后的被测样本成像并记录参数。实验结果显示可降低表层组织散射强度，提高探测深度（约为原探测深度的 2 倍）和组织结构的可读性，并初步验证混合高渗剂 (σ=0.038) 较单一高渗制剂 (σ=0.084) 具有更均一的人体敏感性。

关键词：光学相干层析成像，折射率，散射系数，探测深度，高渗制剂

54. 光纤激光器透镜耦合系统的优化设计

邓元龙, 姚建铨, 阮双琛, 王鹏

激光杂志, 2005, 26(5): 42–43.

摘要：准直—聚焦双透镜系统是光纤激光器中常用的一种泵浦耦合技术，但是对于大功率 LD 泵浦源尾纤输出的多模类高斯光束，难以解析表达其中高阶模式的透镜变换规律，根据此设计透镜耦合系统过程也很复杂。本文用几何光学的方法保证数值孔径的匹配，以类高斯光束光斑尺寸作为基模高斯光束束腰，用基模高斯光束传播规律来简化大功率光纤激光器透镜耦合系统的优化设计。最后给出了一个利用 MATLAB 优化工具函数设计双透镜聚焦耦合系统的实例。

关键词：优化设计，高斯光束，透镜聚焦，耦合系统，光纤激光器

55. 高功率光子晶体光纤激光器及关键技术

邓元龙,姚建铨,阮双琛,王鹏

激光技术, 2005, 29(6): 596–598.

摘要: 与常规双包层光纤相比,空气包层大模面积光子晶体光纤更适用于高功率激光器的研制。介绍了高功率光子晶体光纤激光器研究的最新进展,分析了耦合系统和谐振腔设计中所存在的不利于功率提高的因素,指出低损耗的熔接技术是光子晶体光纤激光器达到更高功率的关键。

关键词: 光纤激光器,光子晶体光纤,高功率,大模面积,熔接

56. 基于周期极化晶体的 Solc 型滤波器透射谱的研究

纪峰,张百钢,路洋,张铁梨,王鹏,姚建铨

激光杂志, 2005, 26(6): 67–67.

摘要: 利用外加电场极化法制作周期极化晶体时占空比一般存在着较大误差,因此本文分析了基于周期极化铌酸锂晶体(PPLN)的 Solc 型滤波器,在非理想占空比下 ($D \neq 1/2$) 的滤波特性,用耦合波理论推导了 D 为任意值时的透过率公式。指出通过增加电压,非理想占空比下中心波长的透过率与理想占空比 $D = 1/2$ 时的情况相同,仍然可以达到 100%(忽略损耗),同时两者的透射谱也相同。

关键词: Solc 型滤波器,PPLN,占空比

57. 大功率激光光纤透镜耦合系统设计

邓元龙,姚建铨,阮双琛,孙秀泉

光电子技术应用, 2005, 20(3): 7-9.

摘要: 在不考虑像差的情况下,采用傍轴光线传输 ABCD 定律,结合混合模系数 M(或 M^2) 的定义,研究了大功率 LD 尾纤输出的混合模类高斯光束经透镜之后束腰及束腰位置和发散角的变化,并在此基础上,利用 MATLAB 软件中的优化设计工具,提出一种设计大功率激光光纤准直—聚焦双透镜耦合系统的简单方法,最后给出了一个光子晶体光纤激光器透镜耦合系统设计实例。

关键词: 耦合系统,类高斯光束,光纤激光器,优化设计

58. 光声技术在医疗成像中的应用

刘英杰,苏翼雄,姚建铨,王瑞康

医疗卫生装备, 2005, 26(8): 26-28.

摘要: 现在的医学成像手段不能很好地探测乳腺癌中的小块肿瘤。对于癌症检测来说,光声成像技术是一种前景光明的成像技术。这一成像手段既有光学成像的高对比度,又有声学成像的高穿透深度,结合了二者的优点。在建立的光声成像系统中,采用 532 nm、脉宽 10 ns 的脉冲激光器作为光源,产生超声信号。为了对系统在癌症检测中的能力进行评估,制作了一个由琼脂和 intralipid 制成的乳房组织模型。成像时,为了更容易探测到超声信号,组织模型和探测器放置在一个透明有机玻璃制成的水容器中。并且对超声信号进行滤波反投影处理,重建组织模型能量吸收的平面分布。实验结果表明,光声成像在早期的乳腺癌和其它癌症的检测中将会占有重要的位置。

关键词: 光声成像,肿瘤检测,图像重建,反投影

59. Two-dimensional photoacoustic imaging of blood vessel networks within biological tissues

Su Yixiong, Ruikang K. Wang, Zhang Fan, Yao Jianquan
Chinese Physics Letters, 2006, 23(2): 512–515.

Abstract: Photoacoustic tomography (PAT) is a powerful imaging technique for medical diagnosis because it combines the merits and most compelling features of light and sound. We describe a PAT experimental system constructed in our laboratory which consists of a Q-switched Nd:YAG pulse laser operating at 532 nm with a 8-ns pulse width to generate the photoacoustic signals from a biological sample. Two-dimensional photoacoustic imaging of blood vessel networks 1 cm below the tissue surface is achieved. We also successfully demonstrate that the system is capable of imaging the blood vessels over the ex vivo rat brain with skull and skin intact.

Key words: Two-dimensional photoacoustic imaging, Blood vessel networks, Biological tissues, medical diagnosis, Q-switched laser

60. Omnidirectional zero-ñ gap in symmetrical Fibonacci sequences composed of positive and negative refractive index materials

H. Y. Zhang, Y. P. Zhang, T. Y. Shang, Y. Zheng, G. J. Ren, P. Wang, J. Q. Yao
European Physical Journal B, 2006, 52(1): 37–40.

Abstract: The band structures of symmetrical Fibonacci sequences (SFS) composed of positive and negative refractive index materials are studied with a transfer matrix method. A new type of omnidirectional zero-ñ gaps is found in the SFS. In contrast to the Bragg gaps, such an omnidirectional zero-ñ gap is insensitive to the incident angles and polarization, and is invariant upon the change of the ratio of the thicknesses of two media. It is found that omnidirectional zero-ñ gap exists in all the SFS, and it is rather stable and independence of the structure sequence.

Key words: Symmetrical Fibonacci sequences, Band structures, Transfer matrix, Omnidirectional zero- gaps

61. Dispersion compensation methods for ultrahigh-resolution optical coherence tomography

Qiang Gong, Jingying Jiang, Ruikang K Wang, Fan Zhang, Jianquan Yao

Proceedings of SPIE, 2006, 6047: 60471S.

Abstract: Optical Coherence Tomography (OCT) has been developed for more than one decade. With the optimum of system configuration such as light source, the imaging elements, the imaging quality has been improved to a higher level. However, many ideal assumptions including dispersion cancellation in the study of OCT system have become inapplicable. Actually, dispersion, can lead to a wavelength dependent phase distortion in sample arm, and finally result in a degrading in image resolution. Therefore, many dispersion compensation methods have been presented by researchers to correct the distorted image. In this paper, the principle of dispersion in OCT imaging system is discussed, and we demonstrate how it affects image quality. Then, with respect to the compensation methods as our knowledge, we classify them into hardware compensation and software compensation and present the detailed procedures and their characteristics, respectively. At last, a detailed discussion has been made to conclude that novel algorithms which can perform higher order compensation with depth variant are necessary and uniform evaluating criteria as well.

Key words: OCT, Dispersion compensation

62. Fiber-optic bending sensor for cochlear implantation

Enbang Li, Jianquan Yao

Proceedings of SPIE, 2006, 6047: 60473N.

Abstract: Cochlear implantation has been proved as a great success in treating profound sensorineural deafness in both children and adults. Cochlear electrode array implantation is a complex and delicate surgical process. Surgically induced damage to the inner wall of the scala tympani could happen if the insertion angle of the electrode is incorrect and an excessive insertion force is applied to the electrode. This damage could lead to severe degeneration of the remaining neural elements. It is therefore of vital importance to monitor the shape and position of the electrode during the implantation surgery. In this paper, we report a fiber-optic bending sensor which can be integrated with the electrode and used to guide the implantation process. The sensor consists of a piece of optical fiber. The end of the fiber is coated with aluminum layer to form a mirror. Bending the fiber with the electrode introduces loss to the light transmitting in the fiber. By detecting the power of the reflected light, we can determine the bending happened to the fiber, and consequently measure the curved shape of the electrode. Experimental results show that the proposed fiber sensor is a promising technique to make in-situ monitoring of the shape and position of the electrode during the implantation process.

Key words: Cochlear implant, Optical fiber sensing, Bending loss

63. Reconstruction algorithm in photoacoustic tomography

Zhiyuan Song, Jingying Jiang, Ruikang K Wang, Fan Zhang, Jianquan Yao
Proceedings of SPIE, 2006, 6047: 60470P.

Abstract: The photoacoustic (PA) effect refers to the generation of acoustic waves by the modulated optical radiation. A novel tissue imaging technique, photoacoustic imaging is using the acoustic waves made in the PA effect to reconstruct the sample. It whose characteristics is combining the advantages of pure optical imaging and pure ultrasound imaging can map the high contrast and high spatial resolution tissue image. The PA imaging reconstruction algorithm performing the signal filtering operation first and then to reduce the signal data to the image, has the important influence on the quality of the image made by the experiment of PA. With the laser appearance in 1960s, the PA imaging technique made great advance and the reconstruction algorithm gains quick development. The prevalent PA reconstruction algorithms include Kruger's 3D inverse Radon transform, Frenz's Fourier transform, Lihong Wang's method based on the analytic solution. Nowadays, PA imaging technique develops to the real-time PA image. So to choose a suitable fast algorithm is significant to the PA imaging system's application. We will reiew the current PA imaging reconstruction algorithm and compare them in the aspect of the tissue imaging spatial resolution and so on.
Key words: PA imaging, Reconstruction algorithm, Spatial resolution, Tissue

64. Photoacoustic imaging of blood-vessel networks of biotissue

Su Yixiong, Ruikang K Wang, Zhang Fan, Yao Jianquan
Proceedings of SPIE, 2006, 6085: 60850L.

Abstract: Photoacoustic tomography(PAT) is a powerful medical imaging technique for medicinal diagnosis in that it combines the merits and most compelling features of light and sound to the biological tissue. It can be potentially used for the detection of the first-stage breast cancers and the blood vessel networks in the deep depth of tissue. In this paper, a PAT experimental system constructed in our laboratory is presented by the use of 532 nm wavelength light as an excitation source. By using this system, we demonstrated that it is feasible to image blood vessel networks in highly scattering ex vivo and in vivo tissue samples.
Key words: Photoacoustic tomography(PAT) imaging, Blood-vessel, Biotissue

65. Spectral optical coherence tomography using three-phase shifting method

Zhenhe Ma, Jingying Jiang, Ruikang K Wang, Fan Zhang, Jianquan Yao
Proceedings of SPIE, 2006, 6047: 60470R.

Abstract: Spectral OCT (SOCT), with high acquisition speed and high dynamic range. has been implemented by many research groups in the world. However, SOCT image inherently has virtual image. including auto-correlation noise and mirror image. The existence of the virtual image may deteriorate the quality of the image. In order to eliminate those virtual images, some methods have been demonstrated effective, such as differential SOCT and complex SOCT. In this paper. A novel method is proposed i.e. three-phase shifting method. The path length of the reference arm is changed for certain distance by PZT controller. Three phase shifted coherence spectra are recorded for A-line. The reconstruction algorithm can eliminate both auto-correlation noise and minor image, thus improve the signal-to-noise ratio of the SOCT image. Furthermore, this method is also able to amplify the measuring range of SOCT by a factor of 2. An intact porcine cornea tissue in vitro is further used to show the potential of this method for high-resolution biological imaging.
Key words: Spectral OCT, Three-phase shifting, CCD camera, Biological tissues

66. Improving dynamic response of a temperature only FBG sensor

Li Enbang, Yao Jianquan, Zhang Weigang
Optoelectronics Letters, 2006, 2(2): 0101–0103.

Abstract: We report a method for improving temporal response of FBG-based temperature sensors. It has been demonstrated that by filling thermal conductive pastes between a sensing FBG and its package, the temporal response of the FBG-based temperature sensor can be significantly improved while isolating the strain and vibration.
Key words: FBG, Dynamic response, Temperature sensor

67. Double-wavelength Fano resonance and enhanced coupled-resonator-induced transparency in a double-microcavity resonator system

Ying Lu, Lijuan Xu, Yizhong Yu, Peng Wang, Jianquan Yao

Journal of the Optical Society of America A, 2006, 23(7): 1718–1721.

Abstract: We present a theoretical study on resonance control in a double-microcavity resonator system coupled to a waveguide with gain in one microresonator and loss in the other. We demonstrate the variation of the output spectra in the waveguide from initial double-wavelength-wide symmetric resonance to double-wavelength sharper asymmetric Fano profiles when gain is introduced in one of the two microresonators, as well as the inversion of the Fano resonance pattern when gain is introduced in the other microresonator. We also investigate the enhanced coupled-resonator-induced-transparency effect in such a system.

Key words: Resonance control, Double-wavelength, Fano resonance, Coupled-resonator-induced, Double-microcavity resonator

68. Photoacoustic imaging: its current status and future development

Tao Lu, Jingying Jiang, Yixiong Su, Ruikang K Wang, Fan Zhang, Jianquan Yao

Proceedings of SPIE, 2006, 6047: 60470Q.

Abstract: Photo-acoustic tomography (PAT) is a new ultrasound-mediated biomedical imaging technology which combines the advantages of high optical contrast and high ultrasonic resolution. In theory, PAT can image object embedded several centimeters under the surface of sample with the resolution oftens of microns. In this paper, several representative image reconstruction algorithms are discussed. Because the PA signal is wide band signal, it is hard to get the whole frequency spectrum due to the tremendous calculation needed. Therefore, the most applicable reconstruction algorithms are all performed in time domain such as "delay-and-sum" and "back projection". The current research methods have been focused on optical detecting and piezoelectric detecting. The optical method has the advantage of high spatial sensitivity due to the short wavelength of the probe laser beam. PA signal detecting using piezoelectric sensor has two main modes, i.e. using unfocused transducer or transducer array or using focused transducer array or linear transducer array. When a focused transducer array is used, the "delay-and-sum" method is often used for image reconstruction. The advantage of the method is that its data acquisition time can be reduced to several minutes or even several seconds by employing the phase control linear scan technique. The future development in PAT research and its potential clinic application is also presented.

Key words: Photo-acoustic tomography (PAT) imaging, Imaging reconstruction algorithm, Time-resolved

69. Simulation study on sensitive detection of small absorbers in photoacoustic tomography

Yixiong Su, Fan Zhang, Jianquan Yao, Ruikang K Wang

Proceedings of SPIE, 2006, 6086: 60861V.

Abstract: Photoacoustic tomography (PAT), which reconstructs the distribution of light-energy deposition in the tissue, is becoming an increasingly powerful imaging tool. For example, the technique has potential applications in the early stage breast cancer sensing and the functional imaging of small animal brain. In PAT, the system signal-to-noise ratio (SNR) and the number of measurement positions (NMP) are the two main factors which affect the quality of final reconstructed image. Undoubtedly, the increase of SNR or the numbers of measurement positions will improves image quality. However, one has to pay a cost on the imaging speed for such improvement of image quality. In this paper, the factors influencing the imaging performance of PAT are investigated by means of computer simulations. The result shows that the increase of the number of averaging times in acquiring of acoustic signal and the number of measurement positions are efficient ways to improve image quality. However, there exists a turning point at which the further increase of NMP and averaging times makes the improvement of imaging performance negligible. Thus a tradeoff should be made to achieve the optimal reconstructed image according to the system SNR.

Key words: Photoacoustic tomography, Simulation study, Signal-to-noise ratio, Number of measurement positions

70. Arbitrary three-phase shifting algorithm for achieving full range spectral optical coherence tomography

Ma Zhenhe, Ruikang K Wang, Zhang Fan, Yao Jianquan
Chinese Physics Letters, 2006, 23(2): 366–369.

Abstract: An arbitrary three-phase shifting algorithm is introduced in order to achieve full range spectral optical coherence tomography imaging of biological tissue. Theoretical treatment behind this approach is given and experimentally verified. It is shown that this method is capable of eliminating the undesired auto-correlation and complex conjugate images, leading to the un-obscured full range spectral OCT imaging. An intact porcine eye is used to demonstrate the potential of such a method for biological imaging.

Key words: Arbitrary three-Phase shifting algorithm, Optical coherence tomography, Full range spectral, Biological imaging

71. Simulation on sensitive detection of small absorber in photoacoustic tomography

Yixiong Su, Jingying Jiang, Ruikang K Wang, Fan Zhang, Jianquan Yao
Proceedings of SPIE, 2006, 6047: 60470S.

Abstract: Noninvasive photoacoustic tomography (PAT) is a novel technique with great potential in biomedical image applications for it combines the merits and most compelling features of light and sound, and has the advantages of providing high contrast and high resolution images in moderate depth below the surface. When the image depth is on the scale of centimeter, the millimeter-scale resolution images still can be obtained. Thus it is a powerful tool for the early-stage breast cancer sensing. In this paper, photoacoustic tomography is studied by using the simulation method. The results show that: (1) the contrast of image increases linearly with respect to the number of measurement position (NMP); (2) the contrast increases exponentially with respect to noise-to-signal ratio.

Key words: Potoacoustic tomography, Simulation study, Sensitive detection, Small absorber

72. 基于 DPS 的双通道动平衡仪研制与应用

卢涛, 姚建铨, 陶洛文
天津大学学报, 2006, 39(2): 240–244.

摘要：普通单片机在测量过程中很难实现对振动信号的实时处理, 为此, 提出一种以 DPS 为核心的双通道动平衡仪的软、硬件实现方法。原始加速度振动信号经电荷/电压转换、滤波、积分和程控增益放大后形成两路电压信号, DPS 以光电传感器产生的转速脉冲信号为基准通过 DPS 对其进行整周期采样, 进而由离散傅里叶变换精确计算振劫信号的幅值与相位, 并通过影响系数法给出了最终的平衡配重结果。系统结构简单, 实时性强, 满足现场双面动平衡对双通道同步测量的要求, 采用此仪器对有五级叶轮的罗茨风机进行双面动平衡, 平均振动下降率超过 90%。

关键词：仪器仪表, 现场动平衡, 数字信号处理, 影响系数

73. 含负折射率材料一维三元光子晶体的特性研究

张会云, 张玉萍, 尚廷义, 郑义, 任广军, 王鹏, 姚建铨
光电子·激光, 2006, 17(5): 587–590.

摘要：运用光学传输矩阵理论, 研究了含负折射率材料一维三元光子晶体的禁带特性和局域模特性, 发现了一种新型全方位光子带隙。与传统的 Bragg 带隙相比, 这种新型全方位光子带隙的中心频率和带宽对入射角的变化不敏感。引入缺陷后, 全方位光子带隙中出现缺陷模, 它的位置对入射角不敏感, 而且当各层介质厚度做一定比例的缩放时也几乎保持不变。

关键词：光子晶体, 负折射率, 全方位光子带隙, 缺陷模

74. 液晶红外磁控双折射效应的研究

任广军, 姚建铨, 李国华, 赵阶林, 王鹏

光电子·激光, 2006, 17(4): 502-505.

摘要：利用 JG-3 型连续可调谐磁场仪搭建实验装置,红外 1 350 nm 激光器做光源,测量了偏振光通过磁场作用下 BL-009 型向列相液晶的透射光强度和旋转角度,详细分析了磁场对液晶透射比和液晶旋光性能的影响,对实验结果进行了理论分析,得出了液晶透射比和旋光角随磁场方向变化的结论。同时,通过实验测试,对液晶的阈值磁场强度进行了讨论。

关键词：液晶,偏振光,磁控双折射（MCB）

75. 光纤传输的脉冲展宽研究

任广军, 姚建铨, 张强, 王鹏, 张玉萍, 张会云

激光杂志, 2006, 27(6): 152-155.

摘要：利用傅立叶方法解非线性薛定谔方程,对单模光纤传输中非线性色散的脉冲展宽进行了研究,通过详细计算,给出了高斯脉冲的均方根宽度以及展宽因子的表达式。同时,作出了光纤内色散造成的高斯脉冲展宽和展宽因子随距离的变换关系图线。并分析研究了光纤色散对不同宽度脉冲的影响,对脉冲压缩技术的研究具有一定的参考价值。

关键词：光纤,脉冲展宽,非线性薛定谔方程,傅里叶变换

76. 液晶的磁旋光特性

任广军,姚建铨,李国华,王鹏

天津大学学报, 2006, 39(8): 973-977.

摘要: 为深入了解液晶的磁旋光特性,利用矩阵方法分析研究了液晶的旋光效应,导出了液晶旋光的矩阵表示。通过测量磁场作用下向列型液晶的旋光角及透射比,研究了液晶的磁旋光特性。详细分析了磁场对液晶分子轴旋转和透射比的影响,做出了液晶分子轴旋转角和透射比与磁场强度关系的曲线。通过改变磁场方向进行实验测试,发现液晶的旋光方向与垂直入射液晶盒的磁场方向无关。

关键词: 液晶,矩阵,旋光,磁场

77. 三维光子晶体典型结构完全禁带的最佳参数理论分析

刘欢,姚建铨,李恩邦,温午麒,张强,王鹏

物理学报, 2006, 55(1): 230-237.

摘要: 基于平面波展开法,理论分析了晶格结构、填充率、介电常数比等因素对 fcc, diamond, woodpile 三种三维光子晶体典型结构完全禁带的影响。三种结构中,fcc 结构由于高对称性导致的能级简并,只适用于密堆积排列的反蛋白石结构;diamond 结构非常容易产生高带隙率的完全禁带,并且可以通过调节多项参数得到所需的完全禁带;woodpile 结构参数调节范围比较宽,为实验制备带来方便。对于不同的三维光子晶体结构,随着介电常数比的增大,完全禁带的宽度和带隙率也会随着增大。还发现了一些以前未引起注意的现象。

关键词: 三维光子晶体,完全禁带,介电常数比,带隙率,平面波展开法

78. 光声成像技术

宋智源，刘英杰，王瑞康，姚建铨，苏翼雄，张帆
中国激光医学杂志, 2006, 15(2): 127-128.

摘要：光声谱技术的原理是当一束光照射到生物组织上以后，生物组织吸收光能量而产生热膨胀，伴随着热膨胀会产生超声波，吸收光能量的多少决定了产生的超声波的强度。于是不同的组织就会产生不同强度的超声波，可以用来区分正常组织和病变组织。正是由于这一特点，光声谱技术在医学中有着广泛的应用前景。

关键词：光声谱技术，超声波，医学

79. Frequency response in photonic heterostructures consisting of single-negative materials

H. Y. Zhang, Y. P. Zhang, P. Wang, J. Q. Yao
Journal of Applied Physics, 2007, 101(1): 013111.

Abstract: Transmission studies for multiple heterostructures consisting of two kinds of single-negative materials inserted with defects are presented. The results show that multiple-channeled filters can be obtained by adjusting the period number m and thicknesses of defects. These structures provide an excellent way to select useful multiple-channeled optical signals from a stop gap, and it is useful in optical device applications.

Key words: Frequency response, Heterostructures, Crystal defects

80. Analysis of square-structured photonic crystal fibers using localized orthogonal function algorithm

Xiaoling Tan, Jianquan Yao, Youfu Geng, Huiyun Zhang, Yuping Zhang, Peng Wang

Journal of Optoelectronics and Advanced Materials, 2007, 9(8): 2312–2316.

Abstract: A modified localized orthogonal function method is applied to the analysis of square-structured PCFs. Because of difficulty in solving the vector wave equation, a scalar approximation is adopted. Then endlessly single-mode operation, electric field distribution, effective area and chromatic dispersion property are, respectively, evaluated. It is shown that the results agree well with the published data for the investigated wavelength window.

Key words: Photonic crystal fibers, Localized orthogonal function, Effective area

81. A new type ultraflattened dispersion photonic crystal fiber with low confinement loss

Xiaoling Tan, Youfu Geng, Yuping Zhang, Huiyun Zhang, Peng Wang, Jianquan Yao

European Physical Journal-Applied Physics, 2007, 40(2): 175–179.

Abstract: In this paper, a new type PCF with three different air-hole diameters in cladding region is proposed and investigated by using a compact 2-D finite difference frequency domain (FDFD) method with anisotropic perfectly matched layers (PML) absorbing boundary conditions. Through numerical simulation we found that by optimizing the geometrical parameters, our proposed structure with fewer of air-hole rings greatly meets the performance criteria such as ultraflattened dispersion and low confinement loss, which is highly instructive for designing photonic crystal fibers.

Key words: Ultraflattened dispersion, Photonic crystal fiber, Confinement loss

82. Omnidirectional single-negative gap and in fibonacci sequences composed of single-negative materials

Tingyi Shang, Huiyun Zhang, Yuping Zhang, Peng Wang, Jianquan Yao
International Journal of Infrared and Millimeter Waves, 2007, 28(8): 671–676.

Abstract: The band structures of Fibonacci sequence composed of single-negative materials are studied with a transfer matrix method. A new type of omnidirectional single-negative gaps is found in the Fibonacci sequence. In contrast to the Bragg gaps, such an omnidirectional single-negative gap is insensitive to the incident angles and polarization, and is invariant upon the change of the ratio of the thicknesses of two media. It is found that omnidirectional single-negative gap exists in the other Fibonacci sequence, and it is rather stable and independence of the structure sequence.

Key words: Photonic crystals, Negative refractive index materials, Omnidirectional single-negative gaps, Fibonacci sequences

83. 液晶磁致旋光的研究

任广军,姚建铨,王鹏,张强,张会云,张玉萍
物理学报, 2007, 56(2): 994–998.

摘要：利用矩阵方法分析了液晶的旋光效应,导出了液晶旋光的矩阵表示。利用JG-3型连续可调谐磁场仪搭建实验装置,红外1 350 nm激光器做光源,测量了偏振光通过磁场作用下BL-009型向列相液晶的旋光角,详细分析了磁场对液晶旋光性能的影响。通过实验测试,对液晶的阈值磁场强度进行了讨论,同时对实验结果进行了理论上的分析,得出了液晶旋光角随磁场与液晶盒表面夹角而变化的结论,验证了液晶分子轴的旋转方向与磁场的方向无关,这为更好的研究液晶的特性以及液晶器件的设计具有重要的参考价值。

关键词：液晶,矩阵,磁致旋光

84. 掺钕保偏光纤激光器的研究

任广军, 张强, 王鹏, 姚建铨
物理学报, 2007, 56(7): 3917–3922.

摘要：对掺钕光纤激光器输出功率沿光纤的分布以及不同光纤长度下抽运功率和输出功率沿光纤的分布进行了数值模拟。以 808 nm 半导体激光器为抽运源, 掺钕双包层保偏光纤为增益介质, 对保偏光纤激光器进行了探索性的实验研究。分别就光纤不同弯曲形状和弯曲半径对激光器输出功率指标和偏振特性的影响进行了研究, 实验中发现在 1 060 nm 和 1 092 nm 处有两个峰值。在波长 1 060 nm 处得到了 7.35 W 的连续偏振激光输出, 斜率效率为 58.3%。

关键词：激光技术, 光纤激光器, 掺钕保偏光纤, 偏振

85. 基于六角结构二维光子晶体绝对带隙的优化设计研究

钟凯, 张会云, 张玉萍, 李喜福, 王鹏, 姚建铨
物理学报, 2007, 56(12): 7029–7033.

摘要：根据平面波展开法对二维光子晶体的能带结构进行计算, 采用栅格结构连接电介质圆柱体对六角结构的二维光子晶体进行了优化。通过计算栅格宽度和圆柱体半径对绝对带隙的影响, 找到了一组可以获得大带隙二维光子晶体结构的最佳参数。优化后的光子晶体的大带隙对光子晶体制造工艺中介质圆柱体半径的偏离具有很好的稳定性, 因此该结构的二维光子晶体具有很高的实用性。

关键词：光子晶体, 绝对带隙, 六角结构

86. The pulse broadening study of Gauss-chirped pulse in optical fibers

Ren Guangjun, Yao Jianquan, Zhang Yuping, Zhang Huiyun, Wang Peng
Modern Physics Letters B, 2007, 21(6): 349–355.

Abstract: The pulse broadening due to dispersion in optical fiber is studied by solving the nonlinear Schrödinger equation through the Fourier transform method. The expression of pulse width in terms of its root-mean-square and the pulse broadening factor of the Gauss-chirped pulse are given. Meanwhile, the influence of the propagating optical fiber distance on the pulse broadening is given. The influence of the chirped factor on the pulse broadening and the optical fiber dispersion on pulses with different widths are analyzed and discussed.

Key words: Optical fiber, Pulse broadening, Gauss-chirped pulse, Fourier transform

87. Feasibility of photoacoustic tomography for ophthalmology

Tao Lu, Zhiyuan Song, Yixiong Su, Fan Zhang, Jianquan Yao
Chinese Optics Letters, 2007, 5(8): 475–476.

Abstract: For the eyeball composed of membrane and liquid, the contrast of ultrasound imaging is not high due to its small variance in acoustic impedance. As a new imaging modality, photoacoustic tomography combines the advantages of pure optical and ultrasonic imaging together and can provide high resolution, high contrast images. In this paper, the feasibility of photoacoustic tomography for ophthalmology is studied experimentally. A Q-switched Nd:YAG pulsed laser with 7-ns pulse width is used to generate photoacoustic signal of a porcine eyeball in vitro. The two-dimensional (2D) optical absorption image of the entire eyeball is reconstructed by time-domain spherical back projection algorithm. The imaging results agree well with the histological structure of the eyeball and show a high imaging contrast.

Key words: Photoacoustic imaging, Medical and biological imaging, Ophthalmic optics, Ophthalmology

88. Low-cost high-resolution wavelength demodulator for multi-channel dynamic FBG sensing

Li Enbang, Peng Gangding, Yao Jianquan

Chinese Journal of Scientific Instrument, 2007, 28(1): 1–6.

Abstract: A simple and low-cost wavelength demodulator with high resolution for multi-channel dynamic fiber Bragg grating (FBG) sensing applications is presented. Thin-film filter (TFF) based dense wavelength division multiplexing (DWDM) modules have been well developed and widely used in optical communications systems. In this study, we adopt a commercially available TFF DWDM module to simultaneously separate and demodulate multi-channel signals from wavelength-multiplexed FBG sensors, leading to a simple and low-cost FBG sensing system with high resolution, which is suitable for multi-channel dynamic measurements. In the present study, a four-channel FBG sensing system with a strain resolution better than 1 nε/Hz$^{1/2}$ at a frequency of 3 kHz has been demonstrated.

Key words: Fiber Bragg grating, Fiber sensing, Wavelength division multiplexing, Dynamic strain, Vibration

89. 保偏光纤激光器的实验研究

任广军，姚建铨，王鹏，张强

中国激光，2007, 34(9): 1208–1211.

摘要： 从耦合波方程出发，对掺钕光纤激光器输出功率沿光纤的分布进行了数值模拟，并对掺钕光纤激光器所需要光纤的最佳长度进行了分析。以 808 nm 半导体激光器为抽运源，掺钕双包层保偏光纤为增益介质，使用对 808 nm 高透，1 060 nm 高反的二色镜和垂直切割的光纤端面 (4% 的菲涅耳反射) 构成法布里 - 珀罗 (F-P) 光学谐振腔，对保偏光纤激光器进行了实验研究。实验中测量了掺钕光纤的荧光光谱，并就不同抽运电流对激光器输出功率和偏振特性进行了研究，在波长 1 060 nm 处得到了 7.5 W 的激光输出，斜率效率为 56%。

关键词： 激光技术，光纤激光器，掺钕保偏光纤，偏振

90. 液晶磁控偏光特性的研究

任广军,姚建铨,李国华,王鹏

光子学报, 2007, 36(1): 152-155.

摘要：利用偏光干涉理论,通过对 BL-009 型向列相液晶透射比的测试,分析了液晶透射比随磁场的变化情况,并对液晶的磁控双折射效应进行了研究。实验在室温 20℃下用 JG-3 型连续可调磁场仪对液晶盒施加垂直于其表面的磁场,用 CT5A 型特斯拉计准确读出磁场强度数值,使液晶盒光轴方向与起偏镜和检偏镜偏振方向成 45°,分别测出了起偏镜和检偏镜偏振方向平行和垂直时的透射光强度。通过数学函数拟合,得出了液晶的双折射率随磁场的变化规律,即:当磁场强度大于液晶的阈值磁场时,拟合函数能很好地描述液晶磁控双折射率的变化规律。

关键词：液晶,磁控,双折射,偏光

91. 含负折射率材料一维光子晶体的全方位带隙和缺陷模

尚廷义,郑义,张会云,张玉萍,姚建铨

光子学报, 2007, 36(4): 663-665.

摘要：运用光学传输矩阵理论,研究了含负折射率材料一维二元光子晶体的禁带特性和局域模特性,发现了一种新型全方位光子带隙。与传统的 Bragg 带隙相比,这种新型全方位光子带隙的中心频率和带宽对入射角的变化不敏感。讨论了引入缺陷层后,入射角变化和各层介质厚度做一定比例的缩放时对缺陷模位置的影响。这种特性在具有固定带宽的全方位反射器和微波技术中全方位或大入射角滤波器方面有重要的应用价值。

关键词：光子晶体,传输矩阵,负折射率,全方位光子带隙

92. 波片和旋转器复合退偏的矩阵研究

任广军，姚建铨，赵阶林
激光技术，2007, 31(3): 314–316.

摘要：为了探索一种简单有效的复色光退偏方法以及偏光探测技术的发展，利用斯托克斯矢量和米勒矩阵分析了波片和旋转器组合对复色光的退偏效应，导出了斯托克斯矢量的具体形式，阐述了复色光的退偏机理。通过计算讨论了偏振光的偏振方向和旋转器的旋转角对退偏情况的影响。结果表明，这种退偏器其退偏效果更加理想，对退偏器的设计与研究具有一定的参考价值。
关键词：物理光学，偏振光，复合退偏，斯托克斯矢量，米勒矩阵

93. 一种激光打标控制系统的软件研究

李玉翔，蔡彬晶，王涛，姚建铨，王鹏
激光杂志，2007, 28(4): 64–65.

摘要：本文介绍了一种微机控制的激光打标的系统软件设计和实践；采用面向对象的分析和设计方法，程序由 VB 和 VC 混合编写，主要包含图形文件的处理，插补运算，屏幕显示以及并口输出几个模块，端口的输出部分编译成动态链接库 DLL (Dynamic Link Library) 供 VB 程序调用，可实现用画图板、Photoshop 等工具随意制作不同字体、不同字号的文字和图形，都可以转换为打标扫描文件，系统还编辑了支持 3B 文件、bmp、jpg、gif 等多种格式的软件，交互性极强；软件驱动激光打标运转良好，具有打标精度高、运行稳定性好等优点。
关键词：激光打标，图像处理，扫描软件

94. 基于阵列探测方式的时域光声成像系统

卢涛, 王瑞康, 苏翼雄, 宋智源, 张帆, 姚建铨

纳米技术与精密工程, 2007, 5(1): 15-18.

摘要：将阵列超声探头和超声相控技术与光声成像相结合的成像系统,与采用水听器的单探头旋转扫描光声成像系统相比,避免了机械旋转机构给光声信号采集所带来的不稳定性,提高了数据采集速度。时域光声信号由 64 阵元线阵超声探头以电子相控聚焦的方式进行线性扫描采集,然后通过时域后向投影算法进行光声图像的重建。采用波长 532 nm, 重复频率 10 Hz 的脉冲激光,系统可快速重建样品内部光学吸收分部的二维图像,单帧图像数据采集时间小于 200 s, 成像横向分辨率小于 2 mm。实验结果表明,采用此方法可显著提高系统对光声信号的扫描稳定性和成像效率,该系统是一种有潜在临床应用价值的光声成像系统。

关键词：光声成像, 时域, 超声波, 阵列传感器

95. CMI 编译码电路的设计

徐海英, 陆颖, 姚建铨, 王鹏

科学技术与工程, 2007, 7(15): 3905-3910.

摘要：CMI(coded mark inversion, 传号反转码)码是数字光纤通信传输码型中的一种。介绍了 CMI 的编译码原理,并用 Prote199 软件对 CMI 码的编译码进行了电路设计,在此基础上用 Multisim8 软件做了仿真。

关键词：CMI, 信道编码, 信道译码, PCM

96. 微环共振器的开关特性分析

徐海英,陆颖,王占鳌

科学技术与工程, 2007, 7(19): 4876–4878.

摘要：微环共振器结构是近年提出的一种微结构,可以实现波长变换,光开关与逻辑门等功能。因此,在光通信和数字信号处理领域有很好的发展前景和应用价值。主要分析了其光开关特性。

关键词：微环共振器,光开关,耦合

97. Spectral domain polarization sensitive optical coherence tomography based on the two phase method

Fan Chuanmao, Jing Zhijun, Jiang Jingying, Gong Qiang, Ma Zhenhe, Zhang Fan, Yao Jianquan, Wang R. K.

Proceedings of SPIE, 2007, 6439: 64390K.

Abstract: In this talk, a spectral domain polarization sensitive optical coherence tomography (SDPS-OCT) system has been developed so as to obtain high scan speed, high dynamic range and high sensitivity, and simultaneously get birefringence contrast of some biological tissue. To reduce corruption of the DC and autocorrelation terms to images, we introduce the two phase method. The stocks vectors (I, Q, U, and V) of the backscattered light from the specimen have been reconstructed by processing the signals from the two channels which are responsible for detecting the vertical and horizontal polarization state light separately. Further, the phase retardation between the two orthogonal polarization states has been acquired. The results from rabbit eye show that SDPS-OCT system based on the two phase method has great potential to imaging biological tissue.

Key words: SDPS-OCT, Two phase method, Stokes vectors, Retardation

98. Birefringence imaging of biological tissue by spectral domain polarization sensitive optical coherence tomography

Jing Zhijun, Fan Chuanmao, Jiang Jingying, Gong Qiang, Ma Zhenhe, Zhang Fan, Yao Jianquan, Wang R. K.

Proceedings of SPIE, 2007, 6439: 64390J.

Abstract: A spectral domain Polarization sensitive optical coherence tomography (SDPS-OCT) system has been developed to acquire depth images of biological tissues such as porcine tendon, rabbit eye. The Stocks vectors (I, Q, U, and V) of the backscattered light from the biological tissues have been reconstructed. Further, the phase retardation and polarization degree between the two orthogonal polarizing states have been computed. Reconstructed images, i.e. birefringence images, from Stokes parameters, retardation and polarization degree of biological tissues show significant local variations in the polarization state. And the birefringence contrast of biological tissue possibly changes by some outside force. In addition, the local thickness of the birefringence layer determined with our system is significant. The results presented show SDPS-OCT is a potentially powerful technique to investigate tissue structural properties on the basis of the fact that any fibrous structure with biological tissues can influence the polarization state of light.

Key words: SDPS-OCT, Birefringence imaging, Biological tissue, Stocks vectors

99. Signal processing using wavelet transform in photoacoustic tomography

Lu Tao, Jiang Jingying, Su Yixiong, Song Zhiyuan, Yao Jiangquan, Wang Ruikang
Proceedings of SPIE, 2007, 6439: 64390L.

Abstract: In order to improve the imaging contrast and resolution in photoacoustic tomography (PAT), the deconvolution between the transducer impulse response and the recorded photoacoustic (PA) signal of the tissue phantom is often used. The suppression of noise is critical in the deconvolution. Compared with the traditional band-pass filter in Fourier domain, wiener filter is more appropriate for the wide band PA signal. The scaling parameter in wiener filter is hard to determine using the traditional Fourier domain method. To solve the problem, the deconvolution algorithm with wiener filter based on the wavelet transform is presented. The scaling parameter is estimated using discrete wavelet transform (DWT) by its multi-resolution analysis (MRA) ability. The white noise had been effectively suppressed. Both numerical simulation and experimental results demonstrated that the contrast and resolution of PA images had been improved.
Key words: Photoacoustic tomography (PAT), Optoacoustic tomography, Acoustic transducer, Signal processing, Wavelet transform

100. Proposal to produce coupled resonator-induced transparency and bistability using microresonator enhanced Mach-Zehnder interferometer

Lu Ying, Xu Lijuan, Shu Meiling, Wang Peng, Yao Jianquan
IEEE Photonics Techonology Letters, 2008, 20(7), 529–531.

Abstract: We propose a microresonator scheme for a Mach–Zehnder interferometer in which a microresonator is side coupled to one arm and the other is inserted in the other arm, to produce coupled resonator-induced transparency (CRIT) and absorption (CRIA) effects. In this system, CRIT and CRIA effects with one effect next to another over a frequency range when two microresonators are identical and Fano-resonance line shape when the sizes of two microresonators are unequal are achieved. We also theoretically investigate the improving optical bistability characteristic based on CRIT.
Key words: Bistability, Coupled resonator-induced transparency (CRIT) and absorption (CRIA), Microresonator

101. Tunable zero-phase-shift omnidirectional filter consisting of single-negative materials

Tong Zhisong, Zhang Huiyun, Yao Jianquan
Apllied physics B: Lasers and optics, 2008, 91(2): 369–371.

Abstract: We theoretically design a heterostructure with which we can achieve a tunable zero-phase-shift omnidirectional filter. The results show that by simply adjusting the thickness of the defect layer of air, we can achieve adjustability of the tunneling mode. We further prove the omnidirectional tunneling modes in the heterostructure consisting of layered single-negative materials and one layer of defect.
Key words: Heterostructure, Tunable omnidirectional filter, Zero-phase-shift

102. Characterization of bent large-mode-area photonic crystal fiber

Tan Xiaoling, Geng Youfu, Li Enbang, Wang Weineng, Wang Peng, Yao Jianquan
Journal of Optics A: Pure and Applied Optics, 2008, 10(8): 085303.

Abstract: In this paper, the characteristics of bent large-mode-area photonic crystal fibers are investigated comprehensively by using a finite difference mode solver with perfectly matched layers. Numerical results show that in bent large-mode-area photonic crystal fibers, the mode-field distributions are deformed and the effective mode area and the width of the guided region are reduced. The propagation character of bent photonic crystal fibers is determined by normalized propagation number and normalized bend radius. Meanwhile, the bending loss oscillation is observed and the position of the first loss peak is obtained in bent large-mode-area photonic crystal fibers.
Key words: Photonic crystal fiber, Large mode area, Bending loss oscillation

103. Effect of thiol on the holographic properties of TiO_2 nanoparticle dispersed acrylate photopolymer films

Ling Furi, Dan Li, Yao Jianquan, Zhou Hai, Liu Jinsong
Journal of Optics A: Pure and Applied Optics, 2008, 10(7): 075303.

Abstract: In this paper we investigatedthe holographic properties by doping thiol into a TiO_2 nanoparticle dispersed acrylate photopolymer in green light with the recording intensity 80 mW cm^{-2}. It is found that the effect of oxygen inhibition on the polymerization was greatly reduced with the increase in thiol concentration. Additionally, noticeable improvements in the recording dynamics of Δn together with the recording sensitivity by doping thiol into the TiO_2 nanoparticle dispersed acrylate photopolymer were obtained and the influence of the thiol groups on the diffusion of TiO_2 nanoparticles is also investigated.
Key words: Holographic, Polymer

104. Bend-induced distortion in large mode area holey fibre

Tan Xiaoling, Geng Youfu, Zhang Tieli, Wang Weineng, Wang Peng, Yao Jianquan
Chinese Physics Letters, 2008, 25(5): 1661–1663.

Abstract: A simplified scheme of bend-induced mode distortion is introduced into bent holey fibres, the distorted mode distribution and mode effective area reduction are investigated using the finite difference method. Numerical results show that the modes of bent holey fibres with small bend radius shift away from the core and are deformed greatly, and the mode areas drop significantly as the bend radius decreases, which severely affects the fibre laser performance. The propagation characteristics of bent holey fibres at given wavelength are determined by fill factor and normalized bend radius. Finally, the transition normalized bend radius that represents the location of the mode area beginning to fall off is obtained.
Key words: Holey fibre, Bend-induced distortion, Finite difference method

105. Structure and optical damage resistance of near-stoichiometric Zn:Fe:LiNbO₃ crystals

Ling Furi, He Zhihong, Yao Jianquan, Wang Baio, Jiang Shaoji
Optics and Laser Technology, 2008, 40(7): 941–945.

Abstract: A series of near-stoichiometric Zn:Fe:LiNbO₃ crystals were grown by the high-temperature top-seed solution growth (HTTSSG) method from stoichiometric melts doped with 6 mol% K₂O. Infrared (IR) transmission spectra were measured and discussed in terms of the defect structure of the near-stoichiometric Zn:Fe:LiNbO₃ crystals. The results of the transmitted beam pattern distortion method show that the optical damage resistance of the near-stoichiometric Zn:Fe:LiNbO₃ crystals increases rapidly when the ZnO concentration exceeds a threshold value. The threshold value concentration of ZnO of the near-stoichiometric Zn:Fe:LiNbO₃ crystals is much lower than that of the congruent LiNbO₃ crystals. The dependence of the optical damage resistance on the defect structure of the near-stoichiometric Zn:Fe:LiNbO₃ crystals are discussed, and the holographic recording properties of the near-stoichiometric Zn:Fe:LiNbO₃ crystals are investigated.

Key words: Stoichiometric, Optical damage resistance, Zn:Fe:LiNbO₃ crystals

106. Performance comparisons between 10 Gb s⁻¹ hybrid TDM/WDM and WDM systems

Huang Jing, Yao Jianquan
Journal of Modern Optics, 2008, 55(11): 1749–1757.

Abstract: The performances of hybrid time-division multiplexing/wavelength-division multiplexing (TDM/WDM) and wavelength-division multiplexing (WDM) technologies are comparatively studied in 10Gb s⁻¹ systems. Hybrid TDM/WDM modulation is superior compared with WDM and with the increase of distance, this advantage turns bigger. It is an applicable technology for dense 10Gb s⁻¹ signal's ultra-long haul transmissions. A higher pulse power upgrades both system's performances. Cycle duty and channel number have little effect on Q. Both systems can obtain better performances for the cases of wider pulse and less channel number. It is also demonstrated that the impact of channel spacing on XPM intensity modulation between 10Gbs⁻¹ return zero (RZ) pulses is small. The decreases of dispersion and nonlinear coefficients will benefit the systems' operations.

Key words: Hybrid TDM/WDM, WDM, Cross-phase modulation, Q

107. 增益导引和折射率导引在大模场单模光纤设计中的应用

王伟能, 张铁犁, 谭晓玲, 姚建铨

中国激光, 2008, 35(3): 426–429.

摘要：传统的大模场光纤是通过设计光纤结构来获得大模场面积的，可以实现的模场面积只能达到几百平方微米。增益导引和折射率导引相结合是实现大模场单模光纤的一种新方法。通过分析增益因子对折射率以及归一化频率的影响，得到了光纤中各阶模式截止条件与纤芯包层折射率差和增益因子的关系。最后以包层折射率为 1.573 4, 纤芯折射率为 1.568 9, 纤芯半径为 50 μm, 10%(原子数分数)重掺杂钕离子的磷酸盐光纤作为模拟计算对象，当波长为 1.064 μm 时，得到其模场直径大于 90 μm。对于普通光纤，增益导引和负折射率导引相结合的方法对实现大模场单模传输很有前景。

关键词：光纤光学, 光纤激光器, 大模场单模光纤, 增益导引, 折射率导引

108. 八角格子光子晶体光纤的传输特性

谭晓玲, 耿优福, 王鹏, 姚建铨

中国激光, 2008, 35(5): 729–733.

摘要：基于带有各向异性完全匹配层吸收边界条件的紧凑二维频域有限差分法 (2D-FDFD) 对八角格子光子晶体光纤 (O-PCF) 的模式分布、模式截止特性以及色散特性进行了数值模拟。通过计算八角格子光子晶体光纤前 20 个模式分布发现，其模场形状比六角格子光子晶体光纤 (H-PCF) 的好，更接近于圆形; 利用有效面积方法分析了八角格子和六角格子光子晶体光纤基模和二阶模的截止特性，得到了非限制模、基模和多模的相图。比较发现相同填充率和空气孔间距时，八角格子光子晶体光纤的单模运转区域要比六角格子光子晶体光纤的宽，且更易用于色散补偿。

关键词：光纤光学, 光子晶体光纤, 频域有限差分, 模式截止, 有效面积

109. Tailoring optical transmission via the arrangement of compound subwavelength hole arrays

Jianqiang Liu, Mengdong He, Xiang Zhai, Lingling Wang, Shuangchun Wen, Li Chen, Zhe Shao, Qing Wan, B. S. Zou, Jianquan Yao

Optics Express, 2009, 17(3): 1859–1864.

Abstract: The transmission properties of light through metal films with compound periodic subwavelength hole arrays is numerically investigated by using the finite-difference time-domain (FDTD) method. The sharp dips in the transmission bands, together with the suppression of surface plasmon resonance (SPR) (0, 1) peak, are found when two square holes in every unit cell are arranged asymmetrically along the polarization direction of the incident light. However, the shape of transmission spectra is not sensitive to the symmetry if the holes are arranged perpendicular to the propagation direction of surface plasmon polaritons (SPPs). The physics origin of these phenomena is explained qualitatively by the phase resonance of SPPs.

Key words: Optical transmission, Subwavelength, Compound hole arrays, SPRs

110. A method to design transmission resonances through subwavelength apertures based on designed surface plasmons

Jinsong Liu, Lan Ding, Kejia Wang, Jianquan Yao

Optics Express, 2009, 17(15): 12714–12722.

Abstract: A plasmonic metamaterial is proposed for which an array of subwavelength apertures is pierced into a metallic foil whose one flat surface has been made with periodic rectangle holes of finite depth. Designed surface plasmons sustained by the holes are explored when the size and spacing of the holes are much smaller than those of the apertures. The transmission property of electromagnetic waves through the metamaterials is analyzed. Results show that the designed surface plasmons characterized by the holes could support the transmission resonances of the incident wave passing through the subwavelength apertures, and that the peak transmission wavelengths could be designed by controlling the geometrical and optical parameters of the holes. Example is taken at THz regime. Our work proposes a method to design the peak wavelengths, and may affect further engineering of surface plasmon optics, especially in THz to microwave regimes.

Key words: Surface plasmons, Surface waves, Subwavelength structures

111. The leakage current mechanisms in the Schottky diode with a thin Al layer insertion between $Al_{0.245}Ga_{0.755}N$/GaN heterostructure and Ni/Au Schottky contact

Liu Fang, Wang Tao, Shen Bo, Huang Sen, Lin Fang, Ma Nan, Xu Fujun, Wang Peng, Yao Jianquan

Chinese Physics B, 2009,18(4): 1614–1617.

Abstract: This paper investigates the behaviour of the reverse-bias leakage current of the Schottky diode with a thin Al inserting layer inserted between $Al_{0.245}Ga_{0.755}N$/GaN heterostructure and Ni/Au Schottky contact in the temperature range of 25–350 ℃. It compares with the Schottky diode without Aluminium inserting layer. The experimental results show that in the Schottky diode with Al layer the minimum point of I–V curve drifts to the minus voltage, and with the increase of temperature increasing, the minimum point of I–V curve returns the 0 point. The temperature dependence of gate-leakage currents in the novelty diode and the traditional diode are studied. The results show that the Al inserting layer introduces interface states between metal and $Al_{0.245}Ga_{0.755}N$. Aluminium reacted with oxygen formed Al_2O_3 insulator layer which suppresses the trap tunnelling current and the trend of thermionic field emission current. The reliability of the diode at the high temperature is improved by inserting a thin Al layer.

Key words: Gate leakage current, Interface states, Tunnelling current, Thermionic field emission current

112. Thermal annealing behaviour of Al/Ni/Au multilayer on n-GaN Schottky contacts

Liu Fang, Wang Tao, Shen Bo, Huang Sen, Lin Fang, Ma Nan, Xu Fujun, Wang Peng, Yao Jianquan

Chinese Physics B, 2009, 18(4): 1618–1622.

Abstract: Recently GaN-based high electron mobility transistors (HEMTs) have revealed the superior properties of a high breakdown field and high electron saturation velocity. Reduction of the gate leakage current is one of the key issues to be solved for their further improvement. This paper reports that an Al layer as thin as 3 nm was inserted between the conventional Ni/Au Schottky contact and n-GaN epilayers, and the Schottky behaviour of Al/Ni/Au contact was investigated under various annealing conditions by current–voltage (I–V) measurements. A non-linear fitting method was used to extract the contact parameters from the I–V characteristic curves. Experimental results indicate that reduction of the gate leakage current by as much as four orders of magnitude was successfully recorded by thermal annealing. And high quality Schottky contact with a barrier height of 0.875 eV and the lowest reverse-bias leakage current, respectively, can be obtained under 12 min annealing at 450 ℃ in N_2 ambience.

Key words: Schottky contact, Barrier height, Ideality factor, Thermal annealing

113. Study of ultraflattende dispersion square-lattice photonic crystal fiber with low confinement loss

Tan Xiaoling, Geng Youfu, Tian Zhen, Wang Peng, Yao Jianquan

Optoelectronics Letters, 2009, 5(2): 124–127.

Abstract: A new type ultraflattened dispersion square-lattice photonic crystal fiber with two different air-hole diameters in cladding region is proposed and the dispersion is investigated using a compact 2-D finite difference frequency domain method with the anisotropic perfectly matched layers (PML) absorbing boundary conditions. Through numerical simulation and optimizing the geometrical parameters, we find that the photonic crystal fibers proposed can realize ultraflattened dispersion of 0 ± 0.06 ps/(km·nm) in wavelength range of 1.375 μm to 1.605 μm, which is more flat than that of triangular PCF, and the confinement loss is as low as about 0.01 dB/km at wavelength of 1.55 μm.

Key words: Ultraflattende dispersion, Square-lattice, Photonic crystal fiber, Confinement loss

114. Influence of lattice symmetry and hole shape on the light enhanced transmission through the subwavelength hole arrays

Sun Mei, Xu Degang, Yao Jianquan

Optoelectronics Letters, 2009, 5(4): 0317–0320.

Abstract: The golden films with various subwavelength hole arrays on the film surface are designed and fabricated on glass substrate by electron beam lithograply (EBL), focused ion beam (FIB), and reactive ion etching (RIE), respectively. The influences of the hole array symmetry and the hole shape on the light-enhanced transmission through the films are observed and simulated. The experimental results show that when the array lattice constant and the hole diameter are the same in the different array structures which are 1 μm and 350 nm respectively, the square hole arrays exhibit two transmission peaks at 1 170 nm and 1 580 nm with the transmissivities of 3% and 6%, respectively, while the hexagonal hole arrays exhibit an enhanced peak of 14% at 1 340 nm; when the lattice constant and the duty cycle are the same for different array stucture, the transmission peaks are different for different hole shapes, which are at 763 nm with transmissivity of 12% for rectangular holes and at 703 nm with the one of 9%, respectively. The numerical simulation results by using the transfer matrix method (TMM) are consistent with the observed results.

Key words: Subwavelength hole arrays, Lattice constant, Lattice symmetry, Hole shape, Transmission

115. 调 Q 脉冲保偏光纤激光器的研究

任广军，魏臻，姚建铨

物理学报，2009, 58(2): 941–945.

摘要：以 808 nm 半导体激光器为抽运源，掺钕双包层保偏光纤为增益介质，对调 Q 脉冲保偏光纤激光器进行了理论分析和实验研究。利用 TDS5104 型示波器探测输出脉冲激光的波形，并用光谱分析仪得到输出脉冲激光的光谱图。利用 F-P 腔型，在 1 060 nm 处获得平均功率为 2.55 W 的脉冲激光输出，重复频率为 1 kHz 时，输出单脉冲能量为 2.3 mJ，峰值功率为 4.7 kW。改变腔型，把二色镜倾斜放置兼作输出镜，最终获得了平均功率为 3.5 W 的偏振脉冲激光输出，重复频率为 1 kHz 时，输出单脉冲能量为 3.3 mJ，脉冲宽度为 184 ns，其峰值功率达 17.9 KW。

关键词：激光技术，光纤激光器，掺钕保偏光纤，调 Q

116. 掺钕保偏光纤放大器的研究

任广军,魏臻,张强,姚建铨

物理学报, 2009, 58(6): 3897-3902.

摘要：对掺钕双包层光纤放大器中抽运光和信号光沿光纤传播的功率分布进行了数值模拟,以 808 nm 半导体激光器为抽运源,掺钕双包层保偏光纤为增益介质,对种子注入主振荡光纤放大器进行了理论分析和实验研究。利用实验室自制的皮秒锁模激光器为种子源,注入 106 nm 皮秒锁模脉冲,获得了稳定的放大脉冲。小信号时的放大倍数为 300（增益为 25 dB）,获得了平均功率 5 W 的皮秒脉冲。同时利用 TDS5104 型示波器探测信号光放大前后的波形,并用光谱分析仪得到输出脉冲激光的光谱图。

关键词：光纤放大器,掺钕保偏光纤,种子注入,反向抽运

117. 四方格子全固光子带隙光纤带隙特性研究

田振,谭晓玲,耿优福,王鹏,姚建铨

激光杂志, 2009, 30(5): 52-53.

摘要：基于平面波展开法,对四方格子全固光子带隙光纤带隙随结构的变化特性进行了数值模拟,并与具有相同结构参数三角格子全固光子带隙光纤进行了比较,数值结果表明带隙的位置由高折射率棒的结构参数决定,而与高折射率棒的排列方式和棒之间的间距无关,这一结果与反谐振效应非常吻合。最后对理想和实际拉制出的四方格子的全固态光子带隙光纤的带隙进行了比较,得出了在设计光纤时,应利用较低的带隙的结论。

关键词：光纤光学,全固光子带隙光纤,光子带隙效应,平面波展开法

118. 增益导引折射率反导引光纤激光特性的研究

王伟能, 谭晓玲, 田振, 耿优福, 王鹏, 姚建铨

激光技术, 2009, 33(5): 503–505.

摘要: 增益导引和折射率反导引是设计大模场单模光纤的一种新方法。基于此方法,详细分析了具有增益导引和折射率反导引光纤的模场特性,数值模拟了不同增益系数和负折射率差值情况下的模场分布及对模场的影响,得出了单模运转条件,为设计大模场光纤提供了理论依据,具有重要的指导意义。

关键词: 增益导引, 折射率反导引, 大模场单模光纤, 光纤激光器

119. Analyses of the performances of 10 Gb s^{-1} time-division multiplexing and wavelength- division multiplexing signals in single-mode fibers and non-zero dispersion-shifted fibers

Jing Huang, Jianquan Yao

Journal of Optics, 2010, 12(1): 015406.

Abstract: The performances of dense 10 Gb s^{-1} time-division multiplexing (TDM) and wavelength-division multiplexing (WDM) signals ($\Delta\lambda$ = 0.2 nm) are analyzed comparatively for transmission in single-mode fibers (SMFs) and in non-zero dispersion-shifted fibers (NZDSFs). In the ultrahigh capacity and ultralong haul communications, the TDM format is superior to WDM when dispersion is precisely compensated (including third-order dispersion compensation). As far as the dispersion compensation is concerned, fiber Bragg gratings (FBGs) are more advantageous than dispersion compensation fibers (DCFs) in SMFs. But in NZDSFs, when the distance is shorter than 2 000 km, the difference between adopting FBGs and DCFs is not distinct, and with the distance further increasing, DCFs retain their superiority even though the group dispersion ripples of FBGs are ignored.

Key words: TDM, WDM, Cross-phase modulation, Return-to-zero pulse

120. 通信波段液晶光电特性的实验研究

任广军,沈远,姚建铨,白育堃,王新闻

光电子·激光, 2010, 21(10): 1492–1494.

摘要：利用偏光干涉理论,通过对 BL-009 型向列相液晶透射比的测试,分析了液晶透射比随电场的变化情况,对液晶的电控双折射效应进行了研究。在 20℃下,利用岛津 UV-3101PC 分光光度计,通过改变加在液晶盒两端的电压,测出了入射光波长为 1 330 nm 和 1 550 nm 时液晶双折射率随电压变化的关系曲线。实验结果表明：当液晶盒两端加 1 V 电压时,入射光波长从 1 310 nm 到 1 350 nm 有一较稳定的高透射带；加 2 V 电压时,从 1 520 nm 到 1 580 nm 有一比较稳定的高透射带。

关键词：液晶,电控双折射,通信

121. 亚波长环形电磁结构的光学特性研究

孙梅,徐德刚,邢素霞,姚建铨

光学学报, 2010, 30(1): 224–227.

摘要：利用聚焦离子束 (focused ion beam, FIB) 刻蚀方法在 120 nm 厚的金膜上制备了实验测量样品。再用实验的方法测量了在可见光波段及近红外波段的透射曲线,当样品具有相同的晶格常数以及空气孔直径时,环形空气孔结构,不仅具有六重旋转对称性还具有中心对称性,它的最大透射率大约是正方角晶格空气孔的 5 倍。亚波长环形空气孔结构透射增强峰的在可见光波段,而正方晶格空气孔的透射增强峰在近红外波段,样品的透射强度和透射增强峰的位置各不相同,证明了结构的旋转对称性对透射曲线的影响。

关键词：表面光学,聚焦离子束刻蚀方法,环形空气孔结构,旋转对称性,亚波长

122. Loss properties of all-solid photonic band gap fibers with an array of rings

Geng Youfu, Li Xuejin, Tan Xiaoling, Yao Jianquan
Optoelectronics Letters, 2010, 6(6): 0454–0457.

Abstract: The confinement loss and bend loss properties of all-solid photonic band gap fibers with an array of rings doped with high index material are investigated. The calculated results show that for a specific structure, the confinement loss and the critical bend radius are reduced simultaneously in some band gaps by increasing the inner diameter of ring, which provides a useful guide and a theoretical basis for designing large mode area fibers with low loss.
Key words: Optical losses, All-solid photonic band gap fibers, Confinement loss, Bend loss, High index material

123. Dynamical analysis of evanescent field lose based fiber laser sensing

Lei Jing, Jianquan Yao, Ying Lu
Proceedings of SPIE - Photonics Asia, 2010, 7853: 785306.

Abstract: Optical fiber sensors (OFS) play an important role in modern intellectualized sensing system. A novel optical fiber sensor based on single mode fiber laser is proposed in this paper. The basic elements of the novel fiber laser sensor (FLS) is based in the fact that the output power of fiber laser is influenced by the loss which caused by the absorption loss of analyte in evanescent field of the fiber. The action of the fiber laser sensor is theoretical investigated using two-level system rate equations. The function which contacted the output power of the fiber laser and the absorption loss of analyte is built upon the complex refraction index of the analyte and the loss of the resonant cavity of the fiber laser though evanescent field. The relative sensitivity of the fiber laser sensor is given finally.
Key words: Optical fiber sensor, Fiber laser, Evanescent field

124. Study on the supermode and in-phase locking in multicore fiber lasers

Yuan Wang, Jianquan Yao, Yibo Zheng, Wuqi Wen, Rui Zhou

Proceedings of SPIE - Photonics Asia, 2010, 7843: 784319.

Abstract: Multicore fiber lasers have larger mode areas, resulting in higher power thresholds for nonlinear processes such as stimulated Raman scattering and stimulated Brillouin scattering. Because of longer distributed distance of the cores, thermal mechanical effects are decreased compared with those of single-core lasers. Therefore, multicore fiber lasers are proposed as a candidate for the power scaling. The progress of multicore fiber lasers is simply introduced. Optical fields propagating in multicore fibers are coupled evanescently, resulting in what are called supermodes. In this article, the coupled-mode theory for analyzing supermode of fiber transmission is introduced. By mean of the theory, assuming under weak-coupling conditions, the supermodes are approximated as linear superposition of modes of individual cores with appropriate coefficients. The near-field mode distributions of some supermodes are numerically calculated, and the corresponding mode distribution patterns are drawn. For making the multicore fiber laser preferentially operate in a particular supermode so that improving beam quality, an in-phase locking method based on self-imaging Talbot external cavity is introduced.

Key words: Multicore fiber, Coupled-mode theory, Supermode, Self-imaging

125. Supermode analysis in multi-core photonic crystal fiber laser

Yibo Zheng, Jianquan Yao, Lei Zhang, Yuan Wang, Wuqi Wen, Rui Zhou, Zhigang Di, Lei Jing

Proceedings of SPIE - Photonics Asia, 2010, 7843: 784316.

Abstract: In this paper, we report on the near-field distribution of multi-core photonic crystal fiber lasers. The supermodes of photonic crystal fibers with foursquarely and circularly distributed multi-cores are observed. The supermode properties are investigated by using full-vector finite-element method (FEM). The mode operations of our 16-core foursquare-array and 18-core circular-array photonic crystal fiber lasers are simulated by the COMSOL Multiphysics software. The near-field distribution patterns of in-phase supermode are presented.

Key words: Multi-core Photonic crystal fiber, Supermode analysis, In-phase supermode

126. Microstructured-core photonic-crystal fiber for ultra-sensitive refractive index sensing

Bing Sun, Mingyang Chen, Yongkang Zhang, Jichang Yang, Jianquan Yao Haixia Cui

Optics Express, 2011, 19(5): 4091–4100.

Abstract: We propose a novel photonic crystal fiber refractive index sensor which is based on the selectively resonant coupling between a conventional solid core and a microstructured core. The introduced microstructured core is realized by filling the air-holes in the core with low index analyte. We show that a detection limit (DL) of 2.02×10^{-6} refractive index unit (RIU) and a sensitivity of 8 500 nm/RIU can be achieved for analyte with refractive index of 1.33.

Key words: Photonic-crystal fiber, Microstructured-core, Ultra-sensitive, Refractive index sensor

127. Fano resonance and spectral compression in a ring resonator drop filter with feedback

Ying Lu, Xiangyong Fu, Danpng Chu, Wuqi Wen, Jianquan Yao

Optics Communications, 2011, 284: 476–479.

Abstract: We propose a ring resonator drop filter with feedback loop which can form the other ring cavity to produce an asymmetric Fano-resonance line shape and mode compression. By properly adjusting the feedback loop, the asymmetric line shapes of the transmission spectra can be controlled and selectively mode compression of the circuiting intensity in the feedback loop is achieved. Such characteristics are useful for applications in ring resonator-based photonic devices such as all-optical switches, sensors and intracavity frequency selection single-mode laser.

Key words: Ring resonator, Fano resonance, Spectral compression

128. 双端泵浦保偏光纤激光器

任广军, 姚建铨

强激光与粒子束, 2011, 23(4): 863–865.

摘要： 以两台 808 nm 半导体激光器 LD1 和 LD2 为泵浦源, 对光纤激光器双端泵浦进行了研究, 获得了 6.5 W 的激光输出。实验分别测出了 LD1 和 LD2 半导体激光器单端泵浦和双端泵浦时的输出功率, 对双端泵浦输出功率与单端泵浦功率之和进行了比较, 利用双端泵浦提高了泵浦效率和输出激光功率。同时测量了输出激光的偏振度, 通过计算得到双端泵浦输出激光的偏振度为 0.5。

关键词： 激光技术, 光纤激光器, 双端泵浦, 保偏

129. Photonic crystal fiber SERS sensors

Yao Jianquan, Di Zhigang, Jia Chunrong, Lu Ying, Xu Degang, Yang Pengfei, Bing Pibin, Zheng Yibo

Infrared and Laser Engineering, 2011, 40(1): 96–106.

Abstract: Optical fiber surface enhanced Raman scattering (SERS) sensors, which achieve optical fiber sensing based on SERS effect, have several particular advantages. However, the use of conventional fiber has brought optical fiber SERS sensors some drawbacks, such as background absorption and interference, fluorescence effect and Raman scattering from the fiber itself. The adaptive SERS sensors are developed to solve these problems, which is at least partially promised by the newly emerging photonic crystal fibers (PCFs). The PCFs SERS sensors composed of PCFs and SERS, have several advantages such as high sensitivity, interference-immunity, simple geometry, light path flexibility, and little dependence on analyte. This review shows the development and the state-of-art of SERS, PCFs, and PCF sensors based on SERS, mainly including theory, geometries, and biosensors applications. Although the recent developments is fast, there still exist a few problems that should be addressed, and particularly, the trend of PCF sensors based on SERS is discussed.

Key words: Optical fiber sensor, PCF, SERS, Nanoparticle

130. Study on HDPE-PCF evanescent wave sensor

Haixia Cui, Jianquan Yao, Ying Lu, Chunming Wan

Journal of Physics: Conference Series (POEM 2010), 2011, 276: 012148.

Abstract: The photonic crystal fiber (PCF) evanescent sensor has unique superior characteristics among the optical sensors, especially in terahertz (THz) wavelength range, PCF evanescent sensing will have huge application potentials. PCF characteristics and THz propagation characteristics in high-density polyethylene (HDPE) were combined. Finite element method was used to simulate and calculate THz-HDPE-PCF some parameters, at last the design considerations were given.

Key words: Optical sensors, High-density polyethylene, Photonic crystal fiber, Evanescent sensor

131. 表面等离子体共振类熊猫型光子晶体光纤传感器

邵丕彬，姚建铨，黄晓慧，陆颖，邱志刚，杨鹏飞

激光与红外，2011, 41(7): 784–787.

摘要：分析了基于表面等离子体共振类熊猫型的光子晶体光纤传感器，采用基全矢量有限元法（FEM）对光纤模式进行了数值计算，在各向异性完美匹配层（PML）边界条件下，求解模场的有效折射率。讨论了各个参量的尺寸对传感的影响。计算表明，激发的等离子体对环境介质折射率的变化非常敏感，所设计传感器最大光谱灵敏度达到 2 μm/RIU，若光谱仪的分辨率为 10 pm，则传感器的分辨率可以达到 5×10^{-6} RIU。

关键词：表面等离子体共振，光子晶体光纤，传感器，折射率

132. 纳米银基底表面增强拉曼散射效应仿真及优化

邸志刚,姚建铨,张培培,贾春荣,邸丕彬,杨鹏飞,郑一博

激光与红外, 2011, 41(8): 850–855.

摘 要:为了更大发挥拉曼光谱在生物检测及传感领域中的作用,表面增强拉曼散射中信号增强与基底复用之间平衡优化需求一直在激励着新型基底的发展。通过用有限元方法对不同银纳米结构(不同尺寸、不同大小、不同结构)基底的表面增强拉曼散射效应进行设计优化。对三种常见结构基底进行仿真,并对结果进行对比分析,表明仿真结果与已发表的相似基底实验结果一致。

关键词:拉曼散射,表面增强拉曼散射,有限元法,基底,纳米粒子

133. 亚波长周期性排列空气孔的传输特性研究

孙梅,徐德刚,姚建铨

光电工程, 2011, 38(4): 72–76.

摘要:从实验和理论两方面来研究了亚波长样品中周期变化时对透过曲线的响。首先,利用聚焦离子束刻蚀方法和电子束直写系统制备了实验样品:正方和六角晶格空气孔及 H-形空气槽。然后,研究了正方和六角晶格空气孔在可见光及近红外波段周期性对透过曲线的影响,证明表面等离子体对透过曲线的增强作用。测量了 H-形空气槽在近红外波段的透过曲线,通过 H-形空气槽的宽度和周期变化时,透过峰的位置和透过率都有明显变化的实验,进一步证明:表面等离子体对透过光增强起决定性的作用。理论模拟都是基于平面波的传输矩阵方法 (TMM),实验和理论模拟很好的结合起来。

关键词:亚波长空气孔,表面等离子体,传输矩阵方法

134. 物联网产业的形成及智慧城市建设

姚建铨

智慧天津高峰会，2011-10-29.

摘要： 本文节选自天津大学精密仪器与光电子工程学院教授、博士生导师、名誉院长、天津大学激光与光电子研究所所长、国家教育部科技委副主任、武汉光电国家实验室（筹）副主任、中国科学院院士姚建铨在"2011CPS 中国安防论坛 - 智慧天津高峰会"的演讲，标题为编者所加。

关键词： 联网技术，数字城市，智能建筑，城市建设，智慧，智能交通，数字地球，相关产业，通信网络，信息技术

135. Transmission and group delay in a double microring resonator reflector

Ying Lu, Congjing Hao, Bing Lu, Xiaohui Huang, BaoqunWu, Jianquan Yao
Optics Communications, 2012, 285(21–22): 4567–4570.

Abstract: We present a microring resonator scheme for an optical reflector in which two microrings are coupled to two bus waveguides, to produce reflected signal at the input port. In this system, a single narrow reflected peak with maximum reflectivity near 100% in a wide spectral range due to the Vernier effect are achieved. We also investigate versatile reflection, transmission spectra and group delay characteristic. These effects are useful for applications in ring resonator-based photonic devices such as all-optical single wavelength tunable laser end mirrors, reflective wavelength switching, reflective filter and dual-wavelength delay line.
Key words: Microring resonator, Group delay, Reflector

136. Grapefruit fiber filled with silver nanowires surface plasmon resonance sensor in aqueous environments

Ying Lu, Congjing Hao, Baoqun Wu, Xiaohui Huang, Wuqi Wen, Xiangyong Fu, Jianquan Yao

Sensors, 2012, 12: 12016–12025.

Abstract: A kind of surface plasmon resonance sensor based on grapefruit photonic crystal fiber (PCF) filled with different numbers of silver nanowires has been studied in this paper. The surface plasmon resonance modes and the sensing properties are investigated comprehensively using the finite element method (FEM). The simulation results show that the intensity sensitivity is related to nanowire numbers and the distance between two nanowires. The optimum value obtained is 2 400 nm/RIU, corresponding to a resolution of 4.51×10^{-5} RIU with a maximum distance of 2 μm. To a certain extent, the PCF filled with more nanowires is better than with just one. Furthermore, the air holes of grapefruit PCF are large enough to operate in practice. Moreover, the irregularity of the filled nanowires has no effect on sensitivity, which will be very convenient for the implementation of experiments.

Key words: Grapefruit photonic crystal fiber, Surface plasmon resonance, Silver nanowire, Nanowire distance and number

137. Characteristics of bend sensor based on two-notch Mach–Zehnder fiber interferometer

Yinping Miao, Kailiang Zhang, Bo Liu, Wei Lin, Jianquan Yao

Optical Fiber Technology, 2012, 18: 509–512.

Abstract: A compact all-fiber modal Mach–Zehnder interferometer is presented based on two notches at different locations along normal single-mode fibers by CO_2 laser irradiation. Two notches are mode splitter and mode combiner, respectively. This configuration which does not need any troublesome cleaving or aligning process is demonstrated as a bending sensor. The results suggest that the transmission loss varies and the interference fringe shifts with the bending increasing. The fluctuation of temperature dose not affects the shape of the interferometer fringe and transmission. Therefore, the proposed device offers salient advantages of simple fabrication technique and low cost. In addition, a high stability over time was expected since the irradiated notches do not degrade over time or with temperature. Due to its asymmetry, it also possesses higher sensitivity to orientation, which would be a promising unit for sensor applications.

Key words: Optical fiber interferometers, Modal Mach–Zehnder interferometer (MMZI), Fiber bend sensor

138. Interference effect in a dual microresonator-coupled Mach–Zehnder interferometer

Ying Lu, Xiaohui Huang, Xiangyong Fu, Wuqi Wen, Jianquan Yao
Optica Applicata, 2012, 42(1): 23–29.

Abstract: We present a theoretical study of interference effect in a Mach–Zehnder interferometer in which two microresonators are side coupled to both arms of the interferometer. The results show that sharp asymmetric Fano resonance, coupled resonator induced transparency and absorption effects can be created in such a structure. We demonstrate that these effects arise from interference between a resonance mode and a continuing propagating mode with asymmetric phase difference, destructive interference between two overcoupled resonance modes, and constructive interference between an overcoupled resonance mode and an undercoupled mode or a continuing propagating mode with symmetric phase differences, respectively. These effects may offer a better understanding of the analogous effects in atomic medium and also make optical resonators a potential device to utilize these effects.
Key words: Microresonator, Mach–Zehnder interferometer, Fano resonance, Coupled resonator induced transparency and absorption

139. A surface-plasmon-resonance sensor based on photonic-crystal-fiber with large size microfluidic channels

Pibin Bing, Jianquan Yao, Ying Lu, Zhongyang Li
Optica Applicata, 2012, 42(3): 493–501.

Abstract: A surface-plasmon-resonance (SPR) sensor based on photonic-crystal-fiber (PCF) with large size microfluidic channels is proposed. The size of the microfluidic channels with gold coating can be enlarged by reducing the number of the holes in the second layer, so the structure of the PCF is simple and easy to manufacture. It is propitious for metal coating and infiltration of microfluidic. The contact area of sample and metal film is increased, and the interface is closer to the core, so the energy coupling between the plasmon mode and the core-guided mode is easier. Numerical results indicate that the excitation of the plasmon mode is sensitive to the change of the refractive index of adjacent analyte. Sensitivity of the sensor is comparable to the ones of the best existing waveguide sensors. The amplitude resolution is demonstrated to be as low as 3.3×10^{-5} RIU, and the spectral resolution is 5×10^{-6} RIU (where RIU means the refractive index unit). The refractive index of microfluidic can be measured effectively.
Key words: Photonic-crystal-fiber, Surface-plasmon-resonance, Sensor, Finite element method

140. A photonic crystal fiber based on surface Plasmon resonance temperature sensor with liquid core

P. B. Bing, Z. Y. Li, J. Q. Yao, Y. Lu, Z. G. Di

Modern Physics Letters B, 2012, 26(13): 1250082.

Abstract: A Photonic Crystal Fiber based on Surface Plasmon Resonance (PCF-SPR) temperature sensor with liquid core is proposed in this paper. Glycerin liquid with a high refractive index is filled in the central air hole of the hollow core photonic bandgap (PBG) PCF, the transmission type of PCF will change to total internal reflection (TIR), which will significantly broaden its transmission bands. The refractive index of glycerin changes with temperature within a certain temperature range and can be detected by measuring the transmission spectra, thus the accurate ambient temperature can be obtained. Numerical results indicate that the plasmon on the surface of the gold-coated channels containing glycerin liquid can be intensively excited by the core-guided mode and the excitation of the plasmon mode is sensitive to the change of the temperature. Resolution of the PCF-SPR temperature sensor with liquid core is demonstrated to be as low as 4×10^{-6} RIU, where RIU means refractive index unit.

Key words: Photonic crystal fiber, Surface plasmon resonance, Sensor, Finite element method

141. Estimation of the fourth-order dispersion coefficient β_4

Jing Huang, Jianquan Yao

Chinese Optics Letters, 2012, 10(10): 101903.

Abstract: The fourth-order dispersion coefficient of fibers are estimated by the iterations around the third-order dispersion and the high-order nonlinear items in the nonlinear Schordinger equation solved by Green's function approach. Our theoretical evaluation demonstrates that the fourth-order dispersion coefficient slightly varies with distance. The fibers also record β_4 values of about 0.002, 0.003, and 0.000 32 ps^4/km for SMF, NZDSF and DCF, respectively. In the zero-dispersion regime, the high-order nonlinear effect (higher than self-steepening) has a strong impact on the transmitted short pulse. This red-shifts accelerates the symmetrical split of the pulse, although this effect is degraded rapidly with the increase of β_2. Thus, the contributions to β_4 of SMF, NZDSF, and DCF can be neglected.

Key words: Dispersion coefficient, Fourth-order, Estimation, High-order nonlinear effect

142. Three-dimensional thermal analysis of 18-core photonic crystal fiber lasers

Zheng Yibo, Yao Jianquan, Zhang Lei, Wang Yuan, Wen Wuqi, Jing Lei, Di Zhigang

Chinese Physics Letters, 2012, 29(2): 024203.

Abstract: The three-dimensional thermal properties of 18-core photonic crystal fiber lasers operated under natural convection are investigated. The temperature sensing technique based on a fiber Bragg grating sensor array is proposed to measure the longitudinal temperature distribution of a 1.6-m-long ytterbium-doped 18-core photonic crystal fiber. The results show that the temperature decreases from the pump end to the launch end exponentially. Moreover, the radial temperature distribution of the fiber end is investigated by using the full-vector finite-element method. The numerical results match well with the experimental data and the coating temperature reaches 422.7 K, approaching the critical value of polymer cladding, when the pumping power is 40 W. Therefore the fiber end cooling is necessary to achieve power scaling. Compared with natural convection methods, the copper cooling scheme is found to be an effective method to reduce the fiber temperature.

Key words: Photonic crystal fiber lasers, 18-core, Three-dimensional thermal properties, Temperature sensing technique, Bragg grating, Longitudinal temperature distribution

143. 液晶光子晶体光纤电场传感的模式特性

杨杨，戴海涛，孙小卫，姚建铨
中国激光，2012, 39: s105013.

摘要：光子晶体光纤（PCF）的导光特性可通过改变空气孔的结构参数（孔径、间距和排列方式）、材料填充等方法进行调节。由于自身具有电可调性，液晶作为PCF的填充材料具有很大的研究价值，可以用于制作电可调PCF。利用有限元法分析了液晶（E7）填充的光子晶体光纤的基模有效折射率、有效模场面积等参量随占空比、外电场的变化关系，得到了不同占空比下基模的截止电压和一定电压下基模的截止波长。结果表明光子晶体光纤的电压可调范围随占空比增大而增大；占空比一定时，电压越大，波长可调范围越小。这种液晶填充的光子晶体光纤可以应用于电场传感等领域。

关键词：光纤光学，光子晶体光纤，有限元法，液晶

144. Supermode analysis of the 18-core photonic crystal filber laser

Wang Yuan, Yao Jianquan, Zheng Yibo, Wen Wuqi, Lu Ying, Wang Peng
Chinese Journal of Lasers, 2012, 39(9): 0902003.

Abstract: The modal of 18-core photonic crystal fiber laser is discussed and calculated. And corresponding far-field distribution of the supermodes is given by Fresnel diffraction integral. For improving beam quality, the mode selection method based on the Talbot effect is introduced. The reflection coefficients are calculated, and the result shows that an in-phase supermode can be locked better at a large propagation distance.
Key words: Photonic crystal filber laser, 18-core, Supermode analysis

145. Tunable thermo-optic switch based on fluid-filled photonic crystal fibers

Wang Ran, Yao Jianquan, Lu Ying, Miao Yinping, Zhao Xiaolei, Wang Ruoqi
Optoelectronics Letters, 2012, 8(6): 0430–0432.

Abstract: A tunable thermo-optic intensity-modulated switch is investigated theoretically and numerically. It is based on the infiltration of temperature-sensitive mixture liquids into index-guiding photonic crystal fibers (PCFs). The switching function attributes to the thermo-optic effect of the effective refractive index of the cladding. The simulation illustrates that the switch presents a tunable transition point according to the concentration of the mixture liquids, and the on-off switching functionality can be realized within a narrow temperature range of 2 ℃. The switches have wide application for innovative all-in-fiber optical communication and logic devices.
Key words: Thermo-optic switch, Intensity-modulated, Tunable, Photonic crystal fibers, Fluid-filled, Refractive index

146. Research on the transverse mode competition in a Yb-doped 18-core photonic crystal fiber laser

Wang Yuan, Yao Jianquan, Zheng Yibo, Wen Wuqi, Lu Ying, Wang Peng
Optoelectronics Letters, 2012, 8(6): 0426–0429.

Abstract: A model based on propagation rate equations is built up for analyzing the multicore transverse mode gain distribution in an 18-core photonic crystal fiber (PCF) laser. The two kinds of feedback cavities are used for the fiber laser, which are the butt contact mirror and the Talbot cavity. According to the model, the transverse mode competitions in different feedback cavities are simulated numerically. The results show that the Talbot cavity can improve in-phase supermode gain, while suppress other supermodes.

Key words: Photonic crystal fiber laser, Yb-doped, 18-core, Transverse mode competition, Feedback cavity

147. Theoretical study on modulating group velocity of light in photonic crystal coupled cavity optical waveguide

Lu Ying, Huang Xiaohui, Fu Xiangyong, Chu Danping, Yao Jianquan
Optoelectronics Letters, 2012, 8(1): 0025–0028.

Abstract: We present a novel mechanism, which is formed by periodically changing the radii of dielectric rods in the middle row of a photonic crystal, to control and stop light. Using the Bloch theory and coupled-mode theory, the dispersion characteristic of such a photonic crystal coupled cavity optical waveguide is obtained. We also theoretically demonstrate that the group velocity of a light pulse in this system can be modulated by dynamically changing the refractive index or radii of the selected dielectric rods, and the light stopping can be achieved.

Key words: Optical modulation, Group velocity, Photonic crystal coupled cavity optical waveguide, Control light, Stop light, Bloch theory, Coupled-mode theory, Dispersion characteristic

148. Theoretical and experimental researches on a PCF-based SPR sensor

Bing Pibin, Li Zhongyang, Yao Jianquan, Lu Ying, Di Zhigang, Yan Xin
Optoelectronics Letters, 2012, 8(4): 0246–0248.

Abstract: A photonic crystal fiber based surface plasmon resonance (PCF-SPR) sensor is simulated by finite element method and experimentally realized. The calculations show that there is an obvious loss peak in the vicinity of 1.2 μm while the PCF of LMA-8 is used as a sensor. The suspension of silver nanoparticle mixed with hexadecyl trimethyl ammonium bromide (CTAB) is inhaled into the PCF to form a metal film which can be stimulated to generate plasmon in the experiment. A spectrometer is utilized to detect the continuous broadband transmission spectrum from the PCF. The experimental results verify the loss peak. Compared with the theoretical calculations, the offset of loss peak about 40 nm can be acceptable, because the uniformity of the metal coating is difficult to guarantee and the film thickness is difficult to control.

Key words: SPR sensor, Photonic crystal fiber, Hexadecyl trimethyl ammonium bromide, Finite element method

149. Microstructured polymer optical fiber-based surface plasmon resonance sensor

Lu Ying, Wu Baoqun, Fu Xiangyong, Hao Congjing, Huang Xiaohui, Yao Jianquan
Proceedings of SPIE, 2012, 8421: 842175.

Abstract: We propose a surface plasmon resonance sensor based on microstructured polymer optical fiber (mPOF) mad of polymethyl methacrylate (PMMA) with the cladding having only one layer of air holes near the edge of the fiber. In such sensor, the nanoscale silver metal film can be deposited on the outer side of the fiber, which is easily realized. Numerical simulation results show that spectral and intensity sensitivity are in the ranger of $8.3 \times 10^{-5} \sim 9.4 \times 10^{-5}$ RIU.

Key words: Microstructured polymer optical fiber, Surface plasmon resonance, Sensor

150. Surface plasmon resonance sensor based on a novel grapefruit photonic crystal fiber

Zhang Peipei, Yao Jianquan, Jing Lei, Cui Haixia, Lu Ying
Proceedings of SPIE, 2012, 8421: 84217I.

Abstract: We propose a novel surface plasmon resonance sensor design based on a grapefruit photonic crystal fiber. In such a sensor, phase matching between plasmon and a core mode is achieved by introducing microstructure into the fiber core. Using the finite element method, the confinement loss of the fiber is calculated to measure the sensitivity of the sensor. Simulation results show that this kind of sensor has an excellent effect, with the amplitude resolution to be as low as 2.88×10^{-5} RIU and the spectral resolution to be 6.67×10^{-5} RIU.

Key words: Photonic crystal fiber, Surface plasmon resonance, Finite element method, Confinement loss

151. Surface plasmon resonance sensor based on grapefruit fiber filled with silver nanowires

Hao Congjing, Lu Ying, Fu Xiangyong, Yao Jianquan
Proceedings of SPIE, 2012, 8421: 84217C.

Abstract: We study the surface plasmon resonance sensors based on grapefruit Photonic crystal fiber (PCF) filled with different numbers silver nanowires. Numerical results show that the intensity sensitivity may be influenced by the distance between two nanowires and nanowires numbers. A best value is got with a maximum distance of 2 μm. And the PCF filled with more nanowires is better than the one. Moreover, the air holes of grapefruit PCF are large enough to operate in practice.

Key words: Grapefruit photonic crystal fiber, Surface plasmon resonance, Silver nanowire, Silver layer

152. 用非球面透镜制作光纤约 1:1 空间耦合器

温午麒,康建翊,丁欣,陆颖,杨鹏飞,伏祥勇,宁鼎,姚建铨

光子学报, 2012, 41(3): 294–298.

摘 要：针对光纤的泵浦耦合问题,对由两片非球面透镜组成的接近 1∶1 光纤间空间耦合器进行了计算和实验验证。利用高斯光束的变化规律对光路进行了分析研究,并根据二极管输出光相干性不好的特点,对非球面透镜进行了光路追迹的模拟计算。研究发现,在泵浦光波长等因素发生变化时,利用椭球面透镜组成的耦合系统较双曲面透镜有更高的稳定性。实验中选用符合计算结果要求的非球面透镜组成耦合装置,利用一台二极管激光器(尾纤输出端面直径约 200 μm,N.A. 约 0.2)泵浦一段芯径约 200 μm(N.A. 约 0.42)的多模光纤,耦合装置的透过率约 95%,在光纤端面有反射的条件下约 90%的泵浦光耦合进光纤。

关键词：类高斯光束,空间耦合器,非球面透镜,光纤

153. Surface plasmon resonance refractive index sensor based on active photonic crystal fiber

C. J. Hao, Y. Lu, M. T. Wang, B. Q. Wu, L. C. Duan, J. Q. Yao

IEEE Photonics Journal, 2013, 5(6): 4801108–4801108.

Abstract: We propose a surface plasmon resonance (SPR) refractive index (RI) sensor based on an active Yb^{3+}-doped photonic crystal fiber (PCF) in this paper. With the proposed sensor, using the pump light at 976 nm can produce laser at 1 060 nm. In addition, the sensitivity can be influenced obviously by a bit change of the refraction index of analyte in the air holes to achieve the intra-cavity fiber sensing. It is found that not only the different air filling ratios but also the different analyte RIs of $n_a > 1:45$ or $n_a < 1:45$ have different effects on the output power and confinement loss, and lead to different trends. The intra-cavity PCF sensing system has great practical value and significance for their advantages of compact structure and high sensitivity.

Key words: Yb^{3+}-doped PCF, PCF-laser-based sensing, Sensor of refractive index, Surface plasmon resonance (SPR)

154. Ferrofluid-infiltrated microstructured optical fiber long-period grating

Yinping Miao, Kailiang Zhang, Bo Liu, Wei Lin, Hao Zhang, Ying Lu, Jianquan Yao

IEEE Photonics Technology Letters, 2013, 25(3): 306–309.

Abstract: Long-period gratings (LPGs) were successfully inscribed in microstructured optical fibers (MOFs) filled with ferrofluid using a scanning CO_2 laser. Its formation mechanism and magnetic field tunability are investigated. The dispersion curve indicates that cladding modes are more viable to be tuned than the core mode for the ferrofluid-filled MOF. The relationship between the phase-matching curve of MOF-based LPG and the magnetic field intensity is theoretically analyzed, and the magnetic-field responses of the LPG with 660-μm-pitch and a resonance peak of 967.56 nm are also discussed. The results show that the MOF-based LPG reaches a sensitivity of 1.946 nm/Oe for a magnetic range of 0–300 Oe, demonstrating its potential application as a high-sensitivity magnetic-field sensor. The proposed magnetic-field sensor could detect the weak magnetic field with high accuracy. It has several unique advantages, such as compactness, good wavelength selectivity, high integration, ease of coupling, high flexibility, and extensibility. Consequently, the proposed device with tunable magnetic-field sensitivity is promising for future-related applications.

Key words: Long-period grating, Magnetic-field sensor, Magnetic-field tunability, Microstructured optical fiber

155. 11 mJ all-fiber-based actively Q-switched fiber master oscillator power amplifier

Qiang Fang, Yuguo Qin, Bo Wang, Wei Shi

Laser Physics Letters, 2013, 10: 115103

Abstract: We report a high energy all fiber format nanosecond pulsed laser source at ~1 064 nm in master oscillator power amplifier (MOPA) configuration. The seed source is an acousto-optic Q-switched fiber laser with a varied pulse duration and repetition rate. The output average power of the oscillator is ~30 mW and two pre-amplifiers were developed to boost the average power to ~3 W. Pulse energy of >11 mJ for ~660 ns pulses at 3 kHz was achieved in the final power amplifier using a commercial 50/400 μm (core/cladding diameter) double cladding Yb-doped fiber (DCYF).

Key words: Master oscillator power amplifier, All fiber, Actively Q-switched, High energy, Nanosecond pulsed laser, 1 064 nm

156. Agarose gel-coated LPG based on two sensing mechanisms for relative humidity measurement

Yinping Miao, Kaikiang Zhang, Yujie Yuam, Bo Liu, Hao Zhang, Yan Liu, Jianquan Yao

Applied Optics, 2013, 52(1): 90–95.

Abstract: A relative humidity (RH) sensor based on long-period grating (LPG) with different responses is proposed by utilizing agarose gel as the sensitive cladding film. The spectral characteristic is discussed as the ambient humidity level ranges from 25% to 95% RH. Since increment of RH will result in volume expansion and refractive index increment of the agarose gel, the LPG is sensitive to applied strain and ambient refractive index; both the resonance wavelength and coupling intensity present particular responses to RH within two different RH ranges (25%–65% RH and 65%–96% RH). The coupling intensity decreases within a lower RH range while it increases throughout a higher RH range. The resonance wavelength is sensitive to the higher RH levels, and the highest sensitivity reaches 114.7 pm/% RH, and shares the same RH turning point with coupling intensity response. From a practical perspective, the proposed RH sensor would find its potential applications in high humidity level, temperature-independent RH sensing and multiparameter sensing based on wavelength/power hybrid demodulation and even static RH alarm for automatic monitoring of a particular RH value owing to the nonmonotonic RH dependence of the transmission power within the whole tested RH range.

Key words: Relative humidity sensor, Different responses, Long-period grating, Agarose gel, Refractive index

157. 230 W average-power all-fiber-based actively Q-switched fiber master oscillator-power amplifier

Qiang Fang, Yuguo Qin, Bo Wang, Wei Shi

Applied Optics, 2013, 52(27): 6744–6747.

Abstract: We report a high-power all-fiber format pulsed laser source at ~1 064 nm in a master oscillator–power amplifier configuration. The seed source is an acousto-optic Q-switched fiber laser with a varied pulse duration and repetition rate. The output power of the oscillator is ~1 W, and two double-cladding Yb-doped fiber amplifiers were used to boost the average power of the seed. >230 W average power was achieved for ~1.4 μs pulses at 100 kHz repetition rate. The optical-to-optical efficiency of the main amplifier is 72.81%.

Key words: Master oscillator-power amplifier, All fiber laser, Actively Q-switched, Acousto-optic, High power, 1 064 nm

158. Surface plasmon resonance sensor based on polymer photonic crystal fibers with metal nanolayers

Ying Lu, Congjing Hao, Baoqun Wu, Mayilamu Musideke, Liangcheng Duan, Wuqi Wen, Jianquan Yao

Sensors, 2013, 13: 956–965.

Abstract: A large-mode-area polymer photonic crystal fiber made of polymethyl methacrylate with the cladding having only one layer of air holes near the edge of the fiber is designed and proposed to be used in surface plasmon resonance sensors. In such sensor, a nanoscale metal film and analyte can be deposited on the outer side of the fiber instead of coating or filling in the holes of the conventional PCF, which make the real time detection with high sensitivity easily to realize. Moreover, it is relatively stable to changes of the amount and the diameter of air holes, which is very beneficial for sensor fabrication and sensing applications. Numerical simulation results show that under the conditions of the similar spectral and intensity sensitivity of 8.3×10^{-5}–9.4×10^{-5} RIU, the confinement loss can be increased dramatically.

Key words: Polymer photonic crystal fiber, Surface plasmon resonance sensor

159. A reflective photonic crystal fiber temperature sensor probe based on infiltration with liquid mixtures

Ran Wang, Jianquan Yao, Yinping Miao, Ying Lu, Degang Xu, Nannan Luan, Mayilamu Musideke, Liangcheng Duan, Congjing Hao

Sensors, 2013, 13: 7916–7925.

Abstract: In this paper, a reflective photonic crystal fiber (PCF) sensor probe for temperature measurement has been demonstrated both theoretically and experimentally. The performance of the device depends on the intensity modulation of the optical signal by liquid mixtures infiltrated into the air holes of commercial LMA-8 PCFs. The effective mode field area and the confinement loss of the probe are both proved highly temperature-dependent based on the finite element method (FEM). The experimental results show that the reflected power exhibits a linear response with a temperature sensitivity of about 1 dB/℃. The sensor probe presents a tunable temperature sensitive range due to the concentration of the mixture components. Further research illustrates that with appropriate mixtures of liquids, the probe could be developed as a cryogenic temperature sensor. The temperature sensitivity is about 0.75 dB/℃. Such a configuration is promising for a portable, low-power and all-in-fiber device for temperature or refractive index monitoring in chemical or biosensing applications.

Key words: Photonic crystal fiber, Temperature sensor probe, Liquid mixtures, Finite element method, Cryogenic temperature

160. The use of a dual-wavelength erbium-doped fiber laser for intra-cavity sensing

Baoqun Wu, Ying Lu, Lei Jing, Xiaohui Huang, Jianquan Yao

Laser Physics, 2013, 23: 115103.

Abstract: A novel intra-cavity sensor based on a dual-wavelength Er-doped fiber laser is proposed. The output power characteristics of the laser are investigated experimentally, and the sensitivity of power to relative cavity loss with different pump currents is studied by defining a sensitivity enhancement factor (SEF). A measurement of the relative sensitivity enhancement of 158.5 is obtained. Moreover, we used such a sensor to measure the absorption peak of NH_3 with a 2 m long hollow core photonic crystal fiber as a gas cell.

Key words: Erbium-doped fiber laser, Dual-wavelength, Intra-cavity sensor, Output power characteristics, Sensitivity enhancement factor

161. Magneto-optical tunability of magnetic fluid infiltrated microstructured optical fiber

Yinping Miao, Bo Liu, Kailiang Zhang, Hao Zhang, Ran Wang, Yan Liu, Jianquan Yao

Optics & Laser Technology, 2013, 48: 280–284.

Abstract: In this paper, the magneto-optical properties of hexagon-hole microstructured optical fiber filled with magnetic fluid were investigated. The results indicate that the proposed device has excellent magnetic field tunability and could be used in sensing area. The dispersion curves of different modes under various magnetic fields demonstrate the higher the intensity of magnetic field, the lower the refractive index of mode; the tunability of cladding modes is greater than that of core modes, and the tunability of higher order modes is greater than that of lower order modes. It shows that filled MOF could be used for the mode ON–OFF or single/multi-mode filters controlled by magnetic field. The transmission of the fiber decreases exponentially with increase of the magnetic field intensity, which permits filled MOF use as intensity-modulated magnetic-field sensors in high sensitivity magnetic-sensing area.

The proposed magneto-optic tunable devices have several advantages of simple structure, high integration, easy fabrication, etc.

Key words: Microstructured optical fiber, Magnetic fluid, Magneto-optical tunability

162. Thermo-optic characteristics of micro-structured optical fiber infiltrated with mixture liquids

Ran Wang, Yuye Wang, Yinping Miao, Ying Lu, Nannan Luan, Congjing Hao, Liangcheng Duan, Cai Yuan, Jianquan Yao

Journal of the Optical Society of Korea, 2013, 17(3): 231–236.

Abstract: We present both theoretically and experimentally the thermo-optic characteristics of micro-structured optical fiber (MOF) filled with mixed liquid. The performance of MOF depends on the efficient interaction between the fundamental mode of the transmitted light wave and the tunable thermo-optic materials in the cladding. The numerical simulation indicates that the confinement loss of MOF presents higher temperature dependence with higher air-filling ratios d/Λ, longer incident wavelength and fewer air holes in the cladding. For the 4 cm liquid-filled grapefruit MOF, we demonstrate from experiments that different proportions of solutions lead to tunable temperature sensitive ranges. The insertion loss and the extinction ratio are 3~4 dB and approximate 20 dB, respectively. The proposed liquid-filling MOF will be developed as thermo-optic sensor, attenuator or optical switch with the advantages of simple structure, compact configuration and easy fabrication.

Key words: Micro-structured optical fiber, Thermo-optic characteristics, Mixture liquid, Thermo-optic, sensor, Optical switch

163. Effects of heterogeneity on the surface plasmon resonance biosensor based on three-hole photonic crystal fiber

Pibin Bing, Zhongyang Li, Jianquan Yao, Ying Lu

Optical Engineering, 2013, 52(5): 054401.

Abstract: A surface plasmon resonance biosensor based on three-hole photonic crystal fiber (PCF) is analyzed by the finite element method. The results demonstrate that the biosensor will exhibit different loss spectra characteristics under the conditions of nonuniform thicknesses of the auxiliary dielectric layer, the gold layer, and the biolayer in the three-hole PCF, respectively. Furthermore, the sensing properties in both areas of resonant wavelength and intensity detection are discussed. Numerical results show excellent sensing characteristics when the thickness of the auxiliary dielectric layer is s =1 μm and the gold d_{gold} = 40 nm, respectively. The sensor resolution of the biolayer thickness is demonstrated more than 0.05 nm in the vicinity of 0.6 μm with the amplitude-based method.

Key words: Photonic crystal fiber, Surface plasmon resonance, Sensor, Finite element method

164. Hollow-core photonic crystal fiber based on C_2H_2 and NH_3 gas sensor

Baoqun Wu, Ying Lu, Congjing Hao, Liangcheng Duan, Nannan Luan, Zhiqiang Zhao, Jianquan Yao

Applied Mechanics and Materials, 2013, 411–414:1577–1580.

Abstract: In this paper, we propose a new hollow-core photonic crystal fiber, which can be available for gas sensor. In addition, properties of the fiber are analyzed at the wavelength of C_2H_2 and NH_3 absorption peak 1 530 nm and 1 967 nm, respectively. For both wavelengths, relative sensitivity coefficients are higher than 0.95, which makes sense in gas sensing. We also get relationship between relative sensitivity coefficient and radius of fiber core, as well as effective refractive index of the mode field.

Key words: Coefficient sensitivity, HC-PCF, Gas detection

165. Numerical investigation of the microstructured optical fiber-based surface plasmon resonance sensor with silver nanolayer

Nannan Luan, Jianquan Yao, Ran Wang, Congjing Hao, Baoqun Wu, Liangcheng Duan, Ying Lu

Applied Mechanics and Materials, 2013, 411–414: 1573–1576.

Abstract. The surface plasmon resonance (SPR) sensor is proposed based on coating the inner surfaces of an index-guiding microstructured optical fiber (MOF) with a silver layer. Fiber core is surrounded by six large metallized holes which should facilitate the fabrication of the layered sensor structure and the infiltration of the analyte. The relationship between the sensitivity of SPR sensor and the refractive index of MOF material is demonstrated with finite element method (FEM). Numerical simulation results indicate that the sensitivity of SPR sensor decreases as the refractive index of the MOF material increasing and both spectral and intensity sensitivity are estimated to be 6.25×10^{-5} and 6.67×10^{-5} with low refractive index of MOF material n=1.46.

Key words: Surface plasmon resonance, Microstructured optical fiber, Silver nanolayer

166. A surface plasmon resonance sensor based on a multi-core photonic crystal fiber

Zhang Peipei, Yao Jianquan, Cui Haixia, Lu Ying

Optoelectronics Letters, 2013, 9(5): 0342–0345.

Abstract: A surface plasmon resonance (SPR) sensor based on a multi-core photonic crystal fiber (PCF) is presented in this paper. There is only one analyte channel positioned in the center of the PCF cross section, rather than several closely arranged analyte channels around the central core. So the design of this sensor not only reduces the consumption of gold and samples, but also effectively avoids the interference between neighboring analyte channels. Optical field distributions of this fiber at different wavelengths and the sensing properties of this sensor are theoretically analyzed and discussed using finite element method (FEM). Simulation results confirm that both the thickness of metallic layer and the fiber structural parameters have significant effect on sensing performance. The amplitude sensitivity of the sensor is found to be 1.74×10^{-5} RIU, and the spectral sensitivity is 3 300 nm/RIU, corresponding to a resolution of 3.03×10^{-5} RIU. Finally, in order to achieve PCF-SPR sensing characteristics, an experiment design scheme based on spectro-scopic detection method is proposed.

Key words: Surface plasmon resonance, Sensor, Photonic crystal fiber, Multi-core, Finite element method

167. 基于液体填充微结构光纤的新型光子功能器件

姚建铨，王然，苗银萍，陆颖，赵晓蕾，景磊

中国激光, 2013, 40(1): 0101002.

摘要： 基于液体材料填充的微结构光纤光子器件有效地将功能材料在不同外界物理场作用下的物理效应同光纤自身的微纳结构结合起来，具有可调谐、设计灵活、全光纤结构和易于集成等优点，是未来光纤光子器件发展的重要方向。掌握不同填充材料、填充方法及所制作器件的不同特性、功能和应用对这一领域的研究具有重要的指导意义。综合阐述了近年来基于液体材料填充的微结构光纤光子器件的研究进展，分析和归纳了各种液态功能材料的种类、物理特性及填充方法，系统阐述了基于该种方法实现的光开关及衰减器、滤波器、调制器、色散补偿器等可调谐光纤光子器件及光纤传感器件，最后对该领域未来的发展方向和前景进行了展望，为未来新型光纤光子器件的研制提供必要的依据和参考。

关键词： 光学器件，微结构光纤，液体填充，光纤传感器

168. Intra-cavity absorption sensor based on erbium doped fiber laser

Ying Lu, Baoqun Wu, Xiaohui Huang, Liangcheng Duan, Congjing Hao, Mayilamu Musideke, Jianquan Yao

IEEE Eighth International Conference on Intelligent Sensors, Sensor Networks and Information Processing (ISSNIP), 2013: 32–35.

Abstract: The sensitivity of power to relative cavity loss of the intra-cavity sensor based on erbium-doped fiber laser is investigated. Theoretical analysis and simulation show that the coupling ratio of coupler and the pump power have a great effect on sensitivity of ring intra-cavity sensor. And we have experimentally found that the sensitivity can be significantly enhanced as pump power approaches the threshold and the coupling ratio is higher.

Key words: Erbium doped fiber laser, Fiber optical sensor, Intra-cavity absorption sensor

169. 1.55 μm 波长处具有高非线性低限制损耗的八边形实心光子晶体光纤

马依拉木·木斯得克,姚建铨,陆颖

激光与光电子学进展, 2013, 50(4): 67-71.

摘要：为了得到高非线性低损耗光子晶体光纤,设计了八角格子圆形空气孔组成的光子晶体光纤结构。利用全矢量有限元法并结合完美匹配层吸收边界条件,对该光子晶体光纤的纤芯材料折射率、非线性系数和限制损耗进行了数值模拟。数值模拟结果表明,该光纤呈现高非线性、低损耗和较好的模场约束能力。调整光纤参数为 $d_1=0.77$ μm, $d_2=0.86$ μm 时可以得到更好的结果,在波长 1.55 μm 处获得高的非线性系数 37.6 $km^{-1}·W^{-1}$ 和低限制损耗 0.7×10^{-17} dB/km。

关键词：光子晶体光纤,有效模面积,模场分布,高非线性,低限制损耗

170. 光纤光栅温度应变同时测量传感技术研究进展

陈曦,姚建铨,陈慧

传感器与微系统, 2013, 32(9): 1-4.

摘要：介绍了光纤光栅温度应变传感器的基本原理,对当前常用的实验方法进行了分类。简要介绍了参考光纤光栅法、双波长叠栅法、光纤 Bragg 光栅 (FBG) 与长周期光栅相结合的方法、不同包层直径光栅对法、纤芯掺杂法以及光纤光栅 F-P 腔法等方法的机理,并对每一种方案进行了详细的分析,讨论了其优缺点和改进方向。

关键词：光纤光栅,温度,应变,区分测量,交叉敏感

171. 光纤气体传感器及其组网技术综述

陈慧, 姚建铨, 陈曦

传感器与微系统, 2013, 32(9): 9–11.

摘要：综述了光纤气体传感器及其组网方式,分别介绍了波分复用、频分复用、空分复用以及码分复用的技术,并分析了各种方法的优缺点,最后讨论了光纤气体传感器的发展方向。

关键词：气体传感器,光纤传感器,气体传感网,复用技术

172. 物联网与智慧城市的关系

姚建铨

枣庄学院学报, 2013, 30(2): 1–4.

摘要：物联网和智慧城市之间是一种相辅相成和相得益彰的关系,智慧城市是目标,需要物联网技术进行支撑,物联网是途径和技术手段,为智慧城市建设服务。物联网是智慧城市的重要标志,智慧城市成为物联网未来发展的热点应用领域。

关键词：物联网,智慧城市,信息技术

173. 物联网是工具手段 智慧城市是目标

姚建铨

中国信息化周报, 2013, 6: 1-2.

摘要：物联网应该为智慧城市服务，而智慧城市的建设一定要借用物联网、云计算、移动互联网等新一代信息技术作为支撑。物联网包括两大产业，分别是物联网技术、物联网产品和服务，通过感知层、网络层和应用层，将物理世界和网络紧密地结合起来，结合国家物联网和智慧城市规划，构建新型战略产业，以物联网和物联网相关的信息领域技术为手段，推进物联网产业发展、技术进步和人才培养，实现建设智慧城市的国家目标。

关键词：物联网，智慧城市，新型战略产业

174. Low temperature sensitive intensity-interrogated magnetic field sensor based on modal interference in thin-core fiber and magnetic fluid

Jixuan Wu, Yinping Miao, Binbin Song, Wei Lin, Hao Zhang, Kailiang Zhang, Bo Liu, Jianquan Yao

Applied Physics Letters, 2014, 104(25): 252402.

Abstract: A fiber-optic magnetic field sensor based on the thin-core modal interference and magnetic fluid (MF) has been proposed and experimentally demonstrated. The magnetic field sensor is spliced with a thin-core fiber (TCF) between two conventional single-mode fibers immersed into the MF. The transmission spectra of the proposed sensor under different magnetic field intensities have been measured and theoretically analyzed. The results show that the magnetic field sensitivity reaches up to −0.058 dB/Oe with the linear range from 75 Oe to 300 Oe. Due to the small thermal expansion of the TCF material, the attenuation wavelength and the transmission power remain almost unchanged as the temperature varies. The proposed magnetic field sensor has several advantages such as intensity-interrogation, low temperature sensitivity, low cost, compact size, and ease of fabrication. And particularly, the temperature cross-sensitivity could be effectively resolved, which makes it a promising candidate for strict temperature environments. Therefore, it would find potential applications in the magnetic field measurement.

Key words: Fiber-optic magnetic field sensor, Thin-core modal interference, Magnetic fluid, Thin-core fiber, Transmission spectra, Sensitivity

175. Intracavity absorption multiplexed sensor network based on dense wavelength division multiplexing filter

Haiwei Zhang, Ying Lu, Liangcheng Duan, Zhiqiang Zhao, Wei Shi, Jianquan Yao
Optics Express, 2014, 22(20): 24545−24550.

Abstract: We report the system design and experimental verification of an intracavity absorption multiplexed sensor network with hollow core photonic crystal fiber (HCPCF) sensors and dense wavelength division multiplexing (DWDM) filters. Compared with fiber Bragg grating (FBG), it is easier for the DWDM to accomplish a stable output. We realize the concentration detection of three gas cells filled with acetylene. The sensitivity is up to 100 ppmV at 1 536.71 nm. Voltage gradient is firstly used to optimize the intracavity sensor network enhancing the detection efficiency up to 6.5 times. To the best of our knowledge, DWDM is firstly used as a wavelength division multiplexing device to realize intracavity absorption multiplexed sensor network. It make it possible to realize high capacity intracavity sensor network via multiplexed technique.

Key words: Intracavity absorption multiplexed sensor network, Hollow core photonic crystal fiber, Dense wavelength division multiplexing filter

176. 700-kW-peak-power monolithic nanosecond pulsed fiber laser

Qiang Fang, Wei Shi, Jingli Fan
IEEE Photonics Technology Letters, 2014, 26(16): 1676−1678.

Abstract: We report a high-pulse energy, high-peak power, and monolithic nanosecond pulsed fiber laser source at similar to 1 064 nm in master oscillator-power amplifier configuration. The seed source is a directly modulated laser diode, producing nanosecond pulses as short as similar to 3 ns with similar to 1-W peak power and a tunable repetition rate. Three core-pumped fiber amplifier stages and two double-cladding fiber amplifier stages were built to boost the peak power and the pulse energy of the ~3-ns seed pulses to ~697 kW and ~2.3 mJ, respectively. The all-fiber construction of the whole laser system enables compact size, maintenance-free, and robust operation.

Key words: Fiber amplifier, Nanosecond pulsed fiber amplifier, Yb-doped fiber

177. 978 nm single frequency actively Q-switched all fiber laser

Qiang Fang, Wei Shi, Xueping Tian, Bo Wang, Jianquan Yao, Nasser Peyghambarian

IEEE Photonics Technology Letters, 2014, 26(9): 874–876.

Abstract: A single-frequency actively Q-switched all-fiber laser at 978 nm was developed using a 2-cm-long commercial ytterbium-doped silica fiber and a pair of silica-fiber-based Bragg gratings. The Q-switching is enabled by the stress-induced polarization modulation in the short linear fiber laser cavity. The laser generates actively Q-switched single-frequency laser pulses with a pulse repetition rate ranging from tens of kilohertz to hundreds of kilohertz. The pulse duration and the average/peak power have been characterized. This is, to the best of our knowledge, the first demonstration of a single-frequency all-fiber Q-switched laser below 1 μm using commercial highly ytterbium-doped silica fiber.

Key words: Single frequency, 978 nm silica fiber laser, Actively Q-switch

178. Dual-direction magnetic field sensor based on core-offset microfiber and ferrofluid

Jixuan Wu, Yinping Miao, Wei Lin, Kailiang Zhang, Binbin Song, Hao Zhang, Bo Liu, Jianquan Yao

IEEE Photonics Technology Letters, 2014, 26(15): 1581–1584.

Abstract: A fiber-optical sensor is proposed for dual-direction magnetic field measurement by cascading an optical microfiber with a core-offset section. Due to the relative direction difference between magnetic field and fiber axis, the refractive index of the magnetic fluid could be tuned accordingly. The transmission spectral characteristics change with the variation of applied magnetic field. Experimental results show that different relative directions turn out different linear responses. When the magnetic field is applied perpendicular or parallel to the fiber axis, the magnetic field sensitivities are -0.025 34 and 0.011 11 dB/Oe, respectively. The proposed sensor has several advantages, including low cost, ease of fabrication, compact structure, and cost-effective intensity interrogation. It is promising in the measurement of magnetic field vector, as well as multiparameter sensing.

Key words: Magnetic fluid, Optical microfiber, Magnetic field sensors

179. Temperature sensing using photonic crystal fiber filled with silver nanowires and liquid

Y. Lu, M. T. Wang, C. J. Hao, Z. Q. Zhao, J. Q. Yao

IEEE Photonics Journal, 2014, 6(3): 6801307.

Abstract: A temperature sensor based on photonic crystal fiber (PCF) surface plasmon resonance (SPR) is proposed in this paper. We use the dual function of the PCF filled with different concentrations of analyte and silver nanowires to realize temperature sensing. The proposed sensor has been analyzed through numerical simulations and demonstrated by experiments. The results of the simulations and experiments show that a blue shift will be obtained with the temperature increase, and different concentrations will change the resonance wavelength and confinement loss. Temperature sensitivity is as high as 2.7 nm/℃ with the experiment, which can provide a reference for the implementation and application of a PCF-based SPR temperature sensor or other PCF-based SPR sensing.

Key words: Photonic crystal fiber, Temperature sensor, Silver nanowires, Surface plasmon resonance

180. Lensed water-core teflon-amorphous fluoroplastics optical fiber

Chunyin Tang, Gongxun Bai, Kwok Lung Jim, Xuming Zhang, Kin Hung Fung, Yang Chai, Yuen H. Tsang, Jianquan Yao, Degang Xu

Journal of Lightwave Technology, 2014, 32(8): 1538–1542.

Abstract: Specially designed liquid-core optical fiber can convey light with demand wavelength and analytes in the same pathway according to the application of the fiber. Water is one of the most useful and common core fluid due to its non-toxicity and other practical optical properties, and which makes water core optical fiber advantageous for biomedical sensing, optical imaging, non-linear optics and optical transmission. Since the numerical aperture of the liquid core fiber is not able to approach to an infinite small value, thus it is necessary to use expensive solid state lens to couple light into or out of the liquid core fiber. In this paper, we have demonstrated a unique method to fabricate water-core lensed fibers by filling light water and heavy water respectively into hollow Teflon-Amorphous Fluoroplastics fibers, and to control the focal length and the spot size by pumping water into or out of the fiber end. By simulation, the focal length over the range of $f = 3.87\text{-}1.33$ mm has been demonstrated using the distilled water, and $f = 4.95\text{-}1.38$ mm using the heavy water. To further reduce the focal length, we have limited the lens aperture by the fiber core and have demonstrated a focal length over $f = 0.34\text{-}0.27$ mm in response to the change of the heavy water lens volume from 0.68 to 1.54 nL. Further simulation shows that the focused spot size can be reduced to 2-6 μm by adjusting the refractive index and fiber geometry. Compared to other optical focusing methods, such lensed fiber and the tuning of its focal length are far easier to make at a much lower cost.

Key words: Lensed fibers, Variable focal length, Water-core optical fiber

181. Magnetic field tunability of square tapered no-core fibers based on magnetic fluid

Yinping Miao, Jixuan Wu, Wei Lin, Binbin Song, Hao Zhang, Kailiang Zhang, Bo Liu, Jianquan Yao

Journal of Lightwave Technology, 2014, 32(23): 4600–4605.

Abstract: A magnetic-field-tuned photonics device based on magnetic fluid (MF) and a square tapered no-core fiber (NCF) sandwiched between two single-mode fibers (SMFs) has been demonstrated experimentally and theoretically. The enhanced evanescent field effect in the NCF is achieved by tapering the square NCF utilizing a fusion splicer. The spectral dependence of the proposed device on the applied magnetic-field intensity has been investigated. The results indicate that the multimode interference spectrum exhibits a blue-shift with the increment of the magnetic-field intensity. A maximal sensitivity of -18.7 pm/Oe is obtained for a magnetic field strength ranging from 25 to 450 Oe. The proposed tunable device has several advantages, including low cost, ease of fabrication, simple and compact structure, and high sensitivity. Therefore, the magnetic-field-tuned square tapered NCF is expected to find potential applications in the fields of optical fiber sensors, as well as fiber communications.

Key words: Magnetic field, Magnetic fluid, Square tapered no-core fiber

182. Surface plasmon resonance temperature sensor based on photonic crystal fibers randomly filled with silver nanowires

Nannan Luan, Ran Wang, Wenhua Lv, Ying Lu, Jianquan Yao
Sensors, 2014, 14: 16035–16045.

Abstract: We propose a temperature sensor design based on surface plasmon resonances (SPRs) supported by filling the holes of a six-hole photonic crystal fiber (PCF) with a silver nanowire. A liquid mixture (ethanol and chloroform) with a large thermo-optic coefficient is filled into the PCF holes as sensing medium. The filled silver nanowires can support resonance peaks and the peak will shift when temperature variations induce changes in the refractive indices of the mixture. By measuring the peak shift, the temperature change can be detected. The resonance peak is extremely sensitive to temperature because the refractive index of the filled mixture is close to that of the PCF material. Our numerical results indicate that a temperature sensitivity as high as 4 nm/K can be achieved and that the most sensitive range of the sensor can be tuned by changing the volume ratios of ethanol and chloroform. Moreover, the maximal sensitivity is relatively stable with random filled nanowires, which will be very convenient for the sensor fabrication.

Key words: Temperature sensor, Fiber optic sensor, Photonic crystal fiber, Surface plasmon resonance, Silver nanowire, Mixture

183. Fiber lasers and their applications [Invited]

Wei Shi, Qiang Fang, Xiushan Zhu, R. A. Norwood, N. Peyghambarian
Applied Optics, 2014, 53(28): 6554–6568.

Abstract: Fiber lasers have seen progressive developments in terms of spectral coverage and linewidth, output power, pulse energy, and ultrashort pulse width since the first demonstration of a glass fiber laser in 1964. Their applications have extended into a variety of fields accordingly. In this paper, the milestones of glass fiber laser development are briefly reviewed and recent advances of high-power continuous wave, Q-switched, mode-locked, and single-frequency fiber lasers in the 1, 1.5, 2, and 3 μm regions and their applications in such areas as industry, medicine, research, defense, and security are addressed in detail.
Key words: Fiber lasers, Applications

184. Temperature-insensitive optical fiber refractometer based on multimode interference in two cascaded no-core square fibers

Jixuan Wu, Yinping Miao, Binbin Song, Kailiang Zhang, Wei Lin, Hao Zhang, Bo Liu, Jianquan Yao
Applied Optics, 2014, 53(22): 5037–5041.

Abstract: A temperature-insensitive optical fiber refractometer, based on multimode interference in no-core square fibers, has been proposed and experimentally demonstrated. The refractometer is formed by a single mode fiber sandwiched between two segments of no-core square fibers through cleaving and fusion splicing. The transmission spectra characteristic of refractive index (RI) and environmental temperature have been investigated. Experimental results show that a transmission dip exhibits a redshift as large as about 25 nm when the ambient RI increases from 1.342 4 to 1.433 4. Within the RI range of 1.403 3 to 1.433 4, the RI sensitivity reaches 474.818 9 nm/RIU. A temperature sensitivity of 0.006 39 nm/℃ is experimentally acquired between 20℃ and 85℃, showing a low temperature cross-sensitivity of about 1.35×10^{-5} RIU/℃. The proposed refractometer has several advantages, such as low cost, simple structure, and compact size. Therefore, it is also expected to be employed in chemical and multi-parameter sensing applications.
Key words: Optical fiber refractometer, Temperature-insensitive, Multimode interference, Cascaded no-core square fibers

185. Green function method for the time domain simulation of pulse propagation

Jing Huang, Jianquan Yao, Degang Xu, Runhua Li
Applied Optics, 2014, 53(16): 3533–3539.

Abstract: Based on the Green function method, the nonlinear Schrödinger equation is directly solved in the time domain (without Fourier transform). Because the dispersion and nonlinear effects are calculated simultaneously, it does not bring any spurious effect such as the split-step method in which the step size has to be carefully controlled by an error estimation. By this time domain solution, the pulse fission is analyzed, and we obtain the relationship between the minimum T_0 (the half-width at 1/e-intensity point of a pulse) and dispersion coefficients (β_2, β_3, and β_4). Thus the concrete dispersion values, which have an impact on ultrashort pulses (the quantity units is femtosecond or attosecond), are listed. It has been demonstrated that pulse fission occurs in the normal and anomalous dispersion regimes, even though fourth-order dispersion and the fifth-order nonlinear effects are not taken into account.
Key words: Green function method, Pulse propagation, Time domain simulation, Nonlinear Schrodinger equation

186. Magnetic-field sensor based on core-offset tapered optical fiber and magnetic fluid

Jixuan Wu, Yinping Miao, Wei Lin, Binbin Song, Kailiang Zhang, Hao Zhang, Bo Liu, Jianquan Yao
Journal of Optics, 2014, 16(7): 075705.

Abstract: A magnetic field sensor based on an asymmetrical fiber modal Mach–Zehnder interferometer (MMZI) is achieved by cascading tapered fiber with the core-offset structure. The MMZI is sealed by the magnetic fluid and its spectral dependence on magnetic field has been investigated. The results show that the transmission variations of the two dips are about 8 dB and 10 dB for a magnetic intensity range from 0 Oe to 400 Oe, respectively. The highest magnetic sensitivity reaches 0.034 07 dB Oe^{-1}. The proposed sensor based on the intensity demodulation is cost-effective and robust; therefore, the device is beneficial to the magnetic field sensing applications and other magneto-optical tunable photonics devices.
Key words: Microstructured fibers, Fiber modal Mach-Zehnder interferometer, Magnetic fluid

187. 300-W-average-power monolithic actively Q-switched fiber laser at 1 064 nm

Wei Shi, Qiang Fang, Xueping Tian, Jingli Fan
Laser Physics, 2014, 24(6): 065102.

Abstract: We demonstrate a high power monolithic nanosecond pulsed fiber laser source at 1 064 nm. This laser was configured as a master oscillator-power amplifier (MOPA) seeded by an acousto-optic Q-switched fiber laser with varied pulse duration and repetition rate. Over 300 W average power is achieved for ~475 ns pulses at 100 kHz repetition rate with an optical to optical efficiency of 75.4%. The M^2 at both directions is <1.6 measured at the highest average power.

Key words: Q-switched laser, High power, Fiber laser

188. Simulation of surface plasmon resonance temperature sensor based on liquid mixture-filling microstructured optical fiber

Nan–Nan Luan, Ran Wang, Ying Lu, Jianquan Yao
Optical Engineering, 2014, 53(6): 067103.

Abstract: We demonstrate a temperature sensor based on surface plasmon resonances supported by a six-hole microstructured optical fiber (MOF). The air holes of the MOF are coated with a silver layer and filled with a large thermo-optic coefficient liquid mixture (ethanol and chloroform). The use of all six fiber holes and their relatively large size should facilitate the coating of the silver and the filling of the liquid mixture. Temperature variations will induce changes of coupling efficiencies between the core-guided mode and the plasmonic mode, thus leading to different loss spectra that will be recorded. The refractive index of the liquid mixture is close to that of the MOF material, which will enhance the coupling efficiency and the sensitivity. Our numerical results indicate that temperature sensitivity as high as 5.6 nm/K can be achieved and that the most sensitive range of the sensor can be tuned by changing the volume ratios of ethanol and chloroform.

Key words: Fiber optic sensors, Fiber optic applications, Sensors, Surface plasmons, Microstructured fibers

189. A photonic crystal fiber sensor based on differential optical absorption spectroscopy for mixed gases detection

Baoqun Wu, Ying Lu, Congjing Hao, Liangcheng Duan, Mayilamu Musideke, Jianquan Yao

Optik, 2014, 125(12): 2909–2911.

Abstract: A photonic crystal fiber sensor based on differential optical absorption spectroscopy for mixed gas detection is presented. In such sensor, hollow core photonic crystal fiber is utilized as gas cell and the feasibility for gas detection is verified by experiment. The components concentration of mixed gas NH_3 and C_2H_2 are measured and the detection sensitivity is 143 ppmv.

Key words: DOAS, HC–PCF, Gas detection, NH_3, C_2H_2

190. 光谱频移的炸药熔铸温度网络监测系统研究

闻强，连素杰，张陈，赵辉，赵宇，王高，徐德刚，姚建铨

光谱学与光谱分析, 2014, 34(3): 592–596.

摘要： 为了准确、稳定、全面地获得炸药熔铸过程各个位置的温度变化情况，设计了基于布喇格光栅光谱频移的温度实时监测系统。通过光纤组网系统对炸药熔铸过程炸药指定位置的多个点同时进行实时温度监测，根据光栅的布喇格波长与光栅温度之间存在的线性关系，建立光栅布喇格波长线性频移与光栅温度的函数，获取炸药不同位置的准确温度。四个通道通过耦合器共用同一个宽带光源，每一根光纤上的 5 个光栅的布喇格波长相互分开。实验所用的光栅为自己设计封装好的光栅，用保偏熔接机将光栅与光纤熔接上，经解调仪获取温度数据。将获取的温度数据经 Origin 处理绘制时间—温度曲线。结果显示，布喇格光栅测得的温度能很好的满足实验要求。

关键词： 布喇格光栅，炸药熔铸，温度监测，温度梯度，光谱频移

191. 基于两种光纤介质模型的长周期光纤光栅传感特性分析

曹洪星,白育堃,李敬辉,马秀荣,姚建铨

中国激光, 2014, 41(2): 0205001.

摘要: 仿真分析了单轴晶体光纤的两层介质模型与三层介质模型所得纤芯模有效折射率的差别,并基于该差别对长周期光纤光栅的外界环境折射率、温度和轴向应变传感灵敏度进行了仿真分析,结果表明对两层、三层介质模型的纤芯模,尤其对薄包层光纤有效折射率有较大差别。用两层介质模型和/或材料的折变系数计算所得的长周期光纤光栅的外界环境折射率、温度和轴向应变灵敏度有较大的误差。需应用三层介质模型以及模式的有效折边系数进行准确计算,据此计算了上述3种传感灵敏度与光纤包层半径,其包层膜系数的关系,为长周期光纤光栅传感器的分析设计提供了指导。

关键词: 光纤光学,光栅,单轴晶体光纤介质模型,薄包层光纤,长周期光纤光栅,传感器

192. 基于亚波长悬浮芯光纤的高灵敏度气体传感器

栾楠楠,王然,郝丛静,袁偲,陆颖,姚建铨

中国激光, 2014, 41(5), 0514001.

摘要: 数值分析了亚波长悬浮芯光纤在气体传感方面的应用。利用有限元法研究了相对灵敏度、有效模场面积和限制损耗与光纤材料折射率包括纤芯直径和光纤材料折射率之间的关系。结果显示,相对灵敏度和限制损耗随着纤芯直径和光纤材料折射率的降低而增加,随着纤芯直径的减小,有效模场面积出现了先减小后增加的现象。增加包层孔径直径能有效降低限制损耗,而相对灵敏度和有效模场面积保持不变,这些结果证明,亚波长悬浮芯光纤非常适合成为高灵敏度、大有效模场面积、低限制损耗的气体传感器。

关键词: 光纤光学,传感器,微结构光纤,气体传感器,悬浮芯光纤,倏逝场

193. Simultaneous measurement of temperature and magnetic field based on a long period grating concatenated with multimode fiber

Yinping Miao, Hao Zhang, Jichao Lin, Binbin Song, Kailiang Zhang, Wei Lin, Bo Liu, Jianquan Yao

Applied Physics Letters, 2015, 106(13): 132410.

Abstract: A dual-parameter measurement scheme based on a long-period fiber grating (LPFG) concatenated with a multimode fiber (MMF) has been proposed and experimentally demonstrated for simultaneous measurement of magnetic field and temperature. Splicing the LPFG with the etched MMF enables the coupling between the core modes and different cladding modes of the LPFG as well as the interferences between higher-order modes in the MMF. Due to different transmission mechanisms of the LPFG and mode interference, the proposed sensor shows transmission dip wavelength sensitivities of 0.028 78 nm/Oe and -0.040 48 nm/℃ for multi-mode interference (MMI) and -0.002 4 nm/Oe and 0.039 29 nm/℃ for the LPFG, respectively. By monitoring the opposite behaviors of resonance wavelength shift corresponding to the LPFG and MMI, the magnetic field and environmental temperature can be simultaneously measured. The spectral characteristics of the proposed sensor that could be tuned through control of both environmental temperature and applied magnetic field, which would provide a promising candidate for dual-channel filtering applications as well as multi-parameter measurement applications.

Key words: Temperature measurement, Magnetic field measurement, Simultaneous measurement, Long-period fiber grating, Multimode fiber, Spectral characteristics

194. Single-frequency fiber laser at 1 950 nm based on thulium-doped silica fiber

Shijie Fu, Wei Shi, Jichao Lin, Qiang Fang, Quan Sheng, Haiwei Zhang, Jinwei Wen, Jianquan Yao

Optics Letters, 2015, 40(22): 5283–5286.

Abstract: A single-frequency fiber laser operating at 1 950 nm has been demonstrated in an all-fiber distributed Bragg reflection laser cavity by using a 1.9 cm commercially available thulium-doped silica fiber, for the first time, to the best of our knowledge. The laser was pumped by a 793 nm single mode diode laser and had a threshold pump power of 75 mW. The maximum output power of the single longitudinal mode laser was 18 mW and the slope efficiency with respect to the launched pump power was 11%. Moreover, the linewidth and relative intensity noise at different pump power have been measured and analyzed.

Key words: Fiber laser, Single-frequency, Thulium-doped silica fiber, 1 950 nm, Bragg reflection laser cavity

195. Low-temperature cross-talk magnetic-field sensor based on tapered all-solid waveguide-array fiber and magnetic fluids

Yinping Miao, Xixi Ma, Jixuan Wu, Binbin Song, Hao Zhang, Kailiang Zhang, Bo Liu, Jianquan Yao

Optics Letters, 2015, 40(16): 3905–3908.

Abstract: A compact fiber-optic magnetic-field sensor based on tapered all-solid waveguide-array fiber (WAF) and magnetic fluid (MF) has been proposed and experimentally demonstrated. The tapered all-solid WAF is fabricated by using a fusion splicer, and the sensor is formed by immersing the tapered all-solid WAF into the MF. The transmission spectra have been measured and analyzed under different magnetic-field intensities. Experimental results show that the acquired magnetic-field sensitivity is 44.57 pm/Oe for a linear magnetic-field intensity range from 50 to 200 Oe. All-solid WAF has very similar thermal expansion coefficient for high- and low-refractive-index glasses, so mode profile is not affected by thermal drifts. Also, magnetically induced refractive-index changes into the ferrofluid are of the order of $\sim 5 \times 10^{-2}$, while the corresponding thermally induced refractive-index changes into the ferrofluid are expected to be lower. The temperature response has also been detected, and the temperature-induced wavelength shift perturbation is less than 0.3 nm from temperature of 26.9 ℃–44 ℃. The proposed magnetic-field sensor has such advantages as low temperature sensitivity, simple structure, and ease of fabrication. It also indicates that the magnetic-field sensor based on tapered all-solid WAF and MF is helpful to reduce temperature cross-sensitivity for the measurement of magnetic field.

Key words: Fiber-optic magnetic-field sensor, Waveguide-array fiber, Magnetic fluid, Refractive index

196. Passive Q-switching of an all-fiber laser induced by the Kerr effect of multimode interference

Shijie Fu, Quan Sheng, Xiushan Zhu, Wei Shi, Jianquan Yao, Guannan Shi, R. A. Norwood, N. Peyghambarian

Optics Express, 2015, 23(13): 17255–17262.

Abstract: A novel passively Q-switched all-fiber laser using a single mode-multimode-single mode fiber device as the saturable absorber based on the Kerr effect of multimode interference is reported. Stable Q-switched operation of an Er^{3+}/Yb^{3+} co-doped fiber laser at 1 559.5 nm was obtained at a pump power range of 190-510 mW with the repetition rate varying from 14.1 kHz to 35.2 kHz and the pulse duration ranging from 5.69 μs to 3.86 μs. A maximum pulse energy of 0.8 μJ at an average output power of 27.6 mW was achieved. This demonstrates a new modulation mechanism for realizing Q-switched all-fiber laser sources.

Key words: All-fiber laser, Passive Q-switching, Kerr effect, Multimode interference, Co-doped

197. Surface plasmon resonance sensor based on D-shaped microstructured optical fiber with hollow core

Nannan Luan, Ran Wang, Wenhua Lv, Jianquan Yao

Optics Express, 2015, 23(7): 8576–8582.

Abstract: To solve the phase matching and analyte filling problems in the microstructured optical fiber (MOF)-based surface plasmon resonance (SPR) sensors, we present the D-shaped hollow core MOF-based SPR sensor. The air hole in the fiber core can lower the refractive index of a Gaussian-like core mode to match with that of a plasmon mode. The analyte is deposited directly onto the D-shaped flat surface instead of filling the fiber holes. We numerically investigate the effect of the air hole in the core on the SPR sensing performance, and identify the sensor sensitivity on wavelength, amplitude and phase. This work allows us to determine the feasibility of using the D-shaped hollow-core MOFs to develop a high sensitivity, real-time and distributed SPR sensor.

Key words: Surface plasmon resonance, Sensor, Microstructured optical fiber, D-shaped, Hollow core

198. High-power all-fiber single-frequency erbium-ytterbium co-doped fiber master oscillator power amplifier

Xiaolei Bai, Quan Sheng, Haiwei Zhang, Shijie Fu, Wei Shi, JianquanYao
IEEE Photonics Journal, 2015, 7(6): 1–6.

Abstract: We report herein an all-fiber single-frequency master oscillator power amplifier (MOPA) at 1 550 nm with Er/Yb-codoped active fiber and wavelength-stabilized 976-nm LD pump source. A pump-limited maximum continuous-wave output power of 56.4 W was achieved under the pump power of 150 W, with corresponding slope efficiency being 37.0%. Via the self-heterodyne method, the evolution of spectrum linewidth during the amplification was investigated for the high-power MOPA-based single-frequency fiber laser. The linewidth and relative intensity noise at the maximum output power are 4.21 kHz and −110 dBm/Hz, respectively.
Key words: Single-frequency laser, All fiber, Erbium–ytterbium, Laser amplifiers, Fiber laser

199. An exposed-core grapefruit fibers based surface plasmon resonance sensor

Xianchao Yang, Ying Lu, Mintuo Wang, Jianquan Yao
Sensors, 2015, 15(7): 17106–17114.

Abstract: To solve the problem of air hole coating and analyte filling in microstructured optical fiber-based surface plasmon resonance (SPR) sensors, we designed an exposed-core grapefruit fiber (EC-GFs)-based SPR sensor. The exposed section of the EC-GF is coated with a SPR, supporting thin silver film, which can sense the analyte in the external environment. The asymmetrically coated fiber can support two separate resonance peaks (x- and y-polarized peaks) with orthogonal polarizations and x-polarized peak, providing a much higher peak loss than y-polarized, also the x-polarized peak has higher wavelength and amplitude sensitivities. A large analyte refractive index (RI) range from 1.33 to 1.42 is calculated to investigate the sensing performance of the sensor, and an extremely high wavelength sensitivity of 13,500 nm/refractive index unit (RIU) is obtained. The silver layer thickness, which may affect the sensing performance, is also discussed. This work can provide a reference for developing a high sensitivity, real-time, fast-response, and distributed SPR RI sensor.
Key words: Exposed-core grapefruit fibers, Surface plasmon resonance, X- and y-polarized, Silver layer thickness

200. 2 μm actively Q-switched all fiber laser based on stress-induced birefringence and commercial Tm-doped silica fiber

Shijie Fu, Quan Sheng, Wei Shi, Xueping Tian, Qiang Fang, Jianquan Yao

Optics & Laser Technology, 2015, 70: 26–29.

Abstract: An actively Q-switched all-fiber laser operating at the wavelength of 1 920 nm is experimentally demonstrated based on a readily accessible commercial Tm-doped silica fiber. The Q-switching is achieved by polarization modulation through the stress-induced birefringence using a piezoelectric transducer (PZT) as the Q-switcher. The pulsed fiber laser can be operated with the repetition rates ranging from tens of kilohertz to nearly two hundreds kilohertz with milliwatt average output power.

Key words: All-fiber laser, Actively Q-switching, Stress-induced birefringence, Commercial Tm-doped silica fiber, Piezoelectric transducer

201. Fiber ring laser sensor based on hollow-core photonic crystal fiber

Z. Q. Zhao, Y. Lu, L. C. Duan, M. T. Wang, H. W. Zhang, J. Q. Yao

Optics Communications, 2015, 350: 296–300.

Abstract: We report an erbium-doped fiber ring laser intra-cavity sensor. Hollow-core photonic crystal fiber (HCPCF) used as the gas absorption chamber is introduced into the laser cavity. When the HC-PCF is filled with gas, its absorption attenuation changes the cavity loss and the laser output. We use a modular NI PXI platform equipped with a programmable voltage and current source and a LabVIEW program to generate driving voltage for the tunable optical filter (TOF) to ensure high precision. The relationship between the concentrations of acetylene, coupling ratios, pump power and output power is theoretically and experimentally investigated. Different output spectra are measured by the optical spectral analyzer (OSA). A minimum detectable acetylene concentration (MDAC) of 5.4 ppm has been experimentally achieved.

Key words: Fiber laser, Photonic crystal fiber, Intra-cavity sensor

202. Simultaneous measurement of displacement and temperature based on thin-core fiber modal interferometer

Jixuan Wu, Yinping Miao, Binbin Song, Wei Lin, Kailiang Zhang, Hao Zhang, Bo Liu, Jianquan Yao

Optics Communications, 2015, 340: 136–140.

Abstract: An optical-fiber sensor based on a thin core fiber (TCF) has been proposed and experimentally demonstrated for simultaneous measurement of displacement and temperature. This in-line sensor consists of a segment of TCF between two segments of single-mode fibers (SMFs), and the interference between the core mode and the cladding mode of the TCF occurs. The transmission spectral responses to displacement as well as environmental temperature have been investigated. Experimental results show that the displacement sensitivities of −0.010 28 nm/μm and -0.015 35 nm/μm for a displacement range of 0–650 μm have been achieved, and the corresponding temperature sensitivities reach about 0.009 42 nm/℃ and 0.004 93 nm/℃ within a large temperature range of 20–90 ℃, respectively. The proposed sensor exhibits such advantages as simple structure, compact size, ease of fabrication, high sensitivity, etc. Therefore, it has potential applications in accurate multi-parameter measurements.

Key words: Optical-fiber sensor, Displacement, Fiber modal interferometer

203. Surface plasmon resonance sensor based on exposed-core microstructured optical fibres

Nannan Luan, Ran Wang, Wenhua Lv, Jianquan Yao

Electronics Letters, 2015, 51(9): 714–715.

Abstract: A surface plasmon resonance (SPR) sensor based on exposed-core microstructured optical fibres (EC-MOFs) is presented and numerically characterized. The exposed section (analyte channel) of the EC-MOF is coated with a SPR supporting thin film of gold. The asymmetrically coated fibre can support two separate resonance peaks, with orthogonal polarisations (x- and y-polarised peaks). Although the sensitivity of both the polarised peaks is very similar, the narrower linewidth of the y-polarised peak provides a much enhanced signal-to-noise ratio. The SPR sensing performance for a large refractive index (RI) range from 1.33 to 1.42 is evaluated. The sensor has higher sensitivity for detecting the high analyte RI. The feasibility of using EC-MOFs to develop a fast-response, real-time and distributed SPR sensor is determined.

Key words: Surface plasmon resonance, Sensor, Exposed-core microstructured optical fibres

204. Surface plasmon resonance sensor based on hollow-core PCFs filled with silver nanowires

Ying Lu, Xianchao Yang, Mintuo Wang, Jianquan Yao
Electronics Letters, 2015, 51(21): 1675–1677.

Abstract: A surface plasmon resonance (SPR) sensor based on hollow-core photonic crystal fibre (HC-PCF) filled with silver nanowires is designed. The analyte and silver nanowires are full filled in the air holes of the HC-PCF to realise the SPR sensing, which is more convenient than silver coated in operation. The designed sensor is analysed through numerical simulations and demonstrated by experiments. All the results show that a blue-shift is obtained with increase of the analyte refractive index (RI), and the silver nanowires concentration has no effect on spectral sensitivity. The highest average spectral sensitivity of 14 240 nm/RIU is obtained by experiments, which is higher than that previously reported for the same type of sensors. The sensor is useful for detecting small analyte RI changes, and can also provide a reference for the implementation and application of PCF-SPR sensors with high sensitivity.

Key words: Surface plasmon resonance, Sensor, Hollow-core photonic crystal fibre, Silver nanowires

205. Simulation analysis of a temperature sensor based on photonic crystal fiber filled with different shapes of nanowires

Mintuo Wang, Ying Lu, Congjing Hao, Xianchao Yang, Jianquan Yao
Optik, 2015, 126(23): 3687–3691.

Abstract: We propose a photonic crystal fiber (PCF)-based surface plasmons resonance (SPR) temperature sensor in this paper, and using the dual function of PCF filled with different shapes of nanowires and analyte to realize temperature sensing and have designed an experimental setup. Different shapes of nanowires with the mixed liquid are filled into the proposed octagon PCF. The simulation results indicate that not only structure parameters of PCF but also the shapes of filled nanowires play a key role to the sensing. A blue-shift in the SPR resonant peak is shown with the temperature increases. Different loss spectra present that a higher confinement loss for low temperature and the octagon PCF filled with elliptical nanowires shows a more obvious peak shift and higher sensitivity, which is about 500 pm/° C and the corresponding sensor resolution is 2.0×10^{-2} °C. This PCF sensing system has great practical value and significance for its advantage of high sensitivity.

Key words: Temperature sensing, Photonic crystal fibers, Surface plasmons resonance, Fiber optics sensors

206. 基于银纳米颗粒的 HCPCF SERS 传感系统优化设计

邱志刚，贾春荣，姚建铨，陆颖

红外与激光工程, 2015, 44(4): 1317–1322.

摘 要：银纳米颗粒与光子晶体光纤、表面增强拉曼散射效应结合而成的 PCF SERS 传感器得到了科研界的广泛关注。而 PCF 结构、SERS 基底的性能是传感器的重要影响因素。为了进一步提高 SERS PCF 传感器的性能，通过研究对比 PCF 和 SERS 基底结构参数对传感性能的影响，设计出适用于 PCF SERS 传感的空芯 PCF 以及 SERS 基底的结构参数。通过数值计算，设计的空芯 PCF 空气填充率为 56.30%，当激发光波长 785 nm 时存在光子带隙，并能够实现单模传输。而半径为 38 nm 的银纳米球在间距为 0.7 nm 时能够产生最大的 SERS 增强因子。研究证明，设计的空芯 PCF 在 785 nm 输入波长下既能够基模传输激发光，又能够为 SERS 提供理想的活性面积，而且银纳米颗粒的形状、尺寸、间距对 SERS 性能影响严重，而且与入射波长有很强的依赖关系。

关键词：空芯光子晶体光纤，表面增强拉曼散射，光子带隙，银纳米颗粒，数值计算

207. All-fiber passively Q-switched fiber laser based on the multimode interference effect

Shijie Fu, Quan Sheng, Xiushan Zhu, Wei Shi, Jianquan Yao, R. A. Norwood, N. Peyghambarian

CLEO: Science and Innovations Optical Society of America, 2015: STh4L. 2.

Abstract: An all-fiber passively Q-switched Er-Yb co-doped fiber laser (EYDFL) using a single mode-multimode-single mode (SMS) fiber structure based on the multimode interference (MMI) effect was proposed and experimentally demonstrated for the first time.

Key words: All-fiber laser, Er-Yb co-doped, Passively Q-switching, SMS, Multimode interference

208. Multipoint hollow core photonic crystal fiber sensor network based on intracavity absorption spectroscopy

Haiwei Zhang, Ying Lu, Wei Shi, Liangcheng Duan, Zhiqiang Zhao, Jianquan Yao

CLEO: Science and Innovations. Optical Society of America, 2015: SM2O. 8.

Abstract: We firstly demonstrate an automatic intracavity absorption multipoint acetylene sensor network via dense wavelength division multiplexing filters by applying voltage gradient to the F-P tunable filter. The sensitivity is up to 100 ppmV at 1 536.71 nm.

Key words: Photonic crystal fiber, Multipoint acetylene sensor network, Hollow core, Intracavity absorption

209. Linearly polarized narrow linewidth single mode fiber laser and nonlinear phenomena

Wei Shi, Qiang Fang, Jingli Fan, Ting Qu, Xiangjie Meng

IEEE Photonics Conference (IPC), 2015: 232−233.

Abstract: We report a high power, narrow linewidth, linearly polarized, MOPA based fiber laser at 1 064.46 nm with 520 W average power, 30 GHz linewidth, 18 dB polarization extinction ratio, 88.7% slope efficiency, and diffraction-limited beam quality.

Key words: MOPA based fiber laser, Single mode, Narrow linewidth, Linearly polarized, Nonlinear phenomena

210. 基于长周期光纤光栅和 ZigBee 组网技术的无线溶液折射率传感网络

石嘉，徐德刚，严德贤，徐伟，姚建铨

激光与光电子学进展, 2015, 52(3): 030602.

摘要：阐述了一种基于长周期光纤光栅和 ZigBee 组网技术的无线溶液折射率传感网络。整个系统包括了传感节点、路由节点、协调器和中心计算机几个部分。传感节点部分，设计了一种基于强度调制型的低成本的长周期光纤光栅溶液折射率传感器，其中包括了长周期光纤光栅和一种特殊结构的封装。同时，设计了传感器的驱动板，包括激光二极管及其驱动电路、光电探测电路。利用损耗峰两侧近似线性的特性，通过测量单侧损耗峰半峰处固定波长光功率的变化实现了折射率传感。系统搭建了基于 ZigBee 的自组网系统，并且实现了实时的多点的溶液折射率监测。测量了折射率为 1.334 7~1.341 8 的氯化钠溶液，分辨率为 0.000 1，最大误差为 0.000 4，平均误差为 0.000 15。所述系统提供了一种将光纤传感器应用于无线传感网络的方案，其优势在于可以大面积铺设大量低成本传感器并组网，方便地增减传感节点，实现了大范围、多点、高分辨率的环境监测。

关键词：传感器，折射率传感器，Zigbee，长周期光纤光栅，无线网络

211. 从工业物联网到智慧城市

姚建铨

科学导报, 2015, B01 版.

摘要：中国是制造大国，但还不是制造强国！在发达国家蓄势占优、新兴经济体追赶比拼的两头挤压下，大而不强的"中国制造"该何去何从？美国、德国等世界主要发达国家纷纷实施以重振制造业为核心的"再工业化"战略，美国要"确保下一轮制造业革命发生在美国"。美国希望通过国家制造创新网络及 15 个制造创新研究所的构建，振兴美国的制造业，并引发制造技术的变革，这对我们是很大的挑战。理论界和产业界普遍认为，世界正处于新一轮技术创新浪潮引发的新一轮工业革命的开端。中国首先要建立一个合理的创新体系加强产业共性技术研究，才能为核心技术和产品研发提供取之不尽的技术支撑。到 2025 年，中国制造业将迈入制造强国行列；到 2035 年，中国制造业将成为名副其实的制造强国。

关键词：工业化，工业物联网，创新体系，智慧城市

212. Dual-wavelength fiber laser operating above 2 μm based on cascaded single-mode-multimode single-mode fiber structures

Shijie Fu, Guannan Shi, Quan Sheng, Wei Shi, Xiushan Zhu, Jianquan Yao, R. A. Norwood, N. Peyghambarian

Optics Express, 2016, 24(11): 11282.

Abstract: A stable dual-wavelength Tm^{3+}: Ho^{3+} co-doped fiber laser operating above 2 μm based on cascaded single-mode-multimode-single mode (SMS) fiber structures is proposed and experimentally demonstrated. Based on the theoretical analysis of the transmission properties of the SMS fiber structure, two cascaded SMS fiber devices with different multimode fiber (MMF) lengths were used in our laser system, where one acted as a long-pass filter to suppress the competitive laser below 2 μm, and the other worked as a band-pass filter to select the specific operating wavelengths of the laser. Dual-wavelength operation of the fiber laser at 2 002.8 and 2 016.1 nm has been achieved in the experiment with a signal to a noise ratio up to 50 dB.

Key words: Dual-wavelength fiber laser, Cascaded SMS

213. ASE suppression in backward-pumped Er/Yb double-cladding fiber amplifier via cladding feedback

Xiaolei Bai, Quan Sheng, Shijie Fu, Haiwei Zhang, Yang Cao, Zhaoxin Xie, Wei Shi, Jianquan Yao

IEEE Photonics Journal, 2016, 8(6): 7102207.

Abstract: We demonstrate a backward-pumped 1 550 nm narrow-linewidth Erbium-Ytterbium double-cladding fiber amplifier with amplified spontaneous emission (ASE) suppression. Compared with the typical backward-pumped amplifier without feedback, the ASE power can be suppressed by over 2 dB by cladding feedback without other filters. The slope efficiency is 23.2%, which only decreased by 0.4%. The linewidth and relative intensity noise of 4.51 kHz and −92 dBm/Hz are also compared with those of the typical backward-pumped amplifier without feedback.

Key words: Amplified spontaneous emission (ASE) suppression, Cladding feedback, Erbium-Ytterbium, Laser amplifies, Fiber laser

214. A refractive index and temperature sensor based on surface plasmon resonance in an exposed-core microstructured optical fiber

Nannan Luan, Chunfeng Ding, Jianquan Yao
IEEE Photonics Journal, 2016, 8(2): 4801608.

Abstract: A surface plasmon resonance sensor is proposed to simultaneously realize the refractive index (RI) and temperature sensing in an exposed-core microstructured optical fiber. Two orthogonal sensing channels coated with silver layers are designed to distinguish the variations of the analyte RI and the temperature. The exposed section of the fiber as an RI sensing channel supports a y-polarized peak. The orthogonally arranged hole is filled with a large thermooptic coefficient liquid as a temperature sensing channel that supports a x-polarized peak. The two polarized peaks can be shifted independently. By following the shifts of the two polarized peaks, the variations of the RI and temperature can be detected simultaneously.

Key words: Fiber optics sensors, Microstructured fibers, Surface plasmons

215. High refractive index surface plasmon resonance sensor based on a silver wire filled hollow fiber

Nannan Luan, Jianquan Yao
IEEE Photonics Journal, 2016, 8(1): 4800709.

Abstract: A surface plasmon resonance (SPR) sensor based on a hollow fiber (HF) is proposed to realize high analyte refractive index (RI) detection. The hole in the HF as a microfluidic channel for the analyte is filled with a silver wire to replace the metal coating. The sensitivity of the proposed SPR sensor with analyte RI from 1.47 to 1.51 is theoretically investigated both in the wavelength and amplitude interrogation methods. The results show that the sensor can support two disparate resonance peaks, with orthogonal polarizations (x- and y-polarized peaks), and the sensitivity of the y-polarized peak is higher than that of x-polarized peak. Moreover, contrary to the performances of the resonance peaks supported by the low RI SPR sensors, the two polarized peaks shift to shorter wavelength as analyte RI increasing and show a higher sensitivity at low analyte RI. These results and analyses, including the abnormal behaviors of resonance peaks and the coupling condition between the core modes and the plasmon modes, are very helpful for the design and improvement of high RI SPR sensors.

Key words: Fiber optics sensors, Microstructured fibers, Surface plasmons

216. Surface plasmon resonance sensor based on exposed-core microstructured optical fiber placed with A silver wire

Nannan Luan, Jianquan Yao

IEEE Photonics Journal, 2016, 8(1): 4800508.

Abstract: We propose a surface plasmon resonance (SPR) sensor based on the exposed core microstructured optical fiber (EC-MOF) placed with a silver wire. The exposed section of the EC-MOF as a microfluidic channel is placed with the silver wire to avoid the metal coating and is then deposited with the analyte to avert the analyte filling. The proposed SPR sensor can support two polarized resonance peaks (x-polarized and y-polarized) caused by the silver wire. We theoretically investigate the sensitivities of the two polarized peaks both in the wavelength and amplitude interrogation methods and analyze the influences of the silver wire location on the sensing performances. The results show that the sensitivities of the two polarized peaks are similar and relatively stable for random locations of the silver wire. Moreover, the x-polarized peak provides a higher resolution for wavelength sensitivity and needs a shorter silver wire for the maximum amplitude sensitivity. This paper demonstrates that using the ECMOF placed with the silver wire can simultaneously solve the metal coating and analyte filling problems in the other SPR sensors with no sacrifice in sensitivity.

Key words: Fiber optics sensors, Surface plasmons, Microstructured fibers

217. Remote gas pressure sensor based on fiber ring laser embedded with fabry-pérot interferometer and sagnac loop

Jia Shi, Yuye Wang, Degang Xu, Yixin He, Junfeng Jiang, Wei Xu, Haiwei Zhang, Genghua Su, Chao Yan, Dexian Yan, Ying Lu, Jianquan Yao

IEEE Photonics Journal, 2016, 8(5): 6804408.

Abstract: In this paper, we propose a remote gas pressure sensor based on a fiber ring laser embedded with a Fabry–Pérot (F–P) interferometer and a Sagnac loop. A compact optic-fiber F–P interferometer is inserted in the fiber laser and works as the sensing head for gas pressure detection. A wavelength selective filter based on optic-fiber Sagnac loop is used in the fiber laser. The sensitivity of -9.69 nm/kPa is obtained with a narrow 3-dB bandwidth less than 0.02 nm and high signal-to-noise ratio (SNR) ~45 dB. Moreover, the excellent performance of the gas pressure sensor for remote detection is demonstrated.

Key words: Fiber optics sensors, Intracavity sensing, Remote sensing

218. Remote magnetic field sensor based on intracavity absorption of evanescent field

Jia Shi, Yuye Wang, Degang Xu, Genghua Su, Haiwei Zhang, Jiachen Feng, Chao Yan, Shijie Fu, Jianquan Yao

IEEE Photonics Journal, 2016, 8(2): 6801607.

Abstract: In this paper, a remote magnetic field sensor based on intracavity absorption of evanescent field is experimentally demonstrated. A single-mode–no-core–single mode (SNCS) fiber coated with magnetic fluid (MF) is inserted into a fiber ring laser through a circulator and a 3-dB coupler, and it works as the filter and sensing head simultaneously. As the light oscillates in the cavity of the fiber ring laser, intracavity absorption of the evanescent field will be induced in the interface of the no-core fiber and MF. We obtain a magnetic field sensor with a narrow 3-dB bandwidth that is less than 0.05 nm, a high signal-to-noise ratio ~40 dB, and magnetic sensitivity of 52.1 pm/mT and −0.367 9 dBm/mT. Moreover, experiments show that it is convenient to achieve remote sensing by selecting different lengths of the transmission optical fiber connecting the circulator and 3-dB coupler, which has a slight impact on the sensitivity.

Key words: Fiber optics sensors, Remote sensing, Intracavity sensing

219. Temperature sensor based on photonic crystal fiber filled with liquid and silver nanowires

X. C. Yang, Y. Lu, B. L. Liu, J. Q. Yao

IEEE Photonics Journal, 2016, 8(3): 6803309.

Abstract: A temperature sensor based on photonic crystal fiber filled with liquid and silver nanowires using surface plasmon resonance is demonstrated both theoretically and experimentally in this paper. Numerical simulation shows that a blue shift is obtained when temperature increases, and the resonance wavelength and resonance intensity can be tuned effectively by adjusting the volume ratios of the liquid constituents. A large temperature range from 25 ℃ to 60 ℃ at different ratios is detected to investigate the sensor's performance, and the sensitivity −2.08 nm/℃ with the figure of merit of 0.157 2 is obtained by experiment. Moreover, with the all-fiber device with strong mechanical stability, it is easy to realize remote sensing by changing the downlead fiber length, which is promising for developing a high-sensitive, real-time, and distributed temperature sensor.

Key words: Temperature sensor, Photonic crystal fiber, Surface plasmon resonance, Silver nanowires

220. Experimental investigation on spectral linewidth and relative intensity noise of high-power single-frequency polarization-maintained thulium-doped fiber amplifier

Haiwei Zhang, Yang Cao, Wei Shi, Quan Sheng, Xiaolei Bai, Shijie Fu, Jianquan Yao

IEEE Photonics Journal, 2016, 8(3): 7101108.

Abstract: We investigate the spectral linewidth and relative intensity noise (RIN) of an all-fiber single-frequency polarization-maintained master oscillator and power amplifier at 1 925 nm. It is found that the linewidth of the fiber amplifier is related to the power ratio of signal laser to accumulated amplified stimulated emission noises. A 3-dB linewidth of 56 kHz is measured at a maximum output power of 45 W, resulting to a linewidth broadening of 20 kHz compared to the 36-kHz linewidth of the seed laser. The RINs of the preamplifier and the power amplifier are measured to be −108 and −90 dBm/Hz at the relaxation frequency, respectively.

Key words: Single-frequency fiber laser, Thulium-doped fiber amplifier, Linewidth, Delayed self-heterodyne, Relative intensity noise

221. Analysis of graphene-based photonic crystal fiber sensor using birefringence and surface plasmon resonance

Xianchao Yang, Ying Lu, Baolin Liu, Jianquan Yao

Plasmonics, 2016, 12(2): 1–8.

Abstract: We present and numerically characterize a photonic crystal fiber (PCF)-based surface plasmon resonance (SPR) sensor. By adjusting the air hole sizes of the PCF, the effective refractive index (RI) of core-guided mode can be tuned effectively and the sensor exhibits strong birefringence. Alternate holes coated with graphene-Ag bimetallic layers in the second layer are used as analyte channels, which can avoid adjacent interference and improve the signal to noise ratio (SNR). The graphene's good features can not only solve the problem of silver oxidation but also increase the absorption of molecules. We theoretically analyze the influence of the air hole sizes of the PCF and the thicknesses of graphene layer and Ag layer on the performance of the designed sensor using wavelength and amplitude interrogations. The wavelength sensitivity we obtained is as high as 2 520 nm/RIU with the resolution of 3.97×10^{-5} RIU, which can provide a reference for developing a high-sensitivity, real-time, fast-response, and distributed SPR sensor.

Key words: Photonic crystal fiber, Surface plasmon resonance, Graphene-Ag bimetallic layers, Birefringence, Wavelength and amplitude interrogations

222. Platinum-scatterer-based random lasers from dye-doped polymer-dispersed liquid crystals in capillary tubes

Jianlong Wang, Yating Zhang, Mingxuan Cao, Xiaoxian Song, Yongli Che, Haiting Zhang, Heng Zhang, Jianquan Yao

Applied Optics, 2016, 55(21): 5702−5706.

Abstract: The resonance characteristics of platinum-scatter-based random lasers from dye-doped polymer-dispersed liquid crystals (DDPDLCs) in capillary tubes were researched for the first time, to the best of our knowledge. After adding platinum nanoparticles (Pt NPs) into the liquid crystal mixtures, the emission spectra of DDPDLCs revealed a lower lasing threshold in comparison with those of DDPDLCs without Pt NPs due to light scattering of liquid crystal droplets and the local field enhancement around Pt NPs. Furthermore, the full width at halfmaximum (FWHM) and the lasing threshold were determined by the doping density of the Pt NPs. The threshold was decreased by about half from 17.5 μJ/pulse to 8.7 μJ/pulse on the condition that around 1.0 wt. % was the optimum concentration of Pt NPs doped into the DDPDLCs. The FWHM of the peaks sharply decreased to 0.1 nm. Our work provides an extremely simple method to enhance random lasers from DDPDLCs doped with Pt NPs, and it has potential applications in random fiber lasers or laser displays.

Key words: Platinum-scatter-based random lasers, Resonance characteristics, Dye-doped polymer-dispersed liquid crystals, Capillary tubes, Nanoparticles

223. Simultaneous magnetic field and temperature measurement based on no-core fiber coated with magnetic fluid

Genghua Su, Jia Shi, Degang Xu, Haiwei Zhang, Wei Xu, Yuye Wang, Jiachen Feng, Jianquan Yao

IEEE Sensors Journal, 2016, 16(23): 8489−8493.

Abstract: In this paper we propose and demonstrate a compact dual-parameter sensor for magnetic field intensity and temperature based on the singlemode–no-core–singlemode (SNCS) fiber. The no-core fiber (NCF) is coated with the magnetic fluid (MF) to detect the magnetic field and temperature simultaneously. The fundamental mode light launched from the singlemode fiber (SMF) will be decomposed into high-order modes lights in the NCF, and then the high-order modes lights will finally interfere in the exit SMF. Because different interference dips have different sensitivities to the magnetic field intensity and temperature, the cross-sensitivity of the magnetic field intensity and temperature can be effectively eliminated by monitoring two discrete interference dips. For the two dips monitored in the experiment, the proposed sensor has the magnetic field sensitivities of 7.432 88 pm/Oe and 6.200 37 pm/Oe, respectively, and temperature sensitivities of -0.246 9 nm/℃ and -0.286 67 nm/℃, respectively. The proposed sensor is compact and easy to setup, and has good potential in applications.

Key words: No-core fiber, Magnetic field, Temperature, Sensor, Magnetic fluid

224. Analysis of hollow fiber temperature sensor filled with graphene-Ag composite nanowire and liquid

Wei Xu, Jianquan Yao, Xianchao Yang, Jia Shi, Junfa Zhao, Cheng Zhang
Sensors, 2016, 16(10): 1656.

Abstract: A hollow fiber temperature sensor filled with graphene-Ag composite nanowire and liquid is presented and numerically characterized. The coupling properties and sensing performances are analyzed by finite element method (FEM) using both wavelength and amplitude interrogations. Due to the asymmetrical surface plasmon resonance sensing (SPR) region, the designed sensor exhibits strong birefringence, supporting two separate resonance peaks in orthogonal polarizations. Results show that x-polarized resonance peak can provide much better signal to noise ratio (SNR), wavelength and amplitude sensitivities than y-polarized, which is more suitable for temperature detecting. The graphene-Ag composite nanowire filled into the hollow fiber core can not only solve the oxidation problem but also avoid the metal coating. A wide temperature range from 22 ℃ to 47 ℃ with steps of 5 ℃ is calculated and the temperature sensitivities we obtained are 9.44 nm/℃ for x-polarized and 5.33 nm/℃ for y-polarized, much higher than other sensors of the same type.

Key words: Hollow fiber, Birefringence, Graphene-Ag composite nanowire, Surface plasmon resonance (SPR)

225. Temperature sensor based on fiber ring laser with sagnac loop

Jia Shi, Yuye Wang, Degang Xu, Haiwei Zhang, Genghua Su, Liangcheng Duan, Chao Yan, Dexian Yan, Shijie Fu, Jianquan Yao

IEEE Photonics Technology Letters, 2016, 28(7): 794–797.

Abstract: In this letter, a temperature sensor based on a fiber ring laser with a reflective Sagnac loop is proposed. The reflective Sagnac loop inserted into the fiber ring laser through a downlead optical fiber is used as the filter and sensing head to supply high temperature sensitivity. As the temperature varies, the transmission spectrum of Sagnac loop changes, leading to the shift of emission wavelength of a fiber ring laser. We obtain a temperature sensor system with a high temperature sensitivity of 1.739 nm/℃, a narrow 3-dB bandwidth less than 0.05 nm, and a high signal-to-noise ratio ~50 dB. Moreover, it is convenient to achieve the remote sensing by changing the length of the downlead optical fiber.

Key words: Fiber laser sensor, Sagnac loop, Temperature sensors

226. SPR sensor based on exposed-core grapefruit fiber with bimetallic structure

Xianchao Yang, Ying Lu, Mintuo Wang, Jianquan Yao

IEEE Photonics Technology Letters, 2016, 28(6): 649–652.

Abstract: A liquid refractive index (RI) sensor based on an exposed-core grapefruit fiber with a bimetallic structure and a central air hole using surface plasmon resonance (SPR) is presented. The asymmetrically SPR region leading to strong birefringence with x-polarized peak provides much higher peak loss and sensitivity than y-polarized. To match with the plasmon and enhance the mode coupling, we introduce a central air hole to lower the effective RI of the core guided mode. We numerically investigate the effects of the central air hole size, silver layer and gold layer thicknesses on the sensor's performance using wavelength, and amplitude interrogations. A large analyte RI range from 1.33 to 1.42 is evaluated, and the results show that the sensor has a higher sensitivity for high analyte RI change than the low. The extremely high wavelength sensitivity 16 400 nm/RIU at 1.42 is obtained. This letter can provide a reference for developing a high sensitivity, real-time, and distributed SPR sensor.

Key words: Exposed-core grapefruit fiber, Bimetallic structure, Surface plasmon resonance, X- and y-polarized peaks, Central air holes

227. A dual-parameter sensor using a long-period grating concatenated with polarization maintaining fiber in sagnac loop

Jia Shi, Genghua Su, Degang Xu, Yuye Wang, Haiwei Zhang, Shijie Fu, Jiachen Feng, Chao Yan, Wei Xu, Jianquan Yao

IEEE Sensors Journal, 2016, 16(11): 4326–4330.

Abstract: In this paper, a novel dual-parameter measurement scheme using a long-period grating (LPG) concatenated with polarization maintaining fiber (PMF) in a Sagnac loop was proposed and experimentally demonstrated to measure temperature and refractive index (RI) simultaneously. There are two types of dips in the output spectrum of the proposed sensor, which are formed by the LPG and Sagnac loop. Compared with the traditional Sagnac loop sensors using PMF only, the birefringence induced by the LPG changes the phase difference and makes the dips of the Sagnac loop sensitive to surrounding RI. Due to different interference mechanisms, the dual-parameter measurement is realized by Sagnac loop and LPG with different sensitivities to temperature and RI. Experiments show that in the output spectrum of the sensor, the dip of the LPG has the sensitivities of 0.201 nm/℃ and −8.36 nm/RIU, and the dip of the Sagnac loop has the sensitivities of −1.06 nm/℃ and −23.068 nm/RIU. The proposed sensor has a simple structure and great potential in biological, chemical, and ocean fields.

Key words: Dual-parameter sensors, Sagnac loop, Long period grating

228. Low-temperature cross-sensitivity refractive index sensor based on single-mode fiber with periodically modulated taper

Yinping Miao, Yong He, Xixi Ma, Hao Zhang, Binbin Song, Xiaoping Yang, Lifang Xue, Bo Liu, Jianquan Yao

IEEE Sensors Journal, 2016, 16(8): 2442−2446.

Abstract: In this paper, a single-mode fiber with periodically modulated tapers has been fabricated using electric arc discharge with a fiber splicer. Such a fiber with periodic tapers could effectively adjust the light propagation. The refractive index (RI) sensitivity of the proposed microfiber under different ambient RI environments has been analyzed. Experimental results show that it has an RI sensitivity of ~490.9 nm/RIU for a linear RI range from 1.364 2 to 1.401 5. The temperature response has also been experimentally investigated, which showing that this device possesses a low-temperature cross sensitivity, and the temperature-induced wavelength shift perturbation is <0.5 nm for an environmental temperature range of 22.5 ℃−80 ℃. Combining such desirable merits as simple fabrication procedure, compactness, low cost, and low-temperature sensitivity of the periodically modulated microfiber, this sensor is expected to find potential applications in fields, such as multi-parameter measurement and biochemical applications.

Key words: Microfiber, Periodically modulated microstructure, Refractive index tunability, Low temperature cross-sensitivity

229. Low-temperature-sensitive relative humidity sensor based on tapered square no-core fiber coated with SiO₂ nanoparticles

Yinping Miao, Xixi Ma, Yong He, Hongmin Zhang, Hao Zhang, Binbin Song, Bo Liu, Jianquan Yao

Optical Fiber Technology, 2016, 29:59–64.

Abstract: A low-temperature-sensitive relative humidity (RH) sensor based on multimode interference effects has been proposed. The sensor consists of a section of tapered square no-core fiber (TSNCF) coated with SiO_2 nanoparticles which is fabricated by splicing the TSNCF with two single-mode fibers (SMFs). The refractive index of SiO_2 nanoparticles changes with the variation of environmental humidity levels. Characteristics of the transmission spectral have been investigated under different humidity levels. The wavelength shifts up to 10.2 nm at 1 410 nm and 11.5 nm at 1 610 nm for a RH range of 43.6–98.6% have been experimentally achieved, and the corresponding sensitivities reach 456.21 pm/%RH and 584.2 pm/%RH for a RH range of 83–96.6%, respectively. The temperature response of the proposed sensor has also been experimentally investigated. Due to the fact that the sensing head is made of a pure silica rod with a low thermal expansion coefficient and the thermo-optic coefficient, the transmission spectrum shows a low temperature sensitivity of about 6 pm/℃ for an environmental temperature of 20.9–80 ℃, which is a desirable merit to resolve the temperature cross sensitivity. Therefore, the proposed sensor could be applied to breath analysis applications with low temperature fluctuations.

Key words: Relative humidity sensor, Multimode interference effects, No-core fiber, SiO_2 nanoparticles

230. Amplified spontaneous emission in distributed feedback active microcavities fabricated by the sol-gel dip-coating method

Jianlong Wang, Yating Zhang, Xiaoxian Song, Mingxuan Cao, Haiyan Wang, Yongli Che, Jianquan Yao

Journal of Modern Optics, 2016, 63(21): 2180–2185.

Abstract: Amplified spontaneous emission (ASE) is demonstrated in a distributed feedback (DFB) active microcavity, formed by rhodamine B molecules in a Bragg grating (BG). The BG was fabricated by alternately depositing titanium dioxide and silicon dioxide sol–gel thin films. The reflectance spectrum of BG simulated by the transfer matrix method was consistent with experimental results, demonstrating that the BG had good periodic structures. With rhodamine B molecules embedded, the ASE was observed from the DFB active microcavity in optical pumped conditions. The full-widthhalf-maximum and threshold of ASE were 7.5 nm and 0.2 mJ/pulse. The slope efficiency of 3% was measured. The DFB active microcavity is promising for low-cost ASE.

Key words: Dip coating, Distributed feedback, Amplified spontaneous emission, Microcavities

231. Backward-pumped 1 550 nm EYDF amplifier with ASE suppression by cladding feedback

Xiaolei Bai, Haiwei Zhang, Quan Sheng, Shijie Fu, Zhaoxin Xie, Wei Shi, Jianquan Yao

CLEO: Applications and Technology, 2016: JTu5A.117.

Abstract: We demonstrate a backward-pumped 1 550 nm EYDF amplifier with ASE suppression by cladding feedback. Compared with the typical backward-pumped amplifier without feedback, the ASE power can be suppressed by over 2 dB without other filters.

Key words: EYDF amplifier, Backward-pump, ASE suppression, Cladding feedback

232. Temperature distribution of double-cladding high-power thulium-dope fiber amplifier

Haiwei Zhang, Wei Shi, Quan Sheng, Xiaolei Bai, Shijie Fu, Yang Cao, Jianquan Yao

CLEO: Applications and Technology, 2016: JTu5A.126.

Abstract: The temperature distributions of double-cladding high-power Thulium-doped fiber amplifier with different inner-cladding shapes are investigated by solving the thermal conductive equation and rate equations. The offset DCF is demonstrated to be a better choice.

Key words: Temperature distribution, Thulium-doped fiber amplifier, Thermal conductive equation, Rate equations

233. Dual-point automatic switching intracavity-absorption photonic crystal fiber gas sensor based on mode competition

Haiwei Zhang, Liangcheng Duan, Wei Shi, Quan Sheng, Ying Lu, Jianquan Yao

Sensors and Actuators B: Chemical, 247: 124–128.

Abstract: Here we propose a dual-point channel-switchable intracavity absorption hollow-core photonic crystal fiber acetylene sensing system based on the mode-competition phenomenon in a ring fiber laser with a Sagnac loop filter. The experimental system operating at 1 532.83 and 1 534.10 nm is applied to 1% acetylene and shows the sensitivities of 398 and 1 905 ppmv, respectively, at the absorption peaks around two operating wavelengths. The resolution error induced by the power fluctuation is no more than 7.2% for both gas cells and it takes similar to 50s to scan the absorption spectra via applying gradient voltage to the tunable F-P filter. This approach is a potential cost-effective resolution for high-capacity intracavity absorption sensor network compatible hybrid gas detection.

Key words: Fiber optics sensors, Intracavity laser absorption, Mode competition, Photonic crystal fiber

234. Multidimensional microstructured photonic device based on all-solid waveguide array fiber and magnetic fluid

Yinping Miao, Xixi Ma, Yong He, Hongmin Zhang, Xiaoping Yang, Jianquan Yao
Nanophotonics, 2017, 6(1): 357–363.

Abstract: An all-solid waveguide array fiber (WAF) is one kind of special microstructured optical fiber in which the higher-index rods are periodically distributed in a low-index silica host to form the transverse two-dimensional photonic crystal. In this paper, one kind of multidimensional microstructured optical fiber photonic device is proposed by using electric arc discharge method to fabricate periodic tapers along the fiber axis. By tuning the applied magnetic field intensity, the propagation characteristics of the all-solid WAF integrated with magnetic fluid are periodically modulated in both radial and axial directions. Experimental results show that the wavelength changes little while the transmission loss increases for an applied magnetic field intensity range from 0 to 500 Oe. The magnetic field sensitivity is 0.055 dB/Oe within the linear range from 50 to 300 Oe. Meanwhile, the all-solid WAF has very similar thermal expansion coefficient for both high-and low-refractive index glasses, and thermal drifts have a little effect on the mode profile.
Key words: Nanophotonic devices, Photonic crystal fiber, Nanomagnetic fluid

235. Compact CNT mode-locked Ho^{3+}-doped fluoride fiber laser at 1.2 μm

Junfeng Wang, Xiushan Zhu, Yunxiu Ma, Yuchen Wang, Minghong Tong, Shijie Fu, Jie Zong, Kort Wiersma, Arturo Chavez–Pirson, Robert A. Norwood, Wei Shi, Nasser Peyghambarian

IEEE Journal of Selected Topics in Quantum Electronics, 2017, 24(3): 1101205.

Abstract: A diode laser pumped carbon nanotube mode-locked Ho^{3+}-doped fluoride fiber laser at 1.2 μm was demonstrated for the first time. Stable mode-locked pulses with an average power of 1 mW at a repetition rate of 18.47 MHz were obtained at a pump power of 348 mW. The pulse energy and peak power of this mode-locked laser oscillator were about 54 pJ and 12.6 W, and were increased to 2.41 nJ and 0.54 kW, respectively, by using a 15-cm Ho^{3+}-doped fluoride fiber amplifier. The pulse duration was measured to be 4.3 ps by an autocorrelator. Because the operation wavelength is close to the zero-dispersion wavelength of 1.3 μm, this mode-locked fiber laser exhibits unique features.

Key words: Carbon nanotube, Fiber lasers, Holmium, Laser mode locking, Optical pulses, Ultrafast optics

236. Simulation of LSPR sensor based on exposed-core grapefruit fiber with a silver nanoshell

Xianchao Yang, Ying Lu, Baolin Liu, Jianquan Yao

Journal of Lightwave Technology, 2017, 35(21): 4728–4733.

Abstract: A localized surface plasmon resonance (LSPR) sensor based on exposed-core grapefruit fiber (EC-GF) with a silver nanoshell (SNS) is presented. The SNS, composed of a dielectric core coated with a thin silver layer, is placed at the exposed section of the EC-GF as the sensing channel to avoid the metal coating, and then deposited with the analyte to avert the liquid filling. Two orthogonal polarized resonance peaks (x-polarized andy-polarized) can be observed due to the birefringence and each polarization exhibits multipolar plasmon resonances, which can realize the cross reference. Due to the good features of the SNS, the position of the LSPR band can be tuned in a broad range, from 2 100 to 4 020 nm, making the proposed sensor of great importance for biosensing. An extremely high sensitivity 7 903.03 nm/RIU is obtained in the sensing range of 1.33–1.42, almost twice as high as the same type works. The influence of the SNS structure on the sensor's performance is also investigated numerically.

Key words: Birefringence, Exposed-core grapefruit fiber, Localized surface plasmon resonance, Silver nanoshell

237. Humidity sensor based on fabry-perot interferometer and intracavity sensing of fiber laser

Jia Shi, Degang Xu, Wei Xu, Yuye Wang, Chao Yan, Chao Zhang, Dexian Yan, Yixin He, Longhuang Tang, Weihong Zhang, Tiegen Liu, Jianquan Yao

Journal of Lightwave Technology, 2017, 35(21): 4789–4795.

Abstract: A humidity sensor based on Fabry-Perot interferometer (FPI) and intracavity sensing of a fiber ring laser is proposed and experimentally demonstrated. A compact humidity-sensitive fiber-optic FPI is developed and inserted in a fiber ring laser. Because the output power of the fiber laser is modulated by the reflection loss of the FPI at different ambient humidity, intracavity humidity sensing is induced. The experiment shows that the relative output power of the fiber laser has a good linear response to ambient humidity from 25%RH to 95%RH and the humidity sensitivity of 0.202 dB/%RH is obtained with a narrow 3-dB bandwidth and high signal-to-noise ratio. Furthermore, the excellent performance of the sensor for remote humidity detection is demonstrated. The sensor also shows a low temperature cross-sensitivity, fast time response, and good stability. The proposed sensor has a great potential in high-capacity sensor network and remote detections.

Key words: Fiber laser sensor, Fabry-Perot interferometer, Humidity sensors

238. Relative humidity sensor based on no-core fiber coated by agarose-gel film

Wei Xu, Jia Shi, Xianchao Yang, Degang Xu, Feng Rong, Junfa Zhao, Jianquan Yao
Sensors, 2017, 17: 2353.

Abstract: A relative humidity (RH) sensor based on single-mode-no-core-single-mode fiber (SNCS) structure is proposed and experimentally demonstrated. The agarose gel is coated on the no-core fiber (NCF) as the cladding, and multimode interference (MMI) occurs in the SNCS structure. The transmission spectrum of the sensor is modulated at different ambient relative humidities due to the tunable refractive index property of the agarose gel film. The relative humidity can be measured by the wavelength shift and intensity variation of the dip in the transmission spectra. The humidity response of the sensors, coated with different concentrations and coating numbers of the agarose solution, were experimentally investigated. The wavelength and intensity sensitivity is obtained as -149 pm/%RH and -0.075 dB/%RH in the range of 30% RH to 75% RH, respectively. The rise and fall time is tested to be 4.8 s and 7.1 s, respectively. The proposed sensor has a great potential in real-time RH monitoring.

Key words: Relative humidity measurement, Agarose, No-core fiber, Multimode interference

239. Improved numerical calculation of the single-mode-no-core-single-mode fiber structure using the fields far from cutoff approximation

Wei Xu, Jia Shi, Xianchao Yang, Degang Xu, Feng Rong, Junfa Zhao, Jianquan Yao
Sensors, 2017, 17: 2240.

Abstract: Multimode interferometers based on the single-mode-no-core-single-mode fiber (SNCS) structure have been widely investigated as functional devices and sensors. However, the theoretical support for the sensing mechanism is still imperfect, especially for the cladding refractive index response. In this paper, a modified model of no-core fiber (NCF) based on far from cut-off approximation is proposed to investigate the spectrum characteristic and sensing mechanism of the SNCS structure. Guided-mode propagation analysis (MPA) is used to analyze the self-image effect and spectrum response to the cladding refractive index and temperature. Verified by experiments, the performance of the SNCS structure can be estimated specifically and easily by the proposed method.

Key words: Single-mode-no-core-single-mode fiber, Guided-mode propagation analysis, Numerical calculation, Transmission spectra

240. Humidity sensor based on intracavity sensing of fiber ring laser

Jia Shi, Wei Xu, Degang Xu, Yuye Wang, Chao Zhang, Chao Yan, Dexian Yan, Yixin He, Longhuang Tang, Weihong Zhang, Jianquan Yao

Journal of Physics D: Applied Physics, 2017, 50: 425105.

Abstract: A humidity sensor based on the intracavity sensing of a fiber ring laser is proposed and experimentally demonstrated. In the fiber ring laser, a humidity-sensitive fiber-optic multimode interferometer (MMI), fabricated by the single-mode-no-core-single-mode (SNCS) fiber coated with Agarose, works as the wavelength-selective filter for intracavity wavelength modulated humidity sensing. The experiment shows that the lasing wavelength of the fiber laser has a good linear response to ambient humidity from 35% RH to 95% RH. The humidity sensitivity of -68 pm/% RH is obtained with a narrow 3 dB bandwidth less than 0.09 nm and a high signal-to-noise ratio (SNR) similar to 60 dB. The time response of the sensor has been measured to be as fast as 93 ms. The proposed sensor possesses a good stability and low temperature cross-sensitivity.

Key words: Fiber optics sensors, Laser sensor, Intracavity sensing

241. General description and understanding of the nonlinear dynamics of mode-locked fiber lasers

Huai Wei, Bin Li, Wei Shi, Xiushan Zhu, Robert A. Norwood, Nasser Peyghambarian, Shuisheng Jian

Scientific Reports: 2017, 7: 1292.

Abstract: As a type of nonlinear system with complexity, mode-locked fiber lasers are known for their complex behaviour. It is a challenging task to understand the fundamental physics behind such complex behaviour, and a unified description for the nonlinear behaviour and the systematic and quantitative analysis of the underlying mechanisms of these lasers have not been developed. Here, we present a complexity science-based theoretical framework for understanding the behaviour of mode-locked fiber lasers by going beyond reductionism. This hierarchically structured framework provides a model with variable dimensionality, resulting in a simple view that can be used to systematically describe complex states. Moreover, research into the attractors' basins reveals the origin of stochasticity, hysteresis and multistability in these systems and presents a new method for quantitative analysis of these nonlinear phenomena. These findings pave the way for dynamics analysis and system designs of mode-locked fiber lasers. We expect that this paradigm will also enable potential applications in diverse research fields related to complex nonlinear phenomena.

Key words: Dissipative soliton resonance, Transmission filter, Normal-dispersion, mode-locked fiber laser

242. Fiber ring laser temperature sensor based on liquid-filled photonic crystal fiber

Xianchao Yang, Ying Lu, Baolin Liu, Jianquan Yao
IEEE Sensors Journal, 2017. 17(21): 6948–6952.

Abstract: A temperature sensor based on a fiber ring laser embedded with liquid-filled photonic crystal fiber (PCF) is proposed and experimentally demonstrated. High refractive index liquid is infiltrated into an index-guiding PCF to generate band gap-like effect then used as the laser filter and the sensing head simultaneously. With the temperature increasing, the emission wavelength of the fiber laser shifts to the shorter wavelength and the output intensity increases due to band gaps' movement. The temperature sensitivity -1.747 nm/℃ and 0.137 dB/℃ are obtained utilizing the wavelength modulation and the intensity modulation, respectively, with high signal-to-nose ratio (~ 55 dB), narrow 3-dB bandwidth (less than 0.08 nm) and good stability.
Key words: Fiber laser temperature sensor, Liquid-filled photonic crystal fiber, Bandgap-like effect

243. Compact hundred-mW 2 μm single-frequency Thulium-doped silica fiber laser

Shijie Fu, Wei Shi, Quan Sheng, Guannan Shi, Haiwei Zhang, Xiaolei Bai, Jianquan Yao
IEEE Photonics Technology Letters, 2017, 29(11): 853–856.

Abstract: The output characteristics of a 2-μm all-silica-fiber single-frequency distributed Bragg reflector (DBR) laser in-band pumped by a 1 570-nm fiber laser were investigated in this letter. More than 50-mW laser power was obtained in this all-silica-fiber laser system with the linewidth of ~ 36 kHz. Low relative intensity noise was achieved in the experiment due to the superior noise properties of 1 570-nm fiber laser as pump source, especially at the relaxation oscillation frequency, where the noise intensity is measured to be as low as -112 dB/Hz. With another piece of Tm^{3+}-doped fiber spliced directly to the DBR laser to effectively exploit the pump power, a compact hundred-mW single-frequency fiber laser at 1 920 nm was obtained without performance degradation. The compact hundred-mW all-silica-fiber single-frequency laser system we demonstrated here can be a competitive candidate to promote the popularization of 2 μm single-frequency fiber laser for the applications such as free-space optical communication and LIDAR.
Key words: Fiber laser, Low noise, Narrow linewidth, Single frequency

244. Temperature self-compensation high-resolution refractive index sensor based on fiber ring Laser

Jia Shi, Yuye Wang, Degang Xu, Tiegen Liu, Wei Xu, Chao Zhang, Chao Yan, Dexian Yan, Longhuang Tang, Yixin He, Jianquan Yao

IEEE Photonics Technology Letters, 2017, 29(20): 1743–1746.

Abstract: In this letter, a temperature self-compensation high-resolution refractive index (RI) sensor based on intracavity intensity-modulated sensing in a fiber ring laser is demonstrated. A fiber-optic multimode interferometer based on single-mode-no-core-single-mode fiber structure is cascaded with a fiber Bragg grating and used as a reflected sensing head to enhance intensity-modulated depth. It is inserted in a fiber ring laser as a wavelength selective filter and the intracavity intensity-modulated RI sensing is induced for the output intensity of the fiber laser. Furthermore, because the lasing wavelength of the sensor system is sensitive to temperature but insensitive to external RI, the temperature self-compensation measurement can be realized. The RI sensitivity is measured to be -4.98 mW/RIU from 1.334 9 to 1.366 5. Correspondingly, the relative intensity sensitivity achieves -196.1 dB/RIU from 1.334 9 to 1.354 4 and -744.6 dB/RIU from 1.354 4 to 1.366 5. The resolution of the fiber laser sensor is obtained as high as 2×10^{-10} RIU with the signal-to-noise ratio more than 55 dB.

Key words: Fiber optics sensors, Refractive index sensors, Laser sensors

245. Dynamic propagation of initially chirped airy pulses in a quintic nonlinear fiber

Yu Yu, Yating Zhang, Xiaoxian Song, Haiting Zhang, Mingxuan Cao, Yongli Che, Heng Zhang, Jianquan Yao

IEEE Photonics Journal, 2017, 9(3): 1–7.

Abstract: We numerically simulate dynamic propagation of finite-energy Airy pulses in anomalous and normal regions of optical fiber and analyze the effects of quintic nonlinearity and initial chirp on evolution properties. Numerical results show that the effects of quantic nonlinearity on finite-energy Airy pulse imposed by initial chirp in two regions are entirely different. In anomalous dispersion region, soliton pulses shed from all finite-energy Airy pulses whether they are chirped or not, and quintic nonlinearity has profound effects on the propagation dynamics of the soliton pulse. However, in normal dispersion region, none soliton pulses are generated from Airy pulses. Depending on the peak intensity varying with the propagation distance, differing from the cases in the abnormal dispersion region, the effect of quintic nonlinearity has a slight impact on the evolution of Airy pulse.

Key words: Quintic nonlinearity, Initial chirp, Finite-energy Airy pulse

246. High sensitivity hollow fiber temperature sensor based on surface plasmon resonance and liquid filling

X. C. Yang, Y. Lu, B. L. Liu, J. Q. Yao

IEEE Photonics Journal, 2017, 10(2): 1–9.

Abstract: A high sensitivity hollow fiber (HF) temperature sensor based on surface plasmon resonance (SPR) and liquid core is designed and analyzed using finite element method (FEM). Toluene with high refractive index (RI) and large thermo-optic coefficient is sealed into the hollow-core of the fiber to guarantee the total internal reflection (TIR) and works as the sensing medium. One single air hole near the fiber core is coated with gold to generate surface plasmons (SPs) and two orthogonal polarization core modes (HE_{11}^x and HE_{11}^y) can be supported due to the asymmetrical SPR region. Contrary to the blue-shift of other liquid filled SPR temperature sensors, the resonance peak of the designed sensor shifts to the longer wavelength with temperature increasing. A large temperature range from 20 ℃ to 100 ℃ is calculated and the extremely high sensitivity 6.51 nm/℃ in good linear relationship is obtained. Moreover, the structure parameters and liquid factors are also discussed to optimize the sensor's performance.

Key words: Hollow fiber, Temperature sensor, Surface plasmon resonance, Liquid core

247. Design of a tunable single-polarization photonic crystal fiber filter with silver-coated and liquid-filled air holes

Xianchao Yang, Ying Lu, Baolin Liu, Jianquan Yao

IEEE Photonics Journal, 2017, 9(4): 1−8.

Abstract: A tunable single-polarization filter is designed by filling high index liquid into the silver-coated air holes of photonic crystal fiber and the polarization characteristics are analyzed utilizing the finite element method. Results show that the filter can be used to realize, to our best knowledge, the filtering of an incoming signal at 1 310 and 1 550 nm bands simultaneously in two orthogonal polarization states with narrowband of full width half maximum only 9 nm (at 1 550 nm bands). The confinement losses of unwanted polarized mode can reach up to 53.1 and 305.1 dB/cm at the two aforementioned communication windows, with other polarized mode losses as low as 0.8 and 2.4 dB/cm, respectively. For a propagation distance of 1 mm, the crosstalk of x-or y-polarized (x-or y-pol) mode can reach a value of 45.4 and -262.9 dB at 1 310 and 1 550 nm bands, and the 20 dB bandwidth at 62 and 133 nm. By adjusting the silver-coated air hole diameter, the silver layer thickness, or the filling liquid, the filtering wavelength can be precisely tuned in a broad band range, from 1 351 to 1 931 nm of x-pol and from 1 208 to 1 392 nm of y-pol, covering almost all the communication wavelength.

Key words: Tunable single-polarization filter, Photonic crystal fiber, Crosstalk

248. 5 kW near-diffraction-limited and 8 kW high-brightness monolithic continuous wave fiber lasers directly pumped by laser diodes

Qiang Fang, Jinhui Li, Wei Shi, Yuguo Qin, Yang Xu, Xiangjie Meng, Robert A. Norwood, Nasser Peyghambarian

IEEE Photonics Journal, 2017, 9(5): 1–7.

Abstract: Tandem pumping technique are traditionally adopted to develop > 3-kW continuous-wave (cw) Yb^{3+}-doped fiber lasers, which are usually pumped by other fiber lasers at shorter wavelengths (1 018 nm e.g.). Fiber lasers directly pumped by laser diodes have higher wall-plug efficiency and are more compact. Here we report two high brightness monolithic cw fiber laser sources at 1 080 nm. Both lasers consist of a cw fiber laser oscillator and one laser-diode pumped double cladding fiber amplifier in the master oscillator-power amplifier configuration. One laser, using 30-μm-core Yb^{3+}-doped fiber as the gain medium, can produce > 5-kW average laser power with near diffraction-limited beam quality (M^2 < 1.8). The slope efficiency of the fiber amplifier with respect to the laughed pump power reached 86.5%. The other laser utilized 50-μm-core Yb^{3+}-doped fiber as the gain medium and produced > 8-kW average laser power with high beam quality (M^2: ~4). The slope efficiency of the fiber amplifier with respect to the launched pump power reach 83%. To the best of our knowledge, this is the first detailed report for > 5-kW near-diffraction-limited and > 8-kW high-brightness monolithic fiber lasers directly pumped by laser diodes.

Key words: Diode-pumped lasers, Fiber lasers, High power, Near-diffraction-limited, Ytterbium-doped

249. Linewidth-narrowed, linear-polarized single-frequency Thulium-doped fiber laser based on stimulated Brillouin scattering effect

Shijie Fu, Wei Shi, Haiwei Zhang, Quan Sheng, Guannan Shi, Xiaolei Bai, Jianquan Yao

IEEE Photonics Journal, 2017, 9(4): 1–7.

Abstract: A 2-μm linear-polarized hybrid Brillouin-Thulium fiber laser (BTFL) was experimentally investigated for linewidth narrowing. The low threshold for the BTFL in terms of Brillouin pump of similar to 200 mW, benefits from the gain of active fiber in the ring cavity for both Brillouin pump and generated Stokes light. More than 205 mW single-frequency Stokes laser was obtained with linewidth reduction ratio of similar to 8 times, from 36 to 4.6 kHz. Linear-polarized output was achieved with a polarization extinction ratio of 23.8 dB.

Key words: Fiber laser, Narrow linewidth, Single frequency, Stimulated Brillouin scattering

250. A hollow-core photonic crystal fiber-based SPR sensor with large detection range

Nannan Luan, Jianquan Yao

IEEE Photonics Journal, 2017, 9(3): 1–7.

Abstract: We propose a large detection range surface plasmon resonance (SPR) sensor based on hollow-core photonic crystal fiber in this paper. The sensor consists of an analyte channel in the core hole and a silver nanowire in the cladding holes. We investigate the resonance properties between the core modes and the surface plasmon polariton (SPP) modes excited on the nanowire surface in a large refractive index (RI) range from 1.33 to 1.5. Numerical results show that the resonance between the core mode and the higher order SPP modes can occur at particular wavelengths, thus exciting resonance peaks that shift to short wavelengths as RI increases. By tracking most sensitive peak, which is formed by the resonance combination of the x-polarized core mode and the second-order SPP mode, our sensor can measure large RI ranges of the analyte either higher or lower than that of the fiber material.

Key words: Optical fiber sensors, Fiber optics

251. Polarization characteristics of high-birefringence photonic crystal fiber selectively coated with silver layers

Xianchao Yang, Ying Lu, Baolin Liu, Jianquan Yao

Plasmonics, 2017, 13(11): 1–8.

Abstract: The polarization characteristics of high-birefringence photonic crystal fiber (HB-PCF) selectively coated with silver layers are numerically investigated using the full-vector finite element method (FEM). The fundamental mode coupling properties and polarization splitting effect are discussed in detail. Results show that the resonance wavelength, resonance strength, and splitting distance between two polarized modes can be adjusted significantly by changing the fiber structure, the diameter of silver rings, and the thickness of silver layers. A single-polarization filter at 1 310 nm bands is proposed with the corresponding loss 500 dB/cm and full width half maximum (FWHM) only 23 nm. This work is very helpful for further studies in polarization-dependent wavelength-selective applications or other fiber-based plasmonic devices.

Key words: Full-vector finite element method, Mode coupling, Polarization splitting effect

252. Tunable surface plasmon resonance sensor based on photonic crystal fiber filled with gold nanoshells

Baolin Liu, Ying Lu, Xianchao Yang, Jianquan Yao
Plasmonics, 2017: 1–8.

Abstract: We present a photonic crystal fiber (PCF)-based surface plasmon resonance (SPR) sensor, whose operating wavelength range is tunable. Gold nanoshells, consisting of silica cores coated with thin gold shells, are designed to be the functional material of the sensor because of their attractive optical properties. It is demonstrated that the resonant wavelength of the sensor can be precisely tuned in a broad range, 660 nm to 3.1 μm, across the visible and near-infrared regions of the spectrum by varying the diameter of the core and the thickness of the shell. Furthermore, the effects of structural parameters of the sensor on the sensing properties are systematically analyzed and discussed based on the numerical simulations. It is observed that a high spectral sensitivity of 4 111.4 nm/RIU with the resolution of 2.45×10^{-5} RIU can be achieved in the sensing range of 1.33–1.38. These features make the sensor of great importance for a wide range of applications, especially in biosensing.

Key words: Sensor, Surface plasmon resonance, Photonic crystal fiber, Gold nanoshells

253. Generation of 2.5 μm and 4.6 μm dispersive waves in kagome photonic crystal fiber with plasma production

Tianqi Zhao, Meng Li, Dong Wei, Xin Ding, Guizhong Zhang, Jianquan Yao
Chinese Physics Letters, 2017, 34(11), 114202.

Abstract: We report our numerical simulation on dispersive waves (DWs) generated in the Kr-filled Kagome hollow-core photonic crystal fiber, by deploying the unidirectional pulse propagation equation. Relatively strong dispersive waves are simultaneously generated at 2.5 μm and 4.6 μm. It is deciphered that the interplay between plasma currents due to Kr ionization and nonlinear effects plays a key role in DW generation. Remarkably, this kind of DW generation is corroborated by the plasma-corrected phase-matching condition.
Key words: Dispersive waves, Numerical simulation, Kagome photonic crystal fiber, Plasma production

254. Fiber lasers and their applications: introduction

Wei Shi, Axel Schulzgen, Rodrigo Amezcua, Xiushan Zhu, Shaif-Ul Alam
Journal of the Optical Society of America B, 2017, 34(3): FLA1.

Abstract: Fiber lasers have more and more applications in industrial, military, medical diagnosis, and scientific research. This issue features the progress in the area of fiber lasers, ranging from new developed gain materials to fiber laser systems and their applications.
Key words: Fiber lasers, Applications, Gain material

255. Review of recent progress on single-frequency fiber lasers [invited]

Shijie Fu, Wei Shi, Yan Feng, Lei Zhang, Zhongmin Yang, Shanhui Xu, Xiushan Zhu, R. A. Norwood, N. Peyghambarian

Journal of the Optical Society of America B, 2017, 34(3): A49–A62.

Abstract: Single-frequency fiber lasers have drawn intense attention for their extensive applications from high-resolution spectroscopy and gravitational wave detection to materials processing due to the outstanding properties of low noise, narrow linewidth, and the resulting long coherence length. In this paper, the recent advances of single-frequency fiber oscillators and amplifiers are briefly reviewed in the broad wavelength region of 1–3 μm. Performance improvements in laser noise and linewidth are addressed with the newly developed physical mechanisms. The solution to achieving higher power/energy is also discussed, accompanied by the start-of-the-art results achieved to date.

Key words: Fiber laser, Single-frequency, Amplifier

256. Simultaneous measurement of refractive index and temperature based on SPR in D-shaped MOF

Xianchao Yang, Ying Lu, Baolin Liu, Jianquan Yao

Applied Optics, 2017, 56(15): 4369–4374.

Abstract: A surface plasmon resonance (SPR) sensor based on D-shaped microstructured optical fiber (MOF) is proposed to realize the simultaneous measurement of refractive index (RI) and temperature. The D-shaped flat surface coated with a gold layer is in direct contact with analyte as a sensing channel of RI, and one of the air holes near the fiber core is filled with chloroform to detect temperature. Two separate channels and birefringence caused by the asymmetric structure can distinguish the variations of RI and temperature independently, thus completely solving the cross-sensitivity problem. This is the first time to realize the simultaneous measurement of multiple parameters without matrix equations, to the best of our knowledge. Results show that the y-polarized peak supported by channel I only shifts with RI variation and is unaffected by the temperature floating. Similarly, the x-polarized peak supported by channel II is only influenced by the change of temperature in the external environment. The effect of gold layer thickness is investigated numerically, and the sensor sensitivity is identified both in wavelength and amplitude interrogations. This work is very helpful for the design and implementation of a highly sensitive, real-time, and distributed SPR sensor for multi-parameter measurement applications.

Key words: Surface plasmon resonance, Sensor, Microstructured optical fiber, Refractive index measurement, Temperature measurement, Simultaneous measurement

257. Hollow-fiber-based surface plasmon resonance sensor with large refractive index detection range and high linearity

Liangcheng Duan, Xianchao Yang, Ying Lu, Jianquan Yao

Applied Optics, 2017, 56(36): 9907–9912.

Abstract: A surface plasmon resonance (SPR) sensor based on the gold-coated hollow fiber (HF) is proposed to realize the detection of refractive index (RI). The large hollow core of the HF is utilized as the microfluidic channel of the analyte, and gold is coated inside the air holes surrounding the core to generate surface plasmons. Benefitting from the unique structure of HF and liquid filling into the core, the sensor can work in a large RI range from 1.27 to 1.45 at the near-infrared region, overcoming the limitation of 1.42, as other side-hole-filled photonic crystal fiber-based SPR sensors reported. Moreover, the relations between the resonance wavelength and RI are highly linear and more suitable for standardization, while almost all similar works show higher sensitivity for the high RI change than for the low. The average spectral sensitivity 5 653.57 nm/RIU is obtained. Other structure parameters, such as gold-coated air hole numbers and gold layer thicknesses, are also investigated.

Key words: Photonic crystal fiber, Surface plasmons

258. Surface plasmon resonance sensor based on photonic crystal fiber filled with gold-silica-gold multilayer nanoshells

Baolin Liu, Ying Lu, Xianchao Yang, Jianquan Yao

Optics Communications, 2017, 405: 281–287.

Abstract: We present a surface plasmon resonance sensor based on photonic crystal fiber filled with gold-silica-gold (GSG) multilayer nanoshells for measurement of the refractive index of liquid analyte. The GSG multilayer nanoshells, composed of a silica-coated gold nanosphere surrounded by a gold shell layer, are designed to be the functional material of the sensor because of their attractive optical properties. Two resonant peaks are obtained due to the hybridization of nanosphere plasmon modes and nanoshell plasmon modes. It is demonstrated that the resonant wavelength of the two peaks can be precisely tuned in 560-716 nm and 849-2 485 nm, respectively, by varying the structural parameters of the GSG multilayer nanoshells in a compact, sub-200 nm size range. The excellent spectral tunability makes the sensor attractive in a wide range of applications, especially in biosensing in near-infrared region. Furthermore, the influences of the parameters on the performance of the sensor are systematically simulated and discussed. It is observed that the spectral sensitivities of 1 894.3 nm/RIU and 3 011.4 nm/RIU can be achieved respectively by the two resonant peaks in the sensing range of 1.33-1.38. The existence of two loss peaks also provides the possibility to realize self-reference in the sensing process.

Key words: Surface plasmon resonance, Sensor, Photonic crystal fiber, Gold-silica-gold multilayer nanoshells, Refractive index measurement

259. Megawatt-peak-power picosecond all-fiber-based laser in MOPA using highly Yb^{3+}-doped LMA phosphate fiber

Guannan Shi, Shijie Fu, Quan Sheng, Jinhui Li, Qiang Fang, Huixian Liu, Arturo Chavez-Pirson, N. Peyghambarian, Wei Shi, Jianquan Yao

Optics Communications, 2017, 411: 133–137.

Abstract: A megawatt-peak-power picosecond all-fiber-based laser in master oscillator power amplifier (MOPA) is experimentally demonstrated. Only 34-cm-long highly Yb^{3+}-doped large mode area (LMA) phosphate fiber was used as the gain fiber in the amplification stage to alleviate nonlinearity and achieve high peak power. Picosecond pulses with single pulse energy of 21.2 μJ and peak power of 0.96 MW were achieved at the repetition rate of 500 kHz. Evident spectral degradation can be observed as the peak power approached 1 MW, and a stimulated Raman scattering (SRS) free peak power of 0.51 MW was obtained in the experiment. Moreover, the output power under different repetition rates was investigated.

Key words: Doped fiber, Amplifier

260. Tunable polarization filter based on high-birefringence photonic crystal fiber filled with silver wires

Xianchao Yang, Ying Lu, Baolin Liu, Jianquan Yao

Optical Engineering, 2017, 56(7): 077108.

Abstract: A tunable single polarization filter based on high-birefringence photonic crystal fiber with silver wires symmetrically filled into cladding air holes is designed. The confinement loss of the unwanted polarized mode (x-polarized mode) at 1 310- and 1 550-nm bands are 371 and 252 dB/cm, whereas another mode confinement loss (y-polarized mode) at the corresponding wavelength as low as 14 and 10 dB/cm, respectively. Moreover, the 20-dB bandwidth can reach 179 (at the 1 310-nm band) and 71 nm (at the 1 550-nm band) for a propagation distance of 1 mm. The dispersion relations and polarization characteristics are analyzed in detail utilizing the finite element method. Numerical results show that by adjusting the pitch between two adjacent air holes, the diameters of cladding air holes or silver wires near the fiber core, the resonance wavelength and resonance strength can be tuned effectively, which is beneficial for tunable polarization filter devices in the communication wave bands.

Key words: Tunable polarization filter, high-briefringence photonic crystal fiber, Finite element method, X- and Y- polarizations

261. 97-μJ single frequency linearly polarized nanosecond pulsed laser at 775 nm using frequency doubling of a high-energy fiber laser system

Qiang Fang, Xuelong Cui, Zhuo Zhang, Liang Qi, Wei Shi, Jinhui Li, Guoqing Zhou

Optical Engineering, 2017, 56(8): 086112.

Abstract: We report a high-energy (~97 μJ), high-peak power (~20 kW), single-frequency, linearly polarized, near diffraction-limited (M^2<1.2) ~4.8-ns pulsed laser source at 775 nm with a 260-Hz repetition rate. This laser was achieved by frequency doubling of a high-energy linearly polarized all-fiber-based master oscillator–power amplifier, seeded by a single-frequency semiconductor distributed feedback laser diode at 1 550 nm. The frequency doubling is implemented in a single-pass configuration using a periodically poled lithium niobate crystal, and a conversion efficiency of 51.3% was achieved.

Key words: Single frequency, Frequency doubling, High-energy fiber laser

262. A novel variable baseline visibility detection system and its measurement method

Meng Li, Lihui Jiang, Xinglong Xiong, Guizhong Zhang, JianQuan Yao

Optical Review, 2017, 24(5): 634–641.

Abstract: As an important meteorological observation instrument, the visibility meter can ensure the safety of traffic operation. However, due to the optical system contamination as well as sample error, the accuracy and stability of the equipment are difficult to meet the requirement in the low-visibility environment. To settle this matter, a novel measurement equipment was designed based upon multiple baseline, which essentially acts as an atmospheric transmission meter with movable optical receiver, applying weighted least square method to process signal. Theoretical analysis and experiments in real atmosphere environment support this technique.

Key words: Visibility, Multiple baseline, Weighted least square

263. 高功率双包层掺铒光纤放大器温度分布特性

张海伟, 盛泉, 史伟, 白晓磊, 付士杰, 姚建铨

红外与激光工程, 2017, 46(6): 622004.

摘要：通过对普适于不同内包层边界条件下的热传导方程进行推导和求解,得到了不同内包层形状的双包层增益光纤所对应的掺铒光纤放大器的三维热分布。计算结果表明,双包层光纤不同内包层形状可导致纤芯处的温度差高达 107 K。同时,信号光与泵浦光功率的比值决定了温度最高点和熔接点的距离,在泵浦光功率为 100 W、信号光功率为 10 mW 的情况下,两者之间的距离可达 30 cm。通过分析不同内包层形状的双包层光纤的径向与轴向的热分布情况发现,相较于其他内包层形状的双包层光纤,偏芯型双包层掺铒光纤因其具有较低的最高温度、较高的泵浦效率和高斯型横截面热分布而较适用于掺铒光纤放大器。

关键词：掺铒光纤放大器, 热效应, 双包层光纤, 解析模型

264. 激光雷达用高性能光纤激光器

史伟, 房强, 李锦辉, 付士杰, 李鑫, 盛泉, 姚建铨

红外与激光工程, 2017, 46(8): 802001.

摘要：系统研究了窄线宽低噪声单频连续光纤激光器、高能量纳秒长脉冲单频光纤激光器以及高峰值功率纳秒短脉冲光纤激光器三类高性能光纤激光器:实现了工作于 1、1.5 及 2 μm 波段的单频连续光纤激光器,典型光谱线宽小于 3 kHz,强度噪声接近于散粒噪声极限;实现了高能量单频光纤激光器,脉冲能量超过 200 μJ,重复频率 20 kHz,脉冲宽度 100~500 ns,激光波长位于 1.5 μm 波段;实现了高峰值功率纳秒短脉冲光纤激光器,峰值功率超过 700 kW,重复频率 10 kHz,脉冲宽度 3 ns;同时还实现了高重频高峰值功率纳秒短脉冲光纤激光器,峰值功率超过 200 W,重复频率 3 MHz,脉冲宽度 1~5 ns。文中阐述了以上几类高性能光纤激光器在激光雷达探测系统中的应用前景。

关键词：光纤激光器, 单频, 纳秒激光脉冲, 高能量, 窄线宽, 高峰值功率

265. High power all fiber-based ultrafast lasers

Zhenhua Yu, Wei Shi, Xinzheng Dong, Jinhui Li, Yizhu Zhao
The 8th International Symposium on Ultrafast Phenomena and Terahertz Waves, 2016: IW1C-4.

Abstract: All fiber-based hundred-watt level average power picosecond laser and mega-watt level peak power picosecond laser are demonstrated based on ultra-large core (Diameter: 50 μm) ytterbium-doped silica fiber and high gain phosphate fiber respectively.
Key words: Ultrafast lasers, All fiber, High power, Picosecond laser

266. Automatic channel-switched intracavity-absorption acetylene sensor based on mode-competition via Sagnac loop filter

Haiwei Zhang, Wei Shi, Liangcheng Duan, Ying Lu, Quan Sheng, Jianquan Yao
Proceedings of SPIE, 2017, 10090: 100901G.

Abstract: We demonstrate an automatic channel-switched intracavity-absorption acetylene sensor via Sagnac loop filter based on the mode-competition in a ring fiber laser. When the photonic crystal fiber gas cell is filled with 1% acetylene, the corresponding absorption intensity can be similar to 14.0 dB and similar to 7.2 dB at 1 532.83 nm and 1 534.01 nm, respectively. Compared with the single transmission pass method, the sensitivity can be improved up to more than 10 times. It spends 50 seconds in scanning the absorption spectra through applying gradient voltage to the tunable F-P filter.
Key words: Optical sensing and sensors, Fiber laser, Multisensor methods

267. High-resolution temperature sensor through measuring the frequency shift of single-frequency Erbium-doped fiber ring laser

Haiwei Zhang, Wei Shi, Liangcheng Duan, Shijie Fu, Quan Sheng, Jianquan Yao
Proceedings of SPIE, 2017, 10090: 1009016.

Abstract: We propose a principle to achieve a high-resolution temperature sensor through measuring the central frequency shift in the single-frequency Erbium-doped fiber ring laser induced by the thermal drift via the optical heterodyne spectroscopy method. We achieve a temperature sensor with a sensitivity about 9.7 pm/℃ and verify the detection accuracy through an experiment. Due to the narrow linewidth of the output single-frequency signal and the high accuracy of the optical heterodyne spectroscopy method in measuring the frequency shift in the single-frequency ring laser, the temperature sensor can be employed to resolve a temperature drift up to similar to 5.5×10^{-6} ℃ theoretically when the single-frequency ring laser has a linewidth of 1 kHz and 10-kHz frequency shift is achieved from the heterodyne spectra.
Key words: Single frequency, Fiber laser, Linewidth narrowing

268. Investigation of ASE and SRS effects on 1 018 nm short-wavelength Yb^{3+}-doped fiber laser

Zhaoxin Xie, Wei Shi, Quan Sheng, Shijie Fu, Qiang Fang, Haiwei Zhang, Xiaolei Bai, Guannan Shi, Jianquan Yao
Proceedings of SPIE, 2017, 10083: 1008327.

Abstract: 1 018 nm short wavelength Yb^{3+}-doped fiber laser can be widely used for tandem-pumped fiber laser system in 1 μm regime because of its high brightness and low quantum defect (QD). In order to achieve 1 018 nm short wavelength Yb^{3+}-doped fiber laser with high output power, a steady-state rate equations considering the amplified spontaneous emission (ASE) and Stimulated Raman Scattering (SRS) has been established. We theoretically analyzed the ASE and SRS effects in 1 018 nm short wavelength Yb^{3+}-doped fiber laser and the simulation results show that the ASE is the main restriction rather than SRS for high power 1 018 nm short wavelength Yb^{3+}-doped fiber laser, besides the high temperature of fiber is also the restriction for high output power. We use numerical solution of steady-state rate equations to discuss how to suppress ASE in 1 018 nm short wavelength fiber laser and how to achieve high power 1 018 nm short-wavelength fiber laser.
Key words: Short wavelength fiber laser, ASE, SRS, Heat load, Temperature

269. Linewidth-narrowed 2-μm single-frequency fiber laser based on stimulated Brillouin scattering effect

Shijie Fu, Wei Shi, Haiwei Zhang, Quan Sheng, Xiaolei Bai, Jianquan Yao
Proceedings of SPIE, 2017, 10083: 10083OL.

Abstract: A 2-μm linear-polarized single-frequency Brillouin-Thulium fiber laser (BTFL) has been experimentally investigated for linewidth narrowing. The threshold for the Brillouin pump is around 200 mW, and more than 205 mW single-frequency Stokes laser was achieved with the 793 nm pump power of 8.5W. The linewidth of the fiber laser has been narrowed for ~ 8 times, from 34 to 4.6 kHz. The measured RIN of the BTFL is <-150 dB/Hz for frequency above 2 MHz, which approaches the shot noise limit.
Key words: Single frequency, Fiber laser, Linewidth narrowing

270. Extended linear cavity 2 μm single-frequency fiber laser using Tm-doped fiber saturable absorber

Shijie Fu, Wei Shi, Quan Sheng, Guannan Shi, Haiwei Zhang, Jianquan Yao
CLEO: Science and Innovations, 2017: JW2A. 64.

Abstract: An extended linear cavity single-frequency fiber laser at 2 μm was investigated using Tm-doped fiber as saturable absorber. More than 60 mW laser power was achieved with the linewidth of ~40 kHz.
Key words: Fiber laser, Single-frequency, Extended linear cavity, Tm-doped fiber

271. 空心 Kagome 光子晶体光纤中等离子诱导产生的色散波

赵天琪, 魏东, 孙甫, 丁欣, 张贵忠, 姚建铨

中国激光, 2017, 44(5): 0508001.

摘要: 通过在充氪(Kr)气体的空心 Kagome 光子晶体光纤中输入超短脉冲,产生中红外波段的色散波,并利用单向脉冲传输方程对色散波进行数值模拟和分析。主要探讨了两种输入脉冲:中心波长为 1.4 μm 的高斯型脉冲和在此基础上加入其双倍频叠加而产生的双色近锯齿波脉冲。两种情况下均产生了多个色散波,并符合相位匹配条件。为优化色散波,将近锯齿波作为抽运光。当充入光纤的 Kr 气体因抽运光压缩而二次电离时,已产生的色散波发生转换,变为新的更长波长的色散波。这种色散波现象可以通过等离子体修正的相位匹配条件来解释。深入探讨了中红外波段超短脉冲的产生机制以及色散波理论。

关键词: 光纤光学, 光子晶体光纤, 超短脉冲非线性效应, 色散波, 激光等离子体

272. 金属-电介质微盘阵列红外吸收器的光学特性分析

魏东, 张贵忠, 丁欣, 姚建铨

光子学报, 2017, 46(8): 0823005.

摘要: 根据亚波长微结构的异常光谱特性,提出了一种"圆形铝盘-圆形 SiO_2 盘-铝衬底"的二维周期性微结构可调谐红外吸波器。利用有限元算法对其红外光谱反射性质进行了数值模拟,发现该结构对入射的 TE 或 TM 光偏振态、在 0°~60° 大入射角范围内具有良好的吸波效果,共振波长调谐范围 4~11 μm。用局域等离子体激元共振理论解释了红外吸波机理,证明了所提出的多层微盘结构具有更宽的调谐特性和更好的吸收效率。

关键词: 耦合等离激元, 红外, 有限元算法, 反射率, 吸收器, 超材料

273. 基于保偏光纤和 LPFG 的 Sagnac 环温度及环境折射率双参量光纤传感器研究

苏耿华, 徐德刚, 石嘉, 张海伟, 许浩展, 冯佳琛, 严德贤, 徐伟, 姚建铨

光电子·激光, 2017, 28(01): 25-31.

摘要: 报导了一种基于级联保偏光纤 (PMF) 和长周期光纤光栅 (LPFG) 的 Sagnac 环温度和环境折射率双参量传感器。在 Sagnac 干涉环内级联 PMF 和 LPFG, LPFG 的双折射效应使得 Sagnac 干涉的相位差受环境折射率调制。所提出的双参量传感器的透射光谱中, 由 LPFG 形成的透射峰和 Sagnac 干涉形成的干涉峰对温度和环境折射率各自具有不同的灵敏度, 通过对两者特征峰波长漂移量的测量就可实现对温度和环境折射率的同时传感。实验上搭建了温度和环境折射率双参量测试装置, 采用波长解调方法, 受光源功率波动的影响小, 温度灵敏度为 1.2 nm/℃; 环境折射率灵敏度为 15 nm/RIU。该双参量传感方案解决了温度和环境折射率的交叉敏感问题, 结构简单, 采用金属板凹槽结构进行封装有效地保护了 LPFG, 在生物传感和化学传感等领域具有广阔的应用前景。

关键词: 传感器, 温度和折射率传感, 长周期光纤光栅 (LPFG), Sagnac 干涉环

274. Enhancement and modulation of photonic spin Hall effect by defect modes in photonic crystal with grapheme

Jie Li, Tingting Tang, Li Luo, Jianquan Yao.
Carbon, 2018, 134: 293-300.

Abstract: Photonic spin Hall effect (PSHE) holds great potential applications for manipulating photon spins. However, the efficient control of PSHE in a particular optical structure is still difficult to realize. In this paper, we report enhanced and magnetically tunable PSHE of reflected light in one-dimensional photonic crystals (1D-PC) with a defect layer. By inserting monolayer graphene into the defect layer of 1D-PC, large polarization rotation and strong photon spin-orbit interaction of reflected light are obtained. Due to the optical field confinement by defect mode, the Kerr rotation angle of reflected light is near 90° and the spin-dependent splitting is about 15 times the wavelength at a specific incident wavelength. In addition, a static perpendicular magnetic field increasing from 1T to 11T shows greatly modulation effect on Kerr rotation angle and spin-dependent splitting. These results pave the way toward the design of spin-based photonic devices in the future.

Key words: Photonic spin Hall effect, Magneto-optical modulation, Photonic crystals

275. Realization of tunable Goos-Hänchen effect with magneto-optical effect in graphene

Tingting Tang, Jie Li, Ming Zhu, Li Luo, Jianquan Yao, Nengxi Li, Pengyu Zhang

Carbon, 2018, 135: 29–34.

Abstract: Tunable Goos-Hänchen (GH) effect with magneto-optical (MO) effect in a prism-graphene coupling structure is proposed. Based on MO effect of graphene in terahertz region, GH effect can be modulated by applied magnetic fields. The GH shift is calculated by stationary phase method and verified by finite element method (FEM). The physical mechanism of MO modulation for GH effect in the prism-graphene coupling structure is also analyzed based on the interaction between graphene and incident light. GH effect based on Kerr rotation may have great potential in the application of optical rotation displacement modulation and sensing. Meanwhile tunable GH effect in graphene shows a big potential in the measurement of Fermi energy or relaxation time of graphene. It provides us an effective method to facilitate its design and applications in terahertz devices and systems.

Key words: Magneto-optical, Goos-Hanchen effect, Kerr rotation, Prism-graphene coupling structure

276. Compact CNT mode-locked Ho^{3+}-doped fluoride fiber laser at 1.2 μm

Junfeng Wang, Xiushan Zhu, Yunxiu Ma, Yuchen Wang, Minghong Tong, Shijie Fu, Jie Zong, Kort Wiersma, Arturo Chavez-Pirson, Robert A. Norwood, Wei Shi, Nasser Peyghambarian

IEEE Journal of Selected Topics in Quantum Electronics, 2018, 24(3): 1–5.

Abstract: A diode laser pumped carbon nanotube mode-locked Ho^{3+}-doped fluoride fiber laser at 1.2 μm was demonstrated for the first time. Stable mode-locked pulses with an average power of 1 mW at a repetition rate of 18.47 MHz were obtained at a pump power of 348 mW. The pulse energy and peak power of this mode-locked laser oscillator were about 54 pJ and 12.6 W, and were increased to 2.41 nJ and 0.54 kW, respectively, by using a 15-cm Ho^{3+}-doped fluoride fiber amplifier. The pulse duration was measured to be 4.3 ps by an autocorrelator. Because the operation wavelength is close to the zero-dispersion wavelength of 1.3 μm, this mode-locked fiber laser exhibits unique features.

Key words: Carbon nanotube, Fiber lasers, Holmium, Laser mode locking, Optical pulses, Ultrafast optics

277. A highly sensitive magnetic field sensor based on a tapered microfiber

Zelong Ma, Yinping Miao, Yi Li, Hongming Zhang, Bin Li, Yanpeng Cao, Jianquan Yao

IEEE Photonics Journal, 2018, 10(4):1–8.

Abstract: A compact all-fiber magnetic field sensor based on the combination of a magnetic fluid and a nonadiabatic microfiber is experimentally demonstrated. By using a fusion splicer combined with an additional flame brushing, the nonadiabatic microfiber, with an abrupt taper, a small waist diameter and a long region, provides a strong evanescent field and a long sensitive distance. The result indicates that a sensitivity as high as 309.3 pm/Oe can be achieved in the range from 0 to 200 Oe. The designed nonadiabatic microfiber sensing structure suggests potential application as a tunable all-in-fiber photonic and other newly fashioned magneto-optical photonic devices.

Key words: Adiabatic microfiber, Magnetic fluid, Magnetic field sensor, Microfiber, Nonadiabatic microfiber

278. High-resolution temperature sensor based on single-frequency ring fiber laser via optical heterodyne spectroscopy technology

Liangcheng Duan, Haiwei Zhang, Wei Shi, Xianchao Yang, Ying Lu, Jianquan Yao
Sensors, 2018, 18(10): 3245.

Abstract: We demonstrate a high-resolution temperature sensor based on optical heterodyne spectroscopy technology by virtue of the narrow linewidth characteristic of a single-frequency fiber laser. When the single-frequency ring fiber laser has a Lorentzian-linewidth <1 kHz and the temperature sensor operates in the range of 3-85 ℃, an average sensitivity of 14.74 pm/℃ is obtained by an optical spectrum analyzer. Furthermore, a resolution as high as ~5 × 10^{-3} ℃ is demonstrated through optical heterodyne spectroscopy technology by an electrical spectrum analyzer in the range of 18.26–18.71 ℃ with the figure of merit up to 3.1 × 10^5 in the experiment.

Key words: Fiber optics sensors, High resolution, Single-frequency laser, Heterodyne spectroscopy

279. Megawatt-peak-power picosecond all-fiber-based laser in MOPA using highly Yb^{3+}-doped LMA phosphate fiber

Guannan Shi, Shijie Fu, Quan Sheng, Jinhui Li, Qiang Fang, Huixian Liu, Arturo Chavez-Pirson, N. Peyghambarian, Wei Shi, Jianquan Yao
Optics Communications, 2018, 411: 133–137.

Abstract: A megawatt-peak-power picosecond all-fiber-based laser in master oscillator power amplifier (MOPA) is experimentally demonstrated. Only 34-cm-long highly Yb^{3+}-doped large mode area (LMA) phosphate fiber was used as the gain fiber in the amplification stage to alleviate nonlinearity and achieve high peak power. Picosecond pulses with single pulse energy of 21.2 μJ and peak power of 0.96 MW were achieved at the repetition rate of 500 kHz. Evident spectral degradation can be observed as the peak power approached 1 MW, and a stimulated Raman scattering (SRS) free peak power of 0.51 MW was obtained in the experiment. Moreover, the output power under different repetition rates was investigated.
Key words: All-fiber-based laser, Picosecond, MOPA, Highly Yb^{3+}-doped, Large mode area, Megawatt-peak-power

280. Numerical simulation of reflective infrared absorber based on metal and dielectric nanorings

Dong Wei, Guizhong Zhang, Xin Ding, Jianquan Yao
Journal of Modern Optics, 2018, 65(7): 869–878.

Abstract: We propose a subwavelength micro-structure of /metal-ring/dielectric-ring/metal-substrate for infrared absorber, and numerically simulate its spectral reflectance in the infrared regime. Besides its pragmatic fabrication, this nanoring structure is characterized by excellent infrared reflectance, angle and polarization insensitivities and large tunability. Based upon the nanoring structure, a multilayered nanoring structure is demonstrated to be able to further tune the resonance wavelength. We also use an area-corrected plasmon polariton model to decipher the resonance wavelengths.
Key words: Nanoring structure, Infrared absorber, Plasmon polariton

281. Liquid crystal-modulated tunable filter based on coupling between plasmon-induced transparency and cavity mode

Dong Li, Fei Liu, Guang jun Ren, Pan Fu, Jian quan Yao

Optical Engineering, 2018, 57(9): 097101.

Abstract: A liquid crystal (LC)-modulated tunable filter is theoretically studied, consisting of an Au nanorod array and an Au film separated by a dielectric indium tin oxide (ITO) glass layer. It is well established that this system can achieve double absorptive peaks resulting from the strong plasmonic coupling between the plasmon-induced transparency effect of the nanorod array and the cavity mode of the Au microcavity structure. The positions of the double absorptive peaks can be tuned dynamically based on the electro-optic effect of LC. The simulation results reveal that a band range of 160 nm has been confirmed at near-infrared wavelengths by altering the driven voltage of LC from 0 to 8 V. The proposed structure is able to filter the two peaks with the reflection coefficients <5%. Compared with the existing tunable filter, it has many advantages, such as continuous tunability, low tuning voltage, and great degree of tunability.

Key words: Tunable filter, Active plasmonics, Plasmon-induced transparency, Cavity mode, Liquid crystal

282. The research on the design and performance of 7 × 1 pump combiners

Yang Cao, Quan Sheng, Shijie Fu, Haiwei Zhang, Xiaolei Bai, Wei Shi, Jianquan Yao

Components and Packaging for Laser Systems IV. International Society for Optics and Photonics, 2018, 10513: 105131M.

Abstract: The 7 × 1 end-pumped pump combiners employing 105/125 μm multimode fibers as pump fibers are investigated. Based on the results of our theoretical analysis, sufficient taper length (TL) and low refractive index (RI) of the capillary have been adopted to fabricate high transmission efficiency combiners. A 7 × 1 end-pumped pump combiner with an average transmission efficiency of 98.9% and a total return loss of 1.1‰ is fabricated in experiments, which could find its application in high-power fiber laser systems.

Key words: Fiber Laser, Fiber device, Pump combiner

283. Influence of seed power and gain fiber temperature on output linewidth in single-frequency Er^{3+}/Yb^{3+} co-doped fiber amplifier

Xiaolei Bai, Quan Sheng, Haiwei Zhang, Shijie Fu, Wei Shi, Jianquan Yao

Fiber Lasers XV: Technology and Systems. International Society for Optics and Photonics, 2018, 10512: 105122H.

Abstract: Based on the 1 550 nm single-frequency Erbium-Ytterbium co-doped fiber amplifier, the output linewidth with different seed power and gain fiber temperature was experimental investigated. The results demonstrated that, to obtain the same output power, the increment of the seed power was benefit to improve the output signal to noise ratio (SNR), and to reduce the linewidth broadening. The increment of gain fiber temperature can improve slope efficiency and enhance the ASE intensity but broaden linewidth. Meanwhile, the ASE was considered as one of the reason of linewidth variation.

Key words: Erbium-Ytterbium co-doped fiber amplifier, Single frequency, ASE, Linewidth

284. High-energy, 100-ns, single-frequency all-fiber laser at 1 064 nm

Shijie Fu, Wei Shi, Zhao Tang, Chaodu Shi, Xiaolei Bai, Quan Sheng, Arturo Chavez–Pirson, N. Peyghambarian, Jianquan Yao

Fiber Lasers XV: Technology and Systems. International Society for Optics and Photonics, 2018, 10512: 1051219.

Abstract: A high-energy, single-frequency fiber laser with long pulse duration of 100 ns has been experimentally investigated in an all-fiber architecture. Only 34-cm long heavily Yb-doped phosphate fiber was employed in power scaling stage to efficiently suppress the Stimulated Brillouin effect (SBS). In the experiment, 0.47 mJ single pulse energy was achieved in power scaling stage at the pump power of 16 W. The pre-shaped pulse was gradually broadened from 103 to 140 ns during the amplification without shape distortion.

Key words: Single frequency, Fiber laser, High energy

285. Dual-wavelength noise-like pulse generation in passively mode-locked all-fiber laser based on MMI effect

Guannan Shi, Shijie Fu, Quan Sheng, Wei Shi, Jianquan Yao

Fiber Lasers XV: Technology and Systems. International Society for Optics and Photonics. 2018, 10512: 105122B.

Abstract: We report on the generation of dual-wavelength noise-like pulse (NLP) from a passively mode-locked all-fiber laser based on multimode interference (MMI) effect. The theory to evaluate and design transmission spectrum of MMI filter is analyzed. A homemade MMI filter was employed in an Er-doped fiber ring laser with NPE configuration and dual wavelength NLP at 1 530 and 1 600 nm was obtained with 3-dB bandwidth of 18.1 and 41.9 nm, respectively. The output had a signal-to-noise ratio higher than 35 dB and can achieve self-started operation.

Key words: Fiber laser, Noise-like pulse, Dual-wavelength, MMI effect

286. Theoretical study and design of third-order random fiber laser

Zhaoxin Xie, Wei Shi, Shijie Fu, Quan Sheng, Jianquan Yao

Fiber Lasers XV: Technology and Systems. International Society for Optics and Photonics 2018, 10512: 105122I.

Abstract: We present result of achieving a random fiber laser at a working wavelength of 1 178 nm while pumping at 1 018 nm. The laser power is realized by 200 m long cavity which includes three high reflectivity fiber Bragg gratings. This simple and efficient random fiber laser could provide a novel approach to realize low-threshold and high-efficiency 1 178 nm long wavelength laser. We theoretically analyzed the laser power in random fiber lasers at different pump power by changing three high reflectivity fiber Bragg gratings. We also calculated the forward and backward power of 1st-order stokes, 2nd-order stokes, 3rd-order stokes. With the theoretical analysis, we optimize the cavity's reflectivity to get higher laser power output. The forward random laser exhibits larger gain, the backward random laser has lower gain. By controlling the value of angle-cleaved end fiber's reflectivity to 3×10^{-7}, when the high reflectivity increases from 0.01 to 0.99, the laser power increases, using this proposed configuration, the 1 178 nm random laser can be generated easily and stably.

Key words: Third-order stokes, Random fiber laser, Random distributed feedback, Rayleigh scattering

287. Switchable and tunable dual-wavelength Er-doped fiber ring laser with single-frequency lasing wavelengths

Haiwei Zhang, Wei Shi, Xiaolei Bai, Shijie Fu, Quan Sheng, Lifang Xue, Jianquan Yao

Fiber Lasers XV: Technology and Systems. International Society for Optics and Photonics, 2018, 10512: 105122C.

Abstract: We obtain a switchable and tunable dual-wavelength single-frequency Er-doped ring fiber laser. In order to realize single-longitudinal output, two saturable-absorber-based tracking narrow-band filters are formed in 3- meter-long un-pumped Er-doped fiber to narrow the linewidth via using the PM-FBG as a reflection filter. The maximum output power is 2.11 mW centered at 1 550.16 nm and 1 550.54 nm when the fiber laser operates in dual-wavelength mode. The corresponding linewidths of those two wavelengths are measured to be 769 Hz and 673 Hz, respectively. When the temperature around the PM-FBG is changed from 15 ℃ to 55 ℃, the dual-wavelength single-frequency fiber laser can be tuned from 1 550.12 nm to 1 550.52 nm and from 1 550.49 nm to 1 550.82 nm, respectively.

Key words: Single frequency, Dual wavelength, Fiber laser, Linewidth narrowing

288. 1-kW monolithic narrow linewidth linear-polarized fiber laser at 1 030 nm

Yang Xu, Qiang Fang, Xuelong Cui, Bowen Hou, Shijie Fu, Zhaoxin Xie, Wei Shi

Fiber Lasers XV: Technology and Systems. International Society for Optics and Photonics, 2018, 10512: 105122S.

Abstract: We demonstrate an all-fiberized, linear-polarized, narrow spectral linewidth laser system with kilowatts-level output power at 1 030 nm in master oscillator-power amplifier (MOPA) configuration. The laser system consists of a linear polarized, narrow linewidth (~28 GHz) fiber laser oscillator and two stages of linear-polarized fiber amplifiers. A 925 W linear-polarized fiber laser with a polarization extinction ratio (PER) of 15.2 dB and a spectral width of ~60 GHz at the central wavelength of 1 030.1 nm is achieved. Owing to the setting of the appropriate parameters for the laser, no indication of Stimulate Brillouin Scattering (SBS) is observed in the system. Moreover, thanks to the excellent quantum efficiency of the laser and the tightly coiling of the active fiber in the main amplifier, the mode instability (MI) is successfully avoided. As a result, the near diffraction-limited beam quality (M^2<1.3) is achieved.

Key words: Fiber optics amplifiers, Narrow linewidth, Single mode

289. Compact bi-direction pumped hybrid double-cladding EYDF amplifier

Xiaolei Bai, Quan Sheng, Shijie Fu, Zhaoxin Xie, Wei Shi, Jianquan Yao
CLEO: Science and Innovations, 2018: SM2N. 7.

Abstract: The double-cladding EYDF with different core diameter is connected by an asymmetric cladding light striper (CLS) to realize bi-direction pumping and increase the loss of counter-propagating light. The experimental results show that this amplifier has benefit to enhance the SBS threshold.
Key words: Double-cladding EYDF, Fiber optics amplifiers, Cladding light striper, Bi-direction pumping

290. 基于915 nm半导体激光单端前向抽运的单纤准单模2 kW全光纤激光振荡器

许阳,房强,谢兆鑫,李锦辉,崔雪龙,史伟
中国激光, 2018, 45(4): 0401003.

摘要: 设计了一种基于915 nm半导体激光单端抽运的单纤准单模全光纤激光振荡器,其工作波长为1 080 nm,输出功率可达2.02 kW。结合理论和实验,研究了增益光纤长度、受激拉曼散射(SRS)和输出功率之间的关系。通过对增益光纤长度进行优化,在保证大于2 kW激光功率的前提下,实现了高SRS抑制比的激光输出,输出激光中SRS功率占比约为0.8%。180 min内激光器的功率不稳定度小于±1%,光—光转换效率约为70%。通过合理设计光纤盘绕,有效抑制了光纤中光的高阶模式,在满功率输出时成功地获得了准单模激光(光束质量因子$M^2 \approx 1.5$),并对该激光器在激光切割中的应用进行了研究。
关键词: 激光器,激光谐振器,915 nm抽运,单端前向抽运,准单模激光器

291. 基于激光器内腔调制的低探测极限折射率传感系统

张超, 徐德刚, 石嘉, 钟凯, 李绪锦, 熊建波, 王胜年, 任宇琛, 陈霖宇, 李长昭, 刘铁根, 姚建铨

中国激光, 2018, 45(12): 1210003.

摘要：报道了一种基于光纤激光器内腔调制的低探测极限折射率传感系统。将基于单模—无芯—单模的全光纤多模干涉结构作为损耗调制器件插入光纤激光器环形腔内，采用激光器内腔调制技术获得了高灵敏度、高信噪比、窄半峰全宽的传感信号，从而实现了低探测极限的折射率测量。系统的折射率探测极限可达 7.3×10^{-7} RIU。该传感系统具有输出稳定、温度交叉敏感小的特点，在高精度生物化学传感、海洋环境监测等领域具有一定的应用潜力。

关键词：传感器, 折射率传感系统, 探测极限, 内腔调制, 光纤激光器

292. 基于 DOAS 的消防应急救援多气体快速遥感仪

姚建铨, 李润宸, 赵帆, 蔡兆文, 王洪兴, 徐德刚, 张帆, 乔旭

光电子 . 激光, 2018, 29(03):314-317.

摘要：在国内首次报道了可对 30 多种有毒气体连续监测的基于 DOAS(differential optical absorption spectra)技术的车载消防应急救援多气体快速遥感仪, 仪器采用长光程紫外差分吸收光谱法痕量气体分析技术, 光发射器和光接收器共用一个卡塞格林望远镜, 以减少大气紊动对测量数据的影响; 仪器的反射系统可以灵活地设置在任意需要区域, 分析系统设置在救援车内, 在需要长时间监测时, 避免了工作人员近距离暴露在危险环境中。多气体 DOAS 仪器特别适合各类气体泄漏事故的消防应急处理。

关键词：长光程, 差分光学吸收谱, 大气监测, 消防应急救援

293. 有机金属卤化物钙钛矿薄膜中的光诱导载流子动力学和动态带重整效应

赵婉莹，库治良，金钻明，刘伟民，林贤，戴晔，阎晓娜，马国宏，姚建铨

物理学报，2019, 68(1):018401.

摘要： 近年来有机-无机金属卤化物钙钛矿太阳能电池因具有光电能量转换效率高、制备工艺简单等优点，引起了学术界和产业界的广泛关注，其优异的光电特性逐渐在能源领域展现出独特的优越特性。在短短几年内，有机-无机混合物钙钛矿太阳能电池的能量转换效率已经高达23%，发展速度逐步赶上甚至超越了成熟的硅太阳能电池。本文利用飞秒瞬态吸收光谱，对二步法制备的 $(5\text{-AVA})_{0.05}(MA)_{0.95}PbI_3$ 和 $(5\text{-AVA})_{0.05}(MA)_{0.95}PbI_3/\text{Spiro-OMeTAD}$ 有机-无机卤化物钙钛矿薄膜材料的激发态动力学进行了对比研究，详细讨论了两种薄膜样品中的电荷载流子产生与复合机制。通过紫外-可见吸收光谱测得钙钛矿薄膜 $(5\text{-AVA})_{0.05}(MA)_{0.95}PbI_3$ 和 $(5\text{-AVA})_{0.05}(MA)_{0.95}PbI_3/\text{Spiro-OMeTAD}$ 的吸收光谱与 $CH_3NH_3PbI_3$ 钙钛矿薄膜材料的双价带结构相对应。从瞬态吸收光谱中，观察到760 nm附近的光致漂白信号，此时的载流子复合过程符合二阶动力学过程，而在约550—700 nm光谱范围内则是光诱导激发态吸收信号。实验结果表明，$(5\text{-AVA})_{0.05}(MA)_{0.95}PbI_3$ 钙钛矿薄膜样品中光生载流子主要的弛豫途径是自由电子和空穴的复合。抽运光激发样品使价带中的电子跃迁到导带，随着延迟时间的增加，电子和空穴复合，光谱发生红移现象。所观察到的带重整效应可以根据Moss-Burstein效应解释。相比较而言，$(5\text{-AVA})_{0.05}(MA)_{0.95}PbI_3/\text{Spiro-OMeTAD}$ 钙钛矿薄膜样品在光激发后电子和空穴分离，空穴迅速转移到空穴传输层，这将导致样品吸收度增加，漂白信号快速恢复，电子-空穴的复合不再对漂白信号的弛豫动力学起主导作用，同时也削弱了带重整现象。本文的实验结果对半导体有机-无机金属卤化物钙钛矿薄膜在光伏领域的应用具有重要意义，为今后高效、稳定的钙钛矿太阳电池的研究提供了参考。

关键词： 有机卤化钙钛矿，超快瞬态光谱，电子-空穴复合，带重整

294. Highly sensitive dual-wavelength fiber ring laser sensor for the low concentration gas detection

Xianchao Yang, Liangcheng Duan, Haiwei Zhang, Ying Lu, Guangyao Wang, JianquanYao

Sensors and Actuators B: Chemical, 2019, 296: 126637.

Abstract: A highly sensitive low concentration gas sensor based on dual-wavelength Erbium-doped fiber ring laser (EDFRL) is designed and demonstrated. The dual-wavelength lasing output is introduced by a tunable Fabry- Perot (F-P) filter and a fiber Bragg grating (FBG) due to the mode competition as they both share a same EDF gain medium. One wavelength at 1 530.37 nm serving as the sensing element is inserted by a hollow core photonic crystal fiber (HCPCF) gas cell which can bring in much more power change when the cavity loss slightly varies while the other wavelength at 1 532.38 nm acting as a reference. Experimental results show that the sensor can achieve a good linear response (R^2=0.986 4) to acetylene concentration variation and a minimum detection limit (MDL) of 10.42 ppmV at 20 s response time, which is the lowest to our best knowledge and improving 6.44 times than that of single-wavelength EDFRL based. Moreover, the absolute detected error induced by the power fluctuation (< 0.1 dB) is less than 1.78% over more than 300 s of observation.

Key words: Gas sensor, Low concentration acetylene detection, Dualwavelength, Fiber ring laser sensor, Highly-Sensitive

295. High-resolution temperature sensor based on intracavity sensing of fiber ring laser

Jia Shi, Fang Yang, Wei Xu, Degang Xu, Hua Bai, Cuijuan Guo, Yajie Wu, Shanshan Zhang, Tiegen Liu, Jianquan Yao

Journal of Lightwave Technology, 2019.

Abstract: The principle and design of high-resolution sensors are of great significance in the development of fiber-optic sensors. In this paper, a general method based on intracavity sensing of fiber ring laser (FRL) is proposed for high-resolution measurement, and a high-resolution temperature sensing system is experimentally demonstrated. The theoretical model for intracavity sensing of FRL is presented. In the sensing system, a fiber-optic Sagnac loop is inserted into an FRL for temperature measurement, which performs as the filter and sensing head simultaneously. Because the filter characteristics of the Sagnac loop is modulated by temperature, the intracavity intensity-modulated sensing is induced for the output power of the FRL. With the method based on intracavity sensing, the temperature resolution level of the sensing system has been improved from 10^{-3} ℃ to 10^{-6} ℃. The stability and measurement errors are also measured. The proposed method can be extensible to similar fiber-optic sensors to improve measurement resolution.

Key words: Fiber optics sensors, High resolution, Fiber lasers

296. Ultrasharp LSPR temperature sensor based on grapefruit fiber filled with a silver nanoshell and liquid

Xianchao Yang, Lianqing Zhu, Ying Lu, Jianquan Yao.
Journal of Lightwave Technology, 2019

Abstract: A grapefruit fiber (GF) based temperature sensor with ultrasharp localized surface plasmon resonance (LSPR) is designed and characterized by finite element method. The silver nanoshell (SNS), formed by a dielectric core and a silver shell together with chloroform own large thermo-optic coefficient is sealed into one large air hole of the GF as the core sensing region. Benefiting from the plasmon hybridization in SNS and birefringence, the designed sensor can support two separate ultrasharp resonance peaks with full width half maximum (FWHM) only 4.6 nm in orthogonal polarizations, which is the narrowest to our best knowledge and the highest figure of merit (FOM) of 0.43/℃ can be obtained. Results show that by following the shifts of two polarized peaks, dual channel temperature sensing can be realized simultaneously at 5℃-55℃ with the maximum sensitivity of -17.8 nm/℃ and -17.2 nm/℃, about 3-17 times higher than other same type works. Moreover, the influences of structure parameters of SNS like deposition position, number, diameter and length on the sensing performances are discussed in detail.

Key words: Grapefruit fiber, Ultrasharp localized surface plasmon resonance, Silver nanoshell, Dual channel temperature sensing

297. Simultaneous measurement of temperature and relative humidity based on a microfiber Sagnac loop and MoS$_2$

Yuting Bai, Yinping Miao, Hongmin Zhang, Jianquan Yao

Journal of Lightwave Technology, 2019

Abstract: A fiber-optic sensor that simultaneously measures relative humidity (RH) and temperature (T) based on a microfiber Sagnac loop (MSL) and MoS$_2$ nanosheets is proposed. The MSL is made from a microfiber coupler fabricated by hydrogen-oxygen flame heating technology. The sensor responses to RH and T are investigated by theoretical and experimental analyses. In our experiments, the RH sensitivities achieved are 176.6 pm/%RH for the wavelength shift and -0.017 dB/%RH for the intensity in the range from 60.6 to 78.6 %RH, and the T sensitivities are -123.5 pm/℃ and -0.079 dB/℃. The proposed sensor has potential applications in the simultaneous measurement of RH and T.

Key words: Fiber-optic sensor, Sagnac loop, MoS$_2$ nanosheets, Simultaneously sensing humidity and temperature

298. Multi-direction bending sensor based on supermodes of multicore PCF laser

Jia Shi, Fan Yang, Dexian Yan, Degang Xu, Cuijuan Guo, Hua Bai, Wei Xu, Yajie Wu, Jinjun Bai, Shanshan Zhang, Tiegen Liu, Jianquan Yao

Optics Express, 2019, 27(16): 23585−23592.

Abstract: A multi-direction bending sensor based on the 18-core photonic crystal fiber (PCF) is demonstrated in this paper. The design of the sensor is discussed theoretically and experimentally. The PCF serves as the laser gain medium and the sensing medium simultaneously in the fiber laser sensor. The operating wavelength of the proposed PCF laser sensor is about 1 032.32 nm. A CMOS image capture system is used to acquire the distribution of supermodes in the PCF laser. Based on the normalized intensity distribution of supermodes, six directions can be measured. The sensor also shows the ability to measure bending radius within 0.11 m. Then, the thermal effects of the bending sensor have been analyzed and the sensing system shows a low temperature crosstalk.

Key words: Multi-direction bending sensor, Photonic crystal fiber, 18-core

299. All-fiber seawater salinity sensor based on fiber laser intracavity loss modulation with low detection limit

Wei Xu, Xiang Yang, Chao Zhang, Jia Shi, Degang Xu, Kai Zhong, Ke Yang, Xujin Li, Weiling Fu, Tiegen Liu, Jianquan Yao

Optics Express, 2019, 27(2): 1529–1537.

Abstract: An all-fiber seawater salinity sensor based on intracavity loss-modulated sensing in a fiber ring laser is proposed and experimentally demonstrated. An optical fiber multimode interferometer, which is based on single-mode-no-core-single-mode fiber structure, is cascaded with a fiber reflector and used as a reflected sensing head to enhance loss-modulated depth. It is inserted in a fiber ring laser and the intracavity loss-modulated salinity sensing is induced for the fiber laser's output intensity. The salinity sensitivity is measured to be 0.1 W/‰ with a high signal-to-noise ratio more than 49 dB and narrow full width at half maximum less than 40 pm. The temperature cross-sensitivity characteristic and stability are also analyzed. Considering the errors from cross-sensitivity, stability and resolution of the photodetector, the detection limit of the sensor system is 0.002 3‰ (0.000 2 S/m), which is comparable to the most advanced commercial electronic salinity sensor.

Key words: Seawater salinity sensor, All fiber, Intracavity loss modulation, Fiber ring laser

300. Ultrafine frequency linearly tunable single-frequency fiber laser based on intracavity active tuning

Lufan Jin, Yonghua Wu, Mingxuan Cao, Zhenggeng Zhong, Liuchao Xu, Kun Wang, Jixing Lin, Chengyu Cai, Tong Xian, Ruiping Li, Liyuan Niu, Jianquan Yao

IEEE Photonics Journal, 2019.

Abstract: An ultrafine frequency linearly tunable single-frequency fiber laser based on the intracavity active tuning is demonstrated. Single-frequency operation can be achieved by an ultrashort cavity and ultranarrow polarization-maintaining fiber Bragg grating. A sub-nanometer-accuracy piezoelectric ceramic actuator with closed-loop mode has been employed to realize the cavity-length active tuning scheme, which can achieve a linear tuning range of >1 GHz with a 3.2-MHz/nm tuning resolution. The periodic tuning repeatability and repeat accuracy show that the scheme can realize linear frequency tuning with great tuning repeatability and high tuning precision.

Key words: Ultrafine, Linearly Tunable, Fiber Laser

301. A highly sensitive magnetic field sensor based on a tapered microfiber

Zelong Ma, Yinping Miao, Yi Li, Hongming Zhang, Bin Li, Yanpeng Cao, Jianquan Yao

IEEE Photonics Journal, 2018, 10(4): 1–8.

Abstract: A compact all-fiber magnetic field sensor based on the combination of a magnetic fluid and a nonadiabatic microfiber is experimentally demonstrated. By using a fusion splicer combined with an additional flame brushing, the nonadiabatic microfiber, with an abrupt taper, a small waist diameter and a long region, provides a strong evanescent field and a long sensitive distance. The result indicates that a sensitivity as high as 309.3 pm/Oe can be achieved in the range from 0 to 200 Oe. The designed nonadiabatic microfiber sensing structure suggests potential application as a tunable all-in-fiber photonic and other newly fashioned magneto-optical photonic devices.

Key words: Adiabatic microfiber, Magnetic fluid, Magnetic field sensor, Microfiber, Nonadiabatic microfiber

302. Simultaneous measurement of relative humidity and temperature using a microfiber coupler coated with molybdenum disulfide nanosheets

Yuting Bai, Yinping Miao, Hongmin Zhang, Jianquan Yao

Optical Materials Express, 2019, 9(7): 2846–2858.

Abstract: The simultaneous measurement of relative humidity (RH) and temperature with an optical fiber sensor is proposed based on a microfiber coupler (MFC) coated with a single layer of molybdenum disulfide (MoS_2) nanosheets. The MFC is fabricated using flame heating technology and has a waist diameter of approximately 6 μm and a waist length of approximately 3 mm. As the RH increases, the effective refractive index of MoS_2 varies as a result of electric charge transfer, the spectrum dip shifts toward longer wavelengths, and the transmission intensity of the spectrum decreases. As the temperature increases, the refractive index of the cladding of the fiber in the MFC waist region increases due to the thermo-optic effect, the spectrum dip shifts toward shorter wavelengths, and the transmission intensity of the spectrum decreases. The experimental results show that the RH sensitivities are 115.3 pm/%RH and -0.058 dB/%RH in the range of 54.0 - 93.2%RH. The temperature sensitivities are -104.8 pm/℃ and -0.042 dB/℃ in the range of 30 - 90 ℃. The proposed sensor is expected to be used for simultaneous measurement of RH and temperature in the field of biochemical analysis.

Key words: Relative humidity measurement, Temperature measurement, Simultaneous measurement, Microfiber coupler, Molybdenum disulfide nanosheets

303. Opening up dual-core microstructured optical fiber-based plasmonic sensor with large detection range and linear sensitivity

Nannan Luan, Haixia Han, Lei Zhao, Jianfei Liu, Jianquan Yao
Optical Materials Express, 2019, 9(2): 819−825.

Abstract: An opening up dual-core microstructured optical fiber based surface plasmon resonance sensor is numerically investigated for the measurement of a broad refractive index (RI) range. An open sensing channel is designed to facilitate the gold coating and accelerate the analyte infiltration. Results indicate that the sensitivity curve shows a nearly linear feature in two parts, and the maximal sensitivity is 4 900 nm/RIU when the RI of the analyte is close to that of the background material of the fiber. Moreover, the sensitivities in low RI range and the signal to noise ratio can be improved by introducing air holes into the core center.

Key words: Microstructured optical fiber, Opening up dual-core, Surface plasmon resonance sensor, Large detection range, Large linear sensitivity

304. High-sensitivity magnetic field sensor based on a dual-core photonic crystal fiber

Guangyao Wang, Ying Lu, Xianchao Yang, Liangcheng Duan, Jianquan Yao
Applied optics, 2019, 58(21): 5800−5806.

Abstract: A high-sensitivity magnetic field sensor based on a dual-core photonic crystal fiber has been designed with an extremely short device length of 2 000 μm in this paper. The two cores of the fiber are separated by one air hole filled with magnetic fluid. The sensitive properties are investigated by the full-vector finite element method. Simulation results illustrate that the highest sensitivity can reach −442.7 pm/Oe in the magnetic field strength range of 30−520 Oe. The photonic crystal fiber filled with magnetic fluid, serving as an excellent platform for magnetic field sensing, has great potential applications in complex environments, remote sensing, and real-time monitoring fields.

Key words: Magnetic field sensor, Photonic crystal fiber, Dual-core, High-sensitivity

305. Microfiber coupler with a Sagnac loop for water pollution detection

Lijiao Zu, Hongmin Zhang, Yinping Miao, Bin Li, Jianquan Yao
Applied optics, 2019, 58(21): 5859−5864.

Abstract: The measurement of chloride ion concentrations has been studied for the purpose of monitoring the quality of water resources. In this paper, a chloride ion sensor based on a microfiber coupler with a Sagnac loop is proposed. The microfiber coupler, which acts as the sensing unit and has a diameter of 10 μm and a length of 1 mm, is fabricated using the flame-brushing technique, and the two ends are connected to form a Sagnac loop, which acts as a reflector to enhance the reflection in the structure. Experimental results show that the sensitivity reaches a maximum of 423 pm/‰ and that the detection limit for the chloride ion concentration is 0.447‰ at a wavelength of 1 595 nm. The proposed sensor is characterized by a simple and easy manufacturing process, compact structure, and low cost; further, this sensing unit has great potential for applications in marine chloride detection and environmental safety monitoring, especially for monitoring building corrosion and water pollution.

Key words: Water pollution detection, Microfiber coupler, Sagnac loop, Chloride ion concentrations

306. Square-lattice alcohol-filled photonic crystal fiber temperature sensor based on a Sagnac interferometer

Guangyao Wang, Ying Lu, Xianchao Yang, Liangcheng Duan, Jianquan Yao
Applied optics, 2019, 58(8): 2132−2136.

Abstract: A square-lattice alcohol-filled photonic crystal fiber temperature sensor based on a Sagnac interferometer (SI) is designed and analyzed by the finite element method. Alcohol is filled in all air holes in the cladding. The temperature-sensing properties of the proposed fiber sensor are investigated. Simulation results exhibit the transmission spectrums of the fiber SI will shift with the change of temperature, because the birefringence of the alcohol-filled fiber will change under different temperatures. The temperature sensitivity is obtained from the fitting line of the temperature and resonant wavelength. The average sensitivity can reach to 16.55 nm/℃ in the range from 45 ℃ to 75 ℃. This designed fiber temperature sensor has advantages in simple structure and high sensitivity. It can be used to detect temperature in complex environments.

Key words: Temperature sensor, Photonic crystal fiber, Square-lattice, Alcohol-filled, Sagnac interferometer, Finite element method

307. Tunability of Hi-Bi photonic crystal fiber integrated with selectively filled magnetic fluid and microfluidic manipulation

Weiheng Wang, Yinping Miao, Zhen Li, Hongmin Zhang, Bin Li, Xiaoping Yang, Jianquan Yao

Applied optics, 2019, 58(4): 979–983.

Abstract: Optical fiber microfluidics technology can implement the mutual tune of the light field and fluid in micro-nano scale. In this paper, one core of high-birefringence photonic crystal fiber (Hi-Bi PCF) is used as a microfluidic channel. The birefringence of Fe_3O_4 nanofluid is experimentally and theoretically investigated by selectively infiltrating the magnetic fluid into the core of the Hi-Bi PCF. The presence of magnetic fluid alters the birefringence of the original Hi-Bi PCF and can be modulated by the intensity of the external magnetic field. The optical field distribution is simulated, and the birefringence of the Hi-Bi PCF with selective filling is approximately 6.672×10^{-4}. The experimental results show that the structure has a highly linear response to the external magnetic field from 0 Oe to 300 Oe, and the sensitivity is 16.8 pm/Oe with a high resolution of 1.19 Oe. Due to several advantages such as all-fiber compact structure, low transmission loss, and high linear response, this device can find various applications, including weak magnetic field measurement with high accuracy, optical fiber gyroscopes, and magneto-optic modulators. Particularly, it also has important significance to realize the all-fiber microfluidic chip laboratory.

Key words: Optical fiber microfluidics technology, Photonic crystal fiber, High-birefringence, Magnetic fluid, Microfluidic manipulation

308. 1.7-μm thulium fiber laser with all-fiber ring cavity

Junxiang Zhang, Quan Sheng, Shuai Sun, Chaodu Shi, Shijie Fu, Wei Shi, Jianquan Yao

Optics Communications, 2020, 457: 124627.

Abstract: A 1.7-μm thulium-doped all-fiber laser based on a ring-cavity configuration is demonstrated. The long wavelength lasing near the 1.9-μm thulium emission peak was suppressed using a wavelength division multiplexer and single-mode–multimode–single-mode (SMS) fiber device, both of which served as a short-pass filter instead of the grating devices usually used in 1.7-μm thulium fiber lasers. A stable hundred-milliwatt level 1 720-nm laser output with a narrow spectral linewidth on the order of gigahertz was obtained after the optimization of the output coupling, active fiber length, and SMS device. With a cascaded SMS filter, the output power was further scaled to 227 mW under a launched 1 570-nm pump power of 2.8 W, while the slope efficiency was 10.3%. The wavelength coverage could be further extended by using different SMS devices and applying tension on them.

Key words: 1.7 μm, Ring cavity, Short-pass filter

309. All-fiberized single-frequency silica fiber laser operating above 2 μm based on SMS fiber devices

Chaodu Shi, Shijie Fu, Guannan Shi, Shuai Sun, Quan Sheng, Wei Shi, Jianquan Yao

Optik, 2019, 187: 291–296.

Abstract: In this paper, a single-frequency thulium-doped silica fiber laser operating above 2 μm based on cascaded single-mode-multimode-single-mode (SMS) structures incorporating a Sagnac loop was experimentally demonstrated. Based on the analysis of the transmission properties of the SMS fiber structure and the mode selection characteristic of the Sagnac loop, two cascaded SMS fiber devices with the same multimode fiber length of approximately 3.25 cm were used in our laser system, to select the specific lasing wavelength above 2 μm, as well as enhance the suppression of lasing around 1 900 nm. A 6.5-m-long unpumped thulium-doped silica fiber was employed in the Sagnac loop as a saturable absorber to achieve single-longitudinal-mode operation. Single-frequency operation of the fiber laser at 2 004.9 nm has been achieved with a maximum output power of 20.9 mW.

Key words: Fiber laser, Single-frequency, Saturable absorber, Single-mode-multimode-single-mode structure

310. Hundred-watts-level monolithic narrow linewidth linearly-polarized fiber laser at 1 018 nm

Zhaoxin Xie, Qiang Fang, Yang Xu, Xuelong Cui, Quan Sheng, Wei Shi, Jianquan Yao

Optical Engineering, 2019, 58(10): 106106.

Abstract: We theoretically and experimentally demonstrated an all-fiber, hundred-watts-level, linearly-polarized, narrow spectral linewidth laser amplifier at a central wavelength of 1 018 nm based on master oscillator-power amplifier configuration, which is composed of a laser oscillator and one stage of the fiber amplifier. The laser system can generate 104-W output power with 3- and 20-dB spectral linewidth of ~0.073 and ~0.25 nm, respectively, and a higher polarization extinction ratio of ~17.89 dB at 1 018.3 nm was obtained. Theoretical analysis based on the rate equations was used to optimize the parameters of 1 018-nm ytterbium doped fiber laser system for the maximum suppression of amplified spontaneous emission (ASE). The ASE was well depressed based on the optimization for the parameters of the laser system including the seed power, seed spectrum, gain fiber length in the amplifier, etc. And ~27-dB signal-to-noise ratio was achieved at the maximum output power. The slope efficiency for the amplifier stage can reach 79%, and near-diffraction-limited beam quality (M_x^2 ~ 1.617 and M_y^2 ~ 1.635) was obtained.

Key words: 1 018 nm, Narrow linewidth, Linearly polarized, Hundred-watts-level, Amplified spontaneous emission

311. Highly sensitive chloride ion concentration measurement based on a multitaper modulated fiber

Lijiao Zu, Hongmin Zhang, Chengwei Fei, Yinping Miao, Bin Li, Jianquan Yao
Optical Engineering, 2019, 58(8): 086109.

Abstract: A chloride ion sensor based on a multitaper modulated fiber fabricated via arc discharge is proposed and studied experimentally and theoretically. The sensing unit consists of four periodic tapers with a diameter of 28.24 μm, a period of 676 μm, and a total length of 2.7 mm. The results show that the concentration sensitivity values are 170 pm/(g/L) at a wavelength of 1 303.60 nm and 220 pm/(g/L) at a wavelength of 1 389.84 nm when the chloride ion concentration is increased from 0 to 40 g/L. The proposed sensor is characterized by a simple manufacturing process, a compact structure, and a low cost, and this sensing unit has great potential for application in marine chloride detection and environmental safety monitoring, especially for monitoring building corrosion and water pollution.

Key words: Fiber-optic chemical sensor, Concentration measurement, Microfiber

312. A novel photonic crystal fiber sensor with three d-shaped holes based on surface plasmon resonance

Pibin Bing, Jialei Sui, Shichao Huang, Xinyue Guo, Zhongyang Li, Lian Tan, Jianquan Yao
Current Optics and Photonics, 2019, 3(6): 541–547.

Abstract: A novel photonic crystal fiber (PCF) sensor with three D-shaped holes based on surface plasmon resonance (SPR) is analyzed in this paper. Three D-shaped holes are filled with the analyte, and the gold film is deposited on the side of three planes. The design of D-shaped holes with outward expansion can effectively solve the uniformity problem of metallized nano-coating, it is beneficial to the filling of the analyte and is convenient for real-time measurement of the analyte. Compared with the hexagonal lattice structure, the triangular arrangement of the clad air holes can significantly reduce the transmission loss of light and improve the sensitivity of the sensor. The influences of the air hole diameter, the distance between D-shaped holes and core, and the counterclockwise rotation angle of D-shaped holes on sensing performance are studied. The simulation results show that the wavelength sensitivity of the designed sensor can be as high as 10 100 nm/RIU and the resolution can reach 9.9×10^{-6} RIU.

Key words: Photonic crystal fiber, Wavelength sensitivity, Surface plasmon resonance

313. Analysis of a photonic crystal fiber sensor with reuleaux triangle

Pibin Bing, Shichao Huang, Xinyue Guo, Hongtao Zhang, Lian Tan, Zhongyang Li, Jianquan Yao

Current Optics and Photonics, 2019, 3(3): 199−203.

Abstract: The characteristics of a photonic crystal fiber sensor with reuleaux triangle are studied by using the finite element method. The wavelength sensitivity of the designed optical fiber sensor is related to the arc radius of the reuleaux triangle. Whether the core area is solid or liquid as well as the refractive index of the liquid core contributes to wavelength sensitivity. The simulation results show that larger arc radius leads to higher sensitivity. The sensitivity can be improved by introducing a liquid core, and higher wavelength sensitivity can be achieved with a lower refractive index liquid core. In addition, the specific channel plated with gold film is polished and then analyte is deposited on the film surface, in which case the position of the resonance peak is the same as that of the complete photonic crystal fiber with three analyte channels being filled with analyte. This means that filling process becomes convenient with equivalent performance of designed sensor. The maximum wavelength sensitivity of the sensor is 10 200 nm/RIU and the resolution is 9.8×10^{-6} RIU.

Key words: Photonic crystal fiber, Wavelength sensitivity, Liquid core

314. Research of laser-induced underwater communication zoom optical system

Yueqi Wang, Yuan Hu, Yating Zhang, Jianquan Yao, Zhiliang Chen, Yifan Zhang

14th National Conference on Laser Technology and Optoelectronics (LTO 2019). International Society for Optics and Photonics, 2019, 11170: 111702B.

Abstract: Based on the loss of laser-induced acoustic signal caused by the undulation of sea level, which affects the communication quality of laser-induced underwater communication system, a laser-induced underwater communication zoom optical system with a focusing position of 10 km and a focusing position variation range of ±500 m is designed. The theoretical relationship between the adjustment distance and the focusing position is established and the far-field distribution of the Gaussian beam focusing system is simulated. The results show that the energy density of the far-field spot is concentrated, and the spot energy is the highest at the focusing position. The sea level condition was simulated, and the indoor pool experiment was carried out to verify the communication performance of the laser-induced acoustic communication system. The experimental results show that the waveform of the laser-induced acoustic signal and the rising edge of the electrical signal are completely consistent, there is no loss of the laser-induced acoustic signal, and the underwater communication quality is good. It provides a viable reference for laser-induced underwater communication systems with adaptive zoom.

Key words: Laser-induced underwater communication, Adaptive zoom, Gaussian beam far field distribution, Optical design

315. Numerical investigation of high efficiency random fiber lasers at 1.5 μm

Zhaoxin Xie, Wei Shi, Quan Sheng, Jianquan Yao

Fiber Lasers XVI: Technology and Systems. International Society for Optics and Photonics, 2019, 10897: 108970P.

Abstract: High power and high efficiency random fiber laser working at 1.5 μm with mixed Erbium-Raman gain at different pump schemes were theoretically discussed. The numerical calculations based on rating equations were established to analysis the laser performance of Erbium-Raman gain random fiber laser by different pump schemes. The optical-to-optical conversion efficiency reached 90% when pump power is 100 W by forward pump scheme in the 1.5 μm regime. In our simulations results we found that forward pump scheme is better than backward pump scheme and bidirectional pump scheme in high power 1.5 μm random laser, besides forward pump scheme is better than the other two pump schemes because of its simple, high efficiency and compact, we also found that long Er-doped fiber(EDF) fiber length had good effect on the laser efficiency, the fiber length of single mode fiber(SMF) from 200 m to 1km had little effect on laser efficiency. This theoretical analysis of 1.5 μm random fiber laser can be a useful design setup for experiment and various laser wavelength could be achieved by the seed of 1.5 μm random fiber laser. Our work provided an effective thinking for high efficiency and high-power random fiber laser at 1.5 μm which could be used for fiber sensing, research field and laser lidar.

Key words: Random fiber laser, High power, High efficiency, Erbium-doped gain, Forward pump scheme

316. Thulium doped silica fiber laser operating in single-longitudinal-mode at a wavelength above 2 μm

Chaodu Shi, Shijie Fu, Guannan Shi, Wei Shi, Quan Sheng, Jianquan Yao

Fiber Lasers XVI: Technology and Systems. International Society for Optics and Photonics, 2019, 10897: 1089709.

Abstract: In this paper, a single frequency thulium doped silica fiber laser operating above 2 μm based on cascaded single-modemultimode-single-mode (SMS) fiber structures incorporating a Sagnac loop was experimentally demonstrated. Based on the theoretical analysis of the trans mission properties of the SMS fiber structure and mode selection characteristic of Sagnac loop, two cascaded SMS fiber devices with same multimode fiber lengths around 3.25 cm were used in our laser system to select the specific lasing wavelength above 2 μm as well as enhance the suppression of lasing around 1 900 nm. And 6.5 m unpumped TDF was employed in Sagnac loop as saturable absorber to achieve single-frequency operation. Single-longitudinal-mode (SLM) operation of the fiber laser operating at 2 004.05 nm was achieved with a signal to noise ratio (SNR) more than 60 dB.

Key words: Fiber laser, Single frequency, Thulium doped, Saturable absorber

317. Tunable CW all-fiber optical parametric oscillator based on the cascaded single-mode-multimode-single-mode fiber structures

Zhao Tang, Shijie Fu, Chaodu Shi, Quan Sheng, Wei Shi, Jianquan Yao

Fiber Lasers XVI: Technology and Systems. International Society for Optics and Photonics, 2019, 10897: 108971W.

Abstract: Aiming at the problem of high-ring-cavity losses caused by the use of a high-insertion-loss filter components to tune the output wavelengths in a continuous-wave (CW) all-fiber optical parametric oscillator (FOPO), this paper proposes a low ring-cavity-loss tunable CW FOPO based on the cascaded single-mode-multimode-single-mode (SMS) fiber structures. By applying an axial pulling force to the SMS fiber device to adjust the transmission spectrum of the filter device, the cascaded SMS fiber devices were used to achieve a tunable CW output of the double sidebands in the range of 1 494~1 501 nm and 1 638~1 629 nm. Among them, the insertion losses of the cascaded SMS fiber devices at the selected wavelengths are less than 1 dB, and the total FOPO ring-cavity losses are no more than 5 dB.

Key words: Lasers, fiber, Fiber optics amplifiers and oscillators, Nonlinear optics, Four-wave mixing, MMI

318. Single-frequency, ultra-narrow linewidth hybrid brillouin-thulium fiber laser based on In-band pumping

Chaodu Shi, Shijie Fu, Quan Sheng, Wei Shi, Jianquan Yao

CLEO: Science and Innovations, 2019: SM1L. 3.

Abstract: A hybrid Brillouin/thulium fiber laser with an ultra-narrow linewidth of 0.93 kHz was demonstrated, with the output coupling, cavity Q factor and pumping scheme optimized for the narrow linewidth.

Key words: Hybrid Brillouin-thulium fiber laser, Ultra-narrow linewidth, Single-frequency

319. An FBG-based high-resolution temperature sensor through measuring the beat frequency of single-frequency ring fiber laser

Liangcheng Duan, Wei Shi, Haiwei Zhang, Xianchao Yang, Ying Lu, Jianquan Yao

CLEO: Science and Innovations, 2019: JTh2A. 90.

Abstract: A high-resolution temperature sensor based on narrow-linewidth (<1 kHz) single frequency ring fiber laser was investigated using optical heterodyne spectroscopy technology. Temperature resolution of ~5×10^{-3} ℃ was achieved in our experiment.

Key words: Temperature sensor, High-resolution, Single frequency ring fiber laser, Optical heterodyne spectroscopy

320. 有源光纤中稀土离子激光上能级寿命测量的研究

刘恒,张钧翔,付士杰,盛泉,史伟,姚建铨

物理学报, 2019, 68(22): 224202.

摘要：提出了一种基于脉冲光纤激光放大器能量特性测量有源光纤中稀土离子上能级寿命的方法。根据光纤激光器速率方程,能够确定有源光纤中反转粒子数储能随抽运功率和时间的变化关系;实验测量不同种子光脉冲重复频率下放大器输出单脉冲能量的变化,可以反映出反转粒子数随时间的变化情况,进而根据理论模型得到激活离子的激光上能级寿命。实验搭建了 1.06 μm 掺镱 (Yb^{3+}) 光纤激光放大系统对该测量方法理论模型的合理性进行了验证,对几种常见商用掺 Yb^{3+} 有源光纤激光上能级寿命分别进行了多次测量和数据处理,测量结果以及变化趋势与其他相关报道中的结果相符。

关键词：上能级寿命,有源光纤,光纤激光器,光纤放大器

321. 重复频率 1.2GHz 皮秒脉冲全光纤掺镱激光器

韦小乐,魏淮,盛泉,付士杰,史伟,姚建铨

光子学报, 2019, 48(11): 1148015.

摘要：研究实现了基于半导体可饱和吸收体被动锁模的高重频全光纤掺镱皮秒脉冲激光器。种子源采取环形腔结构,当抽运功率为 112 mW 时,获得了稳定的锁模脉冲激光,其中心波长为 1 064.1 nm,3 dB 谱宽为 3.6 nm,脉冲宽度为 4.2 ps,重复频率为 19.2 MHz。受限于谐振腔长度,光纤激光器重复频率很难得到进一步提高。因此设计并搭建了一种基于分束器和延时光纤的全新低损耗高重频脉冲调制器,将种子激光重复频率提高到 1.2 GHz。该设计有效降低了脉冲在耦合过程中的能量损耗,为提高全光纤超短脉冲激光器重复频率提供了新途径。

关键词：掺镱光纤激光器,被动锁模,高重频脉冲调制器,超短脉冲,耗散孤子

322. 基于 HCPCF SERS 传感器的吡啶痕量检测

邸志刚, 王彪, 杨健俊, 贾春荣, 张靖轩, 姚建铨, 陆颖

红外与激光工程, 2019, 48(S2): 857–865.

摘要："民以食为天,食以安为先",在各种农药、食品添加剂广泛应用的背景下,非法食品添加剂对食品安全形势造成严重影响,非法食品添加剂痕量定性检测有重大意义。为实现对吡啶的定性检测,提出了一种基于 HCPCF(Hollow Core Photonic Crystal Fiber)SERS 的传感系统,采用银纳米颗粒制作 SERS 基底,设计同向收集 SERS 信号装置,通过对 HCPCF 拉锥并实现选择性填充,利用光谱仪采集输出信号的检测方案。经对浓度为 0.004 975% 的液体样本进行试验,检测出明显的吡啶特征 SERS 谱。实验结果表明该传感系统能够实现对液体样本中吡啶痕量的定性检测。

关键词：食品安全, 光子晶体光纤传感, 纳米颗粒, SERS, 痕量检测

323. 1mJ 窄线宽掺镱脉冲光纤放大器

石锐, 丁欣, 刘简, 姜鹏波, 孙冰, 白云涛, 王靖博, 赵蕾, 张贵忠, 姚建铨

红外与激光工程, 2019, 48(S1): 874–879.

摘要：通过将固体激光器种子源与光纤放大技术相结合的方式,报道了一个单脉冲能量约为 1 mJ,同时窄线宽保持在约 0.1 nm 的大模场面积的掺镱脉冲光纤放大器。不同于传统固体种子源一级大芯径光纤放大结构,所采用的两级掺镱双包层光纤 (YDCF) 放大结构空间耦合调节更简单。放大过程中采用了多种非线性效应抑制方法,放大输出光中未出现明显的 SBS、SRS 等非线性效应,信噪比超过 30 dB。放大后平均功率 10.07 W,相比 0.38 W 的种子光提高了 14.23 dB。在 10 kHz 重复频率下,单脉冲能量约 1 mJ,输出光脉宽 50 ns,放大输出光束质量 M^2=3.72。所实现的毫焦量级窄线宽脉冲光纤放大器可运用于激光遥感、激光测量及非线性频率变换等应用领域。

关键词：脉冲放大, 掺镱光纤放大器, 固体激光器种子源, 窄线宽大能量

324. 基于 MMI 滤波器的可调谐连续光全光纤 OPO

唐钊, 张钧翔, 付士杰, 白晓磊, 盛泉, 史伟, 姚建铨

红外与激光工程, 2019, 48(5): 520002.

摘要：在连续光全光纤光学参量振荡器 (FOPO) 中，目前主要利用可调滤波器 (TBPF) 等高插入损耗的滤波器件进行边带光输出波长的调谐，这种方式所引起的高环形腔损耗限制了 FOPO 输出性能的进一步提升。为解决此问题，提出了基于多模干涉 (MMI) 滤波器的低腔损耗可调谐连续光 FOPO。通过选取不同长度和纤芯尺寸的多模光纤制作级联单模-多模-单模光纤 (SMS) 作为滤波器件，使其在选定波长处的插入损耗小于 1 dB，FOPO 环形腔的总损耗不大于 5 dB，并通过对 SMS 器件施加轴向拉力的方式调节滤波器件的透射谱，实现了 1 494~1 501 nm 和 1 638~1 629 nm 范围内的双边带可调谐连续光输出。

关键词：光纤光学参量振荡器, 四波混频, 多模干涉效应, 损耗

第三章　太赫兹技术

第一节 太赫兹辐射源

1. THz 辐射的研究和应用新进展（New research progress of THz radiation）

姚建铨,路洋,张百钢,王鹏

光电子·激光, 2005, 16(4): 503–507.

摘要： THz 频段是一个非常具有科学价值但尚未开发的电磁辐射区域，它的研究涉及物理学、光电子学及材料科学等，它在成像、医学诊断、环境科学、信息通信及基础物理研究领域有着广阔的应用前景和应用价值。当今，获得 THz 波的方法及 THz 波的探测研究是 THz 研究领域的前沿，更是重点。本文综述了 THz 波的特点、应用领域及发展状况，阐述了 THz 波的产生方法其探测方法。

关键词： THz 辐射, THz 波的产生, THz 波的探测

2. Analysis of surface-emitted terahertz-wave difference frequency generation in slant-stripe-type MgO-doped periodically poled lithium niobate

Y. Lu, B. G. Zhang, Y. Z. Yu, D. G. Xu, H. Liu, B. Sun, P. Zhao, Z. Wang, P. Wang, J. Q. Yao

2006 Joint 31st International Conference on Infrared Millimeter Waves and 14th International Conference on Teraherz Electronics, 2006: 103–103.

Abstract: The theory of THz-wave generation (3.29 THz) via difference frequency generation (DFG) of near infrared radiations (1 318.8 nm and 1 338.2 nm) by a simultaneous dualwavelength lasing (SDWL) of an Nd:YAG laser in slant-stripetype MgO-doped periodically poled lithium niobate (MgO:PPLN) was presented. It is shown that the slant-stripetype MgO:PPLN is capable of realizing the quasi-phasematching in two perpendicular direction of optical and THzwave propagation.

Key words: THz-wave generation, Difference frequency generation (DFG), MgO:PPLN

3. Study of tunable terahertz-wave generation in isotropic semiconductor crystals based on dual-wavelength KTP-OPO operating near degenerate point

B. Sun, J. Q. Yao, Z. Wang, P. Zhao, Y. Lu, H. Liu, D. G. Xu

Conference digest of the **2006** *Joint* **31**ST *International Conference on Infrared and Millimeter Waves and* **14**TH *International Conference on Terahertz Electronics* , 2006: 107–107.

Abstract: Continuously tunable coherent terahertz-wave generation from 0.1 to 2.7 THz can be achieved in isotropic semiconductor nonlinear crystals by difference frequency mixing. We experimentally demonstrated a high-power, narrow-linewidth, angle-tuned pulsed dual-wavelength KTP-OPO operating near the degenerate point, which was used as the pump source for terahertz-wave generation. We theoretically studied the feasibility of the cross-Reststrahlen band dispersion compensation phase matching in the collinear optical mixing technique in this process. A theoretical analysis of perfect phase-matching conditions and coherence length for the ZnTe crystal was presented, according to its optical dispersion property.

Key words: Tunable terahertz-wave generation, Difference frequency mixing, Nonlinear crystals, KTP-OPO, Phase matching

4. Theoretical investigation of dual-wavelength terahertz wave generation based on slant-stripe-type periodic poled lithium niobate crystal

P. Zhao, B. G. Zhang, Y. Z. Yu, D. G. Xu, B. Sun, Y. Lu, H. Liu, T. L. Zhang, Z. Wang, P. Wang, J. Q. Yao

Conference digest of the 2006 Joint 31ST International Conference on Infrared and Millimeter Waves and 14TH International Conference on Terahertz Electronics, 2006: 109–109.

Abstract: We propose a method for generating dualwavelength terahertz radiation with a surface emitted configuration based on slant-stripe-type periodic poled lithium niobate (PPLN) Crystal. The crystal is designed to be phase reserved for two difference frequency generation processes so that the predesigned dual-wavelength terahertz radiation at 1.43 THz and 2.07 THz can be achieved simultaneously. The estimated effective nonlinear coefficient as a function of the terahertz radiation frequencies is also included in our discussion. Theoretical analysis shows that the central frequency of the predesigned terahertz radiation will shift inconspicuously due to the permitted fabrication inaccuracy of the crystal period.

Key words: Dual-wavelength terahertz wave generation, PPLN, Surface emitted configuration

5. Theoretical study of dual-wavelength PPKTP-OPO as a source of DFG THz-wave

Zhuo Wang, Bo Sun, Yuye Wang, Baigang Zhang, Yizhong Yu, Degang Xu, Huan Liu, Peng Wang, Jianquan Yao

Conference digest of the 2006 Joint 31ST International Conference on Infrared and Millimeter Waves and 14TH International Conference on Terahertz Electronics, 2006: 105–105.

Abstract: In this paper, we theoretically investigate the dual-wavelength periodically poled KTP optical parametric oscillator (PPKTP-OPO) operating near degeneracy. According to calculation, the dual-wavelength operation can be achieved by quasi-phase matching with a periodically poled structure. The tunable property of the grating period of the PPKTP crystal with temperature tuning determines that, quasi-CW or continuous tunable THz-wave generation can be achieved by difference frequency generation using a dual-wavelength PPKTP-OPO. Compared with a bulk KTP-OPO operating near degeneracy in theory, the gain property of PPKTP OPO is in better performance, contributing to a higher output power of THz generation.

Key words: Terahertz wave generation, PPKTP-OPO, Quasi-phase matching

6. Widely tunable, dual-signal-wave optical parametric oscillator for terahertz generation by using two periodically poled crystals

T. L. Zhang, X. Y. Zhu, P. Zhao, F. Ji, Y. Lu, P. Wang, B. G. Zhang, J. Q. Yao

Conference digest of the 2006 Joint 31ST International Conference on Infrared and Millimeter Waves and 14TH International Conference on Terahertz Electronics, 2006: 104–104.

Abstract: Dual-signal-wave optical parametric oscillators (DSW OPOs) around 1.5 μm by using two periodically poled crystals was demonstrated experimentally. The widely tunable wavelength interval from 2.5 nm to 69.1 nm was obtained. It can be used for generation of terahertz (THz) waves from 0.3-8.9 THz by different frequency generation. With the incident pump power of 3 W at 50 kHz, the DSW OPO generated 169.6 mW maximum average power of dual signal waves at 1 489.2 nm and 1 558.3 nm.

Key words: DSW OPO, Tunable terahertz-wave generation, Periodically poled crystal

7. 基于光学方法的太赫兹辐射源

孙博,姚建铨

中国激光, 2006, 33(10): 1349–1359.

摘要：太赫兹波技术在物理、化学、生命科学等基础研究学科,以及医学成像、安全检查、产品检测、空间通信、武器制导等应用学科都具有重要的研究价值和应用前景,而太赫兹辐射源正是太赫兹技术发展的关键部分。概述了基于光学方法产生太赫兹辐射的几种常用方法,着重叙述了利用非线性光学差频技术和基于横向晶格振动光学模受激电磁耦子散射过程的太赫兹参量振荡技术工作原理,以及目前的研究状况,并对这两种方法产生太赫兹波辐射源未来的发展方向进行了展望。

关键词：非线性光学,太赫兹辐射,太赫兹波的产生,非线性光学差频,相位匹配,太赫兹波参量振荡器,电磁耦子

8. 利用各向同性半导体晶体差频产生可调谐 THz 辐射的理论研究

孙博,姚建铨,王卓,王鹏

物理学报, 2007, 56(3): 1390–1396.

摘要：理论研究了利用剩余射线带色散补偿相位匹配原理,在Ⅲ-Ⅴ族和Ⅱ-Ⅳ族光学各向同性的半导体非线性晶体中差频产生可调谐 THz 波的可行性问题。根据这些半导体材料的色散特性,并以近简并点双共振 KTP-OPO 的可调谐相干双波长输出作为差频抽运源,对它们的相位匹配能力、差频增益特性、品质因数以及差频过程中的相干长度进行了理论分析和计算,确定了 ZnTe 晶体是在共线相位匹配情况下较为理想的 THz 波差频晶体,而 InP 晶体则更适合用于非共线相位匹配情况。

关键词：非线性光学,THz 辐射,差频,各向同性半导体晶体

9. 光泵重水气体产生 THz 激光的半经典理论分析

何志红，姚建铨，时华锋，黄晓，罗锡璋，江绍基，王鹏

物理学报，2007, 56(10): 5802–5807.

摘要： 从半经典密度矩阵理论出发，采用三能级系统模型对光泵重水气体产生太赫兹激光进行了理论分析，求解得到了脉冲光泵重水气体分子产生太赫兹激光过程中激光信号增益系数 G_s 和抽运光信号被吸收系数 G_p 的表达式，通过迭代法对太赫兹激光信号的输出光强进行了数值计算，理论计算得到的频谱特性曲线完全符合受激 Raman 辐射的频谱特性，即频谱宽度较大、输出光强随抽运失谐量的改变而变化明显等特征。在脉冲激光抽运受激 Raman 辐射过程中，工作介质 D_2O 气体分子的偶极矩由于受到抽运脉冲光场的扰动发生变化，在频谱特性曲线中表现为受激辐射 THz 信号的谱线发生了分裂。理论计算结果与已报道的实验结果能较好地相符。

关键词： 太赫兹激光，半经典理论，重水气体分子

10. 抽运光强度对光学抽运重水气体产生 THz 激光的影响分析

何志红，姚建铨，时华峰，黄晓，罗锡璋，江绍基，李建荣，王鹏

物理学报，2007, 56(11): 6451–6456.

摘要： 利用半经典理论，对脉冲光学抽运重水气体产生 THz 激光信号过程进行分析，对其中抽运光强度与输出 THz 信号光之间的关系进行了数值计算和求解，结果表明，THz 信号出射光强度跟抽运源入射光强度之间不满足简单的线性关系，而是呈现高阶的非线性关系。在工作介质腔长、气压和工作温度一定的条件下，存在最佳抽运光强度，在一定的抽运光强范围内，THz 信号出射光强度与抽运源入射光强度的关系呈现近似线性的增长关系，抽运光能量和信号光能量之间的转换效率相对较高，当抽运光强超过一定值时，由于瓶颈效应的发生会导致 THz 输出信号的逐步减弱，并产生一定的频率调谐范围。

关键词： 太赫兹（THz），半经典理论，受激 Raman 发射，重水气体分子

11. 紧凑型超辐射光泵重水气体 THz 激光器的研制

何志红, 姚建铨, 任侠, 罗锡璋, 时华峰, 江绍基, 王鹏

光电子·激光, 2008, 19(1): 34–37.

摘要: 研制了一种紧凑型超辐射光泵重水气体 THz 激光器系统, 并通过改变 THz 激光激活室的长度, 对 THz 信号出射光强度与信号增益长度之间的关系进行了实验研究; 利用半经典理论, 对两者之间的关系进行了理论计算和求解。实验和理论计算结果表明, THz 信号出射光强度跟信号增益长度之间存在非线性的关系。在工作气压、温度和泵浦光强度一定的条件下, 激光工作室长度存在最佳值, 此时介质对信号光的增益和吸收达到平衡, 出射的 THz 信号光强最大, 继续延长介质工作长度将会导致 THz 输出信号的逐步减弱。根据以上结果可以设计确定相关 THz 激光工作室的最佳长度。

关键词: 太赫兹激光, 半经典理论, 受激 Raman 发射, 重水气体分子

12. The numerical calculation and analyze of the pulse-laser pumped D₂O gas Terahertz laser

He Zhihong, Yao Jianquan, Shi Huafeng, Luo Xizhang, Li Jianrong, Jiang Shaoji, Wang Peng

2007 Joint 32nd International Conference on Infrared and Millimeter Waves and the 15th International Conference on Terahertz Electronics, 2007: 470–471.

Abstract: By using the semi-classical density matrix theory, the pulse-laser pumped D_2O gas THz laser was calculated numerically by iteration method and analyzed. According to the result, the THz radiation output spectral characteristic curves were obtained, and the relation between the intensity of pumping laser and the output Tera-Hz laser was studied.

Key words: D_2O gas molecule, Semi-classical theory, Stimulated Raman emission, Terahertz

13. Effects of pump source on spectra of optically pumped sub-millimeter wave laser

Yangxiang Bao, Xiao Huang, Zhihong He, Xizhang Luo
International Journal of Infrared and Millimeter Waves, 2006, 27: 1315–1322.

Abstract: The output power intensity of optically pumped sub-millimeter wave laser(OPSMMWL) increased with increasing of power intensity of pump source, till to its maximum; and decreased with increasing of the frequency off-set of pump source, if frequency off-set reached or exceeded its Raman threshold, no SMMWL could be obtained; the infrared pumping laser with collision broadening was considered in our computing model, which made our theoretical calculations agreed with the practical. Spectra of CO_2-9R(16) pumped NH_3 90.4 μm laser was used to be a practical case.
Key words: Optically pumped sub-millimeter wave laser(OPSMMWL), Pump source, Spectra

14. Theoretical study of phase-matching properties for tunable terahertz-wave generation in isotropic nonlinear crystals

Sun Bo, Yao Jianquan, Zhang Baigang, Zhang Tieli, Wang Peng
Optoelectronic Letters, 2007, 3(2): 152–156.

Abstract: Theoretical investigation of the phase-matching properties for the tunable coherent terahertz wave generation in the isotropic semiconductor nonlinear materials of CdTe and GaP is presented according to their optical dispersion properties, based on a pulsed dual-wavelength KTP-OPO operating near degenerate point in the process of the difference frequency mixing. The cross-Reststrahlen band dispersion compensation phase-matching technique involved in this interaction is introduced concisely. The optimum lengths of DFG crystals are decided according to the minimum coherence lengths of the crystals in the given pump wavelength ranges. The results of analysis and calculation provide the useful and necessary theoretical basis and guidance for the future experiments.
Key words: Phase matching, Tunable coherent terahertz wave generation, Dual-wavelength KTP-OPO, Difference frequency mixing

15. ZnGeP$_2$ 晶体差频产生 THz 波的研究

王卓, 姚建铨

科学技术与工程, 2007, 7(13): 3101–3103.

摘要: 为利用差频技术获得窄带、低重复频率的 THz 脉冲输出, 将双波长 KTP-OPO 作为泵浦源, ZnGeP$_2$ 作为差频晶体时, 在满足 type II 相位匹配条件下, 可以实现 (0.5-2.7)THz, 峰值功率为 W 量级的 THz 波输出。

关键词: ZnGeP$_2$ 晶体, 差频, 太赫兹, 双波长 KTP-OPO

16. Study of optimal gas pressure in optically pumped D$_2$O gas terahertz laser

He Zhihong, Yao Jianquan, Ren Xia, Yang Yang, Luo Xizhang, Wang Peng

Photonics Asia 2007, · *Proceedings of SPIE* 2007, 6840: 684004.

Abstract: Heavy water vapor (D$_2$O gas) which owns special structure properties, can generate terahertz (THz) radiation by optically pumped technology, and its 385 μm wavelength radiation can be widely used. In this research, on the base of semi-classical density matrix theory, we set up a three-level energy system as its theoretical model, a TEA-CO$_2$ laser 9R (22) output line (λ=9.26 μm) acted as pumping source, D$_2$O gas molecules were operating medium, the expressions of pumping absorption coefficient G_p and THz signal gain coefficient G_s were deduced. It was shown that the gain of THz signal was related with the energy-level parameters of operating molecules and some operating parameters of the THz laser cavity, mainly including gas pressure, temperature etc.; By means of iteration method, the output power density of THz pulse signal was calculated numerically as its initial power density was known; Changing the parameter of gas pressure and keeping others steady, the relationship curve between the output power intensity (Is) of Tera-Hz pulse laser and the operating D$_2$O gas pressure (P) was obtained. The curve showed that the power intensity (Is) increased with gas pressure (P) in a certain range, but decreased when the pressure (P) exceeded some value because of the bottleneck effect, and there was an optimal gas pressure for the highest output power. We used a grating tuned TEA-CO$_2$ laser as pumping power and a sample tube of 97cm length as THz laser operating cavity to experiment. The results of theoretical calculation and experiment matched with each other.

Key words: Terahertz laser, Semi-classical theory, D$_2$O gas molecule, Gas pressure

17. Tera-Hz radiation source by deference frequency generation (DFG) and TPO with all solid state lasers

Jianquan Yao, Xu Degang, Sun Bo, Liu Huan

THE FIRST SHENZHEN INTERNATIONAL CONFERENCE ON ADVANCED SCIENCE AND TECHNOLOGY – Terahertz Radiation Science and Technology (1st SZCAST' 2007), Shenzhen, China, 2007

Abstract: The terahertz technique has attracted much attention from a variety of applications in fundamental and applied research field, such as physics, chemistry, life sciences, medical imaging, safety inspection, radio astronomy, modern communication, weapon guidance and so on. The technological progress of terahertz radiation source plays an important role in promoting the development of various terahertz technique and the related cross subjects. The generation of high-power, coherent, widely tunable, narrow-band terahertz wave, based on the process of the difference frequency generation [1] and the tera-parametric oscillation [2-5] in a polar crystal respectively, is expected to provide a promising terahertz radiation source with the obvious and exclusive advantages of compactness, simplicity for tuning, operation at room temperature and so on, which causes great research interest among the researchers all over the world. However, the research of this potential THz-wave generation technique is still in its infancy, and few relevant theoretical or experimental studies were reported in recent years.

Key words: Terahertz, Terahertz radiation source

18. 基于 GaSe 和 ZnGeP$_2$ 晶体差频产生可调谐太赫兹辐射的理论研究

刘欢, 徐德刚, 姚建铨

物理学报, 2008, 57(9): 5662-5669.

摘要: 基于非线性光学频率变换理论, 采用已报道的利用非线性光学差频方法产生可调谐太赫兹波的实验条件作为理论分析的实验模型, 计算模拟出在不同相位匹配条件下, GaSe 和 ZnGeP$_2$ 晶体差频的相位匹配角、走离角、允许角和有效非线性系数, 并对计算结果进行了分析比较, 总结出对应输出不同太赫兹波长的最佳相位匹配方式。计算结果为利用非线性晶体差频产生可调谐太赫兹辐射的实验研究提供深入和全面的理论基础。

关键词: 太赫兹波, GaSe 晶体, ZnGeP$_2$ 晶体, 差频

19. 产生太赫兹辐射源的 Nd:YAG 双波长准连续激光器

郑芳华, 刘欢, 李喜福, 姚建铨

中国激光, 2008, 35(2): 200-205.

摘要: 产生太赫兹波辐射的方法可分为电子学和光子学两大类。在光子学领域, 非线性光学差频方法是获取高功率、低成本、便携式、室温运转太赫兹波的主要方法之一。实验研究了激光二极管 (LD) 端面抽运 Nd:YAG 1 319 nm/1 338 nm 双波长准连续线偏振运转激光器, 理论计算了输出双波长在非线性晶体 DAST(4-N,N-dimethylamino-4′-N′-methyl—stilbazolium tosylate) 中差频产生太赫兹辐射的平均功率。在重复频率 50 kHz 时, 双波长激光平均输出功率达到 2.22 W, 斜率效率 12.72%, 线偏振度 0.983, 脉冲宽度 71.91 ns。M^2 因子仅为 1.165, 不稳定性小于 0.487%。根据非线性差频理论, 计算出可在 1mm 厚 DAST 晶体中获得 4.71 mW 的 3.23 THz 高相干性太赫兹波辐射。这两条非常接近的谱线为进一步通过非线性光学差频方法获得高相干性太赫兹波提供了理论基础。

关键词: 激光器, Nd:YAG 双波长激光器, 非线性光学差频, 太赫兹波

20. 紧凑型超辐射光泵重水气体 THz 激光器的研制

何志红,姚建铨,任侠,罗锡璋,时华峰,江绍基,王鹏

光电子·激光, 2008, 19(1): 34–37.

摘要：研制了一种紧凑型超辐射光泵重水气体 THz 激光器系统,并通过改变 THz 激光激活室的长度,对 THz 信号出射光强度与信号增益长度之间的关系进行了实验研究;利用半经典理论,对两者之间的关系进行了理论计算和求解。实验和理论计算结果表明, TH 信号出射光强度跟信号增益长度之间存在非线性的关系。在工作气压、温度和泵浦光强度一定的条件下,激光工作室长度存在最佳值,此时介质对信号光的增益和吸收达到平衡,出射的 THz 信号光强最大,继续延长介质工作长度将会导致 THz 输出信号的逐步减弱。根据以上结果可以设计确定相关 THz 激光工作室的最佳长度。

关键词：太赫兹激光,半经典理论,受激 Raman 发射,重水气体分子

21. Collinear phase-matching study of terahertz-wave generation via difference frequency mixed in GaAs and Inp

Huang Lei, Sun Bo, Yao Jianquan, Wang Peng

Optoelectronics Letters, 2008, 4(3): 234–238.

Abstract: The collinearly phase-matching condition of terahertz-wave generation via difference frequency mixed in GaAs and InP is theoretically studied. In collinear phase-matching, the optimum phase-matching wave bands of these two crystals are calculated. The optimum phase-matching wave bands in GaAs and InP are 0.95~1.38 μm and 0.7~0.96 μm respectively. The influence of the wavelength choice of the pump wave on the coherent length in THz-wave tuning is also discussed. The influence of the temperature alteration on the phase-matching and the temperature tuning properties in GaAs crystal are calculated and analyzed. It can serve for the following experiments as a theoretical evidence and a reference as well.

Key words: Collinear phase-matching, Terahertz-wave generation, Difference frequency mixing

22. The numerical calculation and analyze of the pulse-laser pumped D₂O Gas Terahertz laser

He Zhihong, Yao Jianquan, Shi Huafeng, Luo Xizhang, Li Jianrong, Jiang Shaoji, Wang Peng

2007 Joint 32nd International Conference on Infrared and Millimeter Waves and the 15th International Conference on Terahertz Electronics (IRMMW-THz), 2008: 470–471.

Abstract: By using the semi-classical density matrix theory, the pulse-laser pumped D_2O gas THz laser was calculated numerically by iteration method and analyzed. According to the result, the THz radiation output spectral characteristic curves were obtained, and the relation between the intensity of pumping laser and the output Tera-Hz laser was studied.

Key words: D_2O gas molecule, Semi-classical theory, Stimulated Raman emission, Terahertz

23. Simultaneous all-solid-state multi-wavelength lasers- a promising pump source for generating highly coherent terahertz waves

Liu Huan, Xu Degang, Yao Jianquan

Chinese Physics B, 2009, 18(3): 1077–1085.

Abstract: A diode-end-pumped Nd:YAG dual-wavelength laser operating at 1 319 and 1 338 nm is demonstrated. The maximum average output power of the quasi-continuous wave linearly polarized dual-wavelength laser is obtained to be 2.1 W at a repetition rate of 50 kHz with an output power instability of less than 0.38% and beam quality factor M^2 of 1.45. Using the two lines, the highly coherent and narrow linewidth terahertz radiation of 3.23 THz can be generated in an organic 4-N, N-dimethylamino–methyl-stilbazolium tosylate (DAST) crystal. Meanwhile, the multi-wavelength red laser at 659.5, 664 and 669 nm is generated by frequency doubling and sum frequency processes in a lithium triborate (LBO) crystal. The average red laser output power is enhanced up to 1.625 W at a repetition rate of 15 kHz with an output power instability of better than 0.53% and beam quality factor M^2 of 6.05. Using the three lines, it is possible to generate the multi-wavelength THz radiation of 3.3, 3.43 and 6.73 THz in an appropriate difference frequency crystal.

Key words: End-pumped Nd:YAG laser, Dual-wavelength laser, Acousto-optic Q-switch, Terahertz wave

24. Investigation of pump-wavelength dependence of terahertz-wave parametric oscillator based on LiNbO$_3$

Sun Bo, Liu Jinsong, Li Enbang, Yao Jianquan

Chinese Physics B, 2009 18(7): 2846–2843.

Abstract: This paper investigates the performances of terahertz-wave parametric oscillators (TPOs) based on the LiNbO$_3$ crystal at different pump wavelengths. The calculated results show that TPO characteristics, including the frequency tuning range, the THz-wave gain and the stability of THz-wave output direction based on the Si-prism coupler, can be significantly improved by using a short-wavelength pump. It also demonstrates that a long-wavelength-pump allows the employment of a short TPO cavity due to an enlarged phase-matching angle, that is, an increased angular separation between the pump and oscillated Stokes beams under the THz-wave generation at a specific frequency. The study provides an useful guide and a theoretical basis for the further improvement of TPO systems.

Key words: Terahertz-wave parametric oscillator, THz-wave radiation, Pump-wavelength dependence

25. 基于钽酸锂晶体的太赫兹波参量振荡器运转特性的研究

孙博，刘劲松，凌福日，王可嘉，朱大庆，姚建铨

物理学报，2009, 58(3):1745–1751.

摘要： 基于晶格振动模受激电磁耦子散射过程的基本原理，对由 LiTaO$_3$（LT）晶体组成的太赫兹波参量振荡器（TPO）的输出调谐特性、增益和吸收损耗特性，以及基于硅棱镜阵列耦合装置的 THz 波输出特性等方面进行了详细的理论研究和分析。研究结果表明，基于 LiTaO$_3$ 晶体 A_1 对称性晶格振动的特点以及自身优异的非线性光学特性，通过利用短波长抽运光、适当提高抽运能量以及缩短 TPO 谐振腔腔长等方法，完全可以实现 LT-TPO 的高性能运转，证明了 LiTaO$_3$ 晶体是一种性能优良的 TPO 工作介质。理论计算结果及方法为以后的 LT-TPO 实验工作提供了详实的理论依据和实验指导。

关键词： 非线性光学, THz 辐射, LiTaO$_3$ 晶体, 电磁耦子

26. High-energy, continuously tunable intracavity terahertz-wave parametric oscillator

J. Q. Yao, Y. Y. Wang, D. G. Xu, K. Zhong, Z. Wang, P. Wang

2009 34th International Conference on Infrared, Millimeter, and Terahertz Waves, 2009: 1–2.

Abstract: We have demonstrated an intracavity THz optical oscillator pumped by a diode-side-pumped Q-switched Nd:YAG laser. Based on a noncollinear phase matching geometry in the nonlinear crystal MgO:LiNbO$_3$, high-energy, low-threshold, coherent tunable Stokes light is obtained by changing the angle between the resonated idler wave and the pump wave, which means that the widely tunable, high-energy, coherent THz radiation is also generated. The tuning range for Stokes wave is 1 069.4-1 073.4 nm, corresponding to the THz frequency range of 1.4-2.5 THz. The phenomenon of the coherent tunable second-order Stokes light scattering is also observed.

Key words: Terahertz-wave parametric oscillator, Noncollinear phase matching, MgO:LiNbO$_3$

27. The generation of THz frequency comb via surface-emitted optical rectification of fs-pulses in periodically poled lithium niobate

Degang Xu, Pengxiang Liu, Kai Zhong, Yuye Wang, Jianquan Yao

SICAST **2009**, 2009, 11: 16–18.

Abstract: The frequency comb is a powerful tool for frequency metrology. In the optical region, the fs-pulse traingenerated by a mode-locked laser has a comb-like frequency spectrum, which covers the visible and near-infraredregion. In THz region, THz comb generated from photoconductive emitter excited by a fs-pulse train (fs opticalcomb) has already been reported. In this paper, we propose a new method for THz comb generation via surface-emitted optical rectification of fs-pulses in PPLN.

Key words: THz frequency comb, Surface-emitted optical rectification, PPLN

28. Compact and widely tunable terahertz source based on a dual-wavelength intracavity optical parametric oscillation

Y. Geng, X. Tan, X. Li, J. Yao

Applied Physics B, 2010, 99: 181–185.

Abstract: In this paper, a continuously tunable terahertz (THz) source is obtained using a compact intracavity pumped dual-wavelength optical parametric oscillation operating around 2.1 μm as difference-frequency generation pump source. The tuning range of the THz-wave frequency covers from 0.147 THz to 3.651 THz. Based on the collinear difference-frequency generation in the GaSe crystal, the experiment result shows that our schematic is a good option to construct a compact and portable terahertz source with widely tunable range.

Key words: Tunable terahertz (THz) source, Dual-wavelength optical parametric oscillation

29. Enhancement of terahertz wave difference generation based on a compact walk-off compensated KTP OPO

Kai Zhong, Jianquan Yao, Degang Xu, Zhuo Wang, Zhongyang Li, Huiyun Zhang, Peng Wang

Optics Communications, 2010, 283: 3520–3524.

Abstract: A compact, walk-off compensated dual-wavelength KTP OPO near the degenerate point of 2.128 μm pumped by a Nd:YAG pulsed laser is employed as the pump for terahertz (THz) source based on difference frequency generation (DFG) in a GaSe crystal. Coherent THz radiation that is continuously tunable in the range of 81–1 617 μm (0.186–3.7 THz) is achieved. An enhancement of 76.7% in average for the THz energies at different wavelengths is realized using the walk-off compensated KTP OPO than the common one. Using a 8 mm-long GaSe crystal, the maximum output THz pulse energy is 48.9 nJ with the peak power of 11W, corresponding to the energy conversion efficiency of 5.4×10^{-6} and the photon conversion efficiency of about 0.09%.

Key words: Terahertz wave, Difference frequency generation, Walk-off compensate, GaSe

30. Threshold analysis of THz-wave parametric oscillator

Li Zhongyang, Yao Jianquan, Zhu Nengnian, Wang Yuye, Xu Degang
Chinese Physics Letters, 2010, 27(6): 064202.

Abstract: Parametric gain of a terahertz wave parametric oscillator (TPO) is analyzed. Meanwhile the expression of TPO threshold pump intensity is derived and theoretically analyzed with different factors. The effective interactionlength between the pump wave and Stokes wave is calculated, and particular attention is paid to the couplingefficiency of the pump wave and Stokes wave. Such analysis is useful for the experiments of TPO.

Key words: Terahertz wave, Parametric oscillator, Effective interactionlength, Couplingefficiency

31. 基于闪锌矿晶体中受激电磁耦子产生可调谐太赫兹波的理论研究

李忠洋,姚建铨,李俊,邲丕彬,徐德刚,王鹏
物理学报,2010, 59(9): 6229–6234.

摘要： 基于受激电磁耦子散射原理,采用已报道的利用非线性光学参量振荡方法产生可调谐太赫兹波的实验条件作为理论分析的实验模型,以 GaAs,GaP,InP,ZnTe 晶体为代表,计算分析了在闪锌矿晶体中参量振荡产生太赫兹波的吸收、增益特性,对输出 THz 波的调谐特性给出了详尽分析。分析太赫兹波高效耦合输出的腔型结构,并与掺氧化镁铌酸锂晶体组成的太赫兹波参量振荡器做对比。

关键词： 太赫兹波,太赫兹波参量振荡,电磁耦子,闪锌矿晶体

32. 周期结构 GaAs 晶体 ps 脉冲差频产生窄带 THz 辐射的研究

王卓, 王与烨, 姚建铨, 王鹏
物理学报, 2010, 59(5): 3249–3254.

摘要：计算模拟了在 ps 脉冲抽运周期结构 GaAs 晶体差频产生窄带 THz 波过程中，对于不同波长、不同脉宽的抽运光，周期结构 GaAs 晶体的走离长度和最佳周期长度的变化，并对计算结果进行了理论分析。研究了周期结构 GaAs 晶体的畴数对 THz 波光谱的影响。根据 GaAs 晶体温度色散公式，计算并分析了晶体最佳周期长度的温度调谐特性，研究了温度变化对 THz 波光谱的影响，提出了通过温度调谐实现在周期结构 GaAs 晶体中产生宽调谐、窄带宽 THz 波的新方法，该计算结果为下一步的实验研究提供了理论基础。

关键词：ps 脉冲, 差频产生窄带 THz 波, 周期结构 GaAs 晶体, 温度调谐

33. Study on the generation of high-power terahertz wave from surface-emitted THz-wave parametric oscillator with MgO:LiNbO$_3$ crystal

Zhongyang Li, Jianquan Yao, Degang Xu, Kai Zhong, Pibin Bing, Jingli Wang
Proceedings of SPIE - Photonics Asia, 2010, 7854: 78543H.

Abstract: High-power nanosecond pulsed THz-wave radiation was achieved via a surface-emitted THz-wave parametric oscillator (TPO). The effective parametric gain length under the condition of noncollinear phase matching was calculated to optimize the parameters of the TPO. Only one MgO:LiNbO$_3$ crystal with large volume was used as gain medium. THz-wave radiation from 0.8 to 2.9 THz was obtained. The maximum THz-wave output was 289.9 nJ/pulse at 1.94 THz when pump power density was 211 MW/cm^2, corresponding to the energy conversion efficiency of 3.43×10^{-6} and the photon conversion efficiency of about 0.05%. The far-field divergence angle of THz-wave radiation was 0.020 4 rad at vertical direction and 0.006 8 rad at horizontal direction.

Key words: Terahertz-wave parametric oscillator, Noncollinear phase matching, Frequency tuning output

34. Terahertz difference frequency generation in GaSe based on a doubly-resonant walk-off compensated KTP OPO

Kai Zhong, Jianquan Yao, Degang Xu, Peng Wang

Proceedings of SPIE - Photonics Asia, 2010, 7854: 78543.

Abstract: We have achieved a Terahertz (THz) DFG system based on a walk-off compensated intracavity pumped dualwavelength KTP OPO employing two identical KTP crystals. The KTP OPO is doubly resonant and works near the degenerate point at 2.128 μm, which doubles the quantum efficiency compared with DFG using pump pulses around 1 μm. This THz source is simple and compact, about $10 \times 10 \times 40$ cm^2 in size. Besides lower threshold and better stability, the walk-off compensated KTP OPO greatly improves the pump beam quality and enhances the DFG conversion efficiency. With an 8-mm-long GaSe crystal, the generated THz tuning range is from 0.186 THz to 3.7 THz with the maximum output voltage of 489 V on the bolometer at 1.68 THz. An average enhancement of 76.7% for the THz energies is realized using the walk-off compensated KTP OPO than a common one. The conversion efficiency can be improved with a longer and better GaSe crystal.

Key words: Terahertz, Difference frequency generation, Tunable, Coherent, Walk-off compensate, Compact, Efficient

35. Theoretical study on the generation of THz sub-comb via surface-emitted optical rectification of ultra-short pulse in periodically poled lithium niobate

Pengxiang Liu, Degang Xu, Jianquan Yao

Proceedings of SPIE - Photonics Asia, 2010, 7854: 78543E.

Abstract: In this paper, we proposed a method for THz sub-comb generation based on optical rectification. The result of our calculation indicated that THz pulse train, generated by surface-emitted optical rectification (OR) of femtosecond (fs) laser pulse in a periodically poled lithium niobate (PPLN), has a comb-like spectrum. The theoretical analysis was based on radiating antenna model. The characteristic of this THz sub-comb was analyzed both in frequency and time domain. The mechanism of this phenomenon was explained both by spectral interference between early and late pulses and by high-order quasi phase matching. THz sub-comb generated by this method can cover a large bandwidth and have a wide free spectral range.

Key words: Terahertz wave, Optical rectification, Quasi phase matching

36. Terahertz-wave parametric oscillator with a misalignment-resistant tuning cavity

Bo Sun, Sanxing Li, Jinsong Liu, Enbang Li, Jianquan Yao

Optics Letters, 2011, 36(10): 1845–1847.

Abstract: We demonstrate a terahertz-wave parametric oscillator (TPO) with a corner-cube resonator consisting of a cornercube prism (CCP) and a flat mirror. By using the cavity configuration proposed in this Letter, the generation of tunable monochromatic terahertz (THz) waves can be achieved just by rotating the flat mirror instead of rotating the TPO cavity relative to the pump beam. The THz-wave output intensity and pulse width can be controlled periodically by rotating the CCP around the cavity axis. The TPO stability against cavity misalignment is significantly improved by at least 1 to 2 orders of magnitude compared with the conventional plane–parallel resonator configuration.

Key words: Terahertz-wave parametric oscillator (TPO), Cavity, Cornercube prism (CCP)

37. p-polarized Cherenkov THz wave radiation generated by optical rectification for a Brewster-cut LiNbO$_3$ crystal

Pengxiang Liu, Degang Xu, Changming Liu, Da Lv, Yingjin Lv, PengWang, Jianquan Yao

Journal of Optics, 2011, 13: 085202.

Abstract: In this paper, we investigated p-polarized Cherenkov radiation excited by an ultra-short laser pulse focused into a line in an LiNbO$_3$ crystal. The geometries of p- and s-polarized THz generation were both analyzed. We did further calculations on p-polarized THz radiation and designed a Brewster-cut geometry. The radiated energy and conversion efficiency were roughly estimated. Compared with s-polarized waves radiated from a Cherenkov-cut crystal, p-polarized THz radiation has lower energy and conversion efficiency, but higher intensity and better beam quality. The effect of angular dispersion between the spectral components of the THz pulse after refraction at the Brewster surface was also discussed.

Key words: Optical rectification, Terahertz wave, Cherenkov radiation

38. High-power terahertz radiation from surface-emitted THz-wave parametric oscillator

Li Zhongyang, Yao Jianquan, Xu Degang, Zhong Kai, Wang Jingli, Bing Pibin

Chinese Physics B, 2011, 20(5): 054207.

Abstract: We report a pulsed surface-emitted THz-wave parametric oscillator based on two MgO:LiNbO$_3$ crystals pumped by a multi-longitudinal mode Q-switched Nd:YAG laser. Through varying the phase matching angle, the tunable THzwave output from 0.79 THz to 2.84 THz is realized. The maximum THz-wave output was 193.2 nJ/pulse at 1.84 THz as the pump power density was 212.5 MW/cm^2, corresponding to the energy conversion e–ciency of 2.42×10^{-6} and the photon conversion e–ciency of about 0.037%. When the pump power density changed from 123 MW/cm^2 to 148 MW/cm^2 and 164 MW/cm^2, the maximum output of the THz-wave moved to the high frequency band. We give a reasonable explanation for this phenomenon.

Key words: THz-wave parametric oscillator, Noncollinear phase matching, THz-wave polarization, Frequency tunable output

39. High-power terahertz radiation based on a compact eudipleural THz-wave parametric oscillator

Li Zhongyang, Yao Jianquan, Lv Da, Xu Degang, Wang Jingli, Bing Pibin

Chinese Physics Letters, 2011, 28(6): 064209.

Abstract: Tunable high-power THz-wave radiation is achieved via a compact eudipleural THz-wave parametric oscillator. The maximum THz-wave output is 1.164 V at 1.755 THz when the pump energy is 90 mJ. In the experiments we find that the maximum output of THz-wave moves to the high frequency band as the pump energy increases and this phenomenon is reasonably explained. The polarization characteristics of the THz-wave are analyzed.

Key words: THz-wave radiation, THz-wave parametric oscillator, THz-wave output, Polarization

40. Output enhancement of a THz wave based on a surface-emitted THz-wave parametric oscillator

Li Zhongyang, Yao Jianquan, Xu Degang, Bing Pibin, Zhong Kai

Chinese Physics Letters, 2011, 28(11): 114201.

Abstract: High-power nanosecond pulsed THz-wave radiation is achieved via a surface-emitted THz-wave parametric oscillator. One MgO:LiNbO$_3$ crystal with large volume is used as the gain medium. THz-wave radiation from 1.084 THz to 2.654 THz is obtained. The maximum THz-wave average power is 5.8 μW at 1.93 THz when the pump energy is 84 mJ, corresponding to a energy conversion efciency of 6.9 × 10^{-6}. The polarization characteristics of THz wave are analyzed. During the experiments the radiations of the frst-order and the second-order Stokes wave are observed.

Key words: THz wave, Surface-emitted THz-wave parametric oscillator, Output

41. 级联差频产生太赫兹辐射的理论研究

钟凯，姚建铨，徐德刚，张会云，王鹏

物理学报，2011, 60(3): 034210.

摘要：针对差频产生太赫兹（THz）辐射转换效率低的缺点，提出了级联差频的新机理以提高转换效率，并以 ZnTe 晶体为例，对级联差频产生 THz 辐射的原理和过程进行了理论研究。通过对级联差频耦合波方程组的求解，得出了 ZnTe 晶体中级联差频的最佳抽运条件和 ZnTe 晶体的最佳长度，并且分析了晶体吸收、波矢失配及抽运强度对级联差频的影响。计算结果表明，通过级联差频可以大大提高 THz 波的转换效率，其光子转换效率甚至可以超过 Manley-Rowe 关系的限制。

关键词：太赫兹辐射，差频，级联，ZnTe 晶体

42. 铌酸锂晶体中参量振荡产生高功率可调谐太赫兹波的实验研究

李忠洋, 姚建铨, 徐德刚, 钟凯, 邴丕彬, 汪静丽

中国激光, 2011, 38(4): 0411002.

摘要：利用一块大体积 MgO:LiNbO$_3$ 晶体，采用浅表垂直出射方式构成太赫兹波参量振荡器，实现了高功率可调谐的 THz 波输出，调谐范围为 0.8-2.8 THz。当泵浦功率密度为 197.4 MW/cm^2 时，在 1.73 THz 处每个 THz 脉冲的最大输出能量为 173.9 nJ，对应的能量转化效率为 2.2×10^{-6}。实验过程中观察到了一阶和二阶斯托克斯 (Stokes) 光，一阶 Stokes 光相对于抽运的频移等于产生的 THz 波的频率。

关键词：非线性光学, 太赫兹波, 太赫兹波参量振荡器, 电磁耦子

43. Theoretical analysis and numerical simulation of terahertz wave generation and modulation based on GaAs:O

Jian He, Yinli Jiang, Qingyuan Miao, Jiayuan He, Jinsong Liu, Jianquan Yao

Journal of Physics: Conference Series (POEM 2010), 2011, 276: 012204.

Abstract: In this paper, the generation and modulation of terahertz wave based on Oxygen-implanted GaAs (GaAs:O) is proposed and discussed. The theoretical analyses and numerical simulation results about the modulation characteristics of the terahertz wave using the proposed scheme have been obtained. Through comparing the numerical simulation results we can find that the information carried on the optical light can be transferred onto the generated terahertz wave and the power of the modulation side bands in terahertz frequency can keep stable in the whole modulation bandwidth. The relationship between the terahertz power and carrier lifetime has also been studied.

Key words: THz wave generation, THz wave modulation, Oxygen-implanted GaAs (GaAs:O)

44. Analysis on characteristic and application of THz frequency comb and THz sub-comb

Pengxiang Liu, Degang Xu, Jianquan Yao

Journal of Physics: Conference Series (POEM 2010), 2011, 276: 012218.

Abstract: In this paper, we proposed a method for THz sub-comb generation based on spectral interference. The result of our calculation indicated that the THz pulse train, generated by surface-emitted optical rectification of femtosecond (fs) laser pulse in periodically poled lithium niobate (PPLN), has a comb-like spectrum. The characteristic of this THz sub-comb was analyzed both in frequency and time domain. Compared with the THz frequency comb emitted by a photoconductive antenna (PCA), THz sub-comb has a lower spectral resolution and wider free spectral range. Thus it could be an ideal source for wavelength division multiplexing (WDM) in THz wireless communication system.

Key words: Terahertz wave, Frequency comb, Optical rectification

45. Terahertz generation by optical rectification in zincblende crystals with arbitrary crystal-oriention

Xiaoguang Tian, Furi Ling, Jinsong Liu, Jianquan Yao

Remote Sensing, Environment and Transportation Engineering (RSETE), 2011: 6425-6428.

Abstract: We present a set of equations describing the terahertz generation in the zincblende crystals by optical rectification. These equations can be used for every possible polarization direction of the pump beam and for crystals that have arbitrary crystal-oriention. Our results are in excellent agreement with the previous work. Moreover, we have demonstrated the optimal crystal-oriention and polarization of the pump beam to achieve the max emission of terahertz wave.

Key words: Terahertz generation, Rectification, Crystal-oriention, Zincblende crystal

46. 周期极化 GaAs 晶体中差频产生太赫兹辐射的研究

张成国,姚建铨,钟凯,缪岳洋,孙崇玲,王鹏,李敬辉

激光与红外, 2011, 41(10): 1154–1158.

摘 要:传统的光学差频产生的太赫兹辐射转换效率低,不能获得高功率太赫兹辐射。本文对周期极化 GaAs 晶体中差频产生太赫兹辐射进行了理论计算,通过温度调谐实现了周期极化 GaAs 晶体中差频获得可调谐太赫兹波的输出。为了提高差频过程的增益和量子效率,在准相位匹配基础上引进了级联差频机理,并对最佳晶体长度和最佳泵浦频率进行了计算。结果表明,利用周期极化的 GaAs 晶体可以获得更高能量更高效率的太赫兹波辐射。

关键词:GaAs 晶体,准相位匹配,温度调谐,级联差频,太赫兹

47. 太赫兹波及其常用源

杨鹏飞,姚建铨,邢丕彬,邸志刚

激光与红外, 2011, 41(2): 125–131.

摘要:介绍了太赫兹波的特性及其应用,从宽带、窄带两个方面归类,详细介绍了现阶段常用的几种太赫兹源,并列出了相关重要参数的计算。随着源技术的不断进步,太赫兹电磁波也将象光学和微波波段的电磁波一样,给人类社会的许多方面带来深远的影响。

关键词:太赫兹,源,差频,整流,半导体表面辐射

48. Theory of monochromatic terahertz generation via Cherenkov phase-matched difference frequency generation in LiNbO$_3$ crystal

Pengxiang Liu, Degang Xu, Hao Jiang, Zhuo Zhang, Kai Zhong, Yuye Wang, Jianquan Yao

Journal of the Optical Society of America B, 2012, 29(9): 2425–2430.

Abstract: A theory of Cherenkov phase-matched monochromatic terahertz (THz)-wave generation via difference frequency generation in a nonlinear crystal is developed. An experimental situation (LiNbO$_3$ pumped by dual-wavelength near-infrared nanosecond pulses) is considered. This theory accounts for the finite size of pump beam and allows us to explore the generation of transverse THz wave vector. The output characteristic of this THz source is analyzed based on the analytical expression, including radiation pattern, conversion efficiency, and tuning range. Calculated tuning curves are presented, which reasonably agree with previous experimental results. The influence of divergence of the focused pump beam on total radiated energy is studied in detail. Optimal pump beam size that maximizing generated THz energy is obtained.

Key words: Terahertz-wave generation, Difference frequency generation, Nonlinear crystal

49. High-powered tunable terahertz source based on a surface-emitted terahertz-wave parametric oscillator

Zhongyang Li, Pibin Bing, Jianquan Yao, Degang Xu, Kai Zhong

Optical Engineering, 2012, 51(9): 091605.

Abstract: A high-powered pulsed terahertz (THz)-wave has been parametrically generated via a surface-emitted THz-wave parametric oscillator (TPO). The effective parametric gain length under the noncollinear phase matching condition was calculated for optimization of the parameters of the TPO. A large volume crystal of MgO:LiNbO$_3$ was used as the gain medium. THz-wave radiation covering a frequency range from 0.87 to 2.73 THz was obtained. The average power of the THz-wave was 9.12 μW at 1.75 THz when the pump energy was 94 mJ, corresponding to an energy conversion efficiency of about 9.7×10^{-6} and a photon conversion efficiency of about 0.156%. The THz-wave power in our experiments is high enough for practical applications to spectrum analysis and imaging.

Key words: Terahertz-wave generation, Source, Surface-emitted THz-wave parametric oscillator, Power

50. THz-wave difference frequency generation by phase-matching in GaAs/Al$_x$Ga$_{1-x}$ as asymmetric quantum well

Cao Xiaolong, Wang Yuye, Xu Degang, Zhong Kai, Li Jinghui, Li Zhongyang, Zhu Nengnian, Yao Jianquan
Chinese Physics Letters, 2012, 29(1): 014207.

Abstract: An asymmetric quantum well (AQW) is designed to emit a terahertz (THz) wave by using difference frequency generation (DFG) with the structure of GaAs/Al$_{0.2}$Ga$_{0.8}$As/Al$_{0.5}$Ga$_{0.5}$As under a doubly resonant condition. It is found that the second-order nonlinear susceptibility $\chi^{(2)}$ varies with the two pump wavelengths, and it can reach the peak value of 1.61 μm/V when the wavelengths are given as λ_{p1} = 9.756 μm and λ_{p2} = 10.96 μm, respectively. The numerical results show that the refractive index of one pump wave in the AQW is concerned with not only its own wavelength but also the other wavelength. Phase-matching inside the AQW can be obtained through the tuning of the two pump wavelengths.

Key words: Terahertz wave, Difference frequency generation (DFG), Asymmetric quantum well (AQW), Phase matching

51. THz source based on optical Cherenkov radiation

Yao Jianquan, Liu Pengxiang, Xu Degang, Lv Yingjin, Lv Da
Science China - Information Sciences, 2012, 55(1): 27–34.

Abstract: Terahertz (THz) technique has attracted considerable interest due to its broad application prospects. THz source is a crucial part of THz science and technology. Optical Cherenkov radiation in electro-optic crystals is a promising method of THz generation, because phase-matching is automatically satisfied. In this paper, we introduced two types of THz source based on optical Cherenkov radiation: both broadband and tunable monochromatic. The mechanism of radiation was analyzed and recent development was reviewed in detail. The future of THz source based on optical Cherenkov radiation was also forecasted.

Key words: Terahertz source, Optical rectification, Difference frequency generation, Cherenkov radiation

52. 硅棱镜耦合输出 THz 波参量振荡器的实验研究

李忠洋, 姚建铨, 徐德刚, 邴丕彬, 钟凯

光电子·激光, 2012, 23(3): 425–428.

摘要：利用一块 MgO：LiNbO$_3$ 晶体，采用 Si 棱镜耦合出射方式构成太赫兹（THz）波参量振荡器，实现了高效可调谐的 THz 波输出，调谐范围为 0.95~2.10THz。在 1.80 THz 处，当泵浦能量为 101 mJ 时，产生的 THz 波平均功率为 580 nW，THz 波单脉冲能量为 58 nJ，THz 波能量转换效率为 5.74×10^{-7}；同时产生的 Stokes 能量为 2.7 mJ。实验中，观察到了一阶和二阶 Stokes，一阶 Stokes 光与泵浦光的频差等于产生的 THz 波的频率。

关键词：太赫兹波（THz），THz 波参量振荡器，电磁耦子，非共线相位匹配

53. Design and threshold analysis for a novel intracavity THz-wave parametric oscillator

Li Zhongyang, Li Jiwu, Bing Pibin, Xu Degang, Yao Jianquan

Infrared and Laser Engineering, 2012, 41(9): 2339–2345.

Abstract: A novel surface-emitted intracavity THz-wave parametric oscillator (TPO) was designed to improve the down-conversion efficiency. An intersecting cavity geometry was employed that allowed the nonlinear medium to be placed within the cavity of the pump laser. The surface-emitted scheme was employed to extract the THz-wave from MgO:LiNbO$_3$ crystal. The expression of the effective parametric gain length under the condition of noncollinear phase matching condition was derived. The pump beam with larger radius and the crystal with longer length could increase the effective parametric gain length efficiently. Based on the expression of the effective parametric gain length the expression of TPO oscillation threshold was deduced and was theoretically analyzed under different conditions. Larger beam radius of pump wave, lower frequency of THz-wave, shorter cavity physical length, higher reflectivity of the Stokes output coupler and longer crystal length can effectively reduce the oscillation threshold. Such an analysis is useful for the experiments of TPO.

Key words: Intracavity THz-wave parametric oscillator, Oscillation threshold, Noncollinear phase matching, Effective parametric gain length

54. Design of GaAs/Al$_x$Ga$_{1-x}$As asymmetric quantum wells for THz-wave by difference frequency generation

Cao Xiaolong, Yao Jianquan, Zhu Nengnian, Xu Degang
Optoelectronics Letters, 2012, 8(3): 0229–0232.

Abstract: The energy levels, wave functions and the second-order nonlinear susceptibilities are calculated in GaAs/Al$_{0.2}$Ga$_{0.8}$As/Al$_{0.5}$Ga$_{0.5}$As asymmetric quantum well (AQW) by using an asymmetric model based on the parabolic and non-parabolic band. The influence of non-parabolicity cannot be neglected when analyzing the phenomena in narrow quantum wells and in higher lying subband edges in wider wells. The numerical results show that under double resonance (DR) conditions, the secondorder difference frequency generation (DFG) and optical rectification (OR) generation susceptibilities in the AQW reach 2.501 9 Pm/V and 13.208 Pm/V, respectively, which are much larger than those of the bulk GaAs. Besides, we calculate the absorption coefficient of AQW and find out the two pump wavelengths correspond to the maximum absorption, so appropriate pump beams must be selected to generate terahertz (THz) radiation by DFG.

Key words: Asymmetric quantum well (AQW), double resonance (DR) conditions, Difference frequency generation (DFG), Optical rectification (OR) generation susceptibilities

55. Investigation on phase matching in a THz-wave parametric oscillator

Li Zhongyang, Bing Pibin, Xu Degang, Zhong Kai, Yao Jianquan
Optoelectronics Letters, 2012, 8(1): 029–032.

Abstract: The characteristics of noncollinear phase matching and quasi-phase-matching in the THz-wave parametric oscillator (TPO) are investigated. The expression of the effective parametric gain length under the condition of noncollinear phase matching configuration is deduced. The relationship between the poling period of periodically poled LiNbO$_3$ crystal and the generated THz frequency under the condition of quasi-phase-matching configuration is analyzed. Based on the analyses above we propose a new TPO configuration which ensures the three mixing waves interact collinearly. The effects of operation temperature on phase matching are analyzed.

Key words: Noncollinear phase matching, THz-wave parametric oscillator (TPO), Periodically poled LiNbO$_3$ crystal, Quasi phase matching

56. Coupled-mode theory for Cherenkov-type guided-wave terahertz generation via cascaded difference frequency generation

Pengxiang Liu, Degang Xu, Hong Yu, Hao Zhang, Zhongxiao Li, Kai Zhong, Yuye Wang, Jianquan Yao

IEEE Journal of Lightwave Technology, 2013, 31(15): 2058–2064.

Abstract: A scheme for monochromatic terahertz (THz) generation via cascading enhanced Cherenkov-type difference frequency generation (DFG) in a sandwich-like waveguide is proposed. The novel scheme has the potential to overcome the quantum-defect limit and to provide an efficient output coupling. This process is elucidated by developing a coupled-mode theory and taking into account the pump depletion, waveguide mode properties, and THz output coupling. The effect of cascading enhancement is analyzed by comparing with non-cascaded DFG situation. It is predicted that THz power can be boosted by nearly 8-fold with a 400 MW/cm^2 pump in a 40-mm-long $Si-LiNbO_3-Si$ waveguide.

Key words: Cascaded frequency down-conversion, Cherenkov phase-matching, Difference frequency generation, Optical waveguide, Terahertz generation

57. Widely tunable, monochromatic THz generation via Cherenkov-type difference frequency generation in an asymmetric waveguide

Pengxiang Liu, Degang Xu, Weipeng Wang, Changming Liu, Sujia Yin, Xin Ding, Yuye Wang, Jianquan Yao

IEEE Journal of Quantum Electronics, 2013, 49(2): 179–185.

Abstract: Cherenkov-type monochromatic terahertz (THz) generation via difference frequency generation in a general asymmetric sandwich-like waveguide is investigated theoretically. An efficient geometry—conversion from a transverse electric (TE) pump mode to a TE THz mode is mainly considered. The general expression for a Cherenkov THz field is obtained by solving the nonlinear wave equation analytically. Numerical simulation is performed on specific devices with three typical substrate materials: 1) nondoped $LiNbO_3$; 2) BK7 glass; and 3) quartz. Characteristics of this type of THz source and parametric dependence of conversion efficiency are discussed. It is predicted that the conversion efficiency would be increased by one order by optimizing the waveguide structure.

Key words: Cherenkov phase matching, Difference frequency generation, Optical waveguide, Terahertz generation

58. Efficient continuous-wave 1053-nm Nd:GYSGG laser with passively Q-switched dual-wavelength operation for terahertz generation

Kai Zhong, Chongling Sun, Jianquan Yao, Degang Xu, Xinyi Xie, Xiaolong Cao, Qingli Zhang, Jianqiao Luo, Dunlu Sun, Shaotang Yin

IEEE Journal of Quantum Electronics, 2013, 49(3): 375–379.

Abstract: Research on an efficient continuous-wave Nd:GYSGG laser at 1 053 nm with excellent stability is demonstrated. The maximum output power is 4.17 W, corresponding to the conversion efficiency of 33.9% and the slope efficiency of 42.92%. Using a Cr:YAG absorber, pulsed dual-wavelength operation at 1 053 and 1 058.4 nm is obtained, of which the maximum single pulse energy and peak power are 172.1 µJ and 26.1 kW, respectively, when the pulse width is 6.6 ns and the repetition rate is 4.3 kHz. A polarization property is observed, owing to the anisotropy of Cr:YAG. This stably Q-switched dual-wavelength laser is a good pump source for the generation of a terahertz wave at 1.53 THz.

Key words: Diode pumped laser, Nd:GYSGG, Passively Q-switched, Terahertz

59. A high-energy, low-threshold tunable intracavity terahertz-wave parametric oscillator with surface-emitted configuration

Y Y Wang, D G Xu, H Jiang, K Zhong, J Q YAO
Laser Physics, 2013, 23: 055406.

Abstract: A high-energy, low-threshold THz-wave output has been experimentally demonstrated with an intracavity terahertz-wave parametric oscillator based on a surface-emitted configuration, which was pumped by a diode-side-pumped Q-switched Nd:YAG laser. Different beam sizes and repetition rates of the pump light have been investigated for high-energy and high-efficiency THz-wave generation. The maximum THz-wave output energy of 283 nJ/pulse was obtained at 1.54 THz under an intracavity 1 064 nm pump energy of 59 mJ. The conversion efficiency was 4.8×10^{-6}, corresponding to a photon conversion efficiency of 0.088%. The pump threshold was 12.9 mJ/pulse. A continuously tunable range from 0.75 to 2.75 THz was realized.

Key words: Intracavity terahertz-wave parametric oscillator, Surface-emitted configuration, THz-wave output

60. Frequency tuning characteristics of a THz-wave parametric oscillator

Zhongyang Li, Pibin Bing, Degang Xu, Jianquan Yao
Journal of the Optical Society of Korea, 2013, 17(1): 97–102.

Abstract: Frequency tuning characteristics of a THz-wave by varying phase-matching angle and pump wavelength in a noncollinear phase-matching THz-wave parametric oscillator (TPO) are analyzed. A novel scheme to realize the tuning of a THz-wave by moving the cavity mirror forwards and backwards is proposed in a noncollinear phase-matching TPO. The parametric gain coefficients of the THz-wave in a $LiNbO_3$ crystal are explored under different working temperatures. The relationship between the poling period of periodically poled $LiNbO_3$ (PPLN) and the THz-wave frequency under the condition of a quasi-phasematching configuration is deduced. Such analyses have an impact on the experiments of the TPO.

Key words: THz-wave parametric oscillator, Noncollinear phase-matching, Quasi-phase-matching, Tuning THz-wave

61. Intersubband absorption properties of GaAs/Al$_x$Ga$_{1-x}$As asymmetric quantum well based on optical difference frequency

Xiaolong Cao, Jianquan Yao, Kai Zhong, Degang Xu
Optical Engineering, 2013, 52(1): 014001.

Abstract: The intersubband absorptions between the conduction ground state and excited states in Al$_{0.5}$Ga$_{0.5}$As/GaAs/Al$_{0.2}$Ga$_{0.8}$As multiple asymmetric quantum well (AQW) have been investigated based on optical difference frequency in the 9 to 11 μm region. Under an intense resonant excitation from a dual-wavelength CO$_2$ laser, the saturation intensity of intersubband absorption for pump waves is estimated to be 0.3 MW/cm^2. As the well width variation, the position of absorption peak for pump waves and the absorption of terahertz (THz) wave by DFG show concomitant changes. For an AQW of 7 nm deep well-width and 27 nm total well-width, the maximum of absorption for the THz wave is 6.01×10^5 m^{-1} when the two pump wavelengths are 9.69 and 10.64 μm, respectively. These manipulative transitions in AQW can be applied to tunable optical semiconductor devices and implemented in THz wave devices to achieve additional functionalities.
Key words: Intersubband absorption, Asymmetric quantum well (AQW), Optical difference frequency generation, Tetrahertz-wave

62. Theoretical analysis of cascaded optical parametric oscillations generating tunable terahertz waves

Zhongyang Li, Anfu Zhu, Ningke Zuo, Pibin Bing, Degang Xu, Kai Zhong, Jianquan Yao
Optics Engineering, 2013, 52(10): 106103.

Abstract: Cascaded optical parametric oscillations generating a tunable terahertz (THz) wave are analyzed to solve the problem of low quantum conversion efficiency in a THz-wave parametric oscillator. The forward and backward optical parametric processes are theoretically analyzed based on periodically poled lithium niobate (PPLN) as an example. Tuning characteristics of the THz wave that relate to the parameters of the pump wavelength, the grating period of PPLN, and the working temperature are numerically simulated. The gain and absorption characteristics of the generating THz wave are deduced in the situation of quasiphase-matching configuration at different working temperatures.
Key words: Cascaded optical parametric oscillations, Periodically poled lithium niobate (PPLN), Tuning, THz generation

63. High-power tunable terahertz generation from a surface-emitted THz-wave parametric oscillator based on two MgO:LiNbO$_3$ crystals

Zhongyang Li, Pibin Bing, Degang Xu, Jianquan Yao
Optik, 2013, 124: 4884–4886.

Abstract: A high-powered tunable terahertz wave (THz-wave) has been parametrically generated via a surfaceemitted THz-wave parametric oscillator (TPO) pumped by a multi-longitudinal-mode Q-switched Nd:YAG laser. The effective parametric gain length was enlarged by employing two MgO:LiNbO$_3$ crystals. The tunable THz-wave radiation from 0.8 to 2.8 THz was realized via varying phase-matching angle between the pump wave and the Stokes wave. The maximum THz-wave radiation was 173.9 nJ/pulse at 1.7 THz as the pump energy was 82 mJ, corresponding to an energy conversion efficiency of about 2.12×10^{-6} and a photon conversion efficiency of about 0.035%. The first-order, the second-order and the third-order Stokes waves were observed during the experiments.

Key words: THz-wave parametric oscillator, Noncollinear phase matching, Tunable THz-wave

64. High energy terahertz parametric oscillator based on surface-emitted configuration

Xu Degang, Zhang Hao, Jiang Hao, Wang Yuye, Liu Changming, Yu Hong, Li Zhongyang, Shi Wei, Yao Jianquan
Chinese Physics Letters, 2013, 30(2): 024212.

Abstract: We experimentally demonstrate a high power nanosecond pulsed terahertz (THz)-wave parametric oscillator (TPO) by using a wide pump beam. A surface emitted cavity configuration is employed to reduce the THz absorption in MgO:LiNbO$_3$ crystal. The THz wave can be tuned from 1 THz to 3 THz. A maximum THz output energy of 438 nJ/pulse is achieved at 1.56 Hz using a 4.5-mm-diameter pump beam with a pulse energy of 226 mJ pump energy with the repetition of 10 Hz, corresponding to the energy conversion efciency of 1.94×10^{-6}.

Key words: Terahertz (THz)-wave parametric oscillator (TPO), Surface emitted cavity, THz output

65. 级联参量振荡产生太赫兹辐射的理论研究

李忠洋,邴丕彬,徐德刚,曹小龙,姚建铨

物理学报,2013, 62(8): 084212.

摘要:针对光学参量振荡产生太赫兹波转换效率低的缺点,提出了级联参量振荡产生太赫兹波的新机理以提高转换效率。以周期极化铌酸锂晶体为例,对级联参量振荡产生太赫兹波的原理和过程进行了理论研究。分析了抽运光波长、周期极化铌酸锂晶体极化周期和工作温度对产生一阶、二阶闲频光频率的影响。推导了三波共线相互作用条件下太赫兹波的增益特性和吸收特性。计算结果表明,通过级联参量振荡可以有效提高太赫兹波的转换效率,并可以得到宽调谐的太赫兹波输出。基于分析结果,设计了周期极化铌酸锂晶体级联参量振荡产生高效率、宽调谐、窄线宽、连续太赫兹波的实验。

关键词:太赫兹波,太赫兹波参量振荡,级联参量振荡

66. 小型化外腔可调谐 THz 参量振荡器

徐德刚,蒋浩,张昊,王与烨,李忠洋,钟凯,赵刚,杨闯,高恒,姚建铨

强激光与粒子束,2013, 25(6): 1465–1468.

摘要:基于 $LiNbO_3$ 晶体垂直表面输出技术,设计了一台小型化外腔 THz 参量振荡器。利用小型化灯 1 064 nm 脉冲激光器泵浦 $MgO:LiNbO_3$,通过优化设计三波非共线相位匹配的光学参量振荡腔结构,实现 THz 垂直晶体表面输出,减少 $LiNbO_3$ 晶体对 THz 波的吸收,提高了 THz 波输出光束质量。当在泵浦光能量为 128 mJ、重复频率为 10 Hz 时,获得 THz 波的调谐范围为 0.69~3.01 THz,在 1.6 THz 处获得 THz 波最大平均功率为 10.8 μW,脉冲宽度为 10 ns,对应 THz 波能量转换效率为 8.43×10^{-6}。

关键词:光学太赫兹,光学参量振荡,可调谐,表面垂直输出

67. 表面出射太赫兹波参量振荡器的设计与增强输出

李忠洋, 李继武, 邴丕彬, 徐德刚, 姚建铨

红外与激光工程, 2013, 42(4): 935–939.

摘要：利用一台多纵模调 Q Nd:YAG 激光器泵浦的浅表垂直发射太赫兹波参量振荡器参量产生了高功率可调谐太赫兹波辐射。推导了非共线相位匹配条件下的有效参量增益长度以优化太赫兹波参量振荡器参数。实验测得太赫兹波的调谐范围为 0.77~2.83 THz。当泵浦功率密度为 222.3 MW/cm^2 时，在 1.78 THz 处太赫兹波的最大输出能量为 347.8 nJ/pulse, 对应的能量转化效率为 3.91×10^{-6}。太赫兹波在垂直方向上的远场发散角为 0.020 4 rad, 在水平方向上为 0.006 8 rad。

关键词：太赫兹波, 太赫兹波参量振荡器, 非共线相位匹配, 有效参量增益长度

68. The widely tunable THz generation in QPM-GaAs crystal pumped by a near-degenerate dual-wavelength KTP OPO at around 2.127 μm

Degang Xu, Wei Shi, Kai Zhong, YuYe Wang, Pengxiang Liu, Jianquan Yao

Materials, Devices, and Applications XII, 2013, 8604: 86040E.

Abstract: We report a widely tunable terahertz source by using QPM-GaAs crystals pumped by a near-degenerate dual-wavelength KTP OPO around 2.127 μm, based on difference frequency generation (DFG). The tunable THz radiation from 0.06 THz to 3.34 THz has been achieved in QPM-GaAs crystal with coherence length of 650 μm. The maximum output THz energy is 45 nJ with the peak power of 10 W at 1.68 THz, corresponding to the energy conversion efficiency of 5×10^{-6} and the photon conversion efficiency of about 0.08%.

Key words: Widely tunable, Terahertz source, QPM-GaAs, Dual-wavelength KTP OPO

69. High-energy, tunable intracavity terahertz-wave parametric oscillator with surface-emitted configuration

Y. Y. Wang, D. G. Xu, H. Jiang, J. Q. Yao

2013 38th International Conference on Infrared, Millimeter, and Terahertz Waves (IRMMW-THz), 2013: 1–2.

Abstract: A high-energy, low threshold THz-wave output has been experimentally demonstrated with intracavity terahertz-wave parametric oscillator based on surface-emitted configuration. The maximum THz-wave output energy of 283 nJ/pulse was obtained at 1.54 THz under the pump threshold of 12.9 mJ/pulse in the cavity. The continuously tunable range from 0.75 to 2.75 THz was realized.

Key words: Tunable, Intracavity terahertz-wave parametric oscillator, Surface-emitted configuration, THz-wave output

70. Monochromatic Cherenkov THz source pumped by a singly resonant optical parametric oscillator

Pengxiang Liu, Degang Xu, Jiaqi Li, Chao Yan, Zhongxiao Li, Yuye Wang, Jianquan Yao

IEEE PHOTONICS TECHNOLOGY LETTERS, 2014, 26: 494–496.

Abstract: We report an improvement of terahertz (THz) output energy via Cherenkov-type difference frequency generation by developing the dual-wavelength pump source, which is based on a singly resonant near-degenerated optical parametric oscillator. Total utilization efficiency of pump energy is considerably increased by recycling the residual fundamental wave. The highest energy of Cherenkov-type monochromatic THz generation is achieved to be 1.58 nJ/pulse. A tuning range of 0.1-3.2 THz is demonstrated.

Key words: Cherenkov phase matching, Difference frequency generation, Terahertz-wave generation

71. Widely tunable and monochromatic terahertz difference frequency generation with organic crystal DSTMS

Pengxiang Liu, Degang Xu, Yin Li, Xinyuan Zhang, Yuye Wang, Jianquan Yao, Yicheng Wu

Europhysics Letters, 2014, 106(6):60001.

Abstract: Widely tunable and monochromatic terahertz (THz) difference frequency generation with organic crystal DSTMS was investigated experimentally. A double-pass $KTiOPO_4$ optical parametric oscillator, pumped by a frequency-doubled Nd:YAG laser, was employed as the source of 1.3-1.6 μm dual wavelength. The THz tuning spectrum covered a range of 0.88-19.27 THz. The dependence of the THz output on the crystal thickness and pump wavelength was studied. The results were well explained based on theoretical calculations. With a pump energy of 2.47 mJ, the output energy reached 85.3 nJ/pulse at 3.80 THz, which corresponded to a peak power of 17.9 W and photon conversion efficiency of 3.6‰.
Key words: Widely tunable, Difference frequency generation, Organic crystal

72. A study of the multi-mode pumping of terahertz parametric oscillators

J Q Li, Y Y Wang, D G Xu, Z X Li, C Yan, P X Liu, J Q Yao

Laser Physics, 2014, 24: 105401.

Abstract: We experimentally study the influence of multi-transverse-mode pumping on the output characteristics of terahertz parametric oscillators (TPO). We show in our experiments that the quality of the pumping beam affected the output power significantly. The terahertz output energy varied from 172 nJ to 17 nJ when the M^2 value of the pump beam varied from 4.21 to 11.1 under the same pumping energy of 120 mJ/pulse. The experimental results were explained by the gain enhancement effect in stimulated Raman emission under multi-mode pumping.
Key words: Terahertz-wave parametric oscillator, Pump mode, Output characteristics

73. Energy scaling of a tunable terahertz parametric oscillator with a surface emitted configuration

Y Y Wang, Z X Li, J Q Li, C Yan, T N Chen, D G Xu, W Shi, H Feng, J Q Yao
Laser Physics, 2019, 24: 125402.

Abstract: A high-energy THz-wave output has been experimentally demonstrated with a terahertz-wave parametric oscillator based on a surface-emitted configuration. Through optimizing the cavity length and pump beam size, the maximum THz-wave output energy of 854 nJ pulse^{-1} was obtained at 1.62 THz. The conversion efficiency was 0.57×10^{-5}, corresponding to the photon conversion efficiency of 0.099%. The THz beam profile was measured with a THz imager, which had a Gaussian profile. The measured beam diameter sizes were 423 μm and 258 μm in the horizontal and vertical directions, respectively. A wide tunable range from 0.75 to 2.81 THz was realized.

Key words: Terahertz-wave parametric oscillator, High energy, Surface-emitted configuration

74. Investigation of a terahertz-wave parametric oscillator using LiTaO$_3$ with the pump-wavelength tuning method

Bo Sun, Xianpeng Bai, Jinsong Liu, Jianquan Yao
Laser Physics, 2014, 24: 035402.

Abstract: We investigate theoretically the performance of a terahertz parametric oscillator (TPO) using LiTaO$_3$ (LT) with the pump-wavelength tuning method. The frequency tuning accuracy of the LT-TPO is potentially superior to that of a TPO using LiNbO$_3$ (LN) under the same conditions. The variation of the radiation angle of THz waves coupled from the Si prism of the LT-TPO is up to about 20° lower than that for the LN-TPO for a pump-wavelength tuning range of 0.4–1.6 μm. Although the THz-wave parametric gain characteristics of LiTaO$_3$ are somewhat unsatisfactory compared with those of LiNbO$_3$, the LT-TPO with pump-wavelength tuning can still show potential for high-performance operation, with the excellent optical properties of the LiTaO$_3$.

Key words: Terahertz wave, Terahertz-wave parametric oscillator, LiTaO$_3$, Pump-wavelength tuning method, Stimulated polariton scattering

75. 太赫兹波光学参量效应放大特性的理论研究

李忠洋，谭联，袁源，邴丕彬，袁胜，徐德刚，姚建铨

红外与激光工程, 2014, 43(8): 2650–2655.

Abstract: Collinear interaction of pump wave, Stokes wave and terahertz wave was proposed in this letter to solve the problem of intense absorption of terahertz wave in gain medium in terahertz parametric oscillation. The amplification characteristics of forward and backward terahertz wave were theoretically analyzed by solving coupled wave equations under difference approximate conditions based on PPLN crystal as an example. The results indicate that the terahertz wave power can be greatly enhanced by the collinear interaction of the pump wave, the Stokes wave and the terahertz wave as the absorption coefficient are greatly more than the gain coefficient of the terahertz wave. The analysis results in this letter provide comprehensive theoretical foundations for the experiment researches of terahertz wave parametric oscillation in PPLN crystal.

Key words: Terahertz wave, Terahertz wave parametric effect, Quasi-phase matching

76. Study on terahertz parametric oscillator pumped by multi-transverse-mode lasers

Jiaqi Li, Yuye Wang, Degang Xu, Chao Yan, Zhongxiao Li, Pengxiang Liu, Wei Shi, Jianquan Yao

39th International Conference on Infrared, Millimeter, and Terahertz waves (IRMMW-THz), 2014: 1–2.

Abstract: The influence of pumping mode on the output characteristic of terahertz parametric oscillators (TPO) is studied theoretically and experimentally. A multi-transverse mode laser was used as the pump source. With the improvement of the transverse mode of the pump beam, a significant increasing of THz output power was observed. The THz output energy decreased from 172 nJ to 17 nJ when M^2 value of pump beam degraded from 4.21 to 11.1, under the same pumping energy of 120mJ/pulse. The gain enhancement effect under multimode pumping in the TPO is analyzed theoretically, which can illustrate the variation in THz output power under different pumping modes.

Key words: Terahertz parametric oscillators (TPO), Multi-transverse mode laser, THz output

77. 太赫兹参量振荡器研究进展

李惟帆，郭宝山，史伟

激光与光电子学进展，2014, (51), 090005.

摘要：太赫兹(THz)波以其光谱和传输性能等方面的独特特性在基础学科研究,医学成像和无损检测等多领域具有重要的应用前景,其科学研究战略意义重大。根据 THz 波参量振荡器 (TPO) 的结构,主要从产生 THz 波常用的非线性晶体,内腔,外腔及腔增强结构,THz 波输出耦合方式,浅表面输出结构,抽运光参数对 TPO 的影响和种子注入技术几个方面对国内外 THz 参量振荡器的研究进展进行回顾。随着新材料和新结构的研究,TPO 将会在越来越多的领域中发挥作用。

关键词：光谱学,太赫兹参量振荡,铌酸锂晶体,输出耦合,谐振腔结构,种子注入技术

78. Terahertz wave parametric oscillations at polariton resonance using a MgO:LiNbO$_3$ crystal

Zhongyang Li, Pibin Bing, Sheng Yuan, Degang Xu, Jianquan Yao

Applied Optics, 2015, 54(18): 5645–5649.

Abstract: Terahertz wave (THz-wave) parametric oscillations with a noncollinear phase-matching scheme at polariton resonance using a MgO:LiNbO$_3$ crystal with a surface-emitted configuration are investigated. We investigate frequency tuning characteristics of a THz-wave via varying the wavelength of the pump wave and phase-matching angle. The effective parametric gain length under the noncollinear phase-matching condition is calculated. Parametric gain and absorption characteristics of a THz-wave in the vicinity of polariton resonances are analyzed.

Key words: Terahertz parametric oscillations, Noncollinear phase-matching, Surface-emitted configuration, MgO:LiNbO$_3$

79. Investigation on terahertz generation at polariton resonance of MgO:LiNbO$_3$ by difference frequency generation

Zhongyang Li, Pibin Bing, Sheng Yuan, Degang Xu, Jianquan Yao

Optics & Laser Technology, 2015, 69: 13–16.

Abstract: Terahertz (THz) wave generation at polariton resonance of MgO:LiNbO$_3$ with a surface-emitted configuration is investigated. It is shown that by using crystal birefringence of bulk MgO:LiNbO$_3$ crystal THz wave can be efficiently generated by difference frequency generation (DFG) in the vicinity of polariton resonances with Cherenkov phase matching scheme. The frequency tuning characteristics of THz wave via varying wavelength of difference frequency waves, phase matching angle and working temperature are numerically analyzed. Parametric gain coefficient in the low-loss limit and absorption coefficient of THz wave during DFG process in the vicinity of polariton resonances are analyzed.

Key words: Terahertz wave, Difference frequency generation, Cherenkov phase matching

80. Investigation on terahertz parametric oscillators using GaP crystal with a noncollinear phase-matching scheme

Zhongyang Li, Pibin Bing, Sheng Yuan, Degang Xu, Jianquan Yao

Journal of Modern Optics, 2015, 62(4): 302–306.

Abstract: Terahertz parametric oscillator (TPO) using GaP as gain medium with a noncollinear phase-matching scheme is investigated. Frequency-tuning characteristics of the terahertz wave (THz-wave) by varying the phase-matching angle and pump wavelength are analyzed. Gain and absorption characteristics of the THz-wave are investigated. The characteristics of GaP and LiNbO$_3$ (LN) are compared when the two crystals are used as gain medium for TPO. The analyses indicate that GaP is more suitable than LN to be used as gain medium for TPO.

Key words: Terahertz parametric oscillator, Noncollinear phase-matching, GaP

81. Investigation of terahertz generation from bulk and periodically poled LiTaO₃ crystal with a cherenkov phase matching scheme

Zhongyang Li, Pibin Bing, Sheng Yuan, Degang Xu, Jianquan Yao

Journal of the Optical Society of Korea, 2015, 19(3): 297–302.

Abstract: Terahertz (THz) wave generation from bulk and periodically poled LiTaO$_3$ (PPLT) with a Cherenkov phase matching scheme is numerically investigated. It is shown that by using the crystal birefringence of bulk LiTaO$_3$ and a grating vector of PPLT, THz waves can be efficiently generated by difference frequency generation (DFG) with a Cherenkov phase matching scheme. The frequency tuning characteristics of the THz wave via varying wavelength of difference frequency waves, phase matching angle, poling period of PPLT and working temperature are theoretically analyzed. The parametric gain coefficient in the low-loss limit and the absorption coefficient of the THz wave during the DFG process in the vicinity of polariton resonances are numerically analyzed. A THz wave can be efficiently generated by utilizing the giant second order nonlinearities of LiTaO$_3$ in the vicinity of polariton resonances.

Key words: Terahertz wave, Difference frequency generation, Cherenkov phase matching

82. Theoretical analysis of terahertz generation in periodically inverted nonlinear crystals based on cascaded difference frequency generation process

Chenfei Hu, Kai Zhong, Jialin Mei, Maorong Wang, Shibei Guo, Wenzhong Xu, Pengxiang Liu, Degang Xu, Yuye Wang, Jianquan Yao

Modern Physics Letters B, 2015, 29(02): 1450263.

Abstract: The characteristics of terahertz (THz) generation based on cascaded difference frequency generation (DFG) process in periodically inverted gallium arsenide (GaAs) and gallium phosphide (GaP) is calculated from coupled wave equations, in which the output enhancement factors are 5.4 and 3.9 in the two crystals, respectively, compared with DFG without cascading. The optimal interaction length, influence of crystal absorption, wave vector mismatch and pump intensity are analyzed. A short discussion on wavelength tuning is also given. The calculated optimal operating parameters and conclusions can provide good directions for the experimental design.

Key words: Terahertz, Difference frequency generation, Quasi-phase-matching, Cascaded.

83. Investigation on frequency mixing effects in terahertz parametricoscillator with a noncollinear phase-matching scheme

Zhongyang Li, Jun Li, Linfei Liu, Pibin Bing, Sheng Yuan, Degang Xu, Jianquan Yao

Optik, 2015, 126(9): 1032–1035.

Abstract: The frequency mixing effects in terahertz parametric oscillator (TPO) with a noncollinear phase-matching scheme based on bulk lithium niobate ($LiNbO_3$), including sum frequency generation (SFG), difference-frequency generation (DFG), optical parametric oscillation and cascaded optical processes are investigated. The analysis results indicate that the terahertz wave (THz-wave) with N-times frequency can be generated simultaneously from frequency mixing effects in TPO with noncollinear phase-matching scheme. To our best knowledge, this is the first study of the TPO generating THz-wave with N-times frequency in bulk $LiNbO_3$. The frequency tuning characteristics of THz-waves via varying pump wavelength, phase-matching angle and working temperature are theoretically analyzed.

Key words: Terahertz wave, Terahertz wave parametric oscillator, Cascaded optical process.

84. 非共线相位匹配太赫兹波参量振荡器级联参量过程的研究

李忠洋, 张云鹏, 邴丕彬, 袁胜, 徐德刚, 姚建铨

红外与激光工程, 2015, 44(3): 990–995.

摘 要: 在铌酸锂晶体非共线相位匹配太赫兹波参量振荡器中观察到了级联光学参量效应。实验中测量到了一阶、二阶和三阶斯托克斯光。通过分析一阶、二阶和三阶斯托克斯光谱发现相邻阶斯托克斯光频率差相等,表明在太赫兹波的产生过程中发生了级联光学参量效应。在高阶级联光学参量过程中,一个泵浦光子可以产生多个太赫兹光子,表明在太赫兹波产生过程中量子转换效率会有效提高。

关键词: 太赫兹波, 太赫兹波参量振荡器, 级联光学参量振荡

85. Terahertz fiber laser based on a novel crystal fiber converter

Pengxiang Liu, Wei Shi, Degang Xu, Xinzheng Zhang, Jianquan Yao, Robert A. Norwood, Nasser Peyghanbarian

Infrared, Millimeter, and Terahertz waves (IRMMW-THz), **2015 40th International Conference on. IEEE**, 2015: 1–1.

Abstract: We proposed a novel GaAs-based crystal fiber converter for efficient THz difference frequency generation, which combines the single-mode THz fiber and the quasi-phase-matching configuration. Calculations were performed on the characteristics of energy conversion and output beam focusing. Theoretical results indicated that the proposed THz fiber laser can provide high power and high brightness THz source.

Key words: Terahertz wave, Difference frequency generation, Crystal fiber converter, GaAs

86. 高能量、快速可调谐太赫兹参量振荡器

王与烨, 李忠孝, 李佳起, 闫超, 徐德刚, 姚建铨

太赫兹科学与电子信息学报, 2015, 13(1): 19–22.

摘 要: 基于受激电磁耦子散射的太赫兹(THz)波参量振荡器是产生高能量、相干 THz 波的有效手段之一, 实验中采用垂直晶体表面出射结构, 在 1.63 THz 处实现了最高单脉冲能量为 634 nJ 的 THz 波输出。利用电控振镜快速改变泵浦光入射到 MgO:LN 晶体中的角度, 实现了 0.75 THz~2.81 THz 范围内的快速调谐, 该辐射源可以满足 THz 波在生物医学、太赫兹通信、环境监测等应用领域的需求。

关键词: 太赫兹波, 参量振荡, 可调谐

87. High-power high-brightness terahertz source based on nonlinear optical crystal fiber

Pengxiang Liu, Wei Shi, Degang Xu, Xinzheng Zhang, Jianquan Yao, Robert A. Norwood, Nasser Peyghambarian

IEEE Journal of Selected Topics in Quantum Electronics, 2016, 22(2): 8500105.

Abstract: A nonlinear crystal fiber converter for terahertz (THz) difference frequency generation (DFG) is proposed. This axisymmetric structure consists of a periodically inverted GaAs core clad with index-matched chalcogenide glass, which functions as a single-mode fiber for THz radiation to prevent its diffraction and achieve effifficient optical-to-THz conversion. The process of DFG between ps-pulses in the crystal fiber is investigated by developing the guided-wave coupled-mode equations for quasi monochromatic pulses. The output power, spectral power density, and brightness are analyzed, based on the calculations of the dynamic of energy conversion and the characteristics of the THz beam focusing. High output power as well as excellent beam quality contributes to the high brightness of the presented THz crystal fiber source.

Key words: Brightness, Difference frequency generation, Single-mode fiber, Terahertz-wave generation

88. Widely tunable and monochromatic terahertz difference frequency generation with organic crystal 2-(3-(4-hydroxystyryl)-5, 5-dime-thylcyclohex-2-enylidene) malononitrile

Pengxiang Liu, Xinyuan Zhang, Chao Yan, Degang Xu, Yin Li, Wei Shi, Guochun Zhang, Xinzheng Zhang, Jianquan Yao, Yicheng Wu

Applied Physics Letters, 2016, 108(1):621–629.

Abstract: We report an experimental study on widely tunable terahertz (THz) wave difference frequency generation (DFG) with hydrogen-bonded crystals 2-(3-(4-hydroxystyryl)-5,5-dime-thylcyclohex-2-enylidene) malononitrile (OH1). The organic crystals were pumped by a ~1.3 μm double-pass $KTiOPO_4$ optical parametric oscillator. A tuning range of 0.02–20 THz was achieved. OH1 crystals offer a long effective interaction length (also high output) for the generation below 3 THz, owing to the low absorption and favorable phase-matching. The highest energy of 507 nJ/pulse was generated at 1.92 THz with a 1.89-mm-thick crystal. Comprehensive explanations were provided, on the basis of theoretical calculations. Cascading phenomenon during the DFG process was demonstrated. The photon conversion effifficiency could reach 2.9%.

Key words: Widely tunable terahertz, Difference frequency generation (DFG), Organic crystal

89. Green laser induced terahertz tuning range expanding in $KTiOPO_4$ terahertz parametric oscillator

Chao Yan, Yuye Wang, Degang Xu, Wentao Xu, Pengxiang Liu, Dexian Yan, Pan Duan, Kai Zhong, Wei Shi, Jianquan Yao

Applied Physics Letters, 2016, 108(1): 011107.

Abstract: 532 nm green laser is utilized to achieve terahertz tuning range expanding in $KTiOPO_4$ terahertz parametric oscillator. With the theoretical analysis of the stimulated polariton scattering, an expanded tunability of the $KTiOPO_4$ terahertz parametric oscillator can be realized. A wide terahertz output tuning range from 5.7 to 6.1 THz, from 7.4 to 7.8 THz, from 11.5 to 11.8 THz, and from 13.3 to 13.5 THz was demonstrated in our experiment, and the result well matched the analysis. The maximum terahertz output energy was 1.61 μJ under the pump energy of 140 mJ, corresponding to the maximum THz wave conversion effifficiency of 1.3×10^{-5}, and the threshold pump energy is about 30 mJ.

Key words: Terahertz output, Tuning frequency range, Terahertz parametric oscillator

90. Molecular design on isoxazolone-based derivatives with large second-order harmonic generation effect and terahertz wave generation

Xinyuan Zhang, Xingxing Jiang, Pengxiang Liu, Yin Li, Heng Tu, Zheshuai Lin, Degang Xu, Guochun Zhang, Yicheng Wu, Jianquan Yao

Crystengcomm, 2016, 18(20): 3667–3673.

Abstract: A series of 3-phenyl-5-isoxazolone and 3-methyl-5-isoxazolone-based compounds with different electron donors were synthesized. Fourteen single crystals in this family were obtained by slow evaporation. Single crystal X-ray studies showed that four new 3-phenyl-5-isoxazolone-based crystals, $C_{20}H_{18}N_2O_2$ (PDI), $C_{15}H_{11}NO_2S$ (PTI), $C_{23}H_{15}NO_2$ (PFI), and $C_{17}H_{13}NO_2S$ (PMI-(I)), belong to the noncentrosymmertric space group. The inclinations to achieve noncentrosymmetric space groups in the 3-phenyl-5-isoxazolonebased crystals have been confirmed by first-principles calculations. Powder second harmonic generation (SHG) tests revealed that the abovementioned four acentric crystals exhibited very large SHG intensities that were about 1 to 3 times that of OH1 (a commonly used nonlinear optical crystal) under 2.09 μm light. Their UV-vis absorption, diffuse reflectance spectroscopy, and thermal properties were also characterized. Moreover, widely tunable and monochromatic terahertz difference frequency generation in the 3-phenyl-5-isoxazolone-based crystal $C_{17}H_{13}NO_3$ (MLS) was realized for the first time.

Key words: Terahertz, Second harmonic generation (SHG), Difference frequency generation, Isoxazolone-based derivatives

91. High-energy terahertz wave parametric oscillator with a surface-emitted ring-cavity configuration

Zhen Yang, Yuye Wang, Degang Xu, Wentao Xu, Pan Duan, Chao Yan, Longhuang Tang, Jianquan Yao

Optics Letters, 2016, 41(10): 2262–2265

Abstract: A surface-emitted ring-cavity terahertz (THz) wave parametric oscillator has been demonstrated for high-energy THz output and fast frequency tuning in a wide frequency range. Through the special optical design with a galvanooptical scanner and four-mirror ring-cavity structure, the maximum THz wave output energy of 12.9 μJ/pulse is achieved at 1.359 THz under the pump energy of 172.8 mJ. The fast THz frequency tuning in the range of 0.7–2.8 THz can be accessed with the step response of 600 μs. Moreover, the maximum THz wave output energy from this configuration is 3.29 times as large as that obtained from the conventional surface-emitted THz wave parametric oscillator with the same experimental conditions.

Key words: Surface emitted terahertz, THz output, Fast frequency tuning, Terahertz (THz) wave parametric oscillator

92. Efficient phase-matching for difference frequency generation with pump of Bessel laser beams

Pengxiang Liu, Wei Shi, Degang Xu, Xinzheng Zhang, Guizhong Zhang, Jianquan Yao

Optics Express, 2016, 24(2): 901–906.

Abstract: A type of phase matching for difference frequency generation with Bessel-type pump beams is proposed. In this geometry, the phase matching is achieved in a cone around the laser path by properly controlling the beam profile. An experimental case that 1.5 THz generation with ~2 μm lasers pumped bulk GaAs crystal is considered. Calculations of the energy conversion characteristics are performed based on a semi-analytical model. The results indicate that this configuration could relax the phase matching condition in a wide range of nonlinear crystals and pump wavelengths.

Key words: Phase matching, Difference frequency generation, Semi-analytical model

93. Widely-tunable high-repetition-rate terahertz generation in GaSe with a compact dualwavelength KTP OPO around 2 μm

Jiang Mei, Kai Zhong, Maorong Wang, Yang Liu, Degang Xu, Wei Shi, Yuye Wang, Jianquan Yao, Robert A. Norwood, Nasser Peyghambarian

Optics Express, 2016, 24(20): 23368–23375.

Abstract: A compact efficient high-repetition-rate doubly-resonant dual-wavelength KTP optical parametric oscillator (OPO), with output power up to 3.65 W and tuning ranges of 2.088-2.133 μm/ 2.171-2.122 μm for signal/idler waves, was deployed for terahertz (THz) generation in a GaSe crystal. Based on difference frequency generation (DFG), the THz wave was continuously tunable from 730.9 μm (0.41 THz) to 80.8 μm (3.71 THz), believed to be the first report of a compact high-repetition-rate widely-tunable THz source. The maximum THz average power reached 1.2 μW at 1.54 THz and the corresponding DFG efficiency was 7.8×10^{-7}, entirely suitable for portable applications. The utility of the THz source was also demonstrated through spectroscopy and imaging experiments.

Key words: Optical parametric oscillator (OPO), Difference frequency generation (DFG), Widely tunable, High repetition rate THz source

94. Numerical study of compact terahertz gas laser based on photonic crystal fiber cavity

Dexian Yan, Haiwei Zhang, Degang Xu, Wei Shi, Chao Yan, Pengxiang Liu, Jia Shi, Jianquan Yao

Journal of Lightwave Technology, 2016, 34(14):3373–3378.

Abstract: We report a compact terahertz (THz) gas laser based on the cavity photonic crystal fiber (PCF). This compact structure consists of a large air core PCF constructed by high-density polyethylene tubes. The 2.52 THz wave can be generated in the reaction cell filled with the CH_3OH gas. The reaction cell should shrink the volume of THz sources. The saturation intensity is analyzed based on the rate equations theory. Numerical calculations show that the THz wave can realize the fundamental mode transmission and obtain a lower confinement loss of 5.6 dB/m and the flat bending loss spectrum. The optimum pressure, pump threshold are analyzed and the THz output power can reach the magnitude of 10 mW. We combined the THz pump technology and the excellent transmission characteristics of the PCF together. This compact THz gas laser gives a new method to realize the THz source with high beam quality, low transmission loss, well-distributed bending shape and small size.

Key words: Optical resonators, Photonic crystal fiber, Rate equation terahertz generation

95. Compact high-repetition-rate monochromatic terahertz source based on difference frequency generation from a dual-wavelength Nd:YAG laser and DAST crystal

Kai Zhong, Jialin Mei, Marong Wang, Pengxiang Liu, Degang Xu, Yuye Wang, Wei Shi, Jianquan Yao, Bing Teng, Yong Xiao

Journal of Infrared Millimeter & Terahertz Waves, 2016, 38(1):1−9.

Abstract: Although high-repetition-rate dual-wavelength Nd:YAG lasers at 1 319 and 1 338 nm have been realized for quite a long time, we have employed it in generating monochromatic terahertz (THz) wave in this paper for the first time. The dual-wavelength laser was LD-end-pumped and acousto-optically (AO) Q-switched with the output power of watt level operating at different repetition rates from 5.5 to 30 kHz. Using a 0.6-mm-thick organic nonlinear crystal DAST for difference frequency generation (DFG), a compact terahertz source was achieved at 3.28 THz. The maximum average output power was about 0.58 μW obtained at a repetition rate of 5.5 kHz, corresponding to the conversion efficiency of about 6.4×10^{-7}. The output power scaling is still feasible with higher pump power and a longer nonlinear DFG crystal. Owing to the compactness of the dual-wavelength laser and the nonlinear crystal, a palm-top terahertz source is expected for portable applications such as imaging and so on.

Key words: Terahertz generation, Difference frequency generation (DFG), Parametric down conversion, Dual-wavelength laser, Compact, Nd:YAG

96. High-repetition-rate terahertz generation in QPM GaAs with a compact efficient 2-μm KTP OPO

Jialin Mei, Kai Zhong Maorong Wang, Pengxiang Liu, Degang Xu, Yuye Wang, Wei Shi, Jianquan Yao, Robert A. Norwood, Nasser Peyghambarian

IEEE Photonics Technology Letters, 2016, 28(14):1501−1504.

Abstract: An efficient tunable high-repetition-rate KTiOPO$_4$ (KTP) optical parametric oscillator (OPO) was deployed as a compact terahertz (THz) source for the first time. The KTP OPO was intracavity pumped by an acousto-optical Q-switched Nd:YVO$_4$ laser and emitted two tunable wavelengths near degeneracy. The OPO tuning range extended continuously from 2.068 to 2.191 μm with a maximum output power of 3.29 W at 24 kHz, corresponding to an optical–optical conversion efficiency of 20.69%, believed to be among the highest reported. A periodically inverted quasi-phase-matched GaAs crystal was used to generate the THz wave by the difference frequency generation (DFG) method based on a dual-wavelength laser. The maximum output THz power was 0.6 μW at 1.244 THz, and the DFG energy conversion efficiency was 4.32×10^{-7}. This provides a potential practical palm-top tunable THz sources for portable applications.

Key words: Terahertz (THz), Difference frequency generation (DFG), Optical parametric oscillator (OPO), Gallium arsenide (GaAs), Quasi-phase-matching (QPM)

97. Compact high-repetition-rate terahertz source based on difference frequency generation from an efficient 2-μm dual-wavelength KTP OPO

Jialin Mei, Kai Zhong Maorong Wang, Pengxiang Liu, Degang Xu, Yuye Wang, Wei Shi, Jianquan Yao, Robert A. Norwood, Nasser Peyghambarian

International Conference on Infrared, Millimeter, and Terahertz Waves (IRMMW-THz 2016), 2016, 10030: 100301Q.

Abstract: A compact optical terahertz (THz) source was demonstrated based on an efficient high-repetition-rate doubly resonant optical parametric oscillator (OPO) around 2 μm with two type-II phase-matched KTP crystals in the walk-off compensated configuration. The KTP OPO was intracavity pumped by an acousto-optical (AO) Q-switched Nd:YVO$_4$ laser and emitted two tunable wavelengths near degeneracy. The tuning range extended continuously from 2.068 μm to 2.191 μm with a maximum output power of 3.29 W at 24 kHz, corresponding to an optical-optical conversion efficiency (from 808 nm to 2 μm) of 20.69%. The stable pulsed dual-wavelength operation provided an ideal pump source for generating terahertz wave of micro-watt level by the difference frequency generation (DFG) method. A 7.84-mm-long periodically inverted quasi-phase-matched (QPM) GaAs crystal with 6 periods was used to generate a terahertz wave, the maximum voltage of 180 mV at 1.244 THz was acquired by a 4.2-K Si bolometer, corresponding to average output power of 0.6 μW and DFG conversion efficiency of 4.32×10^{-7}. The acceptance bandwidth was found to be larger than 0.35 THz (FWHM). As to the 15-mm-long GaSe crystal used in the type-II collinear DFG, a tunable THz source ranging from 0.503 THz to 3.63 THz with the maximum output voltage of 268 mV at 1.65 THz had been achieved, and the corresponding average output power and DFG conversion efficiency were 0.9 μW and 5.86×10^{-7} respectively. This provides a potential practical palm-op tunable THz sources for portable applications.

Key words: Terahertz (THz), Optical parametric oscillator (OPO), Difference frequency generation (DFG), Gallium arsenide (GaAs), Gallium selenide (GaSe)

98. Enhanced stimulated polariton scattering in KTiOPO$_4$ terahertz parametric oscillator based on green laser pumping

C. Yan, Y. Y. Wang, D. G. Xu, P. X. Liu, J. Shi, D. X. Yan, H. X. Liu, Y. X. He, M. T. Nie, L. H. Tang, K. Zhong, W. Shi, J. Q. Yao

International Conference on Infrared, Millimeter, and Terahertz Waves (IRMMW-THz 2016), 2016: 1–2.

Abstract: Enhanced stimulated polariton scattering induced by 532 nm green laser in KTiOPO$_4$ terahertz parametric oscillator was demonstrated in this paper. An expanded terahertz wave tuning range from 5.7 to 6.1 THz, from 7.4 to 7.8 THz, from 11.5 to 11.8 THz and from 13.3 to 13.5 THz was achieved. The maximum terahertz output energy was 1.61 μJ at 7.57 THz, additionally, the energy levels of the terahertz wave were relatively stable in this condition of 532 nm green laser pumping.

Key words: Terahertz, Enhanced stimulated polariton scattering, Terahertz parametric oscillator, KTiOPO$_4$

99. Widely-tunable terahertz parametric oscillator based on MgO-doped near-stoichiometric LiNbO$_3$ crystal

Longhuang Tang, Yuye Wang, Degang Xu, Zhen Yang, Chao Yan, Wentao Xu, Pan Duan, Yixin He, Jia Shi, Meitong Nie, Jianquan Yao

SPIE/COS Photonics Asia. International Society for Optics and Photonics, 2016, 10030: 10030O.

Abstract: A widely tunable, high-energy terahertz wave parametric oscillator based on 1 mol. % MgO-doped near-stoichiometric LiNbO$_3$ crystal has been proposed with 1 064 nm nanosecond pulsed laser pumping. The tunable range of 1.16 to 4.64 THz was obtained. Under the pump energy of 150 mJ/pulse, the maximum THz wave output energy of 12.56 μJ was achieved at 1.88 THz, corresponding to the THz wave conversion efficiency of 7.61×10^{-5} and the photon conversion efficiency of 1.14%, respectively. Moreover, the THz half maximum (FWHM) beam diameters of MgO:SLN TPO measured at 4 cm from the output surface were 7.42 mm and 6.06 mm in the vertical and horizontal directions, respectively.

Key words: Nonlinear optics, Terahertz parametric oscillator, Tunability, Laser, Terahertz, Nonlinear crystal, THz source

100. High-repetition-rate, widely tunable terahertz generation in GaSe pumped by a dual-wavelength KTP-OPO

Dexian Yan, Degang Xu, Yuye Wang, Wei Shi, Kai Zhong, Pengxiang Liu, Chao Yan, Quan Sheng, Jialin Mei, Jia Shi, Jianquan Yao

SPIE/COS Photonics Asia. 2016, 10030: 100300B.

Abstract: High-repetition-rate, monochromatic and tunable terahertz (THz) source is demonstrated. We use an orthogonally polarized dual-wavelength intracavity OPO to complete the type-II phase-matched collinear diffferencerequency generation in GaSe. A high average-power 2 μm laser with 12 W output power and good beam quality based on an intracavity KTP OPO is experimentally designed. The KTP OPO is intracavity pumped by an acousto-optical Q-switched side-pumped Nd:YAG with the repetition rate of 10 kHz. Two identical KTP crystals were $7 \times 8 \times 15$ mm^3 in size, cut at θ =51.2°, φ = 0°, which were tuned in the x-z plane to achieve type-II phase-matching. The KTP OPO consists of two identical KTP crystals to reduce the walk-off effect and improve the beam overlap area of the output signal and idler waves. The pulse-width of the 2-μm KTP OPO laser is about 11 ns with the linewidth about 0.8 nm. The focused OPO beam is injected into the uncoated GaSe with the length of 8 mm, and the generated THz wave is detected with a 4.2-K Si-bolometer after focusing with a polyethylene lens. The tunable and coherent radiation from 0.2 to 3 THz has been achieved based on the type-II phase-matching DFG when the two pump waves are in the range of 2.106 4 - 2.127 2 μm and 2.151 6 - 2.130 4 μm while symmetrically tuning the phase-matching angle of the KTPs. The maximum output THz average power can reach μW-level around 1.48 THz.

Key words: Terahertz, Difference Frequency generation, Optical parametric oscillator, Gallium selenide, Highrepetition-rate laser

101. THz wave parametric oscillator with a surface-emitted ring-cavity configuration

Zhen Yang, Yuye Wang, Degang Xu, Longhuang Tang, Wentao Xu, Pan Duan, Chao Yan, Jianquan Yao

SPIE/COS Photonics Asia. International Society for Optics and Photonics, 2016, 10030: 10030C

Abstract: A surface-emitted ring-cavity terahertz (THz) wave parametric oscillator has been demonstrated for high-energy THz output and fast frequency tuning. Through the special optical design with a Galvano optical scanner and four-mirror ringcavity structure, a maximum THz output of 12.9 µJ/pulse is achieved at 1.359 THz under the pump pulse energy of 172.8 mJ with the repetition rate of 10 Hz. A further research on the performance of the SE ring-cavity TPO has done to explore more characteristics of THz output. The THz pulse instability and the influence of cavity loss has analyzed. Moreover, the pump depletion rate of the ring-cavity configuration is much lower than the conventional surface-emitted terahertz wave parametric oscillator at the same experimental conditions.

Key words: Parametric processes, Terahertz, Ring cavity, Nonlinear process

102. 小型化可调谐太赫兹波参量振荡器的研究

聂美彤, 王与烨, 徐德刚, 闫超, 徐文韬, 杨振, 姚建铨

第二届全国太赫兹科学技术学术年会, 2016.

摘要: 太赫兹(THz)辐射由于其自身所具有的独特光学性质,在光谱分析、医学成像、环境监测、无损检测、移动通讯等领域均有重要的科研价值和广阔的应用前景。太赫兹参量振荡器结构紧凑,输出功率高,具有较宽的光谱调谐范围,因此是产生高能量、相干 THz 辐射源的有效手段之一。本文基于 MgO:LiNbO$_3$ 晶体垂直表面出射 THz 参量振荡技术与电控振镜旋转,实现了高能量、快速可调谐的小型化太赫兹源。实验中泵浦源采用自主搭建的 LD 侧面泵浦电光调 Q Nd:YAG 激光器,其输出波长为 1 064 nm,脉冲宽度 10ns,重复频率 10Hz,单脉冲能量可达 240 mJ。非线性晶体采用掺杂浓度为 5%mol 的 MgO:LiNbO$_3$ 泵浦光经望远系统将光束直径压缩为 5 mm,再经电控旋转振镜反射到 Stokes 谐振腔内, Stokes 谐振腔腔镜镀 1 064 nm 高透、1 067~1 075 nm 高反膜,以保证腔内只有 Stokes 光振荡。通过改变电控振镜上的外加电压,可以连续改变入射泵浦光与 Stokes 光的夹角,即改变非共线相位匹配角,从而可实现太赫兹频率的快速调谐。实验结果表明,当泵浦光能量为 155.7 mJ 时,在 1.75 THz 处太赫兹波最大单脉冲能量为 10.5 μJ,平均功率约为 0.105 mW,对应的太赫兹波能量转换效率为 6.7×10^{-5}, TPO 阈值为 44.9 mJ。通过电脑控制改变旋转振镜的外加电压,可以实现对太赫兹频率的快速调谐,实验表明,当泵浦光能量为 106 mJ 时,角度调谐范围 0.6°~2.9°,实现的太赫兹波频率调谐范围为 0.8 THz -2.9 THz。在调谐角度为 1.9° 时,采用 F-P 波长计对太赫兹波输出波长进行直接测量,测量结果与由非共线相位匹配条件计算得出的太赫兹波理论波长误差小于 5%,吻合较好。

关键词: 太赫兹源,光参量振荡,可调谐

103. High-average-power, high-repetition-rate tunable terahertz difference frequency generation with GaSe crystal pumped by 2 μm dual-wavelength intracavity KTP optical parametric oscillator

Dexian Yan, Yuye Wang, Degang Xu, Pengxiang Liu, Chao Yan, Jia Shi, Hongxiang Liu, Yixin He, Longhuang Tang, Jianchen Feng, Jianqin Guo, Wei Shi, Kai Zhong, Yuen H. Tsang, Jianquan Yao

Photonics Research, 2017, 5(2): 82–87.

Abstract: We have demonstrated a high-average-power, high-repetition-rate optical terahertz(THz) source based on difference frequency generation(DFG)in the GaSe crystal by using a near-degenerate 2 μm intracavity KTP optical parametric oscillator as the pump source. The power of the 2 μm dual-wavelength laser was up to 12.33 W with continuous tuning ranges of 1 988.0–2 196.2 nm/2 278.4–2 065.6 nm for two waves. Different GaSe cystal lengths have been experimentally investigated for the DFG THz source in order to optimize the THz output power, which was in good agreement with the theoretical analysis. Based on an 8 mm long GaSe crystal, the THz wave was continuously tuned from 0.21 to 3 THz. The maximum THz average power of 1.66 μW was obtained at repetition rate of 10 kHz under 1.48 THz. The single pulse energy amounted to 166 pJ and the conversion efficiency from 2 μm laser to THz output was 1.68×10^{-6}. The signal-to-noise ratio of the detected THz voltage was 23 dB. The acceptance angle of DFG in the GaSe crystal was measured to be 0.16°.

Key words: Terahertz(THz)source, Difference frequency generation(DFG), Optical parametric oscillator, GaSe

104. Energy scaling and extended tunability of terahertz wave parametric oscillator with MgO-doped near-stoichiometric LiNbO$_3$ crystal

Yuye Wang, Longhuang Tang, Degang Xu, Chao Yan, Yixin He, Jia Shi, Dexian Yan, Hongxiang Liu, Meitong Nie, Jiachen Feng, Jianquan Yao

Optics Express, 2017, 25(8): 8926–8936.

Abstract: A widely tunable, high-energy terahertz wave parametric oscillator based on 1 mol. %MgO-doped near-stoichiometric LiNbO$_3$ crystal has been demonstrated with 1 064 nm nanosecond pulsed laser pumping. The tunable range of 1.16 to 4.64 THz was achieved. The maximum THz wave output energy of 17.49 μJ was obtained at 1.88 THz under the pump energy of 165 mJ/pulse, corresponding to the THz wave conversion efficiency of 1.06×10^{-4} and the photon conversion efficiency of 1.59%, respectively. Moreover, under the same experimental conditions, the THz output energy of TPO with MgO: SLN crystal was about 2.75 times larger than that obtained from the MgO: CLN TPO at 1.60 THz. Based on the theoretical analysis, the THz energy enhancement mechanism in the MgO: SLN TPO was clarified to originate from its larger Raman scattering cross section and smaller absorption coefficient.

Key words: Simulated polariton scattering, Generation, MGO-LiNBO$_3$, Coupler, THz wave Output

105. Compact and stable high-repetition-rate terahertz generation based on an efficient coaxially pumped dual-wavelength laser

Yang Liu, Kai Zhong, Jialin Mei, Chu Liu, Jie Shi, Xin Ding, Degang Xu, Wei Shi, Jianquan Yao

Optics Express, 2017. 25(25): 31988–31996.

Abstract: A compact and stable terahertz (THz) source is demonstrated based on difference frequency generation (DFG) pumped by an efficient dual-wavelength acousto-optic (AO) Q-switched solid-state Nd:YLF laser with composite gain media (a-cut and c-cut) in the coaxial pumping configuration. Optimal power ratio and pulse synchronization of the orthogonal polarized 1 047/1 053 nm dual-wavelength laser could be realized by varying the pump focusing depth and/or pump wavelength. The total power of 2.92 W was obtained at 5 kHz pumped by 10-W laser-diode power at 803 nm. Such an efficient dual-wavelength laser demonstrated good stability and inconspicuous timing jitter benefiting from the suppressed gain competition between two resonating wavelengths. An 8-mm-long GaSe crystal was employed to generate THz waves at 1.64 THz by DFG and the maximum THz average output power was about 0.93 μW. This compact coaxial pumping method can be extended to all kinds of neodymium (Nd) doped laser crystals to produce different dual-wavelength lasers for various THz wavelength generation, which have good prospects for portable and costless applications like imaging, non-destructive inspection, etc.

Key words: Terahertz (THz) source, Difference frequency generation (DFG), Coaxial pumping

106. Compact high-repetition-rate monochromatic terahertz source based on difference frequency generation from a dual-wavelength Nd:YAG laser and DAST crystal

Kai Zhong, Jialin Mei, Maorong Wang, Pengxiang Liu, Degang Xu, Yuye Wang, Wei Shi, Jianquan Yao, Bing Teng, Yong Xiao

Journal of Infrared, Millimeter, and Terahertz Waves, 2016. 38(1): 1–9.

Abstract: Although high-repetition-rate dual-wavelength Nd:YAG lasers at 1 319 and 1 338 nm have been realized for quite a long time, we have employed it in generating monochromatic terahertz (THz) wave in this paper for the first time. The dual-wavelength laser was LD-end-pumped and acousto-optically (AO) Q-switched with the output power of watt level operating at different repetition rates from 5.5 to 30 kHz. Using a 0.6-mm-thick organic nonlinear crystal DAST for difference frequency generation (DFG), a compact terahertz source was achieved at 3.28 THz. The maximum average output power was about 0.58 μW obtained at a repetition rate of 5.5 kHz, corresponding to the conversion efficiency of about 6.4×10^{-7}. The output power scaling is still feasible with higher pump power and a longer nonlinear DFG crystal. Owing to the compactness of the dual-wavelength laser and the nonlinear crystal, a palm-top terahertz source is expected for portable applications such as imaging and so on.

Key words: Terahertz generation, Difference frequency generation (DFG), Parametric down conversion, Dual-wavelength laser, Compact, Nd:YAG

107. Optically pumped terahertz sources

Zhong Kai, Shi Wei, Xu Degang, Liu Pengxiang, Wang Yuye, Mei Jialin, Yan Chao, Fu Shijie, Yao Jianquan

Science China Technological Sciences, 2017, 60(12): 1801–1818.

Abstract: High-power terahertz (THz) generation in the frequency range of 0.1-10 THz has been a fast-developing research area ever since the beginning of the THz boom two decades ago, enabling new technological breakthroughs in spectroscopy, communication, imaging, etc. By using optical (laser) pumping methods with near-or mid-infrared (IR) lasers, flexible and practical THz sources covering the whole THz range can be realized to overcome the shortage of electronic THz sources and now they are playing important roles in THz science and technology. This paper overviews various optically pumped THz sources, including femtosecond laser based ultrafast broadband THz generation, monochromatic widely tunable THz generation, single-mode on-chip THz source from photomixing, and the traditional powerful THz gas lasers. Full descriptions from basic principles to the latest progress are presented and their advantages and disadvantages are discussed as well. It is expected that this review gives a comprehensive reference to researchers in this area and additionally helps newcomers to quickly gain understanding of optically pumped THz sources.

Key words: Terahertz (THz) wave, Photoconductive switch, Optical rectification, Difference frequency generation (DFG), Terahertz parametric oscillator (TPO), Photomixing, THz gas lasers, Phase matching (PM)

108. Bursts of efficient terahertz radiation with saturation effect from metal-based ferromagnetic heterostructures.

Shunnong Zhang, Zuanming Jin, Zhendong Zhu, Weihua Zhu, Zongzhi Zhang, Guohong Ma, Jianquan Yao

Journal of Physics D: Applied Physics, 2018, 51: 034001.

Abstract: We report the broadband terahertz (THz) radiation in the metallic ferromagnetic (FM) heterostructures, upon irradiation of a femtosecond laser pulse at room temperature. The origin of THz generation from FM heterostructures can be interpreted using two terms: the transient demagnetization (a local modification of spin order of the FM metal) and electricdipole radiation resulting from a non-local spin current pulses. Here, we show that the THz emission is dominated by the photo-excited transient charge current, which is converted from the spin current with inverse spin Hall effect. We tailor the metallic heterostructures with different non-magnetic thin layer (Pd or Ru) and FM materials (CoFeB or CoFe), to shape the THz transients. Moreover, we find that a saturation effect of THz radiation for CoFeB/Pd is less compared to CoFeB/Ru. THz emission spectroscopy can be used to qualitatively visualize the spin accumulation in the heterostructures.

Key words: Terahertz (THz) radiation, Spin current, Inverse spin Hall effect

109. Low-threshold terahertz-wave generation based on a cavity phase-matched parametric process in a Fabry–Perot microresonator.

Pengxiang Liu, Feng Qi, Weifan Li, Yelong Wang, Zhaoyang Liu, Hongming Wu, Wei Ning, Wei Shi, Jianquan Yao

Journal of Optical Society of America B, 2018, 35(1): 68–72.

Abstract: A configuration for an optical pumped monochromatic terahertz (THz) source operating in a high repetition rate is developed and based on a resonant parametric process within a Fabry–Perot microcavity. Comprehensive analysis of the spectral selection, cavity phase matching, energy-conversion dynamic, and input–output characteristics is provided. Calculations are performed on a Tm^{3+}-doped fiber laser pumped GaAs sheet. It is indicated that a pump threshold can be scaled down by injection seeding of the signal wave. Efficient THz wave generation (mW level average power) under relatively low pump peak power (65 kW) is predicted.

Key words: Nonlinear wave mixing, Nonlinear optics, Parametric processes, Fabry–Perot microcavity

110. Investigation of stimulated polariton scattering from the B_1-symmetry modes of the $KNbO_3$ crystal

Zhongyang Li, Mengtao Wang, Silei Wang, Bin Yuan, Pibin Bing, Degang Xu, Jianquan Yao

Current Optics and Photonics, 2018, 2(1): 90−95.

Abstract : Stimulated polariton scattering from the B_1-symmetry modes of a $KNbO_3$ crystal to generate a terahertz wave (THz-wave) with a noncollinear phase-matching scheme is investigated. The frequency-tuning characteristics of the THz-wave by varying the phase-matching angle and pump wavelength are analyzed. The expression for the effective parametric gain length under the noncollinear phase-matching condition is deduced. Parametric gain and absorption characteristics of the THz-wave in $KNbO_3$ are theoretically simulated. The characteristics of $KNbO_3$ for a terahertz parametric oscillator (TPO) are compared to those of $MgO:LiNbO_3$. The analysis indicates that $KNbO_3$ is an excellent optical crystal for a TPO, to enhance the THz-wave output.

Key words: Stimulated polariton scattering, $KNbO_3$, Terahertz parametric oscillator

111. High-energy and ultra-wideband tunable terahertz source with DAST crystal via difference frequency generation

Yixin He, Yuye Wang, Degang Xu, Meitong Nie, Chao Yan, Longhuang Tang, Jia Shi, Jiachen Feng, Dexian Yan, Hongxiang Liu, Bing Teng, Hua Feng, Jianquan Yao

Applied Physics B, 2018, 124(1): 16.

Abstract: We have demonstrated a high-energy and broadly tunable monochromatic terahertz (THz) source based on difference frequency generation (DFG) in DAST crystal. A high-energy dual-wavelength optical parametric oscillator with two KTP crystals was constructed as a light source for DFG, where the effect of blue light was first observed accompanying with tunable dual-wavelength pump light due to different nonlinear processes. The THz frequency was tuned randomly in the range of 0.3–19.6 THz. The highest energy of 870 nJ/pulse was obtained at 18.9 THz under the intense pump intensity of 247 MW/ cm^2. The THz energy dips above 3 THz have been analyzed and mainly attributed to the resonance absorption induced by lattice vibration in DAST crystal. The dependence of THz output on the input energy was studied experimentally, and THz output saturation was observed. Furthermore, tests of transmission spectroscopy of four typical samples were demonstrated with this ultra-wideband THz source.

Key words: Terahertz (THz) source, Difference frequency generation (DFG), DAST crystal, Optical parametric oscillator

112. Theoretical study of organic crystal-based terahertz-wave difference frequency generation and up-conversion detection

Pengxiang Liu, Feng Qi, Weifan Li, Zhaoyang Liu, Yelong Wang, Wei Shi, Jianquan Yao

Journal of Infrared, Millimeter, and Terahertz Waves, 2018, 39(10): 1005–1014.

Abstract: Optical pumped quasi-monochromatic and tunable terahertz (THz) wave generation and detection system is investigated theoretically. Calculations are performed on an organic nonlinear medium: 4'-dimethylamino-N-methyl-4-stilbazolium tosylate (DAST), due to the superior properties (large dynamic range and extremely wide tunability). The behavior of pulses during the nonlinear interaction is characterized. Previous experimental results are well explained. Our theoretical model provides an approach to improve the performance of the coherent THz system from two aspects. For the source part, an optimal crystal thickness that maximizes the THz output is obtained. For the detector part, a linear conversion relation (used for calibration) and the parameter dependence of responsivity are provided. It was predicted that a dynamic range of ten orders could be achieved with the DAST-based THz system. The presented model can also be applied to guide the design of THz systems with other nonlinear mediums in collinear phase-matching geometry.

Key words: Nonlinear optics, Parametric process, Terahertz-wave generation, Terahertz-wave detection, Organic nonlinear crystal

113. Energy scaling and extended tunability of a ring cavity terahertz parametric oscillator based on KTiOPO$_4$ crystal

Yuye Wang, Yuchen Ren, Degang Xu, Longhuang Tang, Yixin He, Ci Song, Linyu Chen, Changzhao Li, Chao Yan, Jianquan Yao

Chinese Physics B, 2018, 27(11): 114213.

Abstract: A wide terahertz tuning range from 0.96 THz to 7.01 THz has been demonstrated based on ring-cavity THz wave parametric oscillator with a KTiOPO$_4$ (KTP) crystal. The tuning range was observed intermittently from 0.96 THz to 1.87 THz, from 3.04 THz to 3.33 THz, from 4.17 THz to 4.48 THz, from 4.78 THz to 4.97 THz, from 5.125 THz to 5.168 THz, from 5.44 THz to 5.97 THz, and from 6.74 THz to 7.01 THz. The dual-Stokes wavelengths resonance phenomena were observed in some certain tuning angle ranges. Through the theoretical analysis of the dispersion curve of the KTP crystal, the intermittent THz wave tuning range and dual-wavelength Stokes waves operation during angle tuning process were explained. The theoretical analysis was in good agreement with the experiment results. The maximum THz output voltage detected by Golay cell was 1.7 V at 5.7 THz under the pump energy of 210 mJ, corresponding to the THz wave output energy of 5.47 μJ and conversion efficiency of 2.6×10^{-5}.

Key words: Terahertz parametric oscillator, KTiOPO$_4$ crystal, Extended tunability, Energy scaling

114. Conduction-band nonparabolicity effect on refractive index and phase match in asymmetric quantum wells pumped by two infrared beams

Xiaolong Cao, Yongli Che, Jianquan Yao

International Journal of Modern Physics B, 2018, 32(20):1850216.

Abstract: An asymmetric quantum well (AQW) system that is pumped by two infrared beams is designed to generate terahertz (THz) waves. The refractive index and phase mismatch associated with the intersubband transition of the AQW structure are calculated and analyzed for both present and absent conduction band nonparabolicity. The calculated results reveal that, for increasing wavelengths, the refractive index of the AQW for the short-wavelength pump beam varies more than 0.83 and undergoes a 0.204 μm redshift, when the conduction band nonparabolicity is considered. The variation of the refractive index of the AQW with the long-wavelength pump beam, changes from 0.225 to 0.316 after considering the conduction-band nonparabolicity. In addition, no red shift is observed. Whether the refractive index of one pump beam with its specific wavelength increases is determined mainly by the linear terms. However, for increasing the other pump wavelengths, the refractive index of one pump beam mainly depends on the nonlinear terms. Subband energy-levels and dipole transition matrix elements show noticeable changes due to conduction-band nonparabolicity, which change the refractive index. Phase matching can be achieved by adjusting the wavelength of the two pump beams. However, both phase mismatch and coverage increase when the conduction band nonparabolicity is considered.

Key words: Band nonparabolicity, Refractive index, Phase mismatch, Intersubband transition, Asymmetric quantum wells

115. Stimulated polariton scattering in β-BTM crystal

Hongzhan Qiao, Kai Zhong, Chao Yan, Yang Liu, Jie Shi, Degang Xu, Jianquan Yao, Zeliang Gao, Xutang Tao

Infrared, Millimeter-Wave, and Terahertz Technologies V. International Society for Optics and Photonics, 2018, 10826:108260T.

Abstract: Theoretical simulations were carried out to evaluate the properties of terahertz (THz) generation in **β**-BaTeMo$_2$O$_9$ (**β**-BTM) crystal by stimulated polariton scattering (SPS) process. The effects of different polariton modes on THz generation were analyzed, from which we determined the optimal crystal design and polarizations of the coupled waves. The dispersion and absorption characteristics of these vibration modes were also given based on the first-principle calculation and correlation Raman spectrum. Finally, the angle phase matching property and THz-wave gain were calculated. Simulation results showed that **β**-BTM is a kind of potential material for high-power tunable THz generation.

Key words: Stimulated polariton scattering, Terahertz parametric oscillator, β-BaTeMo$_2$O$_9$ (β-BTM)

116. Synchronous dual-wavelength pulse generation in coaxial pumping scheme and its application in terahertz difference frequency generation

Yang Liu, Kai Zhong, Jialin Mei, Shuo Jin, Meng Ge, Degang Xu, Jianquan Yao

Nonlinear Frequency Generation and Conversion: Materials and Devices XVII. International Society for Optics and Photonics, 2018, 10516: 105160G.

Abstract : A compact and flexible dual-wavelength laser with combined two laser crystals (*a*-cut and *c*-cut Nd:YLF) as the gain media under coaxially laser-diode (LD) end-pumping configuration was demonstrated and μW-level THz wave was generated based on difference frequency generation (DFG) in a GaSe crystal. The dynamics of coaxial pumping dual wavelength laser was theoretically investigated, showing that the power ratio and pulse interval for both wavelengths could be tuned by balancing the gains at both wavelengths via tuning pump focal position. Synchronized orthogonal 1 047/1 053 nm laser pulses were obtained and optimal power ratio was realized with the total output power of 2.92W at 5 kHz pumped by 10-W LD power. With an 8-mm-long GaSe crystal, 0.93 μW THz wave at 1.64 THz (182 μm) was generated. Such coaxially LD end-pumped lasers can be extended to various combinations of neodymium doped laser media to produce different THz wavelengths for costless and portable applications.

Key words: Dual-wavelength laser, Synchronous pulse, Coaxial pumping, Difference frequency generation, Terahertz

117. THz radiation modulated by confinement of transient current based on patterned CoFeB/Pt heterostructures

Shunnong Zhang, Weihua Zhu, Qin Li, Zongzhi Zhang, Ye Dai, Xian Lin, Jianquan Yao, Guohong Ma, Zuanming Jin

2018 43rd International Conference on Infrared, Millimeter, and Terahertz Waves (IRMMW-THz). IEEE, 2018: 1–1

Abstract: We report the efficient broadband terahertz (THz) radiation in metallic ferromagnetic CoFeB/Pt heterostructures, which can be significantly modulated by metal-block-pattern through transient charge current confinement.
Key words: Terahertz wave, Modulation, transient charge current confinement

118. 基于 MgO:SLN 晶体的环形腔太赫兹参量振荡器

徐德刚，李长昭，王与烨，唐隆煌，闫超，贺奕焱，姚建铨

光学学报，2018, 38(11): 1119001.

摘要：基于摩尔分数为 1% 的氧化镁掺杂的近化学计量比铌酸锂晶体，采用环形腔结构的浅表垂直出射方式组成太赫兹波参量振荡器。该振荡器的太赫兹波输出调谐范围为 0.99~3.84 THz 频率调谐响应时间为 600 μs。当抽运能量为 150.30 mJ、太赫兹频率为 1.59 THz 时，太赫兹脉冲的输出能量达到最大值，为 16.28 μJ，对应的能量转换效率为 1.08×10^{-4}。在相同的实验条件下，该环形腔结构太赫兹波参量振荡器输出的最大太赫兹波能量是传统线形腔结构的 2.35 倍，实现了高能量、快速可调谐的太赫兹波输出。
关键词：非线性光学，太赫兹波，环形腔太赫兹参量振荡器，受激电磁耦子

119. 铁磁异质结构中的超快自旋流调制实现相干太赫兹辐射

张顺浓,朱伟骅,李炬赓,金钻明,戴晔,张宗芝,马国宏,姚建铨

物理学报, 2018, 67(19): 197202.

摘要: 利用飞秒激光脉冲在生长于二氧化硅衬底上的 W/CoFeB/Pt 和 Ta/CoFeB/Pt 两类铁磁/非磁性金属异质结构中实现高效、宽带的相干 THz 脉冲辐射。实验中，THz 脉冲的相位随外加磁场的反转而反转，表明 THz 辐射与样品的磁有序密切相关。为了考察三层膜结构 THz 辐射的物理机制，分别研究了构成三层膜结构的双层异质结构（包括 CoFeB/W, CoFeB/Pt 和 CoFeB/Ta）的 THz 辐射。实验结果都与逆自旋霍尔效应相符合，W/CoFeB/Pt 和 Ta/CoFeB/Pt 三层膜结构所辐射的 THz 强度优于同等激发功率下的 ZnTe（厚度 0.5 mm）晶体。此外，还研究了两款异质结构和 ZnTe 的 THz 辐射强度与激发光脉冲能量密度的关系，发现 Ta/CoFeB/Pt 的饱和能量密度略大于 W/CoFeB/Pt 的饱和能量密度，表明自旋电子在 Ta/CoFeB/Pt 中的界面积累效应相对较小。

关键词: 太赫兹波, 超快光谱, 自旋流, 逆自旋霍尔效应

120. Rational structural design of benzothiazolium-based crystal HDB-T with high nonlinearity and efficient terahertz-wave generation.

Jingkai Shi, Fei Liang, Yixin He, Xinyuan Zhang, Zheshuai Lin, Degang Xu, Zhanggui Hu, Jianquan Yao, Yicheng Wu

Chemical Communications, 2019, 55(55): 7950–7953.

Abstract: A new promising nonlinear optical crystal $C_{25}H_{25}NO_4S_2$ (HDB-T, $P2_1$ space group) exhibits significant macroscopic second-order optical nonlinearity about 1.5 times larger than that of the benchmark OH1 crystal benefiting from optimized orientated polar HDB cations. Widely tunable monochromatic terahertz-wave from 0.1 to 20 THz in the HDB-T crystal was realized for the first time.

Key words: Widely tunable monochromatic terahertz-wave, Nonlinear optical crystal, Terahertz-wave generation

121. Injection pulse-seeded terahertz-wave parametric generator with gain enhancement in wide frequency range

Longhuang Tang, Degang Xu, Yuye Wang, Chao Yan, Yixin He, Jining Li, Kai Zhong, Jianquan Yao

Optics express, 2019, 27(16): 22808–22818.

Abstract: An injection pulse-seeded terahertz-wave parametric generator (ips-TPG) has been demonstrated with gain enhancement in wide tuning range. Theoretical analysis denotes that the compensation of initial Stokes energy is favorable to the THz gain enhancement in wide frequency range, which is attributed to the improvement on interaction of stimulated polariton scattering (SPS) and difference frequency generation (DFG) processes. In the experiment, the THz frequency tuning range from 1.04 THz to 5.15 THz was achieved based on near stoichiometric $LiNbO_3$ (SLN) crystal. Compared with the traditional terahertz parametric oscillator (TPO) under the same experimental conditions, a significant enhancement of THz output energy was occurred in high frequency range. As the THz frequency increased from 1.9 THz to 3.6 THz, the enhancement ratios from 1.6 times to 34.7 times were obtained. Besides, the 3dB bandwidth of ips-TPG was measured to be 2.1 THz, which was about 2.6 times that of SLN-TPO. This THz parametric source with a relative flat gain in wide frequency range is suitable to a variety of practical applications.

Key words: Injection pulse-seeded terahertz-wave parametric generator (ips-TPG), Gain enhancement, Wide tuning range

122. Active multifunctional terahertz modulator based on plasmonic metasurface

Jie Ji, Siyan Zhou, Weijun Wang, Chunya Luo, Yong Liu, Furi Ling, Jianquan Yao

Optics express, 2019, 27(3): 2363–2373.

Abstract: An active multifunctional terahertz modulator based on plasmon-induced transparency (PIT) metasurface under the effect of external infrared light was investigated theoretically and experimentally. A distinct transparency window, which resulted from the near-field coupling between two resonators, could be observed in the transmission spectra. Experimental results showed a phenomenon infrared light induced blue shift on the both resonator with increasing optical powers. When the optical power was tuned from 0 mW to 400 mW, the amplitude tunability of transmission at transparency window reached to 34.01%, much larger than that at the two resonance frequencies. Moreover, the phase tunability of the transmission at 0.98 THz reached to 31.35%. Meanwhile, the amplitude variation was limited to 10%. Furthermore, a coupled Lorentz oscillator model was adopted to analyze the nearfield interaction of the resonances. Experimental results were in good agreement with the analytical fitting results.

Key words: Terahertz modulator, Plasmonic metasurface, Near-field coupling, Tunability

123. Tunable dual-color terahertz wave parametric oscillator based on KTP crystal

Longhuang Tang, Degang Xu, Yuye Wang, Chao Yan, Yixin He, Jining Li, Kai Zhong, Jianquan Yao

Optics letters, 2019, 44(23): 5675−5678.

Abstract: A tunable dual-color KTP terahertz (THz) wave parametric oscillator (TPO) pumped by a dual-wavelength laser was proposed in this Letter. Theoretical analysis denotes that the emission of a tunable dual-color THz wave can be achieved by the simultaneous stimulated polariton scattering processes from multiple A1-symmetry phonon modes of the KTP crystal. The tunable dual-color THz wave emitted from KTP TPO was demonstrated in our experiment, where the THz frequencies simultaneously tuned from 3.15 THz to 11.63 THz and from 1.47 THz to 6.03 THz with some gaps. The maximum dual-color THz output energy of 1.31 μJ was obtained under the THz frequencies of 5.94 THz and 4.42 THz. Moreover, at a certain phasematching angle, the THz output energies for the two frequencies were independent, which means that the dualcolor THz wave emission with any energy ratio can be achieved by adjusting the pump energy ratio between a dual-wavelength laser.

Key words: Dual-color terahertz wave parametric oscillator (TPO), Dual-wavelength laser, Simultaneous stimulated polariton scattering processes

124. Enhanced terahertz wave generation via stokes wave recycling in non-synchronously picosecond pulse pumped terahertz source

Chao Yan, Degang Xu, Yuye Wang, Zhaohua Wang, Zhiyi Wei, Longhuang Tang, Yixin He, Jining Li, Kai Zhong, Wei Shi, Jianquan Yao

IEEE Photonics Journal, 2019, 11(5): 1–8

Abstract: Enhanced terahertz wave generation via Stokes wave recycling in a nonsynchronously picosecond pulse pump configuration has been demonstrated. It is theoretically analyzed that under the condition of high pump peak power density, the oscillation of the strong 1st order Stokes wave could benefit the higher order Stokes wave emission and terahertz wave generation. In the experiment, 5.71 times enhancement of terahertz wave generation was obtained via Stokes wave recycling in the non-synchronously picosecond pulse pumped terahertz parametric generator compared with the conventional single-pass terahertz parametric generation. The maximum terahertz wave average power was 61.7 μW under the pump power of 20 W and the cavity length of 170 mm, while the maximum power conversion efficiency was 3.085×10^{-6}.

Key words: Terahertz radiation, Nonlinear optics, Raman scattering, Stimulated polariton scattering, Picosecond phenomena

125. Efficient ring-cavity terahertz parametric oscillator with pump recycling technique

Longhuang Tang, Degang Xu, Yuye Wang, Chao Yan, Yixin He, Zhongcheng Sun, Ci Song, Kai Zhong, Jianquan Yao

IEEE Photonics Journal, 2019, 11(1): 1−9.

Abstract: The energy scaling of THz wave in wide range based on surface-emitted (SE) ringcavity terahertz parametric oscillator (TPO) with pump recycling technique is demonstrated in this paper. The continuously tunable range from 1.24 to 3.77 THz was achieved with full pump beam recycling, whereas the tunable range was 1.24–2.72 THz without pump beam recycling. The terahertz output energy and conversion efficiency increased more than seven times throughout the whole tuning range. The maximum enhancement ratio of the terahertz output energy was 12.8 times at 2.72 THz. Additionally, under the same experimental conditions, THz conversion efficiency was increased from 7.77×10^{-6} to 7.85×10^{-5} and the threshold energy was significantly reduced by 41% at 1.63 THz. Moreover, through the theoretical analysis, the enhancement mechanism of TPO by pump recycling was verified to originate from the threshold reduction of the stimulated polariton scattering (SPS) process and the interaction improvement between the pump and Stokes waves in the difference frequency generation (DFG) process.

Key words: Terahertz radiation, Nonlinear optics, Raman scattering, Optical resonators

126. Efficient terahertz generation via GaAs hybrid ridge waveguides

Jialin Mei, Kai Zhong, Jiaming Xu, Degang Xu, Wei Shi, Jianquan Yao
IEEE Photonics Technology Letters, 2019, 31(20): 1666–1669.

Abstract: As a common semiconductor material with high performance in the terahertz band, GaAs is hardly employed in second-order nonlinear frequency conversion (NFC) directly due to its isotropy. Although thin waveguides can break the symmetry and induce birefringence for phase matching (PM), more research seems to be needed with GaAs-based structures. In this paper, a category of ridge waveguide is demonstrated with a hybrid structure composed of GaAs and SiN in different sizes. By using the difference frequency generation (DFG) technique with 2 μm and 10 μm pump lasers, simulations suggest that the hybrid waveguides are able to generate continuous-wave (CW) monochromatic terahertz waves from 1.59 THz to 2.66 THz. The output power reach 56.16 μW, corresponding to a conversion efficiency of 5.62 × 10^{-5} W^{-1}, which is much higher than that using bulk GaAs crystals. Such integrated hybrid waveguides have great potential in efficient on-chip terahertz systems.

Key words: Difference frequency generation (DFG), Gallium arsenide (GaAs), Ridge waveguides, Terahertz waves.

127. Optically pumped gas terahertz fiber laser based on gold-coated quartz hollow-core fiber

Sun Shuai, Guo Zhang, Wei Shi, Zhaoshuo Tian, Quan Sheng, Yao Zhang, Haiwei Zhang, Jianquan Yao
Applied optics, 2019, 58(11): 2828–2831.

Abstract: An optically pumped gas terahertz fiber laser based on gold-coated quartz hollow-core fiber is demonstrated. When the terahertz fiber laser is filled with methanol gas and pumped by a piezoelectric-ceramic-transducer adjusted transversely excited atmospheric pressure carbon dioxide laser, a continuous 2.52 THz laser with the output power of 110 mW is obtained. Attributed to the adjustment of the piezoelectric ceramic transducer in the carbon dioxide laser, the fluctuation of the terahertz fiber laser output power is controlled within ± 5%. The result confirms the possibility of a compact flexible optically pumped gas terahertz fiber laser.

Key words: Terahertz fiber laser, Gold-coated quartz hollow-core fiber, Optically pumped gas

128. Two parallel polarized terahertz waves generation from quasi-phase-matching stimulated polariton scattering with periodically-inverted GaAs

Lian Tan, Yongjun Li, Bin Yuan, Hongtao Zhang, Pibin Bing, Jianquan Yao, Zhongyang Li

Optik, 2019, 183: 664–669.

Abstract: In this work, we theoretically analyze the generation of two parallel polarized terahertz (THz) waves from coupled optical parametric oscillations (OPOs) with periodically-inverted GaAs by stimulated polariton scattering (SPS). We analyze collinear phase-matching (CPM) conditions for bulk GaAs generating a THz wave and quasi-phase-matching (QPM) conditions for periodically inverted GaAs generating two THz waves. The analysis results indicate that the frequencies of the two THz waves and the frequency differences between the two THz waves are efficiently tuning by varying grating periods of periodically-inverted GaAs. The photon flux densities (PFD) of the two THz waves are calculated. The generating PFD of the two THz waves are efficiently enhanced by injecting a Stokes seed laser or increasing pump intensity. The scheme proposed in this work enable new applications of THz wave in imaging and spectroscopy.

Key words: Terahertz wave, Stimulated polariton scattering, Coupled optical parametric oscillation, GaAs

129. Theoretical investigation on collinear phase matching stimulated polariton scattering generating THz waves with a KTP crystal

Lian Tan, Bin Yuan, Yongjun Li, Silei Wang, Hongtao Zhang, Pibin Bing, Jianquan Yao, Zhongyang Li

Current Optics and Photonics, 2019, 3(4): 342–349.

Abstract: We present a theoretical research concerning terahertz (THz) wave generation with $KTiOPO_4$ (KTP) by collinear phase matching (CPM) stimulated polariton scattering (SPS). Both CPM and corresponding nonzero nonlinear coefficients can be simultaneously realized with s→f+f in yz plane, s→f+s with $\theta < \Omega$ in xz plane and s→f+f with $\theta > \Omega$ in xz plane. The effective nonlinear coefficients including electronic nonlinearities and ionic nonlinearities are calculated. Based on the parameter values of refractive indices, absorption coefficients and effective nonlinear coefficients, we simulate THz wave intensities generated with CPM SPS by solving coupled wave equations and give the relationship among the maximum THz wave intensity, optimal crystal length and the angle θ. The calculation results demonstrate that CPM SPS with KTP can generate THz waves with high intensities and quantum conversion efficiencies.

Key words: Terahertz wave, Stimulated polariton scattering, $KTiOPO_4$

130. Simultaneous generation of two THz waves with bulk LiNbO$_3$ and four THz waves with PPLN by coupled optical parametric generation.

Zhongyang Li, Bin Yuan, Yongjun Li, Lian Tan, Pibin Bing, Hongtao Zhang, Jianyuan Yao

Chinese Physics B. 2019.

Abstract: We present a theoretical research concerning simultaneous generation of two terahertz (THz) waves with bulk LiNbO$_3$ and four THz waves with periodically poled LiNbO$_3$ (PPLN) by coupled optical parametric generation (COPG). First, we investigate collinear phase matching of COPG generating two orthogonally polarized THz waves with two types of phase matching of $o = e + e$ and $o = e + o$ with bulk LiNbO$_3$. The two orthogonally polarized THz waves are generated from stimulated polariton scattering (SPS) with A_1 and E symmetric transverse optical (TO) modes in bulk LiNbO$_3$, respectively. Then, we find that perturbations of phase mismatch for $o = e + e$ and $o = e + o$ can be compensated by a same grating vector of PPLN. As a result, four THz waves are simultaneously generated with a PPLN crystal and a pump laser. We calculate third-order nonlinear optical coefficients of $o = e + o$ generating THz waves from E symmetric TO modes. The intensities of four THz waves are calculated by solving coupled wave equations. The calculation results demonstrate that the COPG generating four THz waves have high photon conversion efficiencies.

Key words: Terahertz wave, Coupled optical parametric generation, Stimulated polariton scattering, Periodically poled LiNbO$_3$

131. High-energy and ultra-wideband tunable monochromatic terahertz source and frequency domain system based on DAST crystal

Yixin He, Yuye Wang, Degang Xu, Jining Li, Chao Yan, Longhuang Tang, Xianli Zhu, Hongxiang Liu, Bing Teng, Jianquan Yao

Terahertz, RF, Millimeter, and Submillimeter-Wave Technology and Applications XII. International Society for Optics and Photonics, 2019, 10917: 109170X.

Abstract: We have demonstrated a high-energy and broadly tunable monochromatic terahertz (THz) source via difference frequency generation (DFG) in DAST crystal. The THz frequency is tuned randomly in the range of 0.3-19.6 THz, which is much wider than the THz source based on the inorganic crystal and the photoconductive antenna. The highest energy of 2.53 μJ/pulse is obtained at 18.9 THz corresponding to the optical-to-optical conversion efficiency of 1.31×10^{-4}. The THz output spectroscopy is theoretically and experimentally explained by DFG process and Raman spectroscopy. Meanwhile, a phenomenon of blue light from the KTP-OPO with tunable and multiple wavelengths was firstly observed and explained. Based on our THz source, an ultra-wideband THz frequency domain system (THz-FDS) with transmission mode is realized to measure the ultra-wideband THz spectroscopies of typical materials in solid and liquid states, such as Si, SiC, White PE, water, isopropyl myristate, simethicone, atonlein and oleic acid, etc.. Furthermore, we have studied the THz spectral characteristic of biomedical tissue in the ultra-wideband THz frequency range of 0.3-15 THz to study the biomedical response in the entire THz frequency range, which contains more abundant spectral information and was rarely focused with the limit of the THz source.

Key words: Terahertz radiation, Terahertz radiation, Ultra-wideband THz spectroscopy

132. A gain-boosted terahertz-wave parametric generator in high frequency tuning range via pulse-seed injection

Longhuang Tang, Degang Xu, Yuye Wang, Chao Yan, Yixin He, Changzhao Li, Jining Li, Kai Zhong, Jianquan Yao

Terahertz, RF, Millimeter, and Submillimeter-Wave Technology and Applications XII. International Society for Optics and Photonics, 2019, 10917: 109170T.

Abstract: A gain-boosted terahertz-wave parametric generator (TPG) in high frequency tuning range based on MgO-doped near stoichiometric LiNbO$_3$ (MgO:SLN) crystal has been demonstrated with 1 064 nm nanosecond pulsed laser pumping. The pulse-seed is provided by nanosecond singly resonant near-degenerated KTP optical parametric oscillator with the wavelength range of 1 068.08 nm to 1 084.76 nm. The terahertz tuning range of 0.97 THz to 4.07 THz was achieved. The maximum THz wave output signal was 4 285 mV at 1.82 THz under the pump energy of 180 mJ and pulse-seed energy of 20.2 mJ. During the frequency range of 1.25 THz to 3.43 THz, the THz output energies were larger than 2 000 mV. Compared with the maximum THz output energy, the THz energy attenuation factors of 0.55 dB, 1.71 dB and 3.31 dB were realized in pulse-seeded TPG at 2.5 THz, 3.0 THz and 3.5 THz, respectively. The significantly increasing of THz gain in high frequency range (>2.5 THz) was achieved.

Key words: Nonlinear optics, Terahertz parametric generator, Laser, Terahertz, THz source

133. 新型有机晶体及超宽带太赫兹辐射源研究进展

徐德刚, 朱先立, 贺奕焮, 王与烨, 姚建铨

中国光学, 2019, 12(3): 535–558.

摘要：非线性光学晶体是非线性光学频率变换技术中的核心器件。近些年，为进一步提高基于非线性光学频率变换技术产生太赫兹波的输出能量、转换效率，拓宽产生太赫兹波的带宽，多种新型有机晶体得以发展，并凭借其更加出色的非线性光学性质，成为产生太赫兹波的理想材料。本文按照晶体类型介绍了目前可产生 THz 波的多种有机晶体的性质，并总结了基于多种有机晶体的超宽带太赫兹辐射源的国内外研究进展，同时结合 THz 光谱检测技术的应用需求分析了基于有机晶体宽带 THz 辐射源的发展趋势以及所面临的关键科学问题。

关键词：太赫兹, 有机晶体, 超宽带, 非线性频率变换

134. 基于 DAST 晶体的高能量超宽带可调谐小型化差频 THz 辐射源研究

贺奕焮,庞子博,朱先立,徐德刚,王与烨,孟大磊,武聪,程虹娟,徐永宽,姚建铨

红外与毫米波学报, 2019, 38(4): 485–492.

摘要：基于双温区法生长的高质量 DAST(4-(4-二甲基氨基苯乙烯基) 甲基吡啶对甲苯酸盐) 晶体,成功搭建了高能量、超宽带可调谐差频 THz 辐射源,系统尺寸 40 cm × 25 cm,调谐范围达到 0.3~19.6 THz,最大输出能量达到 4.02 μJ /pulse@18.6 THz,信噪比最高达到 32.24 dB,结合振镜扫描技术,以 0.1 THz 为步长,超宽带光谱扫描时间小于 1 min。实验中观测到差频产生 THz 波的输出饱和现象并研究了基于 DAST 晶体差频产生 THz 波的偏振特性与传输特性,证明基于 DAST 晶体差频产生的 THz 波消光比达到 0.05,且差频过程满足 0 类相位匹配条件。基于该太赫兹辐射源,对多种固体样品在 2~14THz 范围内的超宽带 THz 光谱信息进行了有效获取。

关键词：DAST,晶体,太赫兹,超宽带可调谐

135. 基于 DAST 晶体差频的可调谐 THz 辐射源

徐德刚,朱先立,王与烨,李吉宁,贺奕焮,庞子博,程红娟,姚建铨

光学学报, 2020, 40(04): 0404001.

摘要：基于自发成核法自行生长的 4-(4-二甲基氨基苯乙烯基) 甲基吡啶对甲苯酸盐（DAST）晶体,实现了宽带可调谐 THz 辐射输出。实验探究了饱和生长溶液浓度对晶体生长形态和光学质量的影响,并结合晶体的拉曼光谱,对其生色团振动和转动特性进行了分析。采用 1.3-1.5 μm 的高能量、可调谐双波长为泵浦源,基于 0 类相位匹配外腔差频和电控振镜快速扫描技术,实现了 0.1-20 THz 的宽带 THz 辐射输出,在 18.9 THz 处,最大输出能量为 3.59 μJ/pulse,转化效率为 2.39×10^{-4}。基于宽带 THz 输出谱分析可知,DAST 晶体对 THz 波的吸收主要由晶格振动导致。

关键词：太赫兹,DAST 晶体,自发成核法,差频

136. 基于负曲率空芯光纤的光泵太赫兹光纤激光器的理论研究

张果, 孙帅, 张尧, 盛泉, 史伟, 姚建铨
红外与激光工程, 2019.

摘 要： 针对紧凑型、高效的光泵太赫兹激光器 (OPTL) 技术，设计了基于负曲率空芯光纤的长腔型光泵太赫兹光纤激光器 (OPTFL) 结构。该 OPTFL 以聚甲基戊烯 (PMP) 材料的空芯光纤为工作气室，填充甲醇气体作为工作物质，采用连续 9P(36) 支 CO_2 激光器为泵浦源。从速率方程出发，分析了影响 OPTFL 输出特性的因素，并对负曲率空芯光纤内部微结构进行了探索，通过调整内部结构，能够实现较低损耗的单模太赫兹激光传输。结合设计的负曲率空芯光纤，对长腔型 OPTFL 的可行性进行了分析，理论计算表明，在最佳工作条件下，通过适当增加谐振腔长度，太赫兹激光输出功率有望达到百毫瓦量级。研究结果为高功率、高性能的 OPTFL 提供了一种新的方法与理论指导。
关键词： 太赫兹激光器，负曲率空芯光纤，速率方程，单模

137. 基于超快电子自旋动力学的太赫兹辐射研究进展

金钻明, 宋邦菊, 李炬赓, 张顺浓, 阮舜逸, 戴晔, 阎晓娜, 林贤, 马国宏, 姚建铨
中国激光, 2019, 46(5): 0508005.

摘要： 回顾了近年来利用超快自旋动力学过程产生太赫兹（THz）辐射的研究进展。介绍了基于逆自旋霍尔效应和逆 Rashba-Edelstein 效应的瞬态自旋流—电荷流转换，指出铁磁/非磁性异质结构已被用于设计低成本、高效率的 THz 辐射源。通过优化膜厚、生长条件、衬底和结构，可进一步提高基于自旋电子学的 THz 发射器的效率和带宽。简述了 THz 发射光谱在研究超快自旋泽贝克效应形成动力学中的应用。
关键词： 非线性光学，太赫兹辐射，超快光谱，自旋流，逆自旋霍尔效应，自旋泽贝克效应

第二节　太赫兹传输及功能器件

1. Propagation characteristics of two-dimensional photonic crystals in the terahertz range

H. Liu, J. Yao, D. Xu, P. Wang

Applied Physics B-Lasers and Optics, 2007, 87(1): 57–63.

Abstract: Based on the plane wave expansion method, the complete band gaps of two-dimensional terahertz photonic crystals with typical structures are optimized through varying structural parameters. During our calculation, two kinds of lattice structures that are very promising for the materials of terahertz components are found and some interesting phenomena that were not reported in the former papers are also presented. Using the finite difference time domain method, we simulated the electromagnetic field distribution of THz waves in a PC splitter. And by the plane wave expansion method, we achieved the dispersion relation and defect modes in a rotated PC waveguide. These results provide a useful guide and a theoretical basis for the developments of THz functional components.

Key words: Terahertz wave, Photonic crystals, Finite difference time domain method

2. Characteristics of photonic band gaps in woodpile three-dimensional terahertz photonic crystals

Huan Liu, Jianquan Yao, Degang Xu, Peng Wang

Optics Express, 2007, 15(2): 695–703.

Abstract: Based on plane wave expansion method, complete photonic band gaps (PBGs) of a woodpile three-dimensional (3-D) terahertz (THz) photonic crystal (PC) with face-centered-tetragonal (fct) symmetry are optimized by varying structural parameters and the highest band gap ratio can reach 26.71%. In order to further optimize the complete PBGs, we propose a novel woodpile lattice with comparatively decreased symmetry and the highest band gap ratio can be increased to 27.61%. The woodpile THz PCs with two different symmetries both have a wide range of filling ratios to gain high quality complete PBGs, making the manufacturing process convenient. Woodpile 3-D PCs will be very promising materials for THz functional components.

Key words: Photonic band gaps, Three-dimensional terahertz photonic crystal, Face-centered-tetragonal symmetry

3. Design of terahertz photonic crystal fibers by finite difference frequency domain method

Youfu Geng, Xiaoling Tan, Peng Wang, Jianquan Yao

Journal of Optics: Pure and Applied Optics, 2007, 9(11): 1019–1023.

Abstract: Terahertz photonic crystal fibers are analyzed and designed using a full-vectorial 2D finite difference frequency domain method with nonsplit-field anisotropic perfectly matched layers absorbing boundary conditions. By analyzing the effective mode areas of fundamental mode and second-order mode, a phase diagram which describes the regions of single-mode and multimode operation as well as the non-confined fundamental mode regime is obtained in the terahertz region. The single-mode terahertz photonic crystal fibers with near-zero flattened dispersion are designed based on this method by optimizing the structure parameters.

Key words: Terahertz photonic crystal fiber, Finite difference frequency domain, Mode cutoff, Group-velocity dispersion

4. A novel woodpile three-dimensional terahertz photonic crystal

Liu Huan, Yao Jianquan, Zheng Fanghua, Xu Degang, Wang Peng

Chinese Physics Letters, 2007, 24(5): 1290–1293.

Abstract: A novel woodpile lattice structure is proposed. Based on plane wave expansion (PWE) method, the complete photonic band gaps (PBGs) of the novel woodpile three-dimensional (3D) terahertz (THz) photonic crystal (PC) with a decreasing symmetry relative to a face-centred-tetragonal (fct) symmetry are optimized by varying some structural parameters and the highest band gap ratio can reach 27.61%. Compared to the traditional woodpile lattice, the novel woodpile lattice has a wider range of the filling ratios to gain high quality PBGs, which provides greater convenience for the manufacturing process. The novel woodpile 3D PC will be very promising for materials of THz functional components.

Key words: Woodpile lattice structure, Plane wave expansion method, Photonic band gap, Terahertz photonic crystal

5. Transmission loss and dispersion in plastic terahertz photonic band-gap fibers

Y. F. GENG, X. L. TAN, P. WANG, J. Q. YAO

Apllied physics B: Lasers and optics, 2008, 91(2): 333–336.

Abstract: We present an investigation into the transmission loss and dispersion of terahertz waves in plastic photonic band-gap fibers having a cladding with a finite number of air hole rings. The leakage loss and absorption loss caused by background material are analyzed by a full-vectorial twodimensional finite difference frequencydomain method and the lowest power transmission loss of 6.126 dB/m at 1.75 THz is realized. Numerical results show that a larger diameter-to-pitch ratio is suitable for lower transmission loss and lower group-velocity dispersion in plastic terahertz photonic band-gap fibers.

Key words: Terahertz wave, Transmission Loss, Transmission Dispersion, Plastic photonic band-gap fiber

6. Novel optical controllable terahertz wave switch

Li Jiusheng, Yao Jianquan

Optics Communications, 2008, 281(23): 5697–5700.

Abstract: We study theoretically and demonstrate experimentally light controllable terahertz wave switch. When the modulated optical excitation source is used to irradiate a high resistivity silicon wafer, a novel controllable terahertz wave switch is achieved. The results show that the ON–OFF response time is less than 150 ms and the attenuation of the novel terahertz wave switch is more than 20 dB at frequency of 0.315 THz.

Key words: Terahertz wave, Optical excitation, Switch

7. Controllable terahertz wave attenuator

Jiusheng Li, Jianquan Yao

Microwave and Optical Technology Letters, 2008, 50(7): 1810–1812.

Abstract: A new type of optically controllable terahertz wave attenuator using high-resistivity silicon wafer is developed and tested. Without optical excitation, the high-resistivity silicon is a lossless dielectric material at terahertz wave region. When the high-resistivity silicon wafer is optically excited, free carriers are generated, and the silicon wafer becomes a lossy dielectric. We study theoretically and demonstrate experimentally light-controllable terahertz wave of the high-resistivity silicon wafer. The results show that more than 10-dB attenuation of the novel terahertz wave attenuator is obtained at frequency of 0.3 THz. The proposed device can be used in future terahertz wave communication systems.

Key words: Terahertz wave attenuator, Attenuation, High resistivity silicon, Optically controllable

8. Low loss plastic Terahertz photonic band-gap fibres

Geng Youfu, Tan Xiaoling, Zhong Kai, Wang Peng, Yao Jianquan

Chinese Physics Letters, 2008, 25(11): 3961–3963.

Abstract: We report a numerical investigation on terahertz wave propagation in plastic photonic band-gap fibres which are characterized by a 19-unit-cell air core and hexagonal air holes with rounded corners in cladding. Using the finite element method, the leakage loss and absorption loss are calculated and the transmission properties are analysed. The lowest loss of 0.268 dB/m is obtained. Numerical results show that the fibres could liberate the constraints of background materials beyond the transparency region in terahertz wave band, and efficiently minimize the effect of absorption by background materials, which present great advantage of plastic photonic band-gap fibres in long distance terahertz delivery.

Key words: Terahertz wave, Low loss, Plastic photonic band-gap fibre

9. 基于法布里-珀罗干涉仪的太赫兹波波长测试方法

曹铁岭，姚建铨，郑义

光学仪器，2008, 30(2): 13–16.

摘要： 太赫兹波波长的测量在其科研和实际应用中显得日益迫切和重要。阐述了由两个平行的金属网栅构成的Fabry-Perot(F-P)干涉仪测量太赫兹波波长的原理和方法，对金属网栅的参数设计原则进行了论述，从理论上分析了此方法的可行性，并根据此原理推出了计算太赫兹波线宽的公式。设计制作了测量 92.9 μm 赫兹波的金属网栅。为太赫兹波波长测量和谐线线宽估算提供了理论参考。

关键词： F-P 干涉仪，金属网栅，太赫兹波，波长测量，谱线线宽

10. 基于法布里-珀罗干涉仪的太赫兹波长测试仪

曹铁岭，姚建铨

现代科学仪器，2008, (2): 36–39.

Abstract: The measurement of terahertz wavelength is more and more urgent and important in the science research and practical application. The fundamental of measuring terahertz wavelength is introduced. The parameter of metal mesh is designed and the process of measuring wavelength is narrated, and the formula of calculating the terahertz wave line width is obtained.

Key words: F-P interferometer, Metal grids, Terahertz wave, Wavelength measurement

11. Terahertz liquid crystal tunable filter

Ren Guangjun, Wang Xinchuang, Ma Xiurong, Yao Jianquan
Photonics and OptoElectronics Meetings **2008**, *Wuhan, China (POEM 2008)*, 2008: 72770L.

Abstract: We construct and characterize a room-temperature tunable Terahertz(THz) liquid crystal(LC) filter. Liquid crystal is a substance which is intervenient between inerratic crystal and isotropy liquid. Because their molecular long-range are not likeness crystal, so they are easily affected by exterior conditions, such as electric field, magnetic field, temperature and stress etc. Accordingly, their optical characters are change. According to the electrooptic effect of liquid crystal and the theory of Lyot filter, tunable Lyot liquid crystal filter is designed and tunable F-P liquid crystal filter is designed too. It is proved that the tunable liquid crystal filter has great tuning range and narrow band.
Key words: Terahertz wave, Lliquid crystal filter, Fabry-Perot (F-P) cavity, Mmagnetic field, Room-temperature operation, Lyot THz LC filter

12. Fe-doped polycrystalline CeO_2 as terahertz optical material

Wen Qiye, Zhang Huaiwu, Yang Qinghui, Li Sheng, Xu Degang, Yao Jianquan
Chinese Physics Letters, 2009, 26(4): 047803-1–047803-4.

Abstract: Fe-doped CeO_2 is synthesized by ceramic method and the effects of Fe doping on the structure and properties are characterized by ordinary methods and terahertz-time domain spectrometer (THz-TDS) technique. Our results show that pure CeO_2 only has a small dielectric constant ε of 4, while a small amount of Fe (0.9 at.%) doping into CeO_2 promotes densification and induces a large ε of 23. From the THz spectroscopy, it is found that for undoped CeO_2 both the power absorption and the index of refraction increase with frequency, while for Fe-doped CeO_2 we measure a remarkable transparency together with a flat index curve. The absorption coefficient of Fe-doped CeO_2 at frequency ranging from 0.2 to 1.8 THz is less than 0.35 cm^{-1}, implying that Fe-doped CeO_2 is a potential THz optical material.
Key words: Fe-doped CeO_2, Terahertz-time domain spectrometer (THz-TDS) technique, THz optical material

13. 太赫兹波在有限电导率金属空芯波导中的传输特性

张玉萍, 张会云, 耿优福, 谭晓玲, 姚建铨

物理学报, 2009, 58(10): 7030-7033.

摘要：研究了太赫兹波在具有有限电导率的金属镀层空芯波导中的传输特性，从其传输的特性方程出发，利用 Newton-Raphson 迭代方法数值模拟了传输损耗和相位常数随太赫兹波频率、波导内径以及波导中金属镀层的电导率的变化关系。结果表明，采用大芯径波导和高电导率的金属镀层能有效降低太赫兹波的传输损耗。

关键词：太赫兹波, 空芯波导, 特征方程

14. The propagation characteristics of THz radiation in hollow circular waveguides coated with different material films

Dongmei Lu, Yong Zhang, Zhidong Zhang, Jianquan Yao

Photonics and Optoelectronics Meetings (POEM) 2008: Terahertz Science and Technology. 2009: 72770Q.

Abstract: In the present study, the propagation characteristics of terahertz radiation in both metallic film-coated and dielectric film-coated hollow waveguides have been theoretically analyzed. The dominantmodes in metallic film-coated and metal/dielectric film-coated hollow waveguides are respectively TE11 and HE11 modes. Theoretical attenuation coefficients of terahertz radiations in hollow waveguides with abore diameter of lmm film-coated with Au, Pb and Ni at different incident frequencies are compared. Thedominant mode, i.e. TE11 mode, as a function of bore diameter in hollow waveguides film-coated withdifferent metals are calculated at a given incident frequency. An additional dielectric film with appropriatethickness on the metal films caneffectively enhance the wave reflection, resulting in decreasedattenuation of the terahertz radiation propagating in hollow waveguides. Calculated results indicate that the attenuation of the terahertz radiation in metal/dielectric film-coated hollow waveguide with a bore diameter of 1 mm for a given incident wavelength of 200 μm is about 4 times lower than that in metal hollow waveguide with the same bore diameter.

Key words: Terahertz radiation; Propagation characteristics; Hollow circular waveguide; Attenuation coefficient

15. Propagation characteristics of THz radiation in hollow elliptical waveguide

Yong Zhang, Zhongya Zhang, Dayun Wang, Wenxia Bao, Ming Liu, Dongmei Lu, Peide Zhao, Zhidong Zhang, Jianquan Yao

Photonics and Optoelectronics Meetings (POEM) 2008: Terahertz Science and Technology, 2009: 72770O.

Abstract: Based on the analytical solutions of the wave equations in the elliptical-cylindrical coordinate system, the propagation characteristics of hollow elliptical waveguide for THz Radiation are investigated, and then the mode characteristic equations are also given. Our results show that the mode characteristic equations of a circular waveguide can be treated as a special case of an elliptic waveguide. The propagation characteristics are numerically simulated depending on the refractive index of dielectrics films and the eccentricity of the elliptic waveguide. The cutoff wavelengths of guided modes in the elliptic waveguide are also presented.

Key words: THz Radiation, Elliptic waveguide, Mode characteristic equation, Propagation characteristics, Cutoff wavelengths

16. Time-dependent theoretical model for terahertz wave detector using a parametric process

C. Y. Jiang, J. S. Liu, B Sun, K. J. Wang, S. X. Li, J. Q. Yao

Optics Express, 2010, 18(17): 18180–18189.

Abstract: We have presented a time-dependent theoretical model to describe the time behavior of a quasi-monochromatic nanosecond terahertz detector reported by Guo et al. [2008 Appl. Phys. Lett. 93, 021106]. The temporal input-output characteristic of the detector is investigated numerically by taking the system parameters close to the experimental ones, and the calculated pulse width for the incident terahertz wave agrees well with the experimental one. Our results demonstrate that the energy and width of an output idler wave pulse are proportional to those of the incident terahertz wave pulse. This study provides a strict theoretical basis and could be used to guide the design and optimization for the highly sensitive coherent terahertz detector.

Key words: Parametric processes, Time-dependent theoretical model, Detector

17. Compact terahertz wave polarizing beam splitter

Jiu Sheng Li, Degang Xu, Jianquan Yao

Applied Optics, 2010, 49(24): 4494–4497.

Abstract: We designed a compact terahertz (THz) wave polarizing beam splitter based on a periodic bilayer structure, which operates over a wide THz wavelength range. Within a short length (about 1 mm), this polarizing beam splitter separates THz wave TE- and TM-polarized modes into orthogonal output waveguides. Results of simulations with the finite-element method show that 99.99% of the TE-polarized THz wave is deflected by the periodic bilayer structure (with 39.9 dB extinction ratio), whereas 99.58% of the TM-polarized THz wave propagates through the structure (with a 23.7 dB extinction ratio). Tolerance analysis reveals a large tolerance to fabrication errors.

Key words: Terahertz wave, Polarizing beam splitter, Periodic bilayer structure

18. Proposal of an electrically controlled terahertz switch based on liquid-crystal-filled dual-metallic grating structures

Yinghao Yuan, Jian He, Jinsong Liu, Jianquan Yao

Applied Optics, 2010, 49(31): 6092–6097.

Abstract: We propose an electrically controlled terahertz switch based on metallic grating–liquid crystal–metallic grating (MG-LC-MG) structures. The switching mechanism is realized by modifying the effective refractive of the LC using different bias electric fields. In our design, the MGs not only support supertransmittance at certain frequencies, but they also act as the electrodes. Simulation results show the proposed structure has the potential to realize an electrically controlled terahertz switch with a high extinction ratio and low insertion loss. A prototype design is also proposed for practical implementation.

Key words: Terahertz wave, Switch, MG-LC-MG structures

19. Steady-state theoretical model for terahertz wave detector using a parametric process

C Jiang, J Liu, B Sun, K Wang, J Yao
Journal of Optics, 2010, 12: 045202.

Abstract: This paper establishes a steady-state theoretical model for the main optical part of the terahertz wave detector based on a parametric process in lithium niobate (LN) crystals reported by Guo et al (2008 Appl. Phys. Lett. 93 021106). The formula connecting the incident terahertz wave intensity with the output idler wave intensity is obtained by solving the coupled-wave equations for the case of an anisotropic crystal with lattice vibration modes. Results demonstrate that the output idler wave intensity is proportional to the incident terahertz wave intensity. The expression of the spectral response for this detector is obtained and the spectral response characteristics are analyzed. The dependencies of the response versus the intensity of the pump beam and the length of the amplifying crystal are investigated. This work builds the theoretical basis for analysis of the steady-state characteristics of the detector.

Key words: Terahertz, Scattering, Frequency conversion, Lithium niobate, Anisotropic crystal

20. Four-wave mixing model solutions for polarization control of terahertz pulse generated by a two-color laser field in air

Zheng Chu, Jinsong Liu, Kejia Wang, Jianquan Yao
Chinese Optics Letters, 2010, 8(7): 697–700.

Abstract: A four-wave mixing (FWM) model is used to analyze the polarization control of terahertz (THz) pulse generated by a two-color laser field in air. The analytic formula for the THz intensity varying with the THz polarizer angle, and the relative phase between the two pulses, are obtained. The corresponding numerical results agree well with both numerical result obtained from a quantum model and measured data reported. Moreover, possible phenomena are predicted for variables not found in other experiments. Compared with the quantum model, the FWM model gives analytic formulas and clear physical pictures, and has the advantage of efficient computing time.

Key words: Four-wave mixing, Polarization control, THz generation

21. THz 辐射大气传输研究和展望

姚建铨,汪静丽,钟凯,王然,徐德刚,丁欣,张帆,王鹏

光电子·激光, 2010, 21(10): 1582–1588.

摘要:THz 辐射在大气中的传输特性是 THz 波空间通信,大气科学及遥感检测等空间应用的基础,掌握不同温度、高度、湿度和压力等条件下 THz 频段大气透过率窗口的位置和宽度对于促进该频段的应用具有重要意义。本文分析了 THz 辐射大气传输的研究概况,包括大气传输的基本原理,辐射传输方程的推导以及对各类 THz 辐射大气传输模型的比较,并指出尚待解决的问题,最后对 THz 辐射大气传输研究给出若干建议和展望。

关键词:THz 辐射,大气传输,辐射传输方程,传输模型,THz 辐射应用

22. A ferroelectric polyvinylidene fluoride-coated porous fiber based surface-plasmon- resonance-like gas sensor in the terahertz region

Jing Lei, Yao Jianquan

Optoelectronics Letters, 2010, 6(5):0321–0324.

Abstract: In this paper, a ferroelectric polyvinylidene fluoride (PVDF)-coated porous polymer fiber based surface plasmon resonance (SPR)-like gas sensor is proposed theoretically in the terahertz (THz) region based on the total internal reflection (TIR). In such a sensor, the phase matching is achieved by changing the fiber parameters and the plasmon-like phenomenon at the interface between the ferroelectric polyvinylidene fluoride (PVDF) layer and the gaseous analyte is discussed. Using a fullvector finite-element method, the core-mode loss of the fiber is calculated to measure the resolution of the sensor. The amplitude resolution is demonstrated to be as low as 1.45×10^{-4} RIU, and the spectral resolution is 1.30×10^{-4} RIU in THz region, where RIU means the refractive index unit.

Key words: Gas sensor, Polyvinylidene fluorid, Surface plasmon resonance, Total internal reflection

23. 基于非对称量子阱的太赫兹波调制器

朱能念, 姚建铨, 徐小燕, 李忠洋, 邴丕彬, 徐德刚

激光与红外, 2010, 40(8): 897–900.

摘要: 主要研究了一个特殊的 GaAs/AlGaAs 非对称量子阱中的线性和三阶非线性太赫兹波吸收系数和介质折射率的相对改变。首先利用量子力学中的密度矩阵算符理论和迭代方法导出了线性和三阶非线性光吸收系数和相对折射率改变的表达式,然后以典型的阶梯型量子阱材料为例做了数值计算。计算表明,基于泵浦光场和偏置电压控制的太赫兹波调制器不仅可以做太赫兹波开关,还可以灵活地调制太赫兹波信号的强度和相位,方便实用。

关键词: 量子阱,太赫兹波,调制器,非线性光学

24. The study on THz wave propagation feature in atmosphere

Haixia Cui, Jianquan Yao, Chunming Wan

Proceedings of SPIE - Photonics Asia, 2010, 7854: 785404.

Abstract: Terahertz (THz) transmission can be divided into passive transmission and active transmission, passive transmission refers to atmospheric propagation, active transmission means not only transmission but also completing a function, so also can be said functional transmission. Knowledge of the transmission of terahertz wave is very important for terahertz technology and its applications. We analysis detailed the atmospheric propagation model of terahertz wave, propagation effects, propagation equation, then carry out atmospheric propagation measurement system based on THz-TDS.

Key words: Terahertz (THz) transmission, Atmospheric propagation model, Decay effects

25. Parameter selection and design considerations with MPOF evanescent wave sensor in the THz wavelength range

Haixia Cui, Jianquan Yao, Ying Lu, Chunming Wan

Proceedings of SPIE - Photonics Asia, 2010, 7853: 78530.

Abstract: Using the finite element method to simulate THz photonic crystal fiber(PCF) transmission and sensing characteristics. According to solid TPCF guidance, the refractive index, attenuation and relativity sensitivity were analyzed. And we analyzed firstly the transformation of PCF hole-shape to the influence of sensing on THz wave region. Then we simulate and calculate THz high density polyethylene PCF (HDPE-PCF) some parameters, at last the design considerations were given.

Key words: Terahertz (THz), Photonic crystal fiber(PCF), High density polyethylene(HDPE), Sensing

26. Low-loss and birefringent terahertz polymer elliptical-tube waveguides

Jingli Wang, Jianquan Yao

Proceedings of SPIE - Photonics Asia, 2010, 7854: 78543.

Abstract: In this work, modal birefringence and loss characteristic for the fundamental mode in terahertz (THz) polymer elliptical-tube waveguides are investigated by using a full vector finite element method (FEM). Numerical results show that this kind of waveguide has high birefringence ($\sim 10^{-2}$) and better loss property as a large part of mode power is trapped in the air-core inside the polymer elliptical-tube. Dependence study of the birefringence on several parameters is also provided and numerical results show the birefringence increases as refractive index and thickness of polymer tube increased.

Key words: THz waveguides, Low-loss, Birefringence, Polymer elliptical-tube

27. A THz modulator use the photo-carrier surface plasma effect

Pengfei Yang, Jianquan Yao, Zhigang Di, Pibin Bing, Peng Wang
Proceedings of SPIE - Photonics Asia, 2010, 7854: 78542M.

Abstract: A design of light modulator for THz amplitude and phase modulations has been presented in this paper. Simplest versus of the Drude model is adopted, in which the collision damping is independent of the carrier energy. In our experiment, we use THz-TDS as THz source and detector. A laser whose wavelength is 808 nm was used to irradiate the intrinsic Si(high-resistance), so as to let it generate the Photo-carriers, and to influence the conductance. The Photo-carriers will change the absorption coefficient of the THz wave and also influence the dielectric of the sample, hence to control the characteristics of the THz wave in the silicon. By changing the light intensity, due to the different photon-generated carrier concentration, the single transmission of the THz wave in the silicon wafer sample is changing remarkable. Theoretically, the modulation depth can be more than 80%. we present our design of light modulator for THz, and show the Digital simulation of our design. Also, according to this design theory, Optical/electronic integrated modulation of THz can be realized, that will be our future work.

Key words: THz, Photo-carrier, Concentration, Absorption, Drude model, Modultor

28. The guidance mechanism and numerical simulation of THz polymer hollow-core photonic crystal fiber

Ran Wang, Jianquan Yao, Degang Xu, Jingli Wang, Kai Zhong, Peng Wang
Proceedings of SPIE - Photonics Asia, 2010, 7854: 78543.

Abstract: With the development of terahertz (THz) technology, an efficient propagation waveguide is essential for the construction of compact THz devices. Hollow core photonic crystal fiber with a large air core at the center and a cladding formed by a periodic arrangement of polymer tubes has been demonstrated in this paper. The guidance mechanism is based on anti-resonant reflection from struts of solid material in the cladding. Since most electromagnetic field is dominated in the air core, hollow core fibers have obvious advantages in lower absorption. The propagation characteristics of the fiber, such as the mode field distribution and the loss coefficient are numerically investigated through the finite element method. The result shows that an effective way to reduce the absorption is to enlarge the central air core and reduce the overlap between the field and material.

Key words: THz radiation, Hollow-core fibers, Anti-resonant reflection, Finite element method

29. Analysis on characteristic and application of THz frequency comb and THz sub-comb

Pengxiang Liu, Degang Xu, Jianquan Yao

Photonics and Optoelectronics Meetings (POEM 2010), 2010: 1–5.

Abstract: In this paper, we proposed a method for THz sub-comb generation based on spectral interference. The result of our calculation indicated that the THz pulse train, generated by surface-emitted optical rectification of femtosecond (fs) laser pulse in periodically poled lithium niobate (PPLN), has a comb-like spectrum. The characteristic of this THz sub-comb was analyzed both in frequency and time domain. Compared with the THz frequency comb emitted by a photoconductive antenna (PCA), THz sub-comb has a lower spectral resolution and wider free spectral range. Thus it could be an ideal source for wavelength division multiplexing (WDM) in THz wireless communication system.

Key words: Terahertz wave, Frequency comb, Optical rectification

30. The physical theory and propagation model of THz atmospheric propagation

R Wang, J Q Yao, D G Xu, J L Wang, P Wang

Photonics and Optoelectronics Meetings (POEM 2010), 2011: 012223.

Abstract: Terahertz (THz) radiation is extensively applied in diverse fields, such as space communication, Earth environment observation, atmosphere science, remote sensing and so on. And the research on propagation features of THz wave in the atmosphere becomes more and more important. This paper firstly illuminates the advantages and outlook of THz in space technology. Then it introduces the theoretical framework of THz atmospheric propagation, including some fundamental physical concepts and processes. The attenuation effect (especially the absorption of water vapor), the scattering of aerosol particles and the effect of turbulent flow mainly influence THz atmosphere propagation. Fundamental physical laws are illuminated as well, such as Lamber-beer law, Mie scattering theory and radiative transfer equation. The last part comprises the demonstration and comparison of THz atmosphere propagation models like Moliere(V5), SARTre and AMATERASU. The essential problems are the deep analysis of physical mechanism of this process, the construction of atmospheric propagation model and databases of every kind of material in the atmosphere, and the standardization of measurement procedures.

Key words: Terahertz (THz) radiation, Propagation model, Atmospheric propagation

31. Dielectric behavior of CaCu$_3$Ti$_4$O$_{12}$ ceramics in the terahertz range

Liang Wu, Furi Ling, Ting Liu, Jinsong Liu, Yebin Xu, Jianquan Yao
Optics Express, 2011, 19(6): 5118–5125.

Abstract: The dielectric properties of 1 050 ℃/12 h sintered CaCu$_3$Ti$_4$O$_{12}$ (CCTO) ceramics have been investigated by using terahertz time domain spectroscopy in the frequency range of 0.2-1.6 THz at room temperature. When applying an external optical field, an obvious variation of dielectric constant was observed and reached up to 7%. However, the dielectric loss does not change appreciably. From the results, we found the change of refractive index has a linear relationship on scale with the applied light intensity. These findings were attributed to the change of spontaneous polarization in the ceramic caused by the excited free carriers.

Key words: Dielectric properties, CCTO, Terahertz wave

32. Directional terahertz beams realized by depth-modulated metallic surface grating structures

Yinghao Yuan, Jinsong Liu, Jian He, Jianquan Yao
Journal of the Optical Society of America, 2011, 28(11): 2674–2679.

Abstract: We propose a subwavelength metal slit with surrounding depth-modulated surface grating structures to realize directional beams in the terahertz regime. The surface gratings consist of two sets of grooves of different depths. The shallow grooves are designed to support the spoof surface plasmons, and the deep grooves are utilized to diffract the terahertz surface wave into free space. Theoretical analysis and numerical simulations based on the finite element method confirm that various beaming effects including splitting, on-axis beaming and off-axis beaming can be realized by controlling the distributions of the deep grooves.

Key words: Terahertz wave, Depth-modulated surface grating structures, Directional beams

33. Far-infrared dispersion of complex dielectric constant in the ferroelectric near-stoichiometric LiNbO$_3$:Fe

Liang Wu, Furi Ling, Xiaoguang Tian, Haitao Zhao, Jinsong Liu, Jianquan Yao
Optical Materials, 2011, 33: 1737 - 174.

Abstract: The dielectric properties of near-stoichiometric LiNbO$_3$:Fe single crystal have been investigated by using a terahertz time domain spectroscopy (THz–TDS) in a frequency range of 0.7-1.6 THz at room temperature. When coupled with an applied external optical field, an obvious photorefractive effect was observed, resulting in the modulation of the complex dielectric constant for near-SLN:Fe. The variation of refractive index $|\Delta n|$ has a linear relationship on scale with the applied light intensity accompanied with a steplike decrease. These findings were attributed to the internal space charge field of photorefraction and the light-induced domain reversal in the crystal.
Key words: Near-SLN:Fe, THz, Photorefraction, Domain-reversal

34. Far-infrared dispersion of the complex dielectric constant in ferroelectric near-stoichiometric LiNbO$_3$:Ce

Liang Wu, Furi Ling, Zhigao Zuo, Jinsong Liu, Jianquan Yao
Journal of Optics, 2011, 13: 10550.

Abstract: The dielectric properties of near-stoichiometric LiNbO$_3$:Ce (near-SLN:Ce) single crystal have been investigated by using terahertz time domain spectroscopy (THz-TDS) in a frequency range of 0.7–1.6 THz at room temperature. When coupled with an applied external optical field, a photorefractive effect was observed, resulting in the modulation of the complex dielectric constant for near-SLN:Ce. The variation of the refractive index $|\Delta n|$ has a linear relationship in scale with the applied light intensity accompanied with an abrupt decrease. These findings were attributed to the internal space charge field of photorefraction and the light-induced domain reversal in the crystal.
Key words: Near-SLN:Ce, THz photorefraction, Domain reversal

35. Ultrahigh birefringent polymer terahertz fiber based on a near-tie unit

Jingli Wang, Jianquan Yao, Heming Chen, Kai Zhong, Zhongyang Li
Journal of Optics, 2011, 13: 055402.

Abstract: A novel highly birefringent polymer terahertz (THz) fiber based on a near-tie unit, which is formed by three tangent circular air holes, is proposed in this paper. The introduction of near-tie units in the fiber core can enhance asymmetry to realize high mode birefringence. The dependence of the birefringence on several parameters (radii of the air holes, unit-to-unit spacing lattice constant and lattice arrangement), the power distribution of the mode power and the relative absorption loss are investigated with a full-vector finite element method (FEM). Simulation results show that the polymer THz fiber exhibits high birefringence on a level of 10^{-2} over a wide frequency range, and an extremely large birefringence ($\approx 5.11 \times 10^{-2}$ at f =1.2 THz) is obtained in a triangular lattice THz fiber. Moreover, compared with a round solid-core fiber, the THz fiber guiding loss caused by polymer material absorption can be reduced effectively as a part of the mode power is trapped in the air holes.
Key words: Birefringence, Terahertz fiber, Polymer, Near-tie unit

36. Simulation of continuous terahertz wave transient state thermal effects on static water

Lv Yingjin, Xu Degang, Liu Pengxiang, Lv Da, Wen Qiye, Zhang Huaiwu, Yao Jianquan
Chinese Physics B, 2011, 20(10): 104205.

Abstract: We report a numerical simulation of continuous terahertz beam induced transient thermal effects on static water. The terahertz wave used in this paper has a Gaussian beam profile. Based on the transient heat conduction equation, the finite element method (FEM) is utilized to calculate the temperature distribution. The simulation results show the dynamic process of temperature change in water during terahertz irradiation. After about 300 s, the temperature reaches a steady state with a water layer thickness of 5 mm and a beam radius of 0.25 mm. The highest temperature increase is 7 K/mW approximately. This work motivates further study on the interaction between terahertz wave and bio-tissue, which has a high water content.
Key words: Terahertz, Transient thermal effect, Finite element method

37. A simple birefringent terahertz waveguide based on polymer elliptical tube

Wang Jingli, Yao Jianquan, Chen Heming, Li Zhongyang
Chinese Physics Letters, 2011, 28(1): 014207.

Abstract: We propose a simple birefringent terahertz (THz) waveguide which is a polymer elliptical tube with a cross section of elliptical ring structure. It can be achieved by stretching a normal circular-tube in one direction. Simulations based on the full-vector finite element method (FEM) show that this kind of waveguides exhibits high birefringence on a level of 10^{-2} over a wide THz frequency range. Moreover, as a majority of modal power is trapped in the air core inside the polymer elliptical tube, the THz waveguide guiding loss caused by material absorption can be reduced effectively.

Key words: Terahertz, Waveguide, Polymer elliptical tube, Finite element method

38. Single mode condition and power fraction of air-cladding total refractive guided porous polymer terahertz fibers

Jing Lei, Yao Jianquan
Chinese Physics Letters, 2011, 28(8): 084202.

Abstract: We investigate the single-mode condition and power fraction of an air cladding total internal reflection (TIR) guided porous polymer THz fiber. It is shown that the single mode condition and a high fraction of power in air holes cannot be simultaneously met in porous fibers. Employing the V parameter and the Lorentz–Lorenz formula, we analyze this problem theoretically.

Key words: Single-mode condition, Power fraction, Porous polymer THz fiber

39. 高双折射的混合格子太赫兹光子晶体光纤的设计与研究

汪静丽, 姚建铨, 陈鹤鸣, 邴丕彬, 李忠洋, 钟凯

物理学报, 2011, 60(10): 104219.

摘要: 提出了一种新型高双折射的混合格子太赫兹光子晶体光纤, 通过对芯区亚波长尺寸的空气孔进行多种格子组合排列, 增加结构的非对称性实现高的模式双折射。全文仿真建模采用专业的有限元计算软件 COMSOL Multiphysics 4.0, 结果表明: 混合格子太赫兹光子晶体光纤在很宽的频率范围内都具有较高的双折射 (达到 10^{-2}) 和低的限制损耗, 且通过改变光纤的某些参数可以灵活地控制其双折射或限制损耗特性。相比于同类光通信波段光纤, 由于太赫兹波波长较大, 能够降低芯区微结构加工的难度, 具有可行性。

关键词: 双折射, 混合格子, 太赫兹光子晶体光纤, 限制损耗

40. THz 波在金属镀层空芯波导中传输的理论和实验研究

谭晓玲, 耿优福, 周骏, 姚建铨

物理学报, 2011, 60(5): 054101.

摘要: 本文基于微扰法求得不同金属镀层空心圆波导中各模式的损耗系数, 对金属镀层空芯波导中 THz 波传输损耗随金属材料、波导结构等参数的变化关系进行了数值模拟。根据数值分析结果, 优化设计并拉制了内径为 1.1 mm 的镀银空芯波导, 实验测得当 THz 波的频率为 2.5 THz 时, 传输损耗为 8.6 dB/m, 实现了 THz 波短距离的有效传输。

关键词: 太赫兹波, 金属空芯波导, 传输损耗

41. THz modulator based on the Drude model

Yang Pengfei, Yao Jianquan, Xu Wei
Optoelectronics Letters, 2011, 7(1): 0019–0021.

Abstract: An amplitude modulator for the terahertz (THz) range is designed. The Drude model is adopted, in which the collision damping is independent of the carrier energy. The Si block with 808 nm laser is illustrated, and it will generate the photocarriers. The injected photo-carriers will change the conductivity and dielectric of the sample, which have direct relationship with the absorption coefficient of the THz wave, hence to control the characteristics of the THz wave in the sample. By changing the light intensity, due to the different photon-generated carrier concentrations, the single transmission of the THz wave in the silicon substrate is changed remarkably.

Key words: Terahertz amplitude modulator, Drude model, THz wave

42. Propagation characteristics of THz radiation in hollow rectangle metal waveguide

Y Zhang, Y Z Zhang, D M Lu, D Y Wang, M Liu, S Y WANG, P D Zhao, Z D ZHANG, J Q Yao
Journal of Physics: Conference Series (POEM 2010), 2011, 276: 012229.

Abstract: The secular equation of medium coating were obtained by considering the particularity of THz, which had some reference value in designing of THz waveguides. The attenuation of medium coating metal hallow waveguides were theoretically calculated, the conclusion is that the power loss is less than the metal waveguides. Also we obtained the mode characters and attenuation in metal waveguides, and we noted that there exists an absorption peak. And we studied the affection of shape and size on propagation and got some useful conclusion. The split rectangular waveguide (SRW) is suitable for THz transmission which is confirmed by experiment. Our secular equation can be considered to be a theoretically discussion on it.

Key words: Terahertz wave, Propagation characteristics, Hollow rectangle metal waveguide

43. The study on THz wave propagation feature in atmosphere

Haixia Cui, Jianquan Yao, Chunming Wan

Journal of Physics: Conference Series (POEM 2010), 2011, 276: 012225.

Abstract: THz wave has many applications, such as space communication, Earth environment observation, atmosphere science, remote sensing and so on. The atmospheric propagation of terahertz waves now rank among the most critical issues in the societal implementation of terahertz technology. In this paper we introduce the constitution of THz wave atmosphere propagation system, decay, then analyze the decay, turbulent and enhancement effects about THz wave atmosphere propagation, then carry out atmospheric propagation measurement system based on THz-TDS.

Key words: Terahertz wave, Propagation feature, Atmospheric propagation

44. Electrically controlled broadband THz switch based on liquid-crystal-filled multi-layer metallic grating structures

Yinghao Yuan, Jian He, Jinsong Liu, Jianquan Yao

Journal of Physics: Conference Series (POEM 2010), 2011, 276: 012228.

Abstract: We propose an electrically controlled broadband THz switch based on the liquidcrystal-filled multilayer metallic grating (MMG) structures. During the THz wave passing thought the device, it is multiple reflected among the MMG layers and suffers loss due to the absorption of the LC. The on and off state of the switch is realized by the different absorption intensity of the LC when its complex refractive index is switched between ordinary no and extraordinary ne value in the presence of an electric field. Simulation results show the proposed device exhibits a respectable switching effect for the THz wave over a wide frequency range.

Key words: Terahertz wave, Switch, Liquidcrystal-filled MMG structures

45. 基于液晶的可调谐太赫兹双折射滤波器的设计

吕英进,徐德刚,刘鹏翔,吕达,王鹏,姚建铨

激光与红外, 2011, 41(4): 450-454.

摘要: 液晶双折射滤波器具有室温下工作,调谐简单方便,带宽窄等优点。为了达到设计不同调谐范围和带宽的这种滤波器的目的,以满足实际应用的需要,扩大其应用范围,采用数值模拟的方法,进行参数计算和设计思路的总结,并设计了一套窄带输出的滤波器实例。结果表明,通过对影响滤波器调谐范围和输出带宽的关键参数的数值模拟和分析,为设计不同调谐范围的液晶双折射滤波器提供了依据;设计实例基本上满足预期的设计要求,调谐范围 0.691~0.866 THz。

关键词: 光学器件,太赫兹,液晶,调谐,双折射滤波器

46. 谐振环左手材料设计参数对太赫兹传输的影响

姚建铨,杨鹏飞,邢丕彬,邱志刚

激光与红外, 2011, 41(8): 825-829.

摘要: 左手材料(LHM)的出现,为新型材料的探究和应用开辟了一个全新的领域。电磁波在左手材料中的传播特性已经被许多研究工作者广泛探索并得到了许多新结果,与此同时,太赫兹由于其独特性质也成为近年来研究的热点。归类总结了太赫兹波在典型 LHM 材料中的传输,并通过比较太赫兹波在不同设计的 LHM 中的传输,得出在 LHM 设计中,周期、分形、基板材料等参数对太赫兹传输的影响,这些传输特性在太赫兹波器件的制备方面有很大的应用前景。

关键词: 太赫兹,左手材料,谐振环,传输

47. Terahertz photonic states in semiconductor-graphene cylinder structures

Yizhe Yuan, Jianquan Yao, Wen Xu

Optics Letters, 2012, 37(5): 960–962.

Abstract: We propose a semiconductor–graphene cylinder that can serve as a terahertz (THz) photonic crystal. In such a structure, graphene plays a role in achieving a strong mismatch of the dielectric constant at the semiconductor–graphene interface due to its two-dimensional nature and relatively low value of the dielectric constant. We find that when the radius of the outer semiconductor layer is about $\rho_1 \sim 100$ μm, the frequencies of the photonic modes are within the THz bandwidth and they can be efficiently tuned via varying ρ_1. Furthermore, the dispersion relation of the photonic modes shows that a semiconductor–graphene cylinder is of excellent light transport properties, which can be utilized for the THz waveguide. This study is pertinent to the application of graphene as THz photonic devices.

Key words: Terahertz photonic crystal, Semiconductor–graphene cylinder, Transport property

48. The study of negative THz conductivity of graphene under the phonon scattering mechanism

Weipeng Wang, Degang Xu, Yuye Wang, Changming Liu, Zhuo Zhang, Hao Jiang, JianquanYao

Optics Communications, 2012, 285: 5410–5415.

Abstract: The variety of optical and electronic properties of graphene attracts enormous interests. The relaxation and recombination mechanism of photogenerated electrons and holes in undoped monolayer graphene are studied with the changing pump intensities. The population inversion in graphene can lead to the negative AC conductivity in THz spectral range. And the conductivity depends on the carrier density, the carrier distribution in energy as well as the effective temperature. It indicates that the negative dynamic conductivity is associated with the THz emission and can be used in graphene-based new THz lasers.

Key words: Graphene, Population inversion, AC conductivity, THz emission

49. The dielectric behaviour of doped near-stoichiometric lithium niobate in the terahertz range

Wu Liang, Ling Furi, Zuo Zhigao, Liu Jinsong, Yao Jianquan
Chinese Physics B, 2012, 21(1): 01780.

Abstract: The dielectric properties of near-stoichiometric $LiNbO_3$:Fe and $LiNbO_3$:Ce single crystals have been investigated using terahertz time domain spectroscopy in a frequency range of 0.7–1.6 THz at room temperature. When coupled with an applied external optical field, obvious photorefractive effects were observed, resulting in a modulation of the complex dielectric constant for the crystals. The variation in refractive index, $|\Delta n|$, had a linear relationship with the applied light intensity, accompanied by a step-like decrease at high intensity. The findings were attributed to the internal space charge field of the photorefraction and the light-induced domain reversal in the crystals.

Key words: Near-stoichiometric $LiNbO_3$, Terahertz, Photorefraction, Domain reversal

50. 基于平行金属双柱的太赫兹波二维左手材料

梁兰菊，闫昕，姚建铨，田贵才，薛冬
光学学报，2012, 32(3): 0316001.

摘要：基于产生负介电常数的周期性金属线单元结构，利用平行金属双柱设计了具有双负通带的两种太赫兹波段二维左手材料。应用时域有限积分算法研究了二维左手材料的传输特性，仿真结果表明，在0.76 THz附近，平行金属双柱的表面电荷振荡与反向平行电流引发了电磁谐振，出现良好的负折射效应。在0.75~0.78 THz之间同时具有负等效磁导率和负等效介电常数，双负通带带宽约为0.03 THz。进一步研究了金属双柱间距、长度及基板厚度等结构参数对双负通带带宽的影响。研究结果为太赫兹波段左手材料的设计和研制提供了参考。

关键词：材料，光电子学，太赫兹波，左手材料，金属双柱，传输特性

51. Optical tuning of dielectric properties of $Ba_{0.6}Sr_{0.4}TiO_3$-$La(Mg_{0.5}Ti_{0.5})O_3$ ceramics in the terahertz range

Liang Wu, Linkun Jiang, Yebin Xu, Xin Ding, Jianquan Yao
Applied Physics Letters, 2013, 103: 191111.

Abstract: The dielectric properties of $0.4Ba_{0.6}Sr_{0.4}TiO_3$-$0.6La(Mg_{0.5}Ti_{0.5})O_3$ ceramics and their tenability under external optical fields are investigated at room temperature by means of terahertz time-domain spectroscopy. Application of the optical field leads to an appreciable tuning of the permittivity, which reaches up to 16%. Meanwhile, the dielectric loss changes about 21%. From the results, we find that the change of refractive index has a linear relationship on scale with the applied light power. These findings are attributed to the internal space charge field in the ceramic caused by the excited free carriers.

Key words: Terahertz wave, Dielectric property, Optical tuning, Terahertz time-domain spectroscopy

52. Optical tuning of dielectric properties of $SrTiO_3$:Fe in the terahertz range

Liang Wu, Linkun Jiang, Quan Sheng, Xin Ding, Jianquan Yao
Optics Letters, 2013, 38(14): 2581–2583.

Abstract: Tuning of the dielectric permittivity spectra of iron-doped strontium titanate ($SrTiO_3$:Fe) single crystals in an external optical field is investigated at room temperature by means of terahertz time-domain spectroscopy. Application of the optical field leads to an appreciable tuning of the permittivity, which reaches up to 3.8%. The observed behavior is interpreted in terms of soft-mode hardening due to the anharmonic character of its potential. We also find that the change of refractive index has a linear relationship on scale with the applied light power. These findings are attributed to the internal space charge field of photorefraction caused by the excited free carriers.

Key words: Terahertz wave, Dielectric property, Optical tuning, $SrTiO_3$:Fe, Terahertz time-domain spectroscopy

53. Hyperfine spectrum measurement of an optically pumped far-infrared laser with a Michelson interferometer

Z G Zuo, F R Ling, P Wang, J S Liu, J Q Yao, C X Weng

Laser Physics Letters, 2013, 10(5): 055004.

Abstract: In this letter, we present a Michelson interferometer for the hyperfine spectrum measurement of an optically pumped far-infrared laser with a highest frequency resolution of 3–5 GHz. CH_3OH gas with a purity of 99.9%, is pumped by the CO_2 9P36 and 9R10 laser lines to generate terahertz lasers with frequencies of 2.52 and 3.11 THz, respectively. Moreover, except for the center frequency, which is in good agreement with theoretical work, some additional frequencies on both sides of the center frequency are obtained at a frequency interval of 0.15 THz. Meanwhile, the mechanism behind the observed experimental results is also investigated.

Key words: Michelson interferometer , Hyperfine spectrum measurement, Far-infrared laser

54. Real propagation speed of the ultraslow plasmonic THz waveguide

Baoshan Guo, Wei Shi

Applied Physics B: Lasers and Optics, 2013, 340: 5550.

Abstract: A graded metallic grating structure can not only work as a wave trapping system, but also work as an ultraslow terahertz (THz) waveguide. The depth of the grating waveguide is partial graded and partial fixed. The real propagation speed of such a system is calculated. Different frequencies of THz waves can be propagated at a designed propagation speed, even close to zero.

Key words: Metallic grating structure, Ultraslow terahertz waveguide, Propagation speed

55. Effect of optical pumping on the momentum relaxation time of graphene in the terahertz range

Zuo Zhigao, Wang Ping, Ling Furi, Liu Jinsong, Yao Jianquan
Chinese Physics B, 2013, 22(9): 097304.

Abstract: The momentum relaxation time of a photoexcited graphene in the THz frequency range has been studied by using terahertz time domain spectroscopy under optical pumping at room temperature. It is found that the momentum relaxation time of the graphene as a function of the optical pumping intensity exhibits a threshold behavior. The features of the momentum relaxation time as a function of the optical pumping intensity are also investigated. The results are useful for understanding the basic underlying physics of graphene scattering as well as finding the possible applications in carbonbased electronics.

Key words: Graphene, Momentum relaxation time, Optical pumping, Terahertz

56. Resonance mode-switching in terahertz metamaterials based on varying gallium arsenide conductivity

Xiaolong Cao, Jianquan Yao, Cai Yuan, Kai Zhong
Optical Engineering, 2013, 52(2): 024001.

Abstract: An asymmetric planar terahertz (THz) metamaterial (MM) is designed to be composed of two different single split-ring resonators (SRR) and its character is simulated in the THz region. Gallium arsenide (GaAs) is inserted between the gap of two separated single SRR, and through modulation of its conductivity (σ_{GaAs}) we achieved the switching of three different resonance modes and researched the influence of σ_{GaAs} on transmission of MM. The resonant structure of MM can be switched from the two fundamental LC-modes to the new LC-mode and dipole resonance mode through the coupling LC-mode with the increasing σ_{GaAs}. Such dynamical control of MM resonances provides an efficient way to manipulate electromagnetic wave to push mode-switching of resonance and could be implemented in terahertz devices to achieve additional functionalities.

Key words: Split-ring resonator, Mode-switching, Transmission, Terahertz, Metamaterials

57. Optical control of terahertz nested split-ring resonators

Cai Yuan, Xiaolei Zhao, Xiaolong Cao, Shilin Xu, Nannan Luan, Jianquan Yao
Optical Engineering, 2013, 52(8): 087111.

Abstract: Two kinds of optical modulation processes are designed based on nested split-ring resonators in the terahertz regime. By photo-conductivity induced mode switching effect, the resonant structure and resonant mode will be changed, and the two contrary tunable resonant properties are obtained. Without illumination, the two designs both have three resonant peaks. As the intensity of illumination increases, for the first design, the first and the third resonant peaks disappear and a broadband resonant peak at the second resonant frequency is formed eventually; however, the second design realizes an exactly opposite modulation process. Moreover, in order to identify the modulation mechanism, we discuss and analyze the distributions of the electric fields and surface currents in detail. Our designs can be implemented in tunable terahertz functional devices and provide important reference value for the design of terahertz metamaterial.

Key words: Terahertz, Metamaterial, Split-ring resonators, Modulator

58. Optical tuning of dielectric properties of $LiNbO_3$: Mg in the terahertz range

Liang Wu, Linkun Jiang, Quan Sheng, Xin Ding, Jianquan Yao
Journal of Infrared, Millimeter and Terahertz Waves, 2013, 34(10): 639–645.

Abstract: Tuning of the dielectric permittivity spectra of congruent $LiNbO_3$:Mg single crystals in an external optical pump is investigated at room temperature by means of terahertz time-domain spectroscopy. Application of the optical pump leads to an appreciable tuning of the permittivity, which reaches up to 2.3%. We find that the change of refractive index has a linear relationship on scale with the applied light power. These findings are attributed to the internal space charge field of photorefraction caused by the excited free carriers.

Key words: Optoelectronics, Ferroelectrics, Spectroscopy, Terahertz

59. 太赫兹频段开口环谐振器的可调谐振模式转换

曹小龙, 姚建铨, 袁偲, 赵晓蕾, 钟凯

强激光与粒子束, 2013, 25(9): 2324-2328.

摘要： 在硅衬底上设计了一种单开口环谐振器,对其太赫兹频段内的透射性质进行了研究。假定通过光注入方式改变衬底硅的电导率,实现了谐振环的双谐振透射率可调。将砷化镓材料生长于该谐振环的开口处,通过光注入方式改变砷化镓材料的电导率,可以实现谐振环的双频 LC 共振和偶极子共振模式与单频闭合环共振模式之间的转换。这种通过光注入改变半导体材料电导率的方法,可以在不破坏原来谐振器件物理结构的前提下,实现谐振环谐振模式的可逆转换。

关键词： 开口环谐振器, 太赫兹, 透射, 超材料

60. 基于太赫兹目标散射特性测试系统的设计与应用

杨洋, 姚建铨, 王力, 张镜水

仪器仪表学报, 2013, 34(5): 975-980.

摘要： 搭建了以卧式自动旋转台、立式自动旋转台、龙门架、运动控制器等组成的太赫兹波目标散射特性测试系统,实现了收发分置与收发同置的低频太赫兹波在粗糙铝表面的散射实验测试工作,表明：在收发分置时金属粗糙铝表面散射范围较大,散射角小于 30° 时散射曲线下降较快,超过 30° 时散射曲线变化变得缓慢,且在 45° 附近出现了一个小的散射峰,在收发同置时金属粗糙铝表面散射信号随散射角度变化明显,当散射角达到 10° 时,散射信号几乎衰减为 0,对于同一入射波源,粗、细铝板散射效果显现出不同的现象,其中在细铝板散射曲线中出现信号强度明显的振荡现象。

关键词： 太赫兹, 目标散射特性, 收发同置, 收发分置, 测试系统

61. 铁磁材料在太赫兹波段的研究进展

蒋霖坤,吴亮,姚建铨

激光与光电子学进展, 2013, 50(8): 080022.

摘要: 近年来,铁磁材料特别是铁磁性微纳材料在太赫兹波段的研究取得了许多具有重要应用前景的成果。介绍了太赫兹辐射在一些铁磁性微纳材料中的产生,以及一些铁磁性材料与太赫兹波相互作用的研究。其中包括:外加磁场和非磁性纳米涂层对于在 Co,Ni 等铁磁性微粒中传输的太赫兹脉冲衰减和延迟的影响,铁磁流体中的法拉第旋转,太赫兹脉冲磁场与晶体磁矩的相互作用和一些铁磁性薄膜在太赫兹波段的负折射率。此外,还介绍了两种由人工设计的铁磁性材料构成的磁控太赫兹功能器件。最后,对铁磁材料在太赫兹波段的应用前景进行了展望。

关键词: 材料,铁磁材料,太赫兹,脉冲,磁场,功能器件

62. Modulation of dielectric properties of $KTaO_3$ in terahertz region via 532 nm continuous-wave laser

L. Wu, H. Li, L. Jiang, C. Ding, Q. Sheng, X. Ding, J. Yao

Optical Materials Express, 2014, 4: 2595–2601.

Abstract: The dielectric constant of potassium tantalate ($KTaO_3$) single crystals at 0.1-0.8 THz is modulated by 532 nm continuous-wave laser at room temperature. The dielectric constant decreases with the increasing laser power, especially the real part, which decreases by 3.5% when the laser power is 600 mW. This property of $KTaO_3$ crystal is attributed to its soft-mode hardening due to the anharmonic character of its potential. It is also found that the refractive index linearly decreases with the increasing laser power, which is interpreted as the linear electro-optic effect induced by internal space charge field of the crystal.

Key words: Terahertz, Modulation, Dielectric properties, $KTaO_3$

63. Effect of an optical pump on the absorption coefficient of $Ba_{0.6}Sr_{0.4}TiO_3$-$La(Mg_{0.5}Ti_{0.5})O_3$ ceramics in the terahertz range

Liang Wu, Linkun Jiang, Yebin Xu, Xin Ding, Jianquan Yao
Journal of Optics, 2014, 16: 105703.

Abstract: The absorption coefficient of $0.4Ba_{0.6}Sr_{0.4}TiO_3$–$0.6La(Mg_{0.5}Ti_{0.5})O_3$ ceramic and its tunability under external optical fields are investigated at room temperature by means of terahertz time-domain spectroscopy. Experimental results show that the absorption coefficient of $0.4Ba_{0.6}Sr_{0.4}TiO_3$–$0.6La(Mg_{0.5}Ti_{0.5})O_3$ ceramic is approximately 50–500 cm^{-1} in the frequency range of 0.1–0.8 THz. Application of the optical field leads to an appreciable tuning of dielectric loss, which reaches up to 20.5%. Further theoretical analysis revealed that the variation of absorption coefficient was related to light-induced carriers and OH^- absorption in this ceramic.
Key words: THz-materials, Optical constants, Ferroelectrics

64. Dynamic trapping of terahertz waves by silicon-filled metallic grating structure

Yinghao Yuan, Jinsong Liu, Jian He, Jianquan Yao
Optics Communications, 2014, 332: 132–135.

Abstract: We investigate the feasibility of dynamic trapping of terahertz waves using a silicon-filled metallic grating structure. Using the dispersion relation analysis and the two-dimensional finite element method simulations, we reveal that, if a graded refractive index distribution in the grooves is optical induced, the device has the ability to dynamic trap terahertz waves of different frequencies at different positions (so-called trapping rainbow). Moreover, we demonstrate that the trapped position of a certain frequency of the terahertz waves can be moved continuously along the grooves in subwavelength scale by ingenious control of the distributions of the refractive indices of silicon filled in the grooves. Our design has the potential for the construction of active plasmonic terahertz devices, such as optical controlled terahertz filter, router and demultiplexer in a broadband terahertz communication system.
Key words: Plasmons, Surface waves, Terahertz waves

65. Guided modes in asymmetric negative-zero-positive index metamaterial waveguide in the terahertz regime

Xiaolei Zhao, Cai Yuan, Shilin Xu, Wenhua Lv, Jianquan Yao

Optical Engineering, 2014, 53(4): 045102.

Abstract: An asymmetric negative-zero-positive index metamaterial (NZPIM) waveguide with two different Dirac points (DPs) operating in the terahertz (THz) regime is proposed. Its propagating characteristics are investigated by using the graphical method. Due to the linear dispersion near the DPs, the asymmetric NZPIM waveguide exhibits unique properties of guided modes, which are different from that in conventional waveguide but similar to that of electron waves in graphene waveguides. It is shown that the guided mode properties can be tuned by adjusting the incident angular frequency. In addition, the properties of surface-guided modes are also discussed. Our work may have potential applications in THz functional devices as well as reference value in investigating the guided modes of electrons in graphene waveguides.

Key words: Terahertz, Dirac points, Negative-zero-positive index metamaterials, Waveguide

66. Effect of optical pump on the dielectric properties of SrTiO$_3$ in terahertz range

Wu Liang, Jiang Linkun, Yuan Cai, Ding Xin, Yao Jianquan

Chinese Physics B, 2014, 23(3): 034212.

Abstract: Tuning the dielectric permittivity spectra of strontium titanate (SrTiO$_3$) single crystals in an external optical field is investigated at room temperature by means of terahertz time-domain spectroscopy. The application of the optical field leads to an appreciable tuning of the permittivity, reaching up to 2.8%, with the dielectric loss changing about 3%. The observed behavior is interpreted in terms of soft-mode hardening due to the anharmonic character of its potential. We also find that the change of the refractive index responds linearly to the applied light power. These findings are attributed to a linear electro-optical effect of the internal space charge field of the crystal.

Key words: THz materials, Optical constants, Ferroelectrics

67. Low-loss terahertz waveguide with InAs-graphene-SiC structure

Xu Degang, Wang Yuye, Yu Hong, Li Jiaqi, Li Zhongxiao, Yan Chao, Zhang Hao, Liu Pengxiang, Zhong Kai, Wang Weipeng, Yao Jianquan

Chinese Physics B, 2014, 23(5): 054210.

Abstract: We demonstrate a low-loss terahertz waveguide based on the InAs-graphene-SiC structure. By analyzing the terahertz waveguide proposed in this paper, we can obtain that it is the characteristic of a low transmission loss coefficient ($\alpha_{loss} \approx 0.55$ dB/m) for fundamental mode (LP_{01}) when the incident frequency is larger than 3.0 THz. The critical radii of the inside and outside cylinders have been found for the high-quality transmission. The large inside radius and the high transmission frequency result in a flat transmission loss coefficient curve. As a strictly two-dimensional material, the double graphene surface rings perform better to improve the quality of transmission mode. These results provide a new idea for the research of the long-distance THz waveguide.

Key words: InAs-graphene-SiC structure, Low-loss terahertz waveguide, Transmission, Critical radii

68. Tunable ultra-wideband terahertz filter based on three-dimensional arrays of H-shaped plasmonic crystals

Yuan Cai, Xu Shilin, Yao Jianquan, Zhao Xiaolei, Cao Xiaolong, Wu Liang
Chinese Physics B, 2014, 23(1): 018102.

Abstract: A face-to-face system of double-layer three-dimensional arrays of H-shaped plasmonic crystals is proposed, and its transmission and filtering properties are investigated in the terahertz regime. Simulation results show that our design has excellent filtering properties. It has an ultra-wide bandgap and passband with steep band-edges, and the transmittance of the passband and the forbidden band are very close to 1 and 0, respectively. As the distance between the two face-to-face plates increases, the resonance frequency exhibits a gradual blueshift from 0.88 THz to 1.30 THz. Therefore, we can dynamically control the bandwidths of bandgap and passband by adding a piezoelectric ceramic plate between the two crystal plates. Furthermore, the dispersion relations of modes and electric field distributions are presented to analyze the generation mechanisms of bandgaps and to explain the location of bandgaps and the frequency shift phenomenon. Due to the fact that our design can provide many resonant modes, the bandwidth of the bandgaps can be greatly broadened. This paper can serve as a valuable reference for the design of terahertz functional devices and three-dimensional terahertz metamaterials.

Key words: Terahertz, Wideband filter, Plasmonic crystal, Metamaterials

69. Reflection-type electromagnetically induced transparency analogue in terahertz metamaterials

Ding Chunfeng, Zhang Yating, Yao Jianquan, Sun Chongling, Xu Degang, Zhang Guizhong

Chinese Physics B, 2014, 23(12): 124203.

Abstract: A reflection-type electromagnetically induced transparency (EIT) metamaterial is proposed, which is composed of a dielectric spacer sandwiched with metallic patterns and metallic plane. Experimental results of THz time domain spectrum (THz-TDS) exhibit a typical reflection of EIT at 0.865 THz, which are in excellent agreement with the full-wave simulations. A multi-reflection theory is adopted to analyze the physical mechanism of the reflection-type EIT, showing that the reflection-type EIT is a superposition of multiple reflection of the transmission EIT. Such a reflection-type EIT provides many applications based on the EIT effect, such as slow light devices and nonlinear elements.

Key words: Electromagnetically induced transparency (EIT), Metamaterial, Multiple-reflection interference

70. Design of the novel steering-wheel micro-structured optical fibers sensor based on evanescent wave of terahertz wave band

L. Zhang, G.J. Ren, J.Q. Yao, Y.M. Zhang

Optik, 2014, 125: 5936–5939.

Abstract: In this paper, we present design of a hollow micro-structured photonic crystal fiber with novel steering-wheel pattern of noncircular large holes in cladding as platform for evanescent-field sensing. Based on simulation, confinement loss is less than 0.007 dB/m, and 72% of light intensity overlaps in noncircular large air holes is obtained when the incident wave frequency is 1 THz, and the nonlinear effects of the short-distance transmission are very small simultaneously. The critical value of confinement losses increases with the structure parameters. As for ultra-low loss and high sensitivity of the model, the novel steering-wheel structured fiber is well suited for evanescent-field sensing and detection of chemical and biological products.

Key words: Photonic crystal fiber, Evanescent wave, Terahertz, Sensitivity

71. Thermal tunability and sensitivity of bandgap photonic crystal fiber of teraherz wave

Wu Peng, Ren Guangjun, Yao Jianquan

OPTOELECTRONICS AND ADVANCED MATERIALS-RAPID COMMUNICATIONS, 2014, 8(7–8): 775–778.

Abstract: The production of filter combined terahertz with photonic crystal fiber is rare. Fill the nematic liquid crystal in the photonic crystal fiber air hole, photonic crystal optical fiber transmission mechanism was tuned by changing the nature of the liquid crystal. Liquid crystal refractive index change with temperature, so that the defect mode frequency will move. Based on this mechanism, threshold switching or filter can be designed. For the terahertz sensing characteristics in this article, mode field area and sensitivity become small as the wavelength increases.

Key words: Liquid Crystal, Temperature, Terahertz wave, Photonic crystal fiber

72. Voltage influence on propagation characteristics of liquid crystal photonic crystal fiber of terahertz wave

Chang Lihua, Ren Guangjun, Chen Zhihong, Yao Jianquan

JOURNAL OF OPTOELECTRONICS AND ADVANCED MATERIALS, 2014, 16(9–10): 1175–1179.

Abstract: This paper presents a novel structure of photonic crystal fiber (PCF). Using the finite element method, a new type of LC PCF terahertz waveguide by voltage modulation is designed which is based on that the holes of PCF filled with nematic LC 5CB. We use COMSOL and MATLAB to simulate and calculate the parameters of PCF of different structures under THz wave band. We made PCF have continuously tunable sensing properties without changing its structure. It provides great convenience for the practical application.

Key words: Voltage, Terabertz wave, Photonic crystal fiber, Liquid crystal

73. Multiband metamaterial absorber at terahertz frequencies

Xu Zongcheng, Gao Runmei, Ding Chunfeng, Zhang Yating, Yao Jianquan
Chinese Physics Letters, 2014, 31(5): 054205.

Abstract: We propose a multi-band metamaterial absorber operating at terahertz frequencies. The design, characterization, and theoretical calculation of the high performance metamaterial absorber are reported. The multi-band metamaterial absorber consists of two metallic layers separated by a dielectric spacer. Theoretical and simulated results show that the metamaterial absorber has four distinct absorption points at frequencies 0.57 THz, 1.03 THz, 1.44 THz and 1.89 THz, with the absorption rates of 99.9%, 90.3%, 83.0%, 96.1%, respectively. Two single band metamaterial absorbers and a dual band metamaterial absorber on the top layer are designed. Some multi-band absorbers can be designed by virtue of combining some single band absorbers. The multiple-reflection theory is used to explain the absorption mechanism of our investigated structures.

Key words: Terahertz, Multi-band metamaterial absorber, Multiple-reflection theory

74. Propagation speed calculation of a plasmonic THz wave trapping system

Baoshan Guo, Wei Shi, Jianquan Yao
Chinese Optics Letters, 2014, 12: S12301.

Abstract: A graded metallic grating structure acts as a wave trapping system. Different frequencies of THz waves are trapped at different positions along this structured metal surface grating. The real wave propagation speed of such a system is reduced gradually from the light speed in vacuum to zero, which is demonstrated by calculation and simulation. Different frequencies of THz waves are propagated at a designed propagation speed by a partial graded grating according to the practical demand.

Key words: Terahertz, Wave trapping system, Propagation speed

75. The study of the fundamental nature and electromagnetic parameter retrieval of reverse nested Split-ring resonators

Xiaolei Zhao, Cai Yuan, Jianquan Yao, Shilin Xu, Wenhua Lv

International Photonics and OptoElectronics Meetings, *OSA Technical Digest Optical Society of America*, 2014: OTh4C.1.

Abstract: We present a design of double reverse nested Split-ring resonators operating in terahertz regime. Its resonance characteristics under normal incidence are investigated. Besides, an improved method is used to retrieve the effective parameters.
Key words: Terahertz, Resonance characteristics, Electromagnetic parameter

76. Mechanisms of THz trapping devices based on plasmonic grating

Baoshan Guo, Wei Shi, Jianquan Yao

Lasers & Electro-optics. IEEE, 2014.

Abstract: The mechanisms of plasmonic THz wave trapping devices is attributed to the transformation from surface modes to cavity modes which have a saturated state. An ultraslow THz waveguide is realized by controlling the modes transformation.
Key words: Terahertz wave, THz trapping device, Plasmonic grating

77. Design of a tunable multiband terahertz waves absorber

Chunya Luo, Dan Li, Qin Luo, Jin Yue, Peng Gao, Jianquan Yao, Furi Ling
Journal of Alloys and Compounds, 2015, 652: 18–24.

Abstract: A thermally tunable multiband terahertz metamaterial absorber comprising a periodic array of closed metallic square ring resonators and four metal bars parallel to the four side of the square ring, fabricated on the low-temperature co-fired ceramic (LTCC) strontium titanate (STO) dielectric layer dielectric substrate has been proposed. The resonance frequencies of the absorber are demonstrated to be continuously tuned in the terahertz regime by increasing the temperature. It is found that in the window between 0.05 and 0.35 THz the absorber has three distinctive absorption peaks at frequencies 0.129 THz, 0.198 THz and 0.316 THz (at the room temperature), whose peaks are attained 99.3%, 99.1% and 94.6% respectively. The tunability is attributed to the temperature-dependent permittivity of the substrate and attained to 67.3% frequency tuning depth at the room temperature, when the temperature varied from 400 K to 200 K. The proposed designs ensure broadband thermally tunable terahertz devices.

Key words: Metamaterial, LTCC, Multiband absorber, Tunable

78. Plasmon-induced transparency in metamaterial based on graphene and split-ring resonators

Xiaolei Zhao, Cai Yuan, Wenhua Lv, Shilin Xu, Jianquan Yao
IEEE Photonics Technology Letters, 2015, 27(12): 1321–1324.

Abstract: We present a design of a terahertz plasmon induced transparency (PIT) metamaterial based on a graphene patch and split-ring resonator (SRR) pair. The PIT metamaterial exhibits a sharp-induced transparency peak resulting from the destructive interference between the direct-excited plasmon resonance in the graphene patch acting as the bright mode and the coupling excited inductive–capacitive resonance in the SRR pair acting as the dark mode. Tuning the Fermi energy in graphene results in the modulation of the PIT window, allowing for the active control of the group index. These results may lead to promising applications in tunable terahertz devices, slow light, and sensing technology.

Key words: Terahertz, Metamaterial, Graphene, Plasmon induced transparency.

79. Dual-band ultrasensitive THz sensing utilizing high quality Fano and quadrupole resonances in metamaterials

Chunfeng Ding, Linkun Jiang, Liang Wu, Runmei Gao, Degang Xu, Guizhong Zhang, Jianquan Yao

Optics Communications, 2015, 350: 103–107.

Abstract: High quality (Q) factor resonances are extremely promising for designing ultrasensitive sensors. In this paper, we proposed a metamaterial functioning in the THz regime, which exhibited low loss, high Q Fano and quadrupole resonances simultaneously. Ultrasensitive THz sensing was performed based on the redshifts of the two high Q resonances when analyte was coated at the surface of the metamaterial. Results indicated that both the Fano and quadrupole resonances can sensitively detect slight changes of the analyte (e.g. the thickness or refractive index). The refractive index sensitivity levels are 2.06×10^4 nm/RIU for the Fano resonances and 5.07×10^3 nm/ RIU for the quadrupole resonance, respectively. Both of the two sensitivities have an order of magnitude higher than schemes based on the low Q dipolar resonances in metamaterials. This planar metamaterial would open new degrees of freedom for designing advanced chemical and biological sensors and detectors in the terahertz regime.

Key words: Metamaterials, Fano resonance, Quadrupole dipole, High quality (Q), Sensing

80. Graphene metamaterial for multiband and broadband terahertz absorber

Runmei Gao, Zongcheng Xu, Chunfeng Ding, Liang Wu, Jianquan Yao

Optics Communications, 2015, 356: 400–404.

Abstract: In this paper, we present the efficient design of functional graphene thin film metamaterial on a metalplane separated by a thick dielectric layer. Perfect absorption is characterized by the complete suppression of incident and reflected light and complete dissipation of incident energy. We investigate the properties of graphene metamaterials and demonstrate multiband absorbers that have five absorption bands, using silicon interlayers, in the 0-2.2 THz range. The absorption rate reached up to 99.9% at a frequency of 1.08 THz, and the quality factor was 6.98 for a 0.14 THz bandwidth. We present a novel theoretical interpretation based on standing wave field theory, which shows that coherent superposition of the incident and reflection rays produce stationary waves, and the field energy localized inside the thick spacers and dissipated through the metal-planes. Thus, light was effectively trapped in the metamaterial absorbers with negligible near-field interactions, causing high absorption. The theory developed here explains all features observed in multiband metamaterial absorbers and therefore provides a profound understanding of the underlying physical mechanisms.

Key words: Metamaterials, Terahertz, Subwavelength structures, Coherent optical effects, Theory and design

81. Photoexited switchable metamaterial absorber at terahertz frequencies

Zongcheng Xu, Runmei Gao, Chunfeng Ding, Liang Wu, Yating Zhang, Degang Xu, Jianquan Yao

Optics Communications, 2015, 344: 125–128.

Abstract: We propose a design and numerical study of an optically switchable metamaterial absorber in the terahertz regime. The metamaterial absorber comprises a periodic array of metallic split-ring resonators sitting back to back with an embedded semiconductor silicon. Filing the gap between the resonator arms with a semiconductor (silicon), leads to easy modification of its optical response through a pump beam which changes conductivity of Si. The conductivity of silicon is a function of incident pump power. Therefore, the resonance frequencies of the metamaterial can be tunable by applying an external pump power. The resonance peak of the absorption spectra shows a shift from 1.17 to 0.68 THz via external optical stimulus, with granting a resonance tuning range on the order of 42%. The optical-tuned absorber has potential applications as a terahertz modulator and switchable device and offer a step forward in filling the "THz gap".

Key words: THz, Metamaterials, Terahertz, Tunable, Absorber

82. Stable terahertz toroidal dipolar resonance in a planar metamaterial

Chunfeng Ding, Linkun Jiang, Chongling Sun, Liang Wu, Degang Xu, Guizhong Zhang, Jianquan Yao

Physica Status Solidi (B), 2015, 252(6): 1388–1393.

Abstract: In this paper, we proposed and fabricated a planar terahertz (THz) metamaterial that is composed of four asymmetric split-ring resonators (ASRRs) coated with two polyimide layers. Simulation and experimental results show that a Fano-shaped toroidal dipolar resonance at 0.42THz is acquired from the metamaterial. Further analysis indicates that the toroidal dipolar resonance originates from the coupling of the four ASRRs, and the polyimide coating layers on both sides of the structure play key role in keeping the resonance frequency unchanged. From the designed metamaterials, a confined electromagnetic field inside the dielectric medium with a subwavelength-scale toroidal geometry is observed, and a strong confined E-field component at the toroidal dipole center with its orientation perpendicular to the H-vortex plane can also be numerically acquired. This planar metamaterial would openup an avenue for potential applications in the terahertz regime.

Key words: Asymmetric split-ring resonators, Coupling, Matematerials, Polyimide coatings, Terahertz resonances, Toroidal dipoles

83. Resonant conversion based on GaAs-metal metamaterials within terahertz range

Xiaolong Cao, Yongli Che, Jianquan Yao
International Journal of Modern Physics B, 2015, 29(21): 1550145.

Abstract: In this paper, by utilizing the variable conductivity with photo-injection in gallium arsenide (GaAs), we have designed an asymmetrical planar terahertz (THz) metamaterial, which is connected with two single-gap split ring resonator (SRR) by GaAs strip and demonstrated the resonant conversion of SRR within the THz range under appropriate optical pumping. As central trailing arm of the structure, GaAs is skillfully inserted between the two cross arms of the THz metamaterial and plays a key role in resonant conversion. Through modulation of its conductivity (σ_{GaAs}), the variable conductivity of GaAs can make one dual-gap SRR into two connective single-gap SRRs in physical structure, at the same time, the state conversion of two different resonances in the THz metamaterial has been achieved. The simulation results show that the resonant states of THz metamaterial can be switched from one LC and one dipole (state 1) to two LC and one new dipole (state 2) through the intermediate state with the increasing σ_{GaAs}. This structural design provides a new example to apply variable conductivity to achieve state conversion of resonance and can be extended to the additional application in THz devices.

Key words: Resonant conversion; Metamaterials; Transmission; Terahertz.

84. Direct thermal tuning of the terahertz plasmonic response of semiconductor metasurface

Chunya Luo, Dan Li, Jianquan Yao, Furi Ling
Journal of Electromagnetic Waves and Applications, 2015, 29(18): 2512–2522.

Abstract: In this study, the tunable depth of an array of semiconductor indium antimonide (InSb) subwavelength strips was studied by isothermally increasing the intrinsic carrier density, and then increasing the plasma frequency and the dielectric constant in the terahertz (THz) regime. It could be found that the tunable depth can attain 79.07% and the quality factor Q can be up to 7.1 in the window between 0 and 1.2 THz from 450 to 250 K. In order to improve the performance of the device, the strip array of semiconductor InSb on a ferroelectric strontium titanate film was also investigated. The results showed that when the temperature varied from 450 to 250 K, the modulation depth increased and attained 84.06%. These findings would facilitate the design of tunable components in the communication region of the THz regime.

Key words: Metasurface; InSb; STO; Semiconductor; THz; Thermal tenability

85. 基于编码超表面的太赫兹宽频段雷达散射截面缩减的研究

闫昕, 梁兰菊, 张雅婷, 丁欣, 姚建铨

物理学报, 2015, 64(15): 158101.

摘要：本文设计了一种柔性、非定向低散射的 1bit 编码超表面，实现了太赫兹宽频带雷达散射截面的缩减。这种设计基于对 "0" 和 "1" 两种基本单元进行编码，其反射相位差在很宽的频段范围内接近 180°，为一种非周期的排列方式，该电磁超表面使入射的电磁波发生漫反射，从而实现雷达散射截面的缩减。全波仿真结果表明，在垂直入射条件下，编码超表面的镜像反射率低于 -10 dB 的带宽频段范围为 1.0-1.4 THz，该带宽内超表面相对同尺寸金属板可将雷达散射截面所减量达到 10 dB 以上，最大缩减量达到 19 dB。把柔性编码表面弯曲在直径为 4 mm 的金属圆柱面上，雷达散射截面的所减量高 10 dB 以上的带宽频段范围为 0.9-1.2 THz，仍然可实现宽频带缩减特性。总之，编码超表面为调控太赫兹波提供一种新的途径，将在雷达隐身、成像、宽带通信等方面具有重要的意义。

关键词：编码超表面, 非定向散射, 雷达散射截面, 太赫兹

86. 太赫兹波在沙尘中衰减特性

许文忠, 钟凯, 梅嘉林, 徐德刚, 王与烨, 姚建铨

红外与激光工程, 2015, 44(2): 523–527.

摘 要：为了研究太赫兹波在沙尘大气中的衰减特性，根据 Mie 散射理论计算了单次散射情况下沙尘粒子不同尺寸参数下的散射效率因子和不同散射角下的散射相函数值。并得到了具有一定尺寸分布的沙尘粒子的单位距离衰减和能见度的关系曲线。还利用 Monte Carlo 方法对太赫兹波在沙尘中的多次散射特性进行了模拟计算，分析了不同能见度和不同沙尘类型对太赫兹波传输过程中能量损耗的影响。结果表明：单次散射条件下沙尘粒子的散射主要受尺寸参数影响；沙尘能见度较低时必须要考虑多次散射的影响。研究结果对太赫兹技术在大气环境监测和烟尘和风沙的探测等方面的应用具 有参考价值。

关键词：太赫兹, 沙尘, Mie 散射理论, Monte Carlo 方法

87. Study of the properties of BaGa$_4$Se$_7$ crystal in the terahertz region

W. T. Xu, Y. Y. Wang, C. Yan, D. G. Xu, J. Y. Yao, F. Fan, P. Duan, Z. Yang, P. X. Liu, J. Shi, H. X. Liu, J. Q. Yao

Infrared, Millimeter, and Terahertz waves (IRMMW-THz), **2015 40th International Conference on. IEEE**, 2015: 1–2.

Abstract: The far-infrared properties of a newly grown crystal, BaGa$_4$Se$_7$, were investigated with a THz-TDS system. Refractive indices and absorption coefficients of this material between 0.1 THz-1THz were obtained. The measured properties predicts potential application of BaGa$_4$Se$_7$ in THz-DFG. THz-DFG using BaGa$_4$Se$_7$ was analyzed theoretically and the phase-matching conditions were calculated. It seems that a promising prospect of THz application can be expected by using BaGa$_4$Se$_7$ as the NLO crystal.
Key words: Terahertz wave, Far-infrared property, THz-TDS system

88. 棋盘型结构在太赫兹宽频段 RCS 缩减中应用研究

梁兰菊，闫昕，姚建铨

枣庄学院学报，2015, 32(5): 18–22.

摘 要： 本文设计了一种太赫兹波段薄的新型棋盘型结构，实现了宽频段 RCS 的缩减。该结构为不同尺寸的人工磁导体耶路撒冷十字型组成，通过优化单元结构，使反射相位差在较宽范围内保持在 180 ± 30 度之间，从而降低后向散射能量，达到宽频段 RCS 的缩减。仿真表明在 0.76-1.26 THz 范围内反射波在法向会相消，对应远场产生四个对称方向波束；该结构相对同尺寸的金属板相应的波段内 RCS 的缩减量达到 10 dB 以上，最大缩减量达到 25 dB，与理论分析一致。总之，棋盘型结构为实现宽频段 RCS 缩减提供一种新的途径，将在成像、雷达隐身等方面具有重要的意义。
关键词： 太赫兹，雷达散射截面缩减，人工磁导体，棋盘型结构

89. A New $Ba_{0.6}Sr_{0.4}TiO_3$–Silicon Hybrid Metamaterial Device in Terahertz Regime

Liang Wu, Ting Du, Ningning Xu, Chunfeng Ding, Hui Li, Quan Sheng, Ming Liu, Jianquan Yao, Zhiyong Wang, Xiaojie Lou, Weili Zhang

Small, **2016**, 12(19):2610–2615.

Abstract: Metamaterials, offering unprecedented functionalities to manipulate electromagnetic waves, have become a research hotspot in recent years. Through the incorporation of active media, the exotic electromagnetic behavior of metamaterials can be dramatically empowered by dynamic control. Many ferroelectric materials such as $BaSrTiO_3$ (abbreviated as BST), exhibiting strong response to external electric field, hold great promise in both microwave and terahertz tunable devices. A new active $Ba_{0.6}Sr_{0.4}TiO_3$–silicon hybrid metamaterial device, namely, a SRR (square split ring resonator)–$BaSrTiO_3$ thin film-silicon three-layer structure is fabricated and intensively studied. The active $Ba_{0.6}Sr_{0.4}TiO_3$ thin film hybrid metamaterial, with nanoscale thickness, delivers a transmission contrast up to ≈79% due to electrically enabled carrier transport between the ferroelectric thin film and silicon substrate. This work has significantly increased the low modulation rate of ferroelectric based devices in terahertz range, a major problem in this field remaining unresolved for many years. The proposed BST metamaterial is promising in developing high performance real world photonic devices for terahertz technology.

Key words: Terahertz, Metamaterial device, Ferroelectric material

90. Graphene-based tunable terahertz plasmon-induced transparency metamaterial

Xiaolei Zhao, Cai Yuan, Lin Zhu, Jianquan Yao

Nanoscale, 2016, 8(33):15273–15280.

Abstract: A novel terahertz plasmon induced transparency (PIT) metamaterial structure consisting of single-layered graphene microstructures was proposed and numerically studied in this study. A pronounced transparency peak was obtained in the transmission spectrum, which resulted from the destructive interference between the graphene dipole and monopole antennas. Further investigations have shown that the spectral location and lineshape of the transparency peak can be dynamically controlled by tuning the Fermi level in graphene. Since the monopole antennas in our designed structure exist in a continuous form, a more convenient method for tunablity is available by applying a gate voltage compared to those structures with discrete graphene patterns. This work may open up new avenues for designing tunable terahertz functional devices and slow light devices.

Key words: Plasmon induced transparency, Metamaterial, Tunable, Terahertz

91. Tunable plasmon-induced transparency in a grating-coupled double-layer graphene hybrid system at far-infrared frequencies

Xiaolei Zhao, Lin Zhu, Cai Yuan, Jianquan Yao

Optics Letters, 2016, 41(23): 5470–5473.

Abstract: A grating-coupled double-layer graphene hybrid system is proposed to investigate the plasmon-induced transparency effect at far-infrared frequencies. Based on the guided mode resonance principle, a diffractive grating is used to couple the normally incident waves and excite the plasmonic resonances on two graphene films separated by a spacer, thereby avoiding the need for patterning graphene. It is found that the origin of the observed transparency window transforms from Autler–Townes splitting to electromagnetically induced transparency with the increase of the separation distance between the two graphene films. The tunability of this hybrid system is also investigated via varying the Fermi energy in graphene. The proposed hybrid system has potential applications in tunable switches, sensors, and slow light devices and may open up new avenues for constructing easy-to-fabricate graphene-based plasmonic devices.

Key words: Plasmon-induced transparency, grating coupled, far-infrared frequencies

92. Investigation of optical pump on dielectric tunability in PZT/PT thin film by THz spectroscopy

Jie Ji, Chunya Luo, Yunkun Rao, Furi Ling, Jianquan Yao

Optics Express, 2016, 24(14): 15212−15221.

Abstract: The dielectric spectra of single-layer $PbTiO_3$ (PT), single-layer $PbZr_xTi_{1-x}O_3$ (PZT) and multilayer PZT/PT thin films under an external optical field were investigated at room temperature by time-domain terahertz (THz) spectroscopy. Results showed that the real part of permittivity increased upon application of an external optical field, which could be interpreted as hardening of the soft mode and increasing of the damping coefficient and oscillator strength. Furthermore, the central mode was observed in the three films. Among the dielectric property of the three thin films studied, the tunability of the PZT/PT superlattice was the largest.

Key words: Terahertz spectroscopy, Thin film, Dielectric spectra, Ferroelectric

93. Dual-band tunable perfect metamaterial absorber in the THz range

Gang Yao, Furi Ling, Jin Yue, Chunya Luo, Jie Ji, Jianquan Yao

Optics Express, 2016, 24(2), 1518−1627.

Abstract: In this paper, a dual-band perfect absorber, composed of a periodically patterned elliptical nanodisk graphene structure and a metal ground plane spaced by a thin SiO_2 dielectric layer, is proposed and investigated. Numerical results reveal that the absorption spectrum of the graphene-based structure displays two perfect absorption peaks in the terahertz band, corresponding to the absorption value of 99% at 35 μm and 97% at 59 μm, respectively. And the resonance frequency of the absorber can be tunned by controlling the Fermi level of graphene layer. Further more, it is insensitive to the polarization and remains very high over a wide angular range of incidence around ±60°. Compared with the previous graphene dual-band perfect absorption, our absorber only has one shape which can greatly simplify the manufacturing process.

Key words: Dual-band perfect absorber, Tunable, Metametrial, Terahertz

94. Reconfigurable hybrid metamaterial waveguide system at terahertz regime

Xiaolei Zhao, Lin Zhu, Cai Yuan, Jianquan Yao
Optics express, 2016, 24(16): 18244–18251.

Abstract: We propose an optically controlled reconfigurable hybrid metamaterial waveguide system at terahertz frequencies, which consists of a two dimensional gold cut wire array deposited on top of a dielectric slab waveguide. Numerical findings reveal that this device is able to realize dynamic transformation from double electromagnetically induced transparency like material to ultra-narrow band guided mode resonance (GMR) filter by controlling the optically excited free carriers in gallium arsenide pads inserted between the gold cut wires. During this reconfiguration process of resonance modes, high quality factors up to ~104 and ~118 for the two EIT-like peaks and up to ~578 for the GMR filter are obtained.
Key words: Optically control, Reconfigurable hybrid waveguide, Metamaterial, Terahertz

95. Effect of electric field on the dielectric properties of the Barium Strontium Titanate film

Chunya Luo, Jie Ji, Furi Ling, Dan Li, Jianquan Yao
Journal of Alloys & Compounds, 2016, 687:458–462.

Abstract: We investigated the tunability of the dielectric characterization of the Ferroelectric Barium Strontium Titanate ($Ba_{0.5}Sr_{0.5}TiO_3$) thin film on the Strontium Titanate substrate by a dc bias electric field through the terahertz time-domain spectroscopy. The tunability of the real part of the permittivity reached up to 74.8% and the imaginary part of the permittivity reached up to 33.6% when the bias electric field up to 30 V. And the results are attributed to the soft mode hardening caused by the electric field.
Key words: $BaSrTiO_3$ film, Dielectric characterization, THz, Electrically tunability

96. Dynamically tunable graphene plasmon-induced transparency in the terahertz region

Gang Yao, Furi Ling, Jin Yue, Qin Luo, Jianquan Yao
Journal of Lightwave Technology, 2016, 34(16):3937–3942.

Abstract: Active control of graphene plasmon-induced transparency (GPIT) metamaterial structures, composed of periodically patterned monopolar graphene and dipolar graphene, are presented and investigated. Numerical results reveal that the resonant frequency of GPIT structures can be dynamically tuned by varying the Fermi level of the T-shape graphene strip through controlling the voltage of the electrostatic gating. Coupled Lorentz oscillator model is applied to explore the physical mechanism of the frequency tunable GPIT. Furthermore, the group index of terahertz light can be controlled to exceed 350 in the THz region. It is also found that the interaction strength between the dipolar graphene and the monopolar graphene can be tuned by changing the distance between the radiative mode and the dark mode as well as the degree of the symmetry breaking. These tunable features of the GPIT devices are significant and may offer new opportunity to design active devices in the THz region, such as ultrasensitive sensors, slow light devices, and spectral filters.
Key words: Graphene, Nanostructure, Plasmon-induced transparency (PIT), Terahertz

97. Dynamically electrically tunable broadband absorber based on graphene analog of electromagnetically induced transparency

Gang Yao, Furi Ling, Jin Yue, Chunya Luo, Qin Luo, Jianquan Yao
IEEE Photonics Journal, 2015, 8(1):1–8.

Abstract: A broadband absorber, which is composed of graphene analog of electromagnetically induced transparency (EIT) and a metal ground plane spaced by a thin SiO_2 dielectric layer, is proposed and investigated. Numerical results reveal that the working bandwidth can be dynamically tuned between broadband absorption and narrowband absorption by varying the Fermi level of graphene through controlling of the electrostatic gating. Furthermore, absorption behaviors can be tuned to the state of off when the electric polarized is rotated to another axis. It is also found that the coupling strength between the radiative element and dark element is tuned by the distance between the disconnected vertical graphene strip and horizontal graphene strip. In addition, this type of graphene-based absorber is very sensitive to the incident angles.
Key words: THz optics, Metamaterials, Subwavelength structures

98. Optical-induced absorption tunability of Barium Strontium Titanate film

Chunya Luo, Jie Ji, Jin Yue, Yunkun Rao, Gang Yao, Dan Li, Ying Zeng, Renkui Li, Longsheng Xiao, Xinxing Liu, Jianquan Yao, Furi Ling
Optical Materials, 2016, 60:383–386.

Abstract: The absorption tunability of 100 nm thickness of ferroelectric Barium Strontium Titanate ($Ba_{0.5}Sr_{0.5}TiO_3$) thin films with different densities of pumped optical field is measured by terahertz time-domain spectroscopy in the range of 0.2 THz -1.2 THz at 19 °C. Experimental results show that the absorption coefficient of BST film is approximately at 5 000 cm^{-1}-20 000 cm^{-1} in the range of 0.2 THz-1.2 THz and the absorption coefficient reached up to 16% when we applied the optical field up to 600 mW. The theoretical calculations reveal that increasing photoexcitation fluences is responsible for the increasing of transmission change in the conduction current density cause the absorption coefficient varied.

Key words: Optical tunability, Absorbtion response, $BaSrTiO_3$, THz

99. Slowing and trapping THz waves system based on plasmonic graded period grating

Baoshan Guo, Wei Shi, Jianquan Yao
Journal of Optics, 2016, 45(1):50–57.

Abstract: Graded and partial graded period grating structures are proposed to trap and slow terahertz (THz) waves. Different frequencies of THz waves can be trapped at different positions along the graded grating. The exact trapping positions of every different frequency can be estimated from a fitting curve. The separation distance between trapping point of every 0.1 THz is about 1~2 mm, which can be applied in a THz waves spectrometer. An ultraslow THz waveguide can be realized by using a partial graded grating. The propagated speed of THz waves can be designed and tuned by adjusting the period of the grating. This structure offers the advantage of reducing the speed of the THz waves and propagation them over an ultrawide spectral band. The physical trapping mechanism of the THz waves is attributed to the transformation from surface modes to cavity modes. The localized cavity modes at the trapping position have a saturated stable state. The extra energy will be reflected back along the graded grating.

Key words: Plasmonics, Slowing THz waves

100. Triple-band high Q factor Fano resonances in bilayer THz metamaterials

Chunfeng Ding, Liang Wu, Degang Xu, Jianquan Yao, Xiaohong Sun
Optics Communications, 2016, 370:116−121.

Abstract: In this paper, we proposed a bilayer THz metamaterials, which is constructed by two sets of asymmetric split-ring resonators (ASRRs) with different sizes. Simulation results show that three high Q Fano resonances are excited in the bilayer metamaterials at 0.268, 0.418 THz, and 25 at 0.560 THz, and the Q values are 33, 42, and 25, respectively. The field distributions show that resonances at 0.268 and 0.560 THz originate from one of ASRRs, whereas the resonance at 0.418 THz originates from the other set of ASRRs. Further analysis indicates that the three high Q Fano resonances results from a combined action of the in-plane coupling and the interlayer coupling in the metamaterials: the in-plane coupling lead to resonances enhanced and the interlayer coupling lead to the eigenmode of each set of the ASRRs split into two discrete Fano resonances. This triple-band high Q factor Fano resonance metamaterials would open new degrees of freedom for designing advanced chemical and biological sensors and detectors in the terahertz regime.

Key words: Triple-band, High quality (Q), Fano resonance, Bilayer metamaterials, Coupling

101. Terahertz wavemeter based on scanning Fabry–Perot interferometer: accuracy and optimum designation

Wentao Xu, Yuye Wang, Degang Xu, Chao Yan, Pengxiang Liu, Jianquan Yao
Journal of Modern Optics, 2016, 63(10): 974−981.

Abstract: The performance of terahertz wavemeter based on scanning Fabry–Perot interferometer is discussed in detail. In order to evaluate the absolute accuracy of this kind of wavemeter, the accuracy dependency on some key parameters of the wavemeter, such as the finesse, step size, interference order and noise level, is discussed quantitatively. Two different methods for determining the unknown wavelength are compared in their uncertainty of measurement results. Based on the above analysis, the optimum use of the wavemeter is discussed.

Key words: Terahertz wavemeter, Fabry-Perot interferometer, Accuracy

102. 太赫兹波段表面等离子体波传播距离的调控

王文伯,郭宝山

光子学报, 2016, 45(2): 0224001.

摘要: 设计刻槽深度渐变的金属光栅,利用截止频率控制太赫兹表面波的传播距离,金属光栅刻槽深度在 60 μm 时的截止频率为 0.6 THz, 120 μm 时的截止频率为 1.1 THz。同时根据半导体材料 InSb 在太赫兹波段的色散特性、载流子浓度和迁移率的经验公式,光栅结构的等效介质理论设计了温控的半导体光栅。在 270~300 K 温度范围内,半导体光栅中表面波的传播距离和温度成正比,300 K 时传播距离为 270 K 时的 2~3 倍。

关键词: 太赫兹,表面等离子体激元,时域有限差分法,金属波导光栅,半导体波导光栅

103. Extracting dielectric parameter based on multiple beam interference principle and FTIR system in terahertz range

M. R. Wang, K. Zhong, D. G. Xu, Y. Y. Wang, W. Shi, J. Q. Yao

International Conference on Infrared, Millimeter, and Terahertz Waves (IRMMW-THz 2016), 2016: 1–2.

Abstract: We demonstrate a novel method for obtaining optical constants of transparent material over a broad terahertz spectral range from 0.7 THz to 7.5 THz at room temperature. Based on the interferogram directly acquired by a Fourier transform infrared spectrometer (FTIR), multiple beam interference principle combining Fresnel's formula is employed to extract the refraction index and the extinction coefficient, which are the basis for calculating permittivity. It avoids the shortcomings using Kramers-Kronig (K-K) relations and overcome the frequency range of terahertz time-domain system (TDS). Moreover, this method has better stability and reserves all useful information for thin samples compared with TDS, which makes it a general processing method for terahertz dielectric measurement.

Key words: Optical constants, FTIR, Multiple beam interference, Terahertz dielectric measurement

104. Analyzing terahertz time-domain transmission spectra with multibeam interference principle

M. R. Wang, K. Zhong, Chu Liu, D. G. Xu, Y. Y. Wang, W. Shi, J. Q. Yao

SPIE/COS Photonics Asia. International Society for Optics and Photonics, 2016, 10030: 100301H.

Abstract: We present a data processing method based on multiple beam interference and Fresnel's formula that extract simultaneously the refraction index and the extinction coefficient from terahertz time domain spectra, and the dielectric coefficient can also be calculated. Typical THz-TDS system working in transmission mode was utilized for direct measurement of the transmission spectra with a frequency accuracy of 7.6 GHz and range from 0.3 THz to 4 THz at room temperature. This method is verified with a double-faced polished 350-μm 100-cut GaAs wafer, and the reasonable average relative error for refractive index in the whole range is less than 0.83% comparing with conventional method, which provides a new approach to process the transmission spectra with oscillations.

Key words: Terahertz, THz-TDS, Multi-beam interference, Optical constants

105. Flexible manipulation of Terahertz wave reflection using polarization insensitive coding metasurfaces

Li Jiusheng, Zhao Zejiang, Yao Jianquan

Optics Express.2017, 24: 29983−29992.

Abstract: In order to extend to 3-bit encoding, we propose notched-wheel structures as polarization insensitive coding metasurfaces to control terahertz wave reflection and suppress backward scattering. By using a coding sequence of "00110011…" along x-axis direction and 16 × 16 random coding sequence, we investigate the polarization insensitive properties of the coding metasurfaces. By designing the coding sequences of the basic coding elements, the terahertz wave reflection can be flexibly manipulated. Additionally, radar cross section (RCS) reduction in the backward direction is less than -10dB in a wide band. The present approach can offer application for novel terahertz manipulation devices.

Key words: Spectroscopy, Terahertz, Scattering, In-field

106. Optically tuned dielectric property of ferroelectric PZT/STO/PT superlattice by THz spectroscopy

Jie Ji, Furi Ling, Siyan Zhou, Dan Li, Chunya Luo, Liang Wu, Jianquan Yao
Journal of Alloys & Compounds, 2017, 703: 517–522.

Abstract: The dielectric spectra of ferroelectric $Pb(Zr_{0.52}Ti_{0.48})O_3/SrTiO_3/PbTiO_3$ (PZT/STO/PT) superlattice under an external optical field were investigated at room temperature by time-domain terahertz (THz) spectroscopy. Results showed that the real and imaginary part of permittivity were increased roughly upon application of an external optical field, and tuned by 78.16% and 85.65% respectively. It could be interpreted as softening of the soft mode and increasing of the damping coefficient and oscillator strength. Furthermore, the central mode was observed in the tricolor thin film, and the frequency of the central mode was also increased along with the optical power.
Key words: Ferroelectric superlattice, THz, Dielectric characterization, Soft mode

107. Linear optical properties of $ZnGeP_2$ in the terahertz range

Kai Zhong, Chu Liu, Maorong Wang, Jie Shi, Bin Kang, Zerui Yuan, Jining Li, Degang Xu, Wei Shi, Jianquan Yao
Optical Materials Express, 2017, 7(10): 3571–3579.

Abstract: We investigated the optical properties of zinc germanium phosphide ($ZnGeP_2$ or ZGP) crystals in a wide terahertz (THz) range from 0.2 THz to 6 THz, and made comparisons between crystals grown by both horizontal gradient freezing (HGF) and vertical gradient freezing (VGF) techniques. THz time-domain spectroscopy (TDS) and Fourier transform infrared spectroscopy (FTIR) systems were used to measure and analyze the transmittance, refractive indices and absorption coefficients. It was found that the HGF grown crystals have different birefringence and absorption in the THz range compared with the VGF grown crystals. The anisotropic absorption in the THz range was observed and the polar phonon modes at 3.6 THz and 4.26 THz were also discussed. The dispersion and absorption data of ZGP given in this report enabled us to know it better in the THz range and optimize its THz applications.
Key words: Spectroscopy, Terahertz, Nonlinear optical materials, Crystal optics

108. Active KTaO₃ hybrid terahertz metamaterial

Liang Wu, Jinglong Liu, Hui Li, Chunfeng Ding, Ningning Xu, Xiaolei Zhao, Zongcheng Xu, Quan Sheng, Jianquan Yao, Jining Li, Xin Ding, Weili Zhang
Scientific reports, 2017, 7(1): 6072.

Abstract: The dielectric properties of an active $KTaO_3$ hybrid metamaterial structure and its tunability under external electric fields are investigated at room temperature by means of terahertz time-domain spectroscopy. Application of the electric field leads to an appreciable tuning of the dielectric loss, which is up to 17%. Meanwhile, the refractive index also changes appreciably. These findings are attributed to the internal space charge field in the crystal caused by the excited free carriers.
Key words: Dielectric properties, $KTaO_3$, Tenability, TDS

109. Terahertz wavefront manipulating by double-layer graphene ribbons metasurface

Hongliang Zhao, Zhihong Chen, Fei Su, Guangjun Ren, Fei Liu, Jianquan Yao
Optics Communications, 2017, 402: 523–526.

Abstract: It was recently presented that the phase gradient metasurface can focus the reflection in terahertz range. However, narrow bandwidth and complex tuning method are still challenges. For instance, the size is difficult to be changed once the device is built. We propose a tunable double-layer graphene ribbons array (DLGRA) metasurface which has great potentials for applications in terahertz wavefront control. By changing the Fermi level of each graphene ribbon independently, the DLGRA separated by a bonding agent and a thin dielectric spacer can achieve nearly 2π phase shift with high reflection efficiency. A reflector which can focus terahertz waves over a broad frequency range is demonstrated numerically by the DLGRA. Intriguingly, through a lateral shift between the nearby graphene ribbons, the variation of coupling induces a shift of focusing frequency. Hence, this approach increases the frequency range to a higher degree than the fixed state. The proposed metasurface provides an effective way for manipulating terahertz waves in a broad frequency range.
Key words: Terahertz wavefront, Graphene, Metasurface, Phase shift

110. Theoretical and experimental study on broadband terahertz atmospheric transmission characteristics

Shibei Guo, Kai Zhong, Maorong Wang, Chu Liu, Yong Xiao, Wenpeng Wang, Degang Xu, Jianquan Yao

Chinese Physics B, 2017, 26(1):019501.

Abstract: Broadband terahertz (THz) atmospheric transmission characteristics from 0 to 8 THz are theoretically simulated based on a standard Van Vleck-Weisskopf line shape, considering 1696 water absorption lines and 298 oxygen absorption lines. The influences of humidity, temperature, and pressure on the THz atmospheric absorption are analyzed and experimentally verified with a Fourier transform infrared spectrometer (FTIR) system, showing good consistency. The investigation and evaluation on high-frequency atmospheric windows are good supplements to existing data in the low-frequency range and lay the foundation for aircraft-based high-altitude applications of THz communication and radar.

Key words: Terahertz atmospheric propagation, Line absorption, Continuum absorption

111. Optical coefficients extraction from terahertz time-domain transmission spectra based on multibeam interference principle

Maorong Wang, Kai Zhong, Chu Liu, Degang Xu, Wei Shi, Jianquan Yao

Optical Engineering, 2017, 56(4): 044101.

Abstract: We demonstrated a data-processing method based on multiple beam interference and Fresnel equations that simultaneously gave the refraction index and absorption coefficient from the raw data of terahertz (THz) time-domain spectroscopy (TDS), which laid the foundation for obtaining the dielectric parameters. This method was independent of phase processing, and complete material information was reserved without having to cut the time-domain signal. The optical coefficients including refractive indices and absorption coefficients of white polyethylene and quartz samples at different thicknesses were obtained. The applicability and accuracy of this method were discussed and verified by comparison with the traditional data-processing method of THz TDS.

Key words: Terahertz time-domain spectroscopy, Multibeam interference, Optical coeffcient

112. Characteristic analysis of a photoexcited metamaterial perfect absorber at terahertz frequencies

Pibin Bing, Shichao Huang, Zhongyang Li, Zhou Yu, Ying Lu, Jianquan Yao

Modern Physics Letters B, 2017, 31(18): 1750207.

Abstract: The absorption characteristics of a photoexcited metamaterial absorber at terahertz frequencies were analyzed in this study. Filling photosensitive semiconductor silicon into the gap between the resonator arms leads to modulation of its electromagnetic response through a pump beam which changes conductivity of silicon. Comparisons of terahertz absorbing properties which were caused by different thicknesses and dielectric constants of polyimide, cell sizes and widths of SRRs, and lengths and conductivities of the photosensitive silicon, were studied by using Finite Difference Time Domain (FDTD) from 0.4 THz to 1.6 THz. The results of this study will facilitate the design and preparation of terahertz modulator, filters and absorbers.

Key words: Split-ring resonators; Metamaterial absorber; Terahertz wave

113. Magneto-optical modulation of photonic spin hall effect of graphene in terahertz region

Tingting Tang, Jie Li, Li Luo, Ping Sun, Jianquan Yao

Advanced Optical Materials, 2018: 6(7):1701212.

Abstract: Magneto-optical (MO) modulation of the photonic spin Hall effect (PSHE) of transmitted light in graphene–substrate system in terahertz region is proposed. The expressions for PSHE shifts of left-handed circularly polarized (LHCP) and right-handed circularly polarized (RHCP) components are derived based on angular spectrum analysis. PSHE shifts and their physical mechanisms are discussed in detail in the presence of different relaxation time and Fermi energy of graphene based on simulation results. The potential applications of MO-modulated PSHE in multichannel switch and barcode encryption are also proposed and discussed. The MO-modulated PSHE in graphene–substrate system provides a new mechanism to realize photonic devices in the terahertz region.

Key words: Graphene–substrate system, Magneto-optical modulation, Photonic spin Hall effect

## 114. The novel hybrid metal-graphene metasurfaces for broadband focusing and beam-steering in farfield at the terahertz frequencies

Zhang Zhang, Xin Yan, Lanju Liang, Dequan Wei, Meng Wang, Yaru Wang, Jianquan Yao

Carbon, 2018, 132: 529–538.

Abstract: A hybrid metal-graphene metasurface was proposed and researched in terahertz (THz) frequencies. It is found that these hybrid structure metasurfaces give rise to a larger phase variation in comparison to bare graphene metasurface. Based on the analysis of electric field of the hybrid structures, it implies that due to the enhanced localized electric field and the Fabry-Perot resonance cause the phase increasing quadratically with the improvement of the graphene Fermi energy (E_F) in hybrid metal-graphene structures. By designing different metal sizes and modulating E_F, the structure shows up to 295° of smooth phase modulation at the frequency of 4.5 THz. Based on the phase profile design, a reflection-type focusing metasurface with near 1 000 μm focal length in a broadband frequency range from 4 to 4.8 THz and 1-bit dynamically tunable coding metasurface to steer the reflected THz waves into multiple beams in farfield were both demonstrated successively. Moreover, the degree of freedom for beam-steering of 2-bit coding metasurfaces was enhanced and the process of manipulation was also qualitatively interpreted by the convolution principle. This work may supply a potentially effective method to instruct the design of desirable metasurfaces with multi-function in the modulation of electromagnetic waves.

Key words: Hybrid metal-graphene, Terahertz frequencies, Phase variation

115. Electrical terahertz modulator based on photo-excited ferroelectric superlattice

Jie Ji, Siyan Zhou, Jingcheng Zhang, Furi Ling, Jianquan Yao

Scientific Reports, 2018, 8(1): 2682.

Abstract: The transmission and dielectric spectra of ferroelectric STO/PT superlattice on Si substrate under simultaneous external optical and electric field were investigated and compared at room temperature. Results found that when with an optical field, the electric field realized an effective modulation on the transmission, which displayed a diode property. In addition, a comprehensive model combined with Debye relaxation and Lorentz model was used to analyze the dielectric spectra, variation of the soft mode with external field was put emphasis on exploring.

Key words: Electrical terahertz modulator, Transmission spectra, Dielectric spectra, Soft mode

116. Electrically tuned transmission and dielectric properties of illuminated and non-illuminated barium titanate thin film in terahertz regime

Jie Ji, Jin Yue, Siyan Zhou, Yue Tian, Jingcheng Zhang, Furi Ling, Huaixing Wang, Chunya Luo, Jianquan Yao

Journal of Alloys and Compounds, 2018,747:629–635.

Abstract: Transmission and dielectric properties of non-illuminated and illuminated barium titanate thin film were investigated by time-domain terahertz spectrometer and tuned by an electric field. Compared with the sample illuminated by 532 nm laser, the electric field could realize a higher modulation in the phase of the transmission for the nonilluminated sample. Results displayed that when the sample was illuminated by green laser, optical field suppressed the effect of electric field on transmission and dielectric properties of BTO thin film, because excited carriers from the substrate and ferroelectric thin film would form a built-in electric field to shield the external electric field. In addition, optical field and electrical field would play different roles in the lattice dynamics of the ferroelectric thin film.

Key words: Ferroelectric thin film, Terahertz modulator, Soft mode, Photoelectrical effect

117. Terahertz magnon and crystal-field transition manipulated by R^{3+}-Fe^{3+} interaction in $Sm_{0.5}Pr_{0.5}FeO_3$

Xiumei Liu, Tao Xie, Jiajia Guo, Senmiao Yang, Yuna Song, Xian Lin, Shixun Cao, Zhenxiang Cheng, Zuanming Jin, Anhua Wu, Guohong Ma, Jianquan Yao
Applied Physics Letters, 2018, 113(2): 022401.

Abstract: We use terahertz (THz) magnetic and electric fields to investigate the magnetic and optoelectronic responses of the $Sm_{0.5}Pr_{0.5}FeO_3$ (SPFO) crystal, respectively, by THz time-domain spectroscopy. It is found that the spin reorientation transition (SRT) in SPFO occurs in the temperature range of 175–210 K. The SRT is not observed in $PrFeO_3$. The quasi-antiferromagnetic magnon frequency has a blue-shift from 0.42 THz ($PrFeO_3$) to 0.46 THz (SPFO) at room temperature, due to the enhanced anisotropy constant. The refractive index of SPFO in the THz frequency decreases around 3% compared with that of the $PrFeO_3$ crystal. In addition, it can be found that the energy scale of crystal-field transitions has a red-shift for the doped single crystal. We expect our results to make rare-earth orthoferrites accessible to potential applications in THz spintronic devices.
Key words: Magnetic responses, Optoelectronic responses, SPFO, SRT

118. Terahertz optical properties of nonlinear optical CdSe crystals.

Dexian Yan, Degang Xu, Jining Li, Yuye Wang, Fei Liang, Jian Wang, Chao Yan, Hongxiang Liu, Jia Shi, Longhuang Tang, Yixin He, Kai Zhong, Zheshuai Lin, Yingwu Zhang, Hongjuan Cheng, Wei Shi, Jianquan Yao, Yicheng Wu
Optical Materials, 2018, 78: 484–489.

Abstract: We investigate the optical properties of cadmium selenide (CdSe) crystals in a wide terahertz (THz) range from 0.2 to 6 THz by THz time-domain spectroscopy (THz-TDS) and Fourier transform infrared spectroscopy (FTIR). The refractive index, absorption coefficient and transmittance are measured and analyzed. The properties are characterized by several absorption peaks which represent the relevant phonon vibrations modes. The experimental results are in agreement with the theoretical results. The dispersion and absorption properties of CdSe crystal are analyzed in THz range. These properties indicate a good potential for THz sources and THz modulated devices.
Key words: Terahertz properties, Nonlinear optical materials, Crystal optics, THz devices, CdSe crystal

119. The effect of optical pump on the absorption coefficient of 0.65CaTiO$_3$ -0.35NdAlO$_3$, ceramics in terahertz range.

Dan Li, Chunya Luo, Yebin Xu, Liang Wu, Huaixing Wang, Songjie Shi, Furi Ling , Jianquan Yao

Optical Materials, 2018, 75:280–284.

Abstract: The absorption coefficient of 0.65CaTiO$_3$-0.35NdAlO$_3$ ceramics under external optical fields was investigated by terahertz time-domain spectroscopy in a frequency range of 0.2 THz to 1 THz at room temperature. It could be found that the variation of the absorption coefficient is approximately from 8.95 cm^{-1} to 26.02 cm^{-1} in the frequency range of 0.2 THz to 1 THz, and the tuning range is about 3.32 cm^{-1} at 0.6 THz which almost reaches up to nearly 23.32%. The micromechanism of these results was attributed to the excited free carriers by the external optical pump intensity.

Key words: Absorption coefficient, 0.65CaTiO$_3$-0.35NdAlO$_3$ ceramics, Terahertz

120. A broadband metamaterial absorber based on multi-layer graphene in the terahertz region

Pan Fu, Fei Liu, Guang Jun Ren, Fei Su, Dong Li, Jianquan Yao

Optics Communications, 2018, 417: 62–66.

Abstract: A broadband metamaterial absorber, composed of the periodic graphene pattern on SiO$_2$ dielectric with the double layer graphene films inserted in it and all of them backed by metal plan, is proposed and investigated. The simulation results reveal that the wide absorption band can be flexibly tuned between the low-frequency band and the high-frequency band by adjusting graphene's Fermi level. The absorption can achieve 90% in 5.50–7.10 THz, with Fermi level of graphene is 0.3 eV, while in 6.98–9.10 THz with Fermi level 0.6 eV. Furthermore, the proposed structure can be switched from reflection (>81%) to absorption (>90%) over the whole operation band, when the Fermi level of graphene varies from 0 to 0.6 eV. Besides, the proposed absorber is insensitive to the polarization and can work over a wide range of incident angle. Compared with the previous broadband absorber, our graphene based wideband terahertz absorber can enable a wide application of high performance terahertz devices, including sensors, imaging devices and electro-optic switches.

Key words: Broadband, Metamaterial absorber, Graphene, Polarization insensitive

121. Effect of optical pumping on the dielectric properties of $0.6CaTiO_3$-$0.4NdAlO_3$ ceramics in the terahertz range

Dan Li, Chunya Luo, Yebin Xu, Jing Zhang, Liang Wu, Huaixing Wang, Songjie Shi, Furi Ling, Jianquan Yao
Applied Optics, 2018, 57(1):1–4.

Abstract: The dielectric properties of $0.6CaTiO_3$-$0.4NdAlO_3$ ceramics under external optical fields were investigated by terahertz time-domain spectroscopy in a frequency range of 0.2 THz to 1 THz at room temperature. It could be found that the variation of the real part of complex permittivity is approximately 0.31 in the frequency range of 0.2 THz to 1 THz. However the imaginary part of the dielectric constant does not change appreciably with the external optical field. The micromechanism of these results was attributed to the built-in electric field caused by the excited free carriers in the ceramics.
Key words: Optoelectronics, Ferroelectrics, Far infrared or terahertz, Spectroscopy, Terahertz

122. Active optical modulator based on a metasurface in the terahertz region

Yue Tian, Jie Ji, Siyan Zhou, Hu Wang, Zhichao Ma, Furi Ling, Jianquan Yao
Applied optics, 2018, 57(27): 7778–7781.

Abstract: The characteristics of the electromagnetically induced transparency (EIT) analog exposed under different illumination powers have been investigated theoretically and experimentally. The EIT analog is composed of a fixed aluminum structure fabricated on the silicon substrate. It was found that the resonance degree of the transparent window displayed a decreasing trend, and a blueshift phenomenon emerged by increasing the powers of the laser. Similarly, the properties of the time delay under different illumination powers have also been researched. The realization of the tuning effect may provide a possible choice for the modulation of the slow light devices.
Key words: Spectroscopy, Terahertz, Metamaterials

123. Optical modulation of BST/STO thin films in the terahertz range

Ying Zeng, Songjie Shi, Ling Zhou, Furi Ling, Jianquan Yao
Journal of Electronic Materials, 2018, 47(7): 3855–3860.

Abstract: The $Ba_{0.7}Sr_{0.3}TiO_3$ (BST) thin film (30.3 nm) deposited on a $SrTiO_3$ (STO) film/ silicon substrate sample was modulated by 532 nm continuous-wave laser in the range of 0.2–1 THz at room temperature. The refractive index variation was observed to linearly increase at the highest 3.48 for 0.5 THz with the pump power increasing to 400 mW. It was also found that the BST/STO sample had a larger refractive index variation and was more sensitive to the external optical field than a BST monolayer due to the epitaxial strain induced by the STO film. The electric displacement–electric field loops results revealed that the increasing spontaneous polarization with the STO film that was induced was responsible for the larger refractive index variation of the BST/STO sample. In addition, the real and imaginary part of the permittivity were observed increasing along with the external field increasing, due to the soft mode hardening.

Key words: Thin film, Ferroelectrics, Optoelectronics, Terahertz, Spectroscopy

124. Dynamically tunable terahertz passband filter based on metamaterials integrated with a graphene middle layer

Maosheng Yang, Lanju Liang, Dequan Wei, Zhang Zhang, Xin Yan, Meng Wang, Jianquan Yao.

Chinese Physics B, 2018, 27(9): 098101.

Abstract: The dynamic tunability of a terahertz (THz) passband filter was realized by changing the Fermi energy (E_F) of graphene based on the sandwiched structure of metal–graphene–metal metamaterials (MGMs). By using plane wave simulation, we demonstrated that the central frequency (f_0) of the proposed filter can shift from 5.04 THz to 5.71 THz; this shift is accompanied by a 3 dB bandwidth (Δf) decrease from 1.82 THz to 0.01 THz as the E_F increases from 0 to 0.75 eV. Additionally, in order to select a suitable control equation for the proposed filter, the curves of Δf and f_0 under different graphene E_F were fitted using five different mathematical models. The fitting results demonstrate that the DoseResp model offers accurate predictions of the change in the 3 dB bandwidth, and the Quartic model can successfully describe the variation in the center frequency of the proposed filter. Moreover, the electric field and current density analyses show that the dynamic tuning property of the proposed filter is mainly caused by the competition of two coupling effects at different graphene E_F, i.e., graphene–polyimide coupling and graphene–metal coupling. This study shows that the proposed structures are promising for realizing dynamically tunable filters in innovative THz communication systems.

Key words: Metamaterial, Terahertz, Filter, Mathematical model, Tunability, Graphene

125. Highly sensitive sensors of fluid detection based on magneto-optical optical Tamm state

Nengxi Li, Tingting Tang, Jie Li, Li Luo, Ping Sun, Jianquan Yao
Sensors & Actuators B Chemical, 2018, 265:644−651.

Abstract: We propose a new magneto-optical optical Tamm state (MOOTS) sensor in fluid detection. The device consists of a dielectric magneto-optical thin film of Ce-doped $Y_3Fe_5O_{12}$ (CeYIG), a metallic film of Ag and a periods of the distributed Bragg reflector (DBR). Finite element method (FEM) and 4 × 4 transfer matrix method are employed to calculate the figure of merit (FOM) of sensitivity. Calculation results indicate that a FOM of 1 224.21/RIU for gas refractive index variation from 1.000 0 to 1.000 6 can be obtained when the incident wavelength is 1 064 nm, which is much larger than the FOM of magneto-optical surface plasmon resonance (MOSPR) device. Meanwhile, our MOOTS sensor is insusceptible to the influence of smoothness of the interface of deposited layers and can provide the great error tolerance, which is more conducive to the manufacture, production and actual performance. In addition, the experiment route to detect the refractive index of air and NaCl solution has been revealed.

Key words: Refractive index sensor, Magneto-optical effect, Optical tamm state, Fluid detection

126. Optically tuned optical properties of ferroelectric superlattice by THz spectroscopy

Jie Ji, Siyan Zhou, Qijun Chen, Furi Ling, Jianquan Yao
The 9th International Symposium on Ultrafast Phenomena and Terahertz Waves. Optical Society of America, 2018.

Abstract: Optical properties of ferroelectric superlattice film with different optical powers were characterized by THz spectroscopy. Tunability of transmission could be tuned by 90.1%. Refractive index and absorption coefficient was also varied along with it.

Key words: Photorefractive materials, Ferroelectrics, Spectroscopy, Terahertz

第三章 太赫兹技术

127. 基于石墨烯编码超构材料的太赫兹波束多功能动态调控

闫昕, 梁兰菊, 张璋, 杨茂生, 韦德泉, 王猛, 李院平, 吕依颖, 张兴坊, 丁欣, 姚建铨

物理学报, 2018, 67(11): 118102.

摘要：提出了一种基于石墨烯带的太赫兹波段的1 bit编码超构材料，可以实现太赫兹波束的数目、频率、幅度等参数多功能动态调控。该结构由金属薄膜、聚酰亚胺、硅、二氧化硅、石墨烯带组成。通过对石墨烯带施加两种不同的电压，可以实现一定频率范围内相位差接近180°的"0"和"1"数字编码单元，进而构成1 bit动态可控的编码超构材料。全波仿真结果表明，不同序列的编码超构材料能够实现波束数目从单波束、双波束、多波束到宽波束的调控。相同序列的编码超构材料，通过施加石墨烯带的不同电压能够实现宽频段波束频率的偏移。对于000000或者111111周期序列的编码超构材料，通过施加石墨烯带的不同电压还能够实现波束幅度的调控。因此这种基于石墨烯带的编码超构材料为灵活调控太赫兹波提供了一种新的途径，将在雷达隐身、成像、宽带通信等方面具有重要的意义。

关键词：石墨烯, 太赫兹, 编码超构材料, 动态调控

128. Manipulation of terahertz wave using coding pancharatnam–berry phase metasurface

Jiusheng Li, Jianquan Yao

IEEE Photonics Journal, 2018, 10(5):1–12.

Abstract: Coding metasurface offers a promising way for manipulating electromagnetic waves using several structural metaparticles or changing the metaparticle geometry parameters. It leads to time-consuming optimization in multibit coding designs. Different from the previous conventional coding metasurface, both Pancharatnam–Berry (PB) phase and the predesigned coding sequences in the coding elements are used to manipulate terahertz wave, which becomes more flexible. In this paper, a multibit digital coding metasurface is constructed using the PB phase based on a same size metaparticle with various orientations. Both theoretically calculated and numerically simulated scattering patterns of the designed coding PB phase metasurfaces demonstrate the expected manipulations. The designed coding metasurfaces having polarization dependance for left circularly, right circularly, and linearly polarized waves are confirmed. The maximum bandwidth of a radar cross section reduction of approaching −10 B is 0.8 THz (range from 0.8 to 1.6 THz).

Key words: Terahertz wave, Pancharatnam-Berry phase, Multi-bit coding metasurfaces

129. Active control of terahertz plasmon-induced transparency in the hybrid metamaterial/monolayer MoS$_2$/Si structure

Jie Ji, Siyan Zhou, Weijun Wang, Furi Ling, Jianquan Yao

Nanoscale, 2019, 11(19): 9429–9435.

Abstract: Active control of terahertz waves is critical to the development of terahertz devices. Two-dimensional materials with excellent optical properties provide more choices for opto-electrical devices due to the advancement in their preparation technology. We proposed a hybrid structure of a metamaterial/monolayer MoS$_2$/Si and investigated its optical properties in the terahertz range. The plasmon-induced transparency (PIT) effect was observed in the transmission spectra, resulting from the near-field coupling of two bright modes. According to the simulated results, this phenomenon confirmed its dependency on the length of the cut wire and the distance between DSSRs. Furthermore, an external optical field supported by a 1 064 nm laser could exert a switch effect on the sample. The resonances of the PIT metamaterial disappeared when the optical power was further increased, as the excited carriers in the MoS$_2$/Si substrate blocked the coupling effect. In addition, the experimental results indicated that the PIT metamaterial enhanced the interaction of infrared light with the monolayer MoS$_2$/Si substrate.

Key words: Active control, Monolayer MoS$_2$/Si, PIT effect, Switch effect

130. Amplitude modulation of anomalously reflected terahertz beams using all-optical active Pancharatnam-Berry coding metasurfaces

Jie Li, Yating Zhang, Jining Li, Xin Yan, Lanju Liang, Zhang Zhang, Jin Huang, Jiahui Li, Yue Yang, Jianquan Yao

Nanoscale, 2019, 11(12): 5746−5753.

Abstract: Pancharatnam–Berry (P–B) metasurfaces introduce geometric phases to circularly polarized electromagnetic waves through geometric rotation of the unit cells, thereby converting spin angular momentum (SAM) to orbital angular momentum (OAM) of photons and achieving flexible modulation of spin-polarized waves. It is highly desirable for dynamically tunable P–B metasurfaces to be actively applied. Here, combining double split-ring resonators (DSRRs) with photosensitive semiconductor germanium (Ge), we propose three types of all-optical active Pancharatnam–Berry coding metasurface for dynamic amplitude modulation of spin waves and vortex beams in the terahertz band. Coupled with signal processing methods such as the convolution operation, optical active P–B coding metasurfaces show a strong regulation effect on terahertz beams. This opens up a broad path for coding metasurface applications such as high-speed wireless terahertz communications.

Key words: DSRRs, Germanium, Pancharatnam–Berry coding metasurface, Spin waves, Vortex beams

131. Microfluidic integrated metamaterials for active terahertz photonics

Zhang Zhang, Ju Gao, Maosheng Yang, Xin Yan, Yuying Lu, Liang Wu, Jining Li, Dequan Wei, Longhai Liu, Jianhua Xie, Lanju Liang, Jianquan Yao

Photonics Research, 2019, 7(12): 1400–1406.

Abstract: A depletion layer played by aqueous organic liquids flowing in a platform of microfluidic integrated metamaterials is experimentally used to actively modulate terahertz (THz) waves. The polar configuration of water molecules in a depletion layer gives rise to a damping of THz waves. The parallel coupling of the damping effect induced by a depletion layer with the resonant response by metamaterials leads to an excellent modulation depth approaching 90% in intensity and a great difference over 210° in phase shift. Also, a tunability of slow-light effect is displayed. Joint time-frequency analysis performed by the continuous wavelet transforms reveals the consumed energy with varying water content, indicating a smaller moment of inertia related to a shortened relaxation time of the depletion layer. This work, as part of THz aqueous photonics, diametrically highlights the availability of water in THz devices, paving an alternative way of studying THz wave–liquid interactions and developing active THz photonics.

Key words: Active modulation, THz, Polar configuration, Joint time-frequency analysis, Active THz photonics

132. Characteristic analysis of a photoexcited tunable metamaterial absorber for terahertz waves

Pibin Bing, Xinyue Guo, Hua Wang, Zhongyang Li, Jianquan Yao.
Journal of Optics, 2019, 48(2): 179–183.

Abstract: The absorption characteristics of a photoexcited tunable perfect metamaterial absorber (PMA) at terahertz (THz) frequencies are analyzed based on the split-ring resonators (SRRs)-dielectrics-metallic structure, which are caused by different thicknesses and dielectric constants of polyimide, cell sizes and widths of SRRs, and lengths and conductivities of the photosensitive silicon. The results manifest that the resonance frequency and absorption strength of the designed PMA can be tuned by adjusting the shape, size, thickness, and properties of the metallic structure and dielectric spacer. The simple design and wide frequency tuning range of the PMA can find potential applications in terahertz detectors, filters, switchers, and absorbers.

Key words: Terahertz, Metamaterials, Absorber, Split-ring resonators

133. Effect of optical pump on $Pb_{0.52}Zr_{0.48}TiO_3$ ultrathin film on $LaNiO_3$/Si substrate in the terahertz region

Ying Zeng, Furi Ling, Jianquan Yao
Optical Materials, 2019, 88: 621–624.

Abstract: PZT ($Pb_{0.52}Zr_{0.48}TiO_3$) ultrathin film (12 nm) was deposited on LNO ($LaNiO_3$) film/silicon substrate by magnetron sputtering. XRD and AFM results show the highly oriented ultrathin PZT film with smooth surface suitable for optical modulation. The figure of merit K was observed increased to highest 0.25 with the 532 nm continuous wave (CW) pump power increasing to 500mw at 0.5 THz. The variation of the tunability of dielectric permittivity and the loss were attributed to the built-in electric field caused by the optical pump. It was also found that the conduction loss was predominated in the entire loss when the external field increased.

Key words: Optical materials and properties, Thin films, Spectroscopy, Ferroelectrics

134. BaTeMo$_2$O$_9$ crystals: optical properties and applications in the terahertz range

Hongzhan Qiao, Kai Zhong, Chao Yan, Yang Liu, Longhuang Tang, Kefei Liu, Zeliang Gao, Xutang Tao, Jining Li, Degang Xu, Yuye Wang, Wei Shi, Jianquan Yao

Optical Materials Express, 2019, 9(11): 4390–4398.

Abstract: The optical properties of the α/β-BaTeMo$_2$O$_9$ (α/β-BTM) crystal in the terahertz range were characterized by the terahertz time domain spectroscopy (TDS) system. Frequency-dependent refractive indices and absorption coefficients of two crystals were measured from 0.2 to 2 THz, and discussions and comparisons were made on birefringence, absorption and phonon modes by referring to their structures. The Sellmeier equations for both crystals were also fitted in their transparent ranges. Based on the mode properties and parameters, a feasible scheme for terahertz generation via stimulated polariton scattering (SPS) with β-BTM was proposed and the angle tuning characteristics were calculated. Simulations show that β-BTM has great potential and should be more advantageous than LiNbO$_3$ in generating high-frequency terahertz waves.

Key words: α/β-BTM, Time domain spectroscopy, Stimulated polariton scattering

135. Tunable characteristics of the SWCNTs thin film modulator in the THz region

Weijun Wang, Wen Xiong, Jie Ji, Yue Tian, Furi Ling, Jianquan Yao

Optical Materials Express, 2019, 9(4): 1776–1785.

Abstract: The modulation characteristics of the polyimide-based film of SWCNTs at room temperature were studied with time-domain terahertz (THz) spectroscopy in th SWCNTs e study. The transmission greatly reduced with an increase in the power of the external optical pump. Under the pump power of 300 mW, the transmission even decreased to 3.4% of that of the original SWCNTs sample without illumination. The modulation depth of the film reached 95.6% at 300 mW, indicating the excellent modulation effect. In addition, the optical pump greatly increased the conductivity and caused a blue shift in the real conductivity peak. In order to explore the electric field modulation characteristic of the polyimide-based SWCNTs film, the results of the conductivity at 0 mW and 300 mW under different voltages were discussed. The change in transmission at 300 mW was much more significant than that at 0 mW, indicating that the modulation effect of voltage was more obvious under the condition of illumination. However, even under the pump power of 300 mW, the modulation depth was only 41.11% at 0.7 THz. In terms of the modulation depth of the optic field and electric field, we believed that the optical modulator worked better for the polyimide-based SWCNTs film.

Key words: Modulation characteristics, SWCNTs, Time-domain terahertz spectroscopy

136. Modulation of terahertz electromagnetically induced absorption analogue in a hybrid metamaterial/graphene structure

Wen Xiong, Weijun Wang, Furi Ling, Wenfeng Yu, Jianquan Yao
AIP Advances, 2019, 9(11): 115314.

Abstract: We proposed a two-layer metamaterial structure with graphene that consists of two H-shaped resonators and one I-shaped resonator. The electromagnetically induced absorption (EIA) analog phenomena were observed in absorption spectra, resulting from the near-field coupling of two bright modes. Furthermore, the absorption peak can be tuned by changing the dimension of the I-shaped resonator or changing the Fermi energy of graphene. The theoretical analysis reveals that the EIA analog arises from magnetic resonance using the coupled Lorentz oscillator model. This hybrid-EIA analog structure may provide a possible choice for designing potential devices for dynamic narrow-band filtering and absorber applications.
Key words: Graphene, H-shaped resonators, I-shaped resonator, EIA, Magnetic resonance

137. Optical-induced dielectric tunability properties of DAST crystal in THz range

Degang Xu, Xianli Zhu, Yuye Wang, Jining Li, Yixin He, Zibo Pang, Hongjuan Cheng, Jianquan Yao
Chinese Physics B, 2019, 28(12): 127701.

Abstract: The optical-induced dielectric tunability properties of DAST crystal in THz range were experimentally demonstrated. The DAST crystal was grown by the spontaneous nucleation method (SNM) and characterized by infrared spectrum. With the optimum wavelength of the exciting optical field, the transmission spectra of the DAST crystal excited by 532 nm laser under different power were measured by terahertz time-domain spectroscopy (THz-TDS) at room temperature. The transmitted THz intensity reduction of 26 % was obtained at 0.68 THz when the optical field was up to 80 mW. Meanwhile, the variation of refractive index showed an approximate quadratic behavior with the exciting optical field, which was related to the internal space charge field of photorefractive phenomenon in the DAST crystal caused by the photogenerated carrier. A significant enhancement of 13.7 % for THz absorption coefficient occurred at 0.68 THz due to the photogenerated carrier absorption effect in the DAST crystal.
Key words: THz wave, Organic DAST crystal, Photogenerated carrier, Dielectric property

138. Effect of optical pumping on the dielectric properties of 0.55SrTiO$_3$-0.45NdAlO$_3$ ceramics in terahertz range

Dan Li, Chunya Luo, Jianquan Yao
Optical Engineering, 2019, 58(7): 077109.

Abstract: The dielectric properties of 0.55SrTiO$_3$-0.45NdAlO$_3$ ceramics under external optical fields were investigated by terahertz time-domain spectroscopy at room temperature. From the experimental results, it could be found that the tunability of permittivity with the external optical pump was reached up to 16% at 0.6 THz. And the change of refractive index has a linear relationship on scale with the applied external light power. These results could be explained by a built-in electric field caused by the excited free carriers in the ceramics.

Key words: Dielectric properties, 0.55SrTiO$_3$-0.45NdAlO$_3$ ceramics, External optical fields, Terahertz

139. Photoexcited blueshift and redshift switchable metamaterial absorber at terahertz frequencies

Zongcheng Xu, Liang Wu, Yating Zhang, Degang Xu, Jianquan Yao
Chinese Physics Letters, 2019, 36(12): 124202.

Abstract: We propose a design and numerical study of an optically blueshift and redshift switchable metamaterial (MM) absorber in the terahertz regime. The MM absorber comprises a periodic array of metallic split-ring resonators (SRRs) with semiconductor silicon embedded in the gaps of MM resonators. The absorptive frequencies of the MM can be shifted by applying an external pump power. The simulation results show that, for photoconductivity of silicon ranging between 1 S/m and 4 000 S/m, the resonance peak of the absorption spectra shifts to higher frequencies, from 0.67 THz to 1.63 THz, with a resonance tuning range of 59%. As the conductivity of silicon increases, the resonance frequencies of the MM absorber are continuously tuned from 1.60 THz to 1.16 THz, a redshift tuning range of 28%. As the conductivity increases above 30 000 S/m, the resonance frequencies tend to be stable while the absorption peak has a merely tiny variation. The optical-tuned absorber has potential applications as a terahertz modulator or switch.

Key words: Blueshift, Redshift, Switchable, Metamaterial absorber, Terahertz

140. Temperature-dependent dielectric characterization of magneto-optical $Tb_3Sc_2Al_3O_12$ crystal investigated by terahertz time-domain spectroscopy

Jugeng Li, Senmiao Yang, Xin Chen, Naifeng Zhuang, Qibiao Zhu, Anhua Wu, Xian Lin, Guohong Ma, Zuanming Jin, Jianquan Yao

Chinese Physics Letters, 2019, 36(4): 044203.

Abstract: Terbium scandium aluminum garnet (TSAG) crystals have been widely used in magneto-optical systems. We investigate the complex refractive index of the TSAG crystal in the terahertz frequency range using terahertz (THz) time-domain spectroscopy in the temperature range 100–300 K. It is observed that the refractive index and the absorption coefficient increase with the THz frequency. The refractive index increases with the temperature. We measure the temperature coefficient of the refractive index of the TSAG crystal in the frequency range 0.4–1.4 THz. Furthermore, the loss tangent, i.e., the ratio of experimental values of the imaginary and real part of the dielectric permittivity, is found to be almost independent of frequency. TSAG is very promising for applications in THz optoelectronics because it has a high dielectric constant, low loss, and low thermal coefficient of the dielectric constant.

Key words: TSAG, Terahertz time-domain spectroscopy, refractive index

141. Ultrafast carrier dynamics and terahertz photoconductivity of mixed-cation and lead mixed-halide hybrid perovskites

Wanying Zhao, Zhiliang Ku, Liping Lv, Xian Lin, Yong Peng, Zuanming Jin, Guohong Ma, Jianquan Yao

Chinese Physics Letters, 2019, 36(2): 028401.

Abstract: Using time-dependent terahertz spectroscopy, we investigate the role of mixed-cation and mixed-halide on the ultrafast photoconductivity dynamics of two different methylammonium (MA) lead-iodide perovskite thin films. It is found that the dynamics of conductivity after photoexcitation reveals significant correlation on the microscopy crystalline features of the samples. Our results show that mixed-cation and lead mixed-halide affect the charge carrier dynamics of the lead-iodide perovskites. In the $(5-AVA)_{0.05}(MA)_{0.95}PbI_{2.95}Cl_{0.05}$/spiro thin film, we observe a much weaker saturation trend of the initial photoconductivity with high excitation fluence, which is attributed to the combined effect of sequential charge carrier generation, transfer, cooling and polaron formation.

Key words: Time-dependent terahertz spectroscopy, Mixed-cation, Mixed-halide, Ultrafast photoconductivity dynamics

142. All-optical switchable vanadium dioxide integrated coding metasurfaces for wavefront and polarization manipulation of terahertz beams

Jie Li, Yue Yang, Jining Li, Yating Zhang, Zhang Zhang, Hongliang Zhao, Fuyu Li, Tingting Tang, Haitao Dai, Jianquan Yao

Advanced Theory and Simulations, 2019, 3(1): 1900183.

Abstract: Electromagnetic metasurfaces are artificial 2D structures with sub-wavelength size and have achieved a variety of functions. Coding metasurfaces or programmable metasurfaces are special phase-manipulation type metasurfaces, which can be used for dynamic electromagnetic modulation and information processing in microwave band. However, programmable or tunable coding metasurfaces in terahertz band are still rare. All-optical switchable coding metasurfaces based on vanadium dioxide (VO_2) hybrid resonators are proposed for dynamic modulation of terahertz waves. The C-shaped metal–VO_2 hybrid resonators are switched from anisotropic structures to quasi-isotropic ones due to insulator–metal transition of VO_2, resulting in great attenuation of conversion efficiency for linearly polarized waves and Pancharatnam–Berry phase for circularly polarized waves. Based on this principle, several coding metasurfaces are shown for efficient and simultaneous switching of wavefront and polarization state of terahertz beams, including linearly and circularly polarized waves. It is worth mentioning that a new method for switchable shared-aperture metasurfaces is also proposed, in which only a part of the unit cells are integrated with VO_2 and tunable. This new scheme promotes the applications of terahertz coding metasurfaces such as high-speed terahertz communication.

Key words: All-optical switching, Coding metasurfaces, Shared aperture, Vanadium dioxide, Wavefront manipulation, Polarization manipulation

143. Investigation of optical tuning on the dielectric properties of $0.3\,Ba_{0.4}Sr_{0.6}TiO_3$–$0.7NdAlO_3$ ceramics in terahertz range

Dan Li, Furi Ling, Jianquan Yao

IEEE Transactions on Terahertz Science and Technology, 2019, 9(5): 505−509.

Abstract: Optical tuning of the dielectric properties of $0.3Ba_{0.4}Sr_{0.6}TiO_3$–$0.7NdAlO_3$ ceramics under external optical fields was investigated in a frequency range from 0.2 to 0.8 THz at room temperature. It could be found that the dielectric constant could be modulated by an external optical pump and the tunability of permittivity reached up to 18% at 0.7 THz. Results showed that the real part of permittivity increased upon application of an external optical field, which could be interpreted as hardening of the soft mode and increasing of the damping coefficient and oscillator strength. The experimental results also showed that the change of the refractive index has a linear relationship on the scale with the applied external light power. And these results were attributed to the built-in electric field caused by the excited free carriers in the ceramics. The results of the present experiment provide a reference for further development of tunable structures controlled by an external optical field. Our results should be a reference for future research of the tunable terahertz devices modulated by external optical field, such as terahertz modulator.

Key words: Dielectric properties, External optical fields, $0.3Ba_{0.4}Sr_{0.6}TiO_3$–$0.7NdAlO_3$ ceramics, Terahertz (THz), Tunability.

144. A novel terahertz beam splitter using ultrathin flexible transmission-type coding metasurface

L. Liang, J. Wang, X. Yan, S. Gao, Y. Lv, L. Xiao, Y. Li, D. Wei, M. Wang, X. Wang, Y. Zhang, J. Yao

Journal of Optoelectronics and Advanced Materials, 2019, 21(7−8): 443−449.

Abstract: A new flexible and simple transmission-type coding metasurface was proposed and characterized, which can generate different transmitted beam patterns by coding sequences of "0" and "1" digital elements. The two digital elements with almost identical transmission amplitude are composed of three metallic layers and three dielectric layers, and their transmission phase difference is approximately 180° at 0.95 THz. The characterizations and working mechanisms of the coding metasurface were investigated through theoretical analysis and electromagnetic (EM) simulation. The simulation and theoretical analysis results show that the normally incident waves will be realized a single beam, two main beams and four main beams, respectively using different transmission-type coding sequences. And the amplitude modulation of transmission beam was realized by an optical pump beam. This work reveals a new and simple route for designing novel terahertz beam splitter and offers widespread applications.

Key words: Beam splitter, Coding metasurface, Transmission-type, Digital elements, Coding sequences

145. Nested anti-resonant hollow core fiber for terahertz propagation

Xianli Zhu, Degang Xu, Yuye Wang, Jining Li, Yixin He, Chao Yan, Longhuang Tang, Jianquan Yao

Infrared, Millimeter-Wave, and Terahertz Technologies VI. International Society for Optics and Photonics, 2019, 11196: 111960N.

Abstract: A novel nested anti-resonant hollow core fiber (NAHF), based on Topas, with low loss and flattened dispersion is proposed for efficient transmission of terahertz wave. Finite element method with an ideally matched layer (PML) boundary condition is used to investigate its guiding properties. A cladding structure of nested anti-resonant elliptical rings is introduced to reduce mode power leakage. The NAHF shows a low confinement loss (< 0.29 cm^{-1}) and a small effective material loss (< 0.019 cm^{-1}) in the frequency range of 0.9-1.5 THz. An ultra-flatted near zero dispersion profile of ±0.029 ps/THz/cm is obtained within a broad frequency range of 0.6-1.5 THz. Furthermore, optimizing the structure parameters in NAHF, higher core power fraction over 80 %, higher effective mode area of ~10^{-6} μm^2 and the bending loss of 3.05 × 10^{-5} cm^{-1} at the bending radius of 10 cm are also achieved.

Key words: Terahertz, Hollow core fiber, Anti-resonant effect, Dispersion

146. Terahertz wave transmission and reflection characteristics in plasma

Xingning Geng, Degang Xu, Jining Li, Kai Chen, Yixin He, Jianquan Yao

Infrared, Millimeter-Wave, and Terahertz Technologies VI. International Society for Optics and Photonics, 2019, 11196: 111961T.

Abstract: After entering the near space, a layer of plasma sheath is formed outside the hypersonic vehicles due to the high-temperature and high-pressure environment. The plasma sheath, which characteristic frequency is similar to microwave, will cause serious impediment to communication signal. This phenomenon is known as the blackout problem. With the rapid development of aerospace industry, plasma sheath blackout has become an urgent problem to be solved. Current research shows that increasing the frequency of electromagnetic wave higher than the plasma characteristic frequency can effectively reduce the shielding effect of plasma. The frequency of terahertz (THz) wave is much higher than microwave, it can propagate through plasma sheath, which provides an effective method to solve the problem of plasma sheath. In this paper, a theoretical model of plasma is established, and the transmission properties of THz wave in plasma is simulated using scattering matrix method. Then a kind of plasma jet is produced in laboratory environment according to dielectric barrier discharge. And the experiments of a broadband THz source and THz time-domain spectrum transmission in this kind of plasma and a 2.52 THz wave reflection imaging of target under plasma shelter are carried out respectively. The transmittance increases with frequency under 0.5 THz and becomes stable at 100% over 0.5 THz, and the result of experiments and simulation are in good agreement. Both theory and experiments show that THz wave has good penetration in plasma jet and can detect targets behind plasma, and this study will lay a theoretical foundation for solving the plasma blackout problem of hypersonic vehicle in near space.

Key words: Terahertz wave, Plasma, Transmission, Reflection

147. Photo active control of plasmon-induced reflection in complementary terahertz metamaterials

Yue Yang, Jining Li

Infrared, Millimeter-Wave, and Terahertz Technologies VI. International Society for Optics and Photonics, 2019, 11196: 1119604.

Abstract: A typical plasmon-induced transmission metamaterial and its corresponding complementary structure is designed and explored. The results show that the transmission spectra of the two structures are complementary, but the reflection spectra of the complementary structure and the transmission spectra of the ordinary structure are in good agreement. Furthermore, this work integrate photosensitive silicon into the complementary metal-based metamaterials, realizing the optical active control of the plasmon-induced reflection and transmission. This research strategy provides a new way for the study of reflective structures. Moreover, the active control of the reflection and transmission play a role in slow light devices and terahertz filter.

Key words: Metamaterial, Terahertz, Optical, Active control

148. 基于超材料的可调谐的太赫兹波宽频吸收器

陈俊，杨茂生，李亚迪，程登科，郭耿亮，蒋林，张海婷，宋效先，叶云霞，任云鹏，任旭东，张雅婷，姚建铨

物理学报，2019, 68(24): 247802.

摘要：随着频谱资源的日益稀缺，太赫兹波技术在近十几年的时间里得到了越来越多的关注，并取得了巨大的进展。由于高吸收、超薄厚度、频率选择性和设计灵活性等优势，超材料吸收器在太赫兹波段备受关注。本文设计了一种"T型"结构的超材料太赫兹吸收器，同时获得了太赫兹多频吸收器和太赫兹波宽频可调谐吸收器。它们结构参数一致，唯一的区别是在太赫兹波宽频可调谐吸收器的顶端超材料层上添加了一块方形光敏硅。这种吸收器都是三层结构，均由金属基板、匹配电介质层以及顶端超材料层组成。仿真结果表明，太赫兹波多频吸收器拥有 6 个吸收率超过 90% 的吸收峰，其平均吸收率高达 96.34%。而太赫兹波宽频可调谐吸收器通过改变硅电导率，可以控制吸收频带的存在与否，同时可以调整吸收峰的频率位置，使吸收峰频率在一个频带宽度大约为 30 GHz 的范围内调整。当硅的电导率为 1 600 S/m 时，吸收率超过 90% 的频带宽度达到 240 GHz，而且其峰值吸收率达到 99.998%。

关键词：太赫兹吸收器，多频，可调控宽频，超材料

149. 稀土正铁氧体中 THz 自旋波的相干调控与强耦合研究进展

金钻明,阮舜逸,李炬赓,林贤,任伟,曹世勋,马国宏,姚建铨
物理学报, 2019, 68(16): 167501.

摘要:基于反铁磁材料的自旋逻辑器件被认为具有更低的能量损耗、更快的速度和更高的稳定性,这使得反铁磁材料的超快自旋动力学成为当前自旋电子学研究的热点。由于反铁磁体具有强的交换耦合和高的共振频率,将在 GHz 甚至 THz 波段得到广泛应用。本文综述了利用太赫兹 (THz) 脉冲的磁场分量与反铁磁自旋序之间的相互作用进行探测与操控。利用 THz 脉冲时域光谱,系统研究了反铁磁性稀土正铁氧体 ($RFeO_3$) 中自旋共振的非热激发及其弛豫动力学。总结了 $RFeO_3$ 的准铁磁和准反铁磁自旋模式的共振频率,以及由 R^{3+}-Fe^{3+} 离子间的相互作用所确定的自旋重取向温区。不仅可以利用具有时间延迟的 THz 双脉冲实现 $DyFeO_3$ 中自旋极化的相干控制,利用材料的各向异性以单个 THz 脉冲也可以实现 $YFeO_3$ 中的自旋波相干调控。在 $Er_xY_{1-x}FeO_3$ 单晶样品中,找到了自旋与真空磁子的关联交换耦合的实验证据,证明了存在以物质-物质相互作用形式的迪克协作耦合。最后,讨论了 THz 波在 $TmFeO_3$ 晶体传播过程中诱导的磁极化子。

关键词:反铁磁,自旋共振,超快太赫兹光谱

150. 可调控的太赫兹多频带吸波器特性研究

初启航,杨茂生,陈俊,曾彬,张海婷,宋效先,叶云霞,任云鹏,张雅婷,姚建铨
中国激光, 2019, 46(12): 1214003.

摘要:基于不同的形状、大小的谐振环对于电磁场具有不同的响应原理,设计了一种对 4 个频带的电磁响应的圆形谐振环结构所组成的太赫兹吸波器。本文通过利用时域有限差分法 (FDTD) 对于该吸收器的特性进行研究,分别通过改变顶层金属环形图案几何尺寸、中间层电介质的厚度以及顶层金属图案圆环处的硅电导变化率,进行了太赫兹多频带吸收器的设计与仿真。耦合后的多频吸收器的吸收峰中低频部分达到完美吸收,高频部分吸收率由 70 % 增长为 94 %。同时,随着电导率变化低频分别从 0.775 THz 和 1.064 THz 移动到 0.697 THz 和 1.017 THz,分别移动了 78 GHz 和 47 GHz,实现了连续频率调谐。

关键词:太赫兹波,超材料吸收器,隐身材料,电导率,调控

151. Ultra-wideband low-loss control of Terahertz scatterings via an all-dielectric coding metasurface

Maosheng Yang, Xin Yan, Zhang Zhang, Ju Gao, Lanju Liang, Xinyue Guo, Jie Li, Dequan Wei, Meng Wang, Yunxia Ye, Xiaoxian Song, Haiting Zhang, Yunpeng Ren, Xudong Ren, Jianquan Yao

*ACS Applied Electronic Material*s, 2020, 2(4), 1122−1129.

Abstract: Despite the growing interests for controlling terahertz (THz) scatterings via metallic metasurfaces, the avoidance of Ohmic loss resulting from the interaction between light and phonons or free electrons remains an arduous task. Herein, we propose a new approach of ultra-wideband low-loss control of THz scatterings via an all-dielectric coding metasurface (DCMF). In our experiments, the DCMFs manipulate THz waves remarkably in an ultra-wideband of 900 GHz. When a THz wave enters into the DCMFs, it emerges split into two beams with a higher amplitude of transmission up to 60% and the lower amplitude about 30%. In our simulations, the high amplitude of transmission may even reach 99%. Besides, the variation of coding sequence enables diversification of manipulation for THz scatterings. Notably, the DCMF with the sequence of "101010..." refracts a beam of 1.3 THz in two directions. The energy of the main beam seems to dissipate in all directions. We also find that as a beam of 1.63 THz passes through the DCMFs consisting of the sequence of "110110...", its energy may scatter in multiple directions, demonstrating an unprecedented control of THz scatterings. Our results not only enrich the strategy of THz modulation but also guide the superior design for THz devices.

Key words: Terahertz, Coding metasurface, Metamaterial, Phase sensitivity, Modulation

152. Position-guided Fano resonance and amended GaussAmp model for the control of slow light in hybrid graphene–silicon metamaterials

Maosheng Yang, Xiaoxian Song, Haiting Zhang, Yunxia Ye, Yunpeng Ren, Xudong Ren, Jianquan Yao

Optics Express, 2020, 28(8): 11933–11945.

Abstact: Position-guided Fano resonance is observed in hybrid graphene–silicon metamaterials. An outstanding application of such resonance is slow-light metadevices. The maximum group delay is 9.73 ps, which corresponds to a group delay in free-space propagation of 2.92 mm. We employ a coupled oscillator model to illustrate anomalous transmission, where the intensity of the Fano peak increases with the Fermi level. Furthermore, we amend the GaussAmp model to serve as a suitable control equation for the group delay. The coefficient of correlation (R^2) is as high as 0.999 98, while the lowest values of the root-mean-square error and sum of squared errors are respectively 0.004 21 and 0.001 56. These results indicate that the amended GaussAmp model accurately controls the trend of the group delay. This work not only clarifies the mechanism of Fano resonance generation but also provides a promising platform for dynamically adjustable optical switches and multidimensional information sensors.

Key words: Position-guide, Fano resonance, Slow-light metadevices, GaussAmp model

153. Plasmon-induced reflection metasurface with dual-mode modulation for multi-functional THz devices

Yue Yang, Jie Li, Jining Li, Jin Huang, Yating Zhang, Lanju Liang, Jianquan Yao

Optics and Lasers in Engineering, 2020, 127:105969.

Abstract: The active controllable metasurfaces based on electromagnetically induced transparency (EIT) have attracted great interest in the research of terahertz functional devices. We demonstrate a dual-mode controllable terahertz metasurface based on complementary plasmon-induced reflection structure. Both photosensitive silicon and monolayer graphene are integrated into the unit cell to realize the active control of the EIT-like reflection window and the whole resonance, modulated by optical pumping and extra electrical voltage, respectively. The excellent properties of the modulation depth and the group delay highlight this metasurfaces in the applications of terahertz modulators and slow light devices. Moreover, this dual-mode modulation in one design provides a good way of commercial terahertz multi-functional devices.

Key words: Multi-function, Dual-mode control, Complementary, Metasurfaces

154. Frequency-switchable VO_2-based coding metasurfaces at the terahertz band

Jiahui Li, Yating Zhang, Jining Li, Jie Li, Yue Yang, Jin Huang, Chengqi Ma, Zhenzhen Ma, Zhang Zhang, Lanju Liang, Jianquan Yao

Optics Communications, 2020, 458: 124744.

Abstract: In this paper, a frequency-switchable terahertz coding metasurface is demonstrated based on the phase transition of vanadium dioxide. Temperature excitation can initiate the insulator–metal transition of vanadium dioxide, thereby changing the resonant frequency of the metal-vanadium dioxide composite unit structure, then changing the operating frequency of the entire coding metasurfaces. We first design a 1-bit coding metasurface whose working frequency can be switched between 1 THz and 1.4 THz, with a switching bandwidth of 0.4 THz. Further more, a 2-bit coding metasurface is designed, whose working frequency can be switched between 1 THz and 1.5 THz with a switching bandwidth of 0.5 THz. This work provides a new design idea for the terahertz active coding metasurfaces which opens up a broad path for coding metasurface applications such as wireless terahertz communications.

Key words: Terahertz, Frequency-switchable coding metasurfaces, Vanadium dioxide

155. Metal-graphene hybrid active chiral metasurfaces for dynamic terahertz wavefront modulation and near field imaging

Jie Li, Jitao Li, Yue Yang, Jining Li, Yating Zhang, Liang Wu, Zhang Zhang, Maosheng Yang, Chenglong Zheng, Jiahui Li, Jin Huang, Fuyu Li, Tingting Tang, Haitao Dai, Jianquan Yao
Carbon, 2020, 163: 34−42.

Abstract: Chiral metasurfaces can achieve giant chiral optical responses and have been expanded from the optical band to other electromagnetic bands. Here, we propose a new method for dynamic terahertz circular dichroism (CD) manipulation in metasurfaces. By introducing a patterned and electrically doped graphene into the metamirror consists of double layer C-shaped split ring resonators (SRRs), efficient terahertz CD modulation is observed. Since the electrical doping of graphene changes the absorption loss of the metasurface cavity, the structure shows switching between a chiral metasurface and an ordinary metal mirror, which can be seen as "on" and "off" states. The calculation results show an efficient modulation of the terahertz CD in a large dynamic range. In addition, we also use the new method to design two metasurfaces for dynamic terahertz wavefront modulation and near-field digital imaging, both of which show a high-performance electrical switching. This method provides a new way for the design of active terahertz devices based on metasurfaces, and also promotes the applications of terahertz chiral metasurfaces in high-speed wireless communication and dynamic imaging.
Key words: Chiral metasurfaces, Doped graphene, Terahertz CD modulation

156. Active controllable bandwidth of THz metamaterial bandpass filter based on vanadium dioxide

Jin Huang, Jining Li, Yue Yang, Jie Li, Jiahui Li, Yating Zhang, Jianquan Yao
Optics Communications, 2020, 465: 125616.

Abstract: A terahertz metamaterial bandpass filter with actively control bandwidth based on vanadium dioxide is proposed. This filter has an initial bandwidth with 0.32 THz without a bias voltage. By applying different bias voltages on different parts of the metal patterns, not only the bandwidth can be continuously adjusted, but also the passband frequency can be shifted. Simulation results show that the ranges of the bandwidth increasing and decreasing are 0.13 THz and 0.12 THz, respectively. And the range of the passband frequency shifting is 0.12 THz. Meanwhile, excellent broadband response and bandpass performance can still be maintained during the phase transition process. The physical mechanism of the bandwidth tunability is elucidated by comparing the surface current and electric field distributions of the insulator and metal phase. In addition, it can keep stable working performance when the incident angle varies up to 40°. This bandwidth tunable terahertz passband filter can eliminate undesired information to obtain useful information more flexibly and has a great potential in terahertz communication, bioimaging and safety detection.

Key words: Metamaterial filter, Terahertz wave, Broadband, Bandwidth tunability, Vanadium dioxide

157. Active controllable dual broadband terahertz absorber based on hybrid metamaterials with vanadium dioxide

Jin Huang, Jining Li, Yue Yang, Jie Li, Jiahui Li, Yating Zhang, Jianquan Yao

Optics Express, 2020, 28(5): 7018−7027.

Abstract: In this paper, we present an active controllable terahertz absorber with dual broadband characteristics, comprised by two diagonal identical patterns of vanadium dioxide in the top layer of the classical three-layer structure of metamaterial perfect absorbers. Simulation results show that two bandwidths of 80% absorption are 0.88 THz and 0.77 THz from 0.56 to 1.44 THz and 2.88 to 3.65 THz, respectively. By using thermal control to change the conductivity of the vanadium dioxide, absorptance can be continuously adjusted from 20% to 90%. The impedance matching theory is introduced to analyze and elucidate the physical mechanism of the perfect absorption. Field analyses are further investigated to get more insight into the physical origin of the dual broadband absorption. In addition, incident polarization insensitivity and wide-angle absorption are also demonstrated. The proposed absorber promises diverse applications in terahertz regime, such as imaging, modulating, sensing and cloaking.

Key words: Active controllable terahertz absorber, Dual broadband, Thermal control, Impedance matching theory

158. All-optical switchable terahertz spin-photonic devices based on vanadium dioxide integrated metasurfaces

Jie Li, Jitao Li, Yating Zhang, Jining Li, Yue Yang, Hongliang Zhao, Chenglong Zheng, Jiahui Li, Jin Huang, Fuyu Li, Tingting Tang, Jianquan Yao

Optics Communications, 2020, 460: 124986.

Abstract: Metasurfaces based on Pancharatnam–Berry (P–B) phase can achieve strong spin angular momentum (SAM) to orbital angular momentum (OAM) conversion of light, which provides a new degree of freedom for light control and opens up a new way for the applications of metasurfaces in classical and quantum optics. With the development of high-speed, large-capacity information transmission and high-definition imaging, demand for multifunctional and tunable P–B phase metasurfaces increases. Here, we propose three switchable terahertz spin-photonic devices based on P–B phase metasurfaces for focusing (divergence), splitting and vortex generation of terahertz beams. Based on photo-induced insulator–metal phase transition of the vanadium dioxide (VO_2) islands in reflective hybrid resonators, switching of the devices function between on- and off- state is obtained, and the amplitude switching efficiency is as high as 90%. This work provides new ideas for the design of active terahertz devices and facilitates the applications of terahertz spin-photonic devices based on metasurfaces.

Key words: Vanadium dioxide, Tunable metasurfaces, Terahertz waves, Spin manipulation

第三节　太赫兹技术的应用

1. Highly precise determination of optical constants olive oil in terahertz time-domain spectroscopy

Li Jiusheng, Yao Jianquan

Proceedings of SPIE, 2008, 7277: 727707.

Abstract: Recently, there has been a remarkable effort in employing terahertz (THz) spectroscopy for investigating material properties. Pulsed THz time-domain spectroscopy is a coherent technique, in which both the amplitude and the phase of a THz pulse are measured. Coherent detection enables direct calculations of both the imaginary and the real parts of the refractive index without using the Kramers-Kronig relations. In this letter, the terahertz absorption spectra and the refractive indices of olive oil were measured by using terahertz time-domain spectroscopy (THz-TDS) in the frequency range extending from 0.2 to 2.5 THz. The terahertz dielectric properties of olive oil were characterized by THz-TDS, and the consistency with the known parameters was identified. A novel iterative algorithm improves the existing data extraction algorithms and further enhances the accuracy of the parameter extraction for terahertz time-domain spectroscopy. The results obtained in this study suggest that the THz-TDS method is a useful tool for vegetable oils characterization in the far infrared region. This method can be applied not only to terahertz time-domain spectroscopy but also to any kind of optical constant measurement in the time domain.

Key words: Terahertz time-domain spectroscopy, Olive oil, Refractive index, Absorption coefficient

2. Terahertz spectrum analysis of leather at room temperature

Li Jiusheng, Yao Jianquan, Li Jianrui

Proceedings of SPIE, 2008, 7277: 727704.

Abstract: Over the past ten years, electromagnetic terahertz (THz) frequencies region from 100 GHz to 10 THz (or wavelengths of 30 μm ~3 mm) have received extensive attention and investigation. Terahertz wave detection enables direct calculations of both the imaginary and the real parts of the refractive index without using the Kramers-Kronig relations. There are many potential applications such as radio astronomy, atmospheric studies, remote sensing, and plasma diagnostics.Shoes, neckties and sofa, etc. are mainly made of skin of animal, imitated skin and artificial leather. It has important practical value to component analysis and quality assessment by measuring absorption, refractive index, and other optical parameters. In this paper, the spectral characteristics of sheepskin, imitated sheepskin and artificial leather have been measured with terahertz time-domain spectroscopy (THz-TDS) in the range of 0.1~2.0THz. The results show that there have not absorption peak in the absorption spectrum of the sheepskin. However, it is found that there are three absorption peaks in the absorption spectrum of the artificial leather at the frequency of 1.13THz, 1.21THz, and 1.36THz, respectively. The potential application of the leather in THz frequency region is also discussed.

Key words: THz-TDS, Leather, Refractive index, Absorption coefficient

3. Terahertz time-domain spectroscopy and application on peanut oils

Li Jiusheng, Yao Jianquan, Li Jianrui

Proceedings of SPIE, 2008, 7277: 727705.

Abstract: Many materials were previously studied using far-infrared Fourier transform spectroscopy (FTS) in transmission and reflection modes. Recently, there has been a remarkable effort in employing terahertz time-domain spectroscopy (THz-TDS) for investigating material properties, including environment pollutants, semiconductors, polymers, explosive materials, and gases, etc. Since the absorption coefficient and the refractive index of the material studied are directly related to the amplitude and phase respectively of the transmitted field, both parts of the complex permittivity can be obtained by THz-TDS. In this letter, the optical properties of peanut oils in the frequency range from 0.2 to 2.5 THz were studied by employing terahertz time-domain spectroscopy. Several peanut oils, such as clean unused peanut oil, peanut oil after five minutes of boiling, and peanut oil after ten minutes of boiling were tested. The time delays of clean unused peanut oil, peanut oil after five minutes of boiling, and peanut oil after ten minutes of boiling are 8.33ps, 8.46ps and 8.46ps, respectively. The refractive indices of the three oil samples show slow a decrease as the terahertz wave frequency increases. The power absorption coefficients increase as the frequency increases within the investigated terahertz wave frequency range.

Key words: Terahertz time-domain spectroscopy, Oils, Absorption coefficient

4. Terahertz-wave Spectrum of Cotton

Li Jiusheng, Yao Jianquan

Proceedings of SPIE, 2008, 7277: 727706.

Abstract: Recently, there has been a remarkable effort in employing terahertz (THz) spectroscopy for investigating material properties. THz-TDS has been employed to investigate a wide variety of materials, including environment pollutants, semiconductors, polymers, explosive materials, and gases, etc. In this letter, the spectral characteristics of cotton in the range of 0.2 ~ 2.5THz have been measured with THz time-domain spectroscopy. Its absorption and refraction spectra are obtained at room temperature in nitrogen atmosphere. It is found that cotton has the spectral response to THz waves in this frequency region. The results provided in this paper will help us to study the THz application to cotton commercial transaction inspection further.

Key words: THz-TDS, Cotton, Absorption spectra

5. 太赫兹通信技术的研究与展望

姚建铨,迟楠,杨鹏飞,汪静丽,崔海霞,李九生,徐德刚,丁欣

中国激光, 2009, 36(9): 2213-2233.

摘要: 太赫兹通信是一个极具应用前景的技术,太赫兹波有非常宽的还没分配的频带,并且具有传输速率高、方向性好、安全性高、散射小及穿透性好等许多特性。发展太赫兹通信技术已成为各发达国家研究的热点。本文分析了国内外太赫兹通信的研究情况;以太赫兹通信系统的整体框架,全面的介绍和分析了太赫兹通信的一些关键技术和最新研究成果;同时,对太赫兹的技术发展趋势和应用前景做了展望,提出了太赫兹技术的发展战略。

关键词: 太赫兹波,太赫兹通信系统,高速短距离无线互联通信,太赫兹空间通信

6. High-quality continuous-wave imaging with a 2.53 THz optical pumped terahertz laser and a pyroelectric detector

Bing Pibin, Yao Jianquan, Xu Degang, Xu Xiaoyan, Li Zhongyang
Chinese Physics Letters, 2010, 27(12): 124209.

Abstract: A cw terahertz (THz) transmission imaging system is demonstrated and a high-quality THz image can be obtained using a pyroelectric detector. The factors that affect the imaging quality, such as the THz wavelength, spot size on the sample surface, step length of the motor, and frequency of the chopper, are theoretically and experimentally investigated. The experimental results show that the maximum resolution of the THz image can reach 0.4mm with the THz wavelength of 118.8 μm, the spot size of 1.8mm and the step length of 0.25mm.

Key words: Terahertz transmission imaging system, Imaging quality, Spot size, Step length

7. High quality continuous-wave imaging at 2.53 THz

Pibin Bing, Jianquan Yao, Degang Xu, Xiaoyan Xu, Zhangyang Li
Proceedings of SPIE, 2010, 7854: 78542E.

Abstract: Low-resolution of terahertz (THz) imaging troubled its applications in the field of medical diagnosis and security inspection. The continuous wave (CW) THz imaging system utilizing a pyroelectric detector has been realized. The two crucial factors were analyzed in theory and verified in experiment; a high-quality THz image with the resolution of 0.4mm was obtained by choosing suitable imaging parameters. In our experiment the THz wave frequency of 2.53 THz, the spot size of 1.8 mm and the step length of 250 μm were selected to achieve high quality THz image. We also image several samples with different materials utilizing this system, and the results were very good.

Key words: Continuous wave imaging, Terahertz wave, Pyroelectric detector, Imaging quality

8. Terahertz imaging technique and application in large scale integrated circuit failure inspection

Zhigang Di, Jianquan Yao, Chunrong Jia, Degang Xu, Pibin Bing, Pengfei Yang, Yibo Zheng

Proceedings of SPIE, 2010, 7854: 78542L.

Abstract: Terahertz ray, as a new style optic source, usually means the electromagnetic whose frequencies lies in between 0.1THz~10THz, the waveband region of the electromagnetic spectrum lies in the gap between microwaves and infrared ray. With the development of laser techniques, quantum trap techniques and compound semiconductor techniques, many new terahertz techniques have been pioneered, motivated in part by the vast range of possible applications for terahertz imaging, sensing, and spectroscopy. THz imaging technique was introduced, and THz imaging can give us not only the density picture but also the phase information within frequency domain. Consequently, images of suspicious objects such as concealed metallic or metal weapons are much sharper and more readily identified when imaged with THz imaging scanners. On the base of these, the application of THz imaging in nondestructive examination, more concretely in large scale circuit failure inspection was illuminated, and the important techniques of this application were introduced, also future prospects were discussed. With the development of correlative technology of THz, we can draw a conclusion that THz imaging technology will have nice application foreground.

Key words: THz, THz imaging, Nondestructive inspection, Spatial resolution

9. Review of explosive detection using terahertz spectroscopy technique

Wang Gao, Xu Degang, Yao Jianquan

2011 International Conference on Electronics and Optoelectronics (ICEOE 2011), 2011, V4: 22–25.

Abstract: With the rapid development of the terahertz spectroscopy technique, it has been widely used in the field of safety inspection, aerospace, life science and chemistry. Most explosives and related compounds have characteristic absorption and many nonmetal and nonpolarity materials are transparent to terahertz wave, so it has shown significant potential for the safety inspection. This article firstly introduces the characteristics of terahertz and the progress of THz spectroscopy technique. Then the authors summarize the abroad research status of explosives detection using terahertz spectroscopy technique, and also sum up the research of our country. Finally this article outlooks the future development trend.

Key words: Terahertz, Characteristic absorption spectrum, Explosive detection

10. 粗糙表面对雷达目标散射截面的影响

杨洋，姚建铨，唐世星，李忠洋，邴丕彬

激光与红外，2011, 41(7): 800–803.

摘 要：粗糙表面对雷达目标信号的散射在红外、微波以及太赫兹波领域是普遍存在的。研究结果表明，在红外领域目标表面粗糙的状况是影响激光雷达散射截面的重要因素，在微波领域这种影响是作为一种随机过程，是一种微扰，雷达散射截面的回波功率是粗糙表面高度散布密度分布函数与平坦表面脉冲响应函数的卷积，通常情况下卷积的效果在数学上相当于作用在平坦表面脉冲响应函数上一个δ函数。

关键词：RCS，LRCS，太赫兹雷达散射截面，雷达方程，粗糙表面，卷积

11. 球型目标在不同波段的雷达散射截面

杨洋,姚建铨,宋玉坤,邴丕彬,李忠洋

激光与红外, 2011, 41(5): 552–556.

摘要：文中对球型目标在微波、红外、太赫兹等不同波段的雷达散射截面进行深入探讨,利用电磁波理论和红外辐射理论得到了理想金属球的微波雷达和朗伯球的激光雷达的散射截面的数学表达式,并在此基础上给出了球型目标太赫兹雷达散射截面的具体研究内容和研究方式,指出选用理想金属朗伯球体的目标作为太赫兹雷达散射截面的标准体,提出了"中值加权修正"的研究方法,并对方法的具体实施方案给予了阐述。

关键词：微波雷达,激光雷达,太赫兹雷达,雷达散射截面,球型目标

12. 太赫兹成像技术在无损检测中的实验研究

邸志刚,姚建铨,贾春荣,邴丕彬,杨鹏飞,徐小燕

激光与红外, 2011, 41(10): 1163–1166.

摘要：太赫兹成像是一种新兴技术,为了将其应用在无损检测领域,提出了一种基于连续扫描的新型太赫兹成像系统。通过对隐藏的硬币和带小孔的金属板等样品进行检测,得到了清晰的图像。实验结果表明通过本成像系统可以对隐藏的金属样品进行鉴别,其空间分辨率低于 0.5 mm。因而本系统可以成功地应用于无损检测。

关键词：激光光学,太赫兹,太赫兹成像,无损检测,空间分辨率

13. 太赫兹光谱技术在气体检测中的应用

吴亮,凌福日,刘劲松,姚建铨

激光与光电子学进展,2009, 46(7): 29-35.

摘要:相关环境污染信息在太赫兹波段内具有较强的吸收特性,使得利用太赫兹光谱技术探测大气中的污染物成为可能。概述了国内外太赫兹时域光谱技术在气体检测方面的研究历史和发展现状,介绍了不同的气体分子在太赫兹波段的吸收谱,以及对混合气体分子和同素异形体分子的识别。对太赫兹时域光谱技术在气体检测方面的研究方向进行了展望。

关键词:太赫兹光谱技术,大气污染,时域光谱系统,转动能级跃迁,吸收谱

14. 爆炸物太赫兹光谱探测技术研究进展

王高,周汉昌,姚宝岱,徐德刚,姚建铨

激光与光电子学进展,2011, 48(1): 013001.

摘要:近年来,随着太赫兹(THz)技术的快速发展,它在安检、航空航天、生命科学、化学等领域展现了巨大的应用前景。在安检领域,由于许多炸药及其相关材料在太赫兹波段具有特征吸收谱,许多非金属、非极性材料对太赫兹波是透明的,因此爆炸物太赫兹光谱探测技术具有巨大的潜力,受到了国内外的高度重视。介绍了国内外利用太赫兹波进行爆炸物光谱探测技术的研究现状及太赫兹光谱探测技术的新进展,综述了在固态爆炸物和气态爆炸物的特征吸收光谱研究方向所取得的成果。最后总结了目前存在的技术难题,展望了未来的发展趋势。

关键词:太赫兹,特征吸收光谱,爆炸物检测

15. 太赫兹波段纳米颗粒表面增强拉曼散射的研究

吴玉登,任广军,郝芸,姚建铨

光谱学与光谱分析, 2013, 33(5): 1230–1233.

摘要：研究了太赫兹波段纳米颗粒的表面增强拉曼散射,证明在太赫兹波段同样存在拉曼增强现象。通过研究表面增强拉曼散射的电磁增强原理,提出利用时域有限差分法仿真模拟纳米颗粒在太赫兹波照射下的表面增强拉曼散射,分析了太赫兹波的增强效果。仿真实验表明,时域有限差分法可以有效精确仿真太赫兹波段纳米颗粒的散射效果,结果使表面增强拉曼散射从可见光和红外波段扩展到太赫兹波段,为太赫兹波与表面增强拉曼散射的结合应用提供了依据。

关键词：太赫兹,表面增强拉曼散射,时域有限差分法,纳米颗粒

16. THz 光谱技术检测 DNAN 炸药含量的研究

王高,徐德刚,姚建铨

光谱学与光谱分析, 2013, 33(4): 886–889.

摘要：为了对一种新型低感熔铸炸药(DNAN)含量进行检测,设计了 THz 光谱相干探测系统,检测确定了 DNAN 的 THz 特征波长位置,又利用比尔朗伯定律求解了 DNAN 的含量。设计了由主控系统控制步进电机实现光电探测器微扫描的时间相干 THz 光谱检测系统,计算推导了 DNAN 含量求解需要的系统参数,获得了 DNAN 的 THz 特征光谱。实验采用三种方法分别对不同 DNAN 含量的爆炸物样品进行检测,结果显示,该系统精度接近目前广泛应用的 THz 光谱检测设备 MINI-Z 太赫兹光谱仪。在此基础上,设计了特征吸收峰优化算法,通过 Origin 软件仿真分析可知,该算法可以进一步提高系统的探测精度及稳定性。

关键词：太赫兹,炸药检测,特征吸收峰,2,4- 二硝基苯甲醚

17. 粗糙铜表面对低频太赫兹波的散射实验

杨洋，姚建铨，张镜水，王力

红外与毫米波学报，2013, 32(1): 36–39.

摘要：搭建了以 0.2 THz 返波管振荡器源、热释电探测器、小型自动旋转光学平台等组成的太赫兹波目标散射特性实验测试系统，对两种不同粗糙度铜盘表面的散射特性进行了测试，表明：太赫兹金属粗糙目标散射中导体表面的感应电流产生电磁散射和粗糙导体表面引起的朗伯散射是同时存在的；在斜入射时这种近似于朗伯体的金属粗糙表面几乎可以被看成镜体，随着目标表面的粗糙度变大，反射变弱，散射增强，主峰向小于反射角的方向偏移；在垂直入射的情况下，散射角小于 40 度时散射曲线下降较快，超过 40 度散射曲线变化变得很缓慢，但在 50 度附近很多材料都会出现一个小的散射峰。

关键词：太赫兹，金属粗糙表面，目标散射特性，实验测试系统

18. Numerical simulation of the thermal response of continuous-wave terahertz irradiated skin

Xu Degang, Liu Changming, Wang Yuye, Wang Weipeng, Jiang Hao, Zhang Zhuo, Liu Pengxiang, Yao Jianquan

Optoelectronics Letters, 2013, 9(1): 73–76.

Abstract: We report a two-layer model to describe the thermal response of continuous-wave (CW) terahertz (THz) irradiated skin. Based on the Pennes bio-heat conduction equation, the finite element method (FEM) is utilized to calculate the temperature distribution. The THz wave with a Gaussian beam profile is used to simulate the photo-thermal mechanism. The simulation results show the dynamic process of temperature increasing with irradiation time and possible thermal damage. The factors which can affect temperature distribution, such as beam radius, incident power and THz frequency, are investigated. With a beam radius of 0.5 mm, the highest temperature increase is 3.7 K/mW.

Key words: Two-layer model, Thermal response, Finite element method, Gaussian beam

19. 太赫兹空间应用研究与展望

姚建铨,钟凯,徐德刚

空间电子技术, 2013, 10(2): 1-16.

摘要:近年来,我国在空间技术领域取得了举世瞩目的成就,多项技术已经达到国际先进甚至领先水平。要保持我国在空间领域的优势地位,必须加强适用于空间技术及应用的新方法、新手段。太赫兹(THz)技术为空间技术的发展提供了新的途径,因此对THz波空间应用技术进行研究非常必要。文章介绍了THz波的性质及特点,综述了THz技术在雷达、通信、遥感及航天工业等方面的应用,并对THz波在空间应用领域进行了展望,希望能够促进THz技术在该领域的研究及应用技术的进步。

关键词:太赫兹(THz),空间应用,雷达,通信,遥感,成像,无损检测

20. 太赫兹波段超材料在生物传感器的应用研究进展

闫昕,张兴坊,梁兰菊,姚建铨

光谱学与光谱分析, 2014, 34(9): 2365-2371.

摘要:超材料对电磁场的局域增强以及对周围环境的介电性质敏感等特性,可用于无标记生物检测,因而越来越受到国内外的学术关注,特别是太赫兹波段的超材料生物传感器。总结了近年来太赫兹波段超材料在生物传感器方面取得的进展。首先介绍了超材料生物传感器的基本原理,接着分析和讨论了衬底材料和厚度的选择、超材料结构对传感器灵敏度的影响。分析表明,通过优化结构、采用低介电常数和损耗低的薄衬底,能进一步提高生物传感器的灵敏度,并且多种物质在太赫兹波段都有直接的电磁响应特征,因此利用太赫兹波段超材料实现无标记生物检测具有很大的应用潜力。最后初步探讨了该生物传感器的发展趋势与前景。

关键词:传感器,太赫兹,时域光谱,超材料,生物,进展

21. 太赫兹雷达散射截面测量中定标体的确定

杨洋,姚建铨,钟凯

激光与红外, 2014, 44(10): 1149–1153.

摘要：围绕赫兹雷达散射截面定标体选定的内容开展了一系列工作,确定了适合作为太赫兹雷达散射截面标准体的工艺要求和加工方式,并先后对6种通过不同工艺加工成的太赫兹雷达散射截面的标准体材料进行了测试,分别测出半球反射率随波长的变化关系,确定了适合作为太赫兹雷达散射截面标准体的工艺要求和加工方式,加工出符合条件的太赫兹雷达散射截面测量中用作定标金属铝球,并利用该定标体对其他目标体在低频太赫兹波段的雷达散射截面进行了初步测量。

关键词：太赫兹雷达散射截面,金属球,收发同置,半球反射率

22. THz 波连续波透射式逐点扫描快速成像实验研究

李忠孝,徐德刚,王与烨,李佳起,闫超,刘鹏翔,姚建铨

光电子·激光, 2015, 26(1): 177–183.

摘要：利用 CO_2 激光抽运太赫兹(THz)激光器,搭建了 THz 连续波透射式快速扫描成像系统。将样品置于平移台上,利用透镜将 THz 光束聚焦于样品表面,通过计算机控制平台连续移动与信号实时采集,大大缩短了成像时间,通过计算机重构获得扫描图像。从扫描速度、成像分辨率的角度出发,对连续 THz 扫描成像系统进行优化研究。实验对 1 cm × 1 cm 的样品,用 0.25 mm 步长进行扫描成像,得到最短成像时间仅需 3 min,在输出频率为 4.3 THz 时,图像最高分辨率达到 0.1 mm。对刀片与树叶进行扫描成像,分析并比较了衍射极限对成像质量的影响。结果表明,在选用高频率输出光源和扫描平台小步长移动情况下可以获得更好的成像质量。

关键词：成像系统,太赫兹(THz)成像,透射,快速扫描

23. 低频太赫兹标准目标雷达散射截面的实验研究

杨洋, 姚建铨, 王力, 张镜水, 刘婧

红外与激光工程, 2015, 44(3): 985−989.

摘 要: 以研究太赫兹雷达散射截面的特性为目的, 选用所搭建低频太赫兹雷达测试系统, 并借助于标准目标法开展了有关太赫兹雷达粗糙铝盘散射截面的实验研究工作。实验结果表明: 在小角度散射中太赫兹雷达散射截面随散射角的增大变化比较明显, 在散射角超过 5° 后太赫兹雷达散射截面随散射角的变化趋向缓慢, 但当散射角超过 12° 后探测信号的强度已衰减到无法测量, 在太赫兹雷达散射截面的测试中没有出现微波雷达散射截面的大小随散射角的变化而剧烈振荡的现象; 将测试结果与同尺寸微波、激光雷达散射截面的结果进行了对比, 得到结论: 在 0° 附近太赫兹雷达散射截面的数值比同尺寸微波雷达散射截面的数值要小两个数量级, 但比同尺寸激光雷达散射截面的数值要高一个数量级。

关键词: 太赫兹雷达散射截面, 标准目标, 收发同置, 目标散射

24. 基于相干层析的太赫兹成像技术研究

黄亚雄, 姚建铨, 凌福日, 李丹

激光与红外, 2015, 45(10): 1261−1265.

摘 要: 介绍了一种结合光学相干层析技术和太赫兹技术的太赫兹三维成像技术——太赫兹相干层析成像技术。该技术利用宽频太赫兹的弱相干原理, 可以实现对待测样品进行高精度的三维成像。实验结果表明太赫兹相干层析成像技术的纵向分辨能力高于 100 μm。在纵向探测精度方面, 该技术相对传统的方案有了较大的提高, 在高精度太赫兹无损探测领域具有巨大的应用前景。

关键词: 太赫兹波, 相干层析成像, OCT 技术, 纵向分辨能力

25. 超高速太赫兹通信系统中调制方式的探讨

孙禹，詹亚锋，陆宇颖，姚建铨

现代电子技术，2015, 38(9): 1–8.

摘 要：针对传输速率为 10 Gb/s 太赫兹通信系统，给出了调制方式优选方案。首先详细介绍毫米波、自由空间激光通信中常用调制方式的特性，包括功率有效性、频带利用率、实现复杂度和峰均比；接着介绍太赫兹通信目前采用的调制方式，然后重点讨论在具体实现时，相位噪声、模/数转换器采样率、功率放大器非线性对调制方式选取的影响；最后在链路预算基础上结合具体器件参数，考虑相位噪声和功放非线性因素，对优选调制方式误码率性能进行仿真。

关键词：太赫兹通信，调制方式，链路预算，相位噪声，非线性功率放大器

26. Single pixel imaging with tunable terahertz parametric oscillator

Pan Duan, Yuye Wang, Degang Xu, Chao Yan, Zhen Yang, Wentao Xu, Wei Shi, Jianquan Yao

Applied Optics, 2016, 55(13): 3670–3675.

Abstract: A method of active terahertz imaging based on compressive sampling is demonstrated. A metal mask structure is designed with all modulation matrices engraved on. The imaging approach based on the mask eliminates the need for imaging object movement in point-wise scanning and shows high sensitivity. A terahertz parametric oscillator with tunability from 0.5 to 2.7 THz was used as the light source. Holes with circular, rectangular, and letter "H" shapes were imaged at 1.75 THz at 20% sampling rate. The influence of sampling rates and averaging times on the image was analyzed. Imaging of the letter "H" at different frequencies from 1.0 to 2.2 THz was tested and evaluated, and recognizable results were obtained in the range of 1.4–2.0 THz.

Key words: Terahertz imaging, Spatial light modulators, Image reconstruction techniques, Parametric oscillators and amplifiers

27. Characterizing the oil and water distribution in low permeability core by reconstruction of terahertz images

Rima Bao, Xinyang Miao, Chengjing Feng, Yingzhi Zhang, Honglei Zhan, Kun Zhao, Maorong Wang, Jianquan Yao

Science China: Physics, Mechanics & Astronomy, 2016, 59(6): 664201.

Abstract: Exploitation of low permeability oilfield has become an interest of the world due to the decline of conventional oil reserves. Flooding has been observed to be able to enhance oil recovery in tight sandstone reservoir. The pore throat diameter of the sandstone described above is in nano/micro scale, and remaining oil and irreducible water exist simultaneously. Therefore, better understanding of reservoir behavior is essential to predict future performance of water flooding process. Terahertz (THz) radiation (wavelengths from 30 μm to 3 mm) has recently emerged as an efficient tool in many fields such as oil and gas characterization, cultural heritage investigation, air pollution detection and design of metamaterials. Applications of THz techniques are promising such as time domain spectroscopy (TDS) in solid-state physics and aqueous chemistry. Response of THz wave is intrinsically associated with low energy events such as molecular torsion or vibration as well as inter and intramolecular hydrogen-bonding, causing the strong attenuation of THz waves by water. Therefore, THz spectroscopy was considered as an effective means for detection of water content and distribution. In this letter, we investigated the distribution of oil and water in low permeability core via reconstruction of THz images.

Key words: Oil recovery, Nano/micro scale, Low permeability, THz images

28. 太赫兹波近场成像综述

刘宏翔,姚建铨,王与烨,徐德刚,贺奕焱

红外与毫米波学报, 2016, 35(3): 300-309.

摘要： 太赫兹波成像作为可见光和微波成像等的拓展,在半导体材料表征、生物组织诊断、无损检测和安检等领域表现出许多独特的优点,得到了越来越广泛的关注。传统太赫兹波成像受长波长对应的衍射极限的限制,分辨率较低。而太赫兹波近场成像是目前突破该限制,获得亚微米甚至是纳米量级高分辨率图像的研究热点之一。首先介绍了近场机制与成像的基本原理；其次总结了太赫兹波近场成像的几种常用方法及其对应研究进展和当前存在的问题,包括孔径型、针尖型、亚波长太赫兹源型和微纳结构调控型等；最后探讨了该方向的发展趋势。

关键词： 太赫兹波,近场成像,衍射极限,隐失波,微纳结构

29. Attenuated total internal reflection imaging with continuous terahertz wave

Hongxiang Liu, Yuye Wang, Degang Xu, Pan Duan, Meitong Nie, Jia Shi, Chao Yan, Yixin He, Tunan Chen, Hua Feng, Jianquan Yao

International Conference on Infrared, Millimeter, and Terahertz Waves (IRMMW-THz), 2016.

Abstract: An attenuated total internal reflection imaging system was demonstrated with continuous terahertz wave for the first time. A drop of water was imaged to validate its efficacy. The approach enables qualitative or even quantitative studies of the surface layer characteristics. The work paves the way for fast terahertz imaging in an in-vivo biomedical diagnosis without complicated sample preparations and specialized data processing.

Key words: Imaging system, Continuous terahertz wave, Surface layer characteristics, Biomedical diagnosis

30. Biomedical diagnosis of cerebral ischemia with continuous-wave THz imaging

Jia Shi, Yuye Wang, Degang Xu, Chao Yan, Longhuang Tang, Pan Duan, Yixin He, Hongxiang Liu, Tunan Chen, Hua Feng, Jianquan Yao

41st *International Conference on Infrared, Millimeter, and Terahertz Waves (IRMMW-THz)*, 2016.

Abstract: The fast label-free detections of the extent and degree of cerebral ischemia have been the difficulties and hotspots for precise and accurate neurosurgery. We experimentally demonstrated the terahertz (THz) transmission imaging of the fresh cerebral tissues at different ischemic stages within 24 hours. With a CO_2 laser pumped continuous-wave THz imaging system operating at 2.52 THz, THz imaging could distinguish the ischemic cerebral tissues from the normal cerebral tissues. It was indicated that the transmittance of the THz wave for ischemic cerebral tissues was lower than that for normal cerebral tissues. As the ischemic time increased, the ischemic areas became larger and diffused. Compared to the traditional triphenyltetrazolium chloride-stained (TTC-stained) images, the ischemic tissues can been detected using THz wave with a high sensitivity as early as the ischemic time of 2 hours. Thus, THz imaging has a great potential in recognition of cerebral ischemia and it may become a new method for intraoperative real-time guidance, recognition in situ, and precise excision.

Key words: Cerebral ischemia, Terahertz transmission imaging, High sensitivity, Recognition

31. Fast terahertz imaging based on compressive sensing

Yuye Wang, Pan Duan, Degang Xu, Jia Shi, Longhuang Tang, Yixin He, Hongxiang Liu, Meitong Nie, Jianquan Yao

The 8th International Symposium on Ultrafast Phenomena and Terahertz Waves, 2016.IW4B.5.

Abstract: We demonstrate an active terahertz imaging method based on compressive sensing. With all modulation matrices engraved on one mask, the fast imaging has achieved. Recognizable result of letter "H" has obtained with PSNR of 10.21.

Key words: Terahertz imaging, Spatial light modulators, Image reconstruction techniques

32. 基于太赫兹参量振荡器的太赫兹压缩感知成像研究

段攀,王与烨,徐德刚,闫超,徐文韬,杨振,姚建铨

第二届全国太赫兹科学技术学术年会, 2016: 213.

摘要：介绍了一种以太赫兹参量振荡器为成像辐射源的太赫兹压缩感知成像方法。太赫兹因其具有低能性、指纹特性和对极性物质的敏感性等特性,使得太赫兹成像技术在安检、质检以及医学领域具有广阔前景并取得了快速发展。常用的太赫兹成像方式为扫描成像和阵列式成像,扫描成像速度慢,阵列成像成本高且需要配备高功率辐射源。压缩感知技术作为一种新型的信号采集方法能够在信号采集的同时实现信号数据的有效压缩,能够大幅减小数据的采集量,缩短成像时间,同时实现较高灵敏度的探测。本方法以基于铌酸锂晶体的太赫兹参量振荡器为主动成像辐射源,其具有覆盖频带宽、室温运转、连续可调谐、易于实现小型化等优点。采用太赫兹垂直出射腔型结构,实现了 0.5-2.7 THz 频率范围的太赫兹辐射输出,当输入 1 064 nm 泵浦光能量为 120 mJ/pulse 时,在 1.7 THz 处得到最大为 780 n/pulse 的太赫兹辐射输出。用部分镂空的金属片实现对 10 mm × 10 mm 区域的太赫兹辐射的空间强度调制,镂空部分最小单元为 0.5 mm × 0.5 mm,所以每个用于辐射调制的掩模板单元为由镂空 (1) 和非镂空 (0) 部分组成的 20 × 20 的矩阵。为了实现掩模板单元的快速更换并减少掩模板制作的复杂程度,相邻掩模单元共用了其掩模矩阵中的 19/20, 80 个 20 × 20 的掩模矩阵压缩为 1 个 20 × 99 的掩模矩阵。金属片制作的掩模板在太赫兹辐射传播方向的垂直方向上移动,实现对太赫兹辐射的不同调制。收集不同掩模板单元对应得到的不同的太赫兹辐射强度,由 TVAL3 算法可以得到对应的恢复图像。在太赫兹输出为 1.7 THz 处,分别对直径为 2 mm 的圆孔、3 mm × 3.5 mm 的矩形、8mm × 10mm 的字母 "H" 进行了成像测试,均得到了辨识度较高的恢复图像。采集一个 20 × 20 图像数据所需时间为 8 秒,TVAL3 算法恢复出该图像的时间约为 0.25 秒。为了减小太赫兹辐射源的不稳定性对成像结果的影响,每次数据采集取 20 个脉冲平均。在此背景下对字母 "H" 在 1-22 THz 范围内做了成像测试,在 1.4-2.0 THz 频率范围内,信噪比较高,恢复出图像效果均可辨认。该方法适用于宽带的太赫兹波成像。

关键词：太赫兹成像,压缩感知,太赫兹参量振荡器

33. Terahertz imaging based on morphological reconstruction

Jia Shi, Yuye Wang, Degang Xu, Chao Yan, Tunan Chen, Yixin He, Longhuang Tang, Meitong Nie, Pan Duan, Dexian Yan, Hua Feng, Jianquan Yao

IEEE Journal of Selected Topics in Quantum Electronics, 2017, 23(4): 6800107.

Abstract: Terahertz (THz) imaging technology is a developing and promising candidate for biological diagnosis, security inspection, and semiconductor wafer examination, due to the low photon energy, the high transparency, and the fingerprint properties of the THz radiation. However, a major encountered bottleneck is the degradation of image quality caused by the power fluctuation of THz source, interference phenomenon, complex environment, and so on. In this paper, we present the mathematical morphology for THz imaging to improve the image quality, taking advantage of morphological reconstructions. Based on the original THz image of a paper with some letters taken from our continuous THz imaging system, the visibility of objects has been successfully enhanced with the suppression of complex background and improvement of peak signal-to-noise ratio using morphological reconstruction. Moreover, morphological reconstruction with proper structuring element parameter was then performed to a THz image of fresh rat cerebral tissue. It presents a satisfactory result with clearer edges and suppressions of the interference fringes and noises. It is suggested that THz imaging based on morphological reconstruction opens a pathway towards automatic techniques for denoising, recognitions, and segmentations in THz biomedical imaging.

Key words: Terahertz imaging, Image reconstruction techniques, Morphological transformations

34. High-sensitivity attenuated total internal reflection continuous-wave terahertz imaging

Hongxiang Liu, Yuye Wang, Degang Xu, Limin Wu, Chao Yan, Dexian Yan, Longhuang Tang, Yixin He, Hua Feng, Jianquan Yao

Journal of Physics D: Applied Physics, 2017, 50(37): 375103.

Abstract: We demonstrate an attenuated total internal reflection imaging system. The surface information of the sample on top of a prism can be acquired by two-dimensionally scanning this prism moving in the vertical plane with horizontally incident continuous terahertz waves at a fixed height. The principles and feasibility of this method are investigated. The effective imaging area on the prism, image resolution and polarization dependence of contrast enhancement and stability improvement are analyzed. Examples including solid agar, distilled water and porcine tissue are presented, demonstrating the method's advantages of high sensitivity and simple sample preparation. The experimental and theoretical results consistently show that p-polarization contributes to enhanced image contrast and more stable intensity of the attenuated total internal reflected signal.

Key words: Terahertz imaging, Total internal reflection, THz, ATR

35. Label-free and reagentless bacterial detection and assessment by continuous-wave terahertz imaging

Xiang Yang, Jia Shi, Ke Yang, Degang Xu, Jianquan Yao, Yuye Wang, Weiling Fu

International Conference on Infrared, Millimeter, and Terahertz Waves (IRMMW-THz), 2017.

Abstract: Rapid and accurate bacterial detection with minimal operation step is critical for food safety control and infectious diseases diagnostics. In this report, we demonstrate a new strategy of bacterial colonies sensing by THz imaging which do not need sample processing. A Continuous-wave THz imaging system in transmission mode was used to detect single bacterial colonies after culture directly. Four species of bacteria could be detected as different THz images; thus the mixed samples could be clearly recognized. In addition, living and dead bacteria (thermally treated) can also be display as different images due to their different hydration levels and structural changes. Our results clearly demonstrated the potential of THz imaging for bacterial detection and viability assessment in a label-free manner.

Key words: Bacterial detection, THz imaging, Hydration levels, Viability assessment

36. Label-free bacterial colony detection and viability assessment by continuous-wave terahertz transmission imaging

Xiang Yang, Jia Shi, Yuye Wang, Ke Yang, Xiang Zhao, Guiyu Wang, Degang Xu, Yunxia Wang, Jianquan Yao, Weiling Fu

Journal of biophotonics, 2018, 11(8): e201700386.

Abstract: Timely and accurate bacterial detection is critical for various health and safety applications, which promotes the continuous development of versatile optical sensors for bacterial investigations. Here, we report a new strategy for bacterial colony sensing using terahertz (THz) imaging with minimal assay procedures. The proposed method utilizes the acute sensitivity of THz wave to the changes in the water content and cellular structures. Single bacterial colonies of 4 bacterial species were directly distinguished using THz imaging by utilizing their differences in THz absorption. In addition, the distribution of mixed bacterial samples has been demonstrated by THz imaging, which demonstrated that the target bacterium could be easily recognized. Furthermore, we investigated the differentiation of bacterial viability, which indicated that bacteria under different living states could be distinguished by THz imaging because of their different hydration levels and cellular structures. Our results suggest that THz imaging has the potential to be used for mixed bacterial sample detection and bacterial viability assessment in a label-free and nondestructive manner.

Key words Bacterial detection, Biosensor, Terahertz imaging, THz absorption

37. Optimization for vertically scanning terahertz attenuated total reflection imaging

Hongxiang Liu, Yuye Wang, Degang Xu, Zhinan Jiang, Jining Li, Limin Wu, Chao Yan, Longhuang Tang, Yixin He, Dexian Yan, Xin Ding, Hua Feng, Jianquan Yao

Optics express, 2018, 26(16): 20744−20757.

Abstract: Terahertz attenuated total reflection imaging has been used to develop preliminary applications without any in-depth analysis of the nature of present systems. Based on our proposed vertically scanning imaging system, an analysis of optimum prism design and polarization selection for error reduction is presented theoretically and experimentally, showing good agreement. By taking the secondary reflection inside the prism and the prism deflection into consideration, p-polarized terahertz waves are recommended for prisms with a base angle below 31°, leading to minimum error. This work will contribute to the development of more practical application of terahertz attenuated total reflection scanning imaging in various fields with enhanced performance.

Key words: Total reflection imaging, Scanning imaging, Secondary reflection, Deflection

38. Automatic evaluation of traumatic brain injury based on terahertz imaging with machine learning

Jia Shi, Yuye Wang, Tunan Chen, Degang Xu, Hengli Zhao, Linyu Chen, Chao Yan, Longhuang Tang, Yixin He, Hua Feng, Jianquan Yao

Optics express, 2018, 26(5): 6371–6381.

Abstract: The imaging diagnosis and prognostication of different degrees of traumatic brain injury (TBI) is very important for early care and clinical treatment. Especially, the exact recognition of mild TBI is the bottleneck for current label-free imaging technologies in neurosurgery. Here, we report an automatic evaluation method for TBI recognition with terahertz (THz) continuous-wave (CW) transmission imaging based on machine learning (ML). We propose a new feature extraction method for biological THz images combined with the transmittance distribution features in spatial domain and statistical distribution features in normalized gray histogram. Based on the extracted feature database, ML algorithms are performed for the classification of different degrees of TBI by feature selection and parameter optimization. The highest classification accuracy is up to 87.5%. The area under the curve (AUC) scores of the receiver operating characteristics (ROC) curve are all higher than 0.9, which shows this evaluation method has a good generalization ability. Furthermore, the excellent performance of the proposed system in the recognition of mild TBI is analyzed by different methodological parameters and diagnostic criteria. The system can be extensible to various diseases and will be a powerful tool in automatic biomedical diagnostics.

Key words: Terahertz imaging, Medical, Biological imaging, Machine learning

39. Terahertz computed tomography of high-refractive-index objects based on refractive index matching

Linyu Chen, Yuye Wang, Degang Xu, Yuchen Ren, Yixin He, Changzhao Li, Chao Zhang, Longhuang Tang, Chao Yan, Jianquan Yao

IEEE Photonics Journal, 2018, 10(6): 5900813.

Abstract: Terahertz (THz) computed tomography (CT) is a promising technique to obtain the internal information of objects due to its properties of nonionizing and high penetration of nonpolar objects. However, THz CT is hindered by the strong refractive effect and Fresnel reflection loss when it is applied to high-refractive-index samples. In this paper, the three-dimensional image of the sample is performed in the low absorption liquid, the refractive index of which is close to the sample. This novel experimental procedure can eliminate the refractive effect and Fresnel reflection loss effectively. Together with the data processing to eliminate the liquid absorption and to restrain the artifacts, the THz CT imaging of high-density polyethylene (n = 1.53) cylinders and cubes is achieved successfully. The internal hole and metallic foreign body are shown accurately. In addition, the defects with different sizes inside the sample are measured with refractive index matching method. It shows that THz CT imaging based on refractive index matching method has high sensitivity to detect the small defect inside the sample as small as 0.5 mm diameter.

Key words Terahertz, Computed tomography, High-refractive-index

40. Interference elimination in terahertz imaging based on inverse image processing

Yuye Wang, Linyu Chen, Tunan Chen, Degang Xu, Jia Shi, Yuchen Ren, Changzhao Li, Chao Zhang, Hongxiang Liu, Limin Wu, Hengli Zhao, Hua Feng, Jianquan Yao

Journal of Physics D: Applied Physics, 2018, 51(32): 325101.

Abstract: We propose a novel image restoration approach to eliminate the interference fringes in terahertz (THz) imaging. By making use of the information in the background of a THz image, inverse image processing is introduced under a priori knowledge about the degradation mechanism to suppress the interference. Based on a simulated interference image and the original THz image of a piece of paper coated with graphite powder, taken from our continuous THz imaging system, the interference fringes are successfully eliminated. The enhancement of the peak signal-to-noise ratio and structural similarity index is achieved, compared with the frequency domain filtering. Moreover, the inverse image processing is performed on a THz image of fresh rat brain tissue. It presents an excellent result with accurate information for the brain tissue after interference elimination.

Key words: Terahertz imaging, Interference elimination, Application in biomedicine

41. High-sensitivity terahertz imaging of traumatic brain injury in a rat model

Hengli Zhao, Yuye Wang, Linyu Chen, Jia Shi, Kang Ma, Longhuang Tang, Degang Xu, Jianquan Yao, Hua Feng, Tunan Chen

Journal of biomedical optics, 2018, 23(3): 036015.

Abstract: We demonstrated that different degrees of experimental traumatic brain injury (TBI) can be differentiated clearly in fresh slices of rat brain tissues using transmission-type terahertz (THz) imaging system. The high absorption region in THz images corresponded well with the injured area in visible images and magnetic resonance imaging results. The THz image and absorption characteristics of dehydrated paraffin-embedded brain slices and the hematoxylin and eosin (H&E)-stained microscopic images were investigated to account for the intrinsic differences in the THz images for the brain tissues suffered from different degrees of TBI and normal tissue aside from water. The THz absorption coefficients of rat brain tissues showed an increase in the aggravation of brain damage, particularly in the high-frequency range, whereas the cell density decreased as the order of mild, moderate, and severe TBI tissues compared with the normal tissue. Our results indicated that the different degrees of TBI were distinguishable owing to the different water contents and probable hematoma components distribution rather than intrinsic cell intensity. These promising results suggest that THz imaging has great potential as an alternative method for the fast diagnosis of TBI.

Key words: Traumatic brain injury, Terahertz imaging, Terahertz spectroscopy

42. Terahertz two-pixel imaging based on complementary compressive sensing

Yuye Wang, Yuchen Ren, Linyu Chen, Ci Song, Changzhao Li, Chao Zhang, Degang Xu, Jianquan Yao

Chinese Physics B, 2018, 27(11): 114204.

Abstract: A compact terahertz (THz) imaging system based on complementary compressive sensing has been proposed using two single-pixel detectors. By using a mechanical spatial light modulator, sampling in the transmission and reflection orientations was achieved simultaneously, which allows imaging with negative mask values. The improvement of THz image quality and anti-noise performance has been verified experimentally compared with the traditional reconstructed image, and is in good agreement with the numerical simulation. The demonstrated imaging system, with the advantages of high imaging quality and strong anti-noise property, opens up possibilities for new applications in the THz region.

Key words: Terahertz imaging, Image reconstruction techniques, Spatial light modulators

43. Broadband terahertz dielectric measurement based on multi-beam interference and Fourier transform infrared spectrometer

Jie Shi, Maorong Wang, Kai Zhong, Chu Liu, Jialin Mei, Degang Xu, Jianquan Yao

Modern Physics Letters B, 2018, 32(25): 1850298.

Abstract: We demonstrate a method for obtaining optical coefficients over a broad terahertz spectral range from 1.5 THz to 16 THz at room temperature. Based on the interferograms directly acquired by a Fourier transform infrared spectrometer (FTIR), multi-beam interference principle combining Fresnel's formula is employed to extract the refraction index and the extinction coefficient, giving the basis for calculating dielectric coefficients. It avoids the uncertainty and phase instability while using Kramers-Kronig (KK) relations and overcomes the limited frequency range of terahertz time-domain spectroscopy (TDS). Moreover, this method has better stability and is needless of cutting useful information between neighboring interference peaks for thin samples compared with TDS, making it a general processing method for interferograms and a good alternative for terahertz dielectric measurement.

Key words: Terahertz spectroscopy, Optical coefficients, Multi-beam interference, FTIR

44. Terahertz computed tomography of high-refractive-index object based on a novel experimental procedure

Linyu Chen, Yuye Wang, Degang Xu, Jia Shi, Yuchen Ren, Changzhao Li, Chao Zhang, Longhuang Tang, Yixin He, Jianquan Yao

Proceedings of SPIE, 2018, 10826: 108260Z.

Abstract: Recently, terahertz (THz) computed tomography (CT) has emerged as a possible effective technique for 3D structural information detection. However, THz-CT is difficult to be applied to high refractive index object, due to the severe refraction phenomenon occurred during the acquisition of raw data. We propose a novel experimental procedure to solve this problem. Including the use of a sink filled with liquid whose refractive index is close to the sample, and a correction algorithm to eliminate the noise of liquid. The proposed method is applied to the high-density polyethylene samples of different shapes.

Key words: Terahertz, Computed tomography, Three-dimensional imaging, High-refractive-index

45. Terahertz reflectometry imaging of traumatic brain injury

Limin Wu, Yuye Wang, Degang Xu, Hengli Zhao, Jia Shi, Jining Li, Tunan Chen, Hua Feng, Jianquan Yao

Proceedings of SPIE, 2018, 10816: 1081616.

Abstract: We set up the terahertz continuous reflectometry imaging system and the spatial resolution of our system was roughly 0.6 × 0.6mm at 2.52 THz. We also demonstrated that the paraffin embedded traumatic brain injury (TBI) in rat model sample can be differentiated clearly. The results show that the THz reflection intensity of the TBI area was lower than that of normal area. These promising results suggest that THz reflection imaging has great potential as an alternative method for the fast diagnosis of TBI.

Key words: Traumatic brain injury, Terahertz imaging, Terahertz spectroscopy, Reflection intensity, Paraffin embedded, THz, Spatial resolution, Continuous terahertz wave, Terahertz technique, Reflection intensity

46. 3.11 THz 标准体雷达散射截面测量

王茂榕, 钟凯, 刘楚, 徐德刚, 姚建铨

红外与激光工程, 2018, 47(2): 0225001.

摘要: 针对太赫兹近场散射特性测量特点,基于 CO_2 激光抽运的太赫兹激光器和双层独立转动平台搭建了一套高频段太赫兹雷达散射截面 (RCS) 测量系统。利用不锈钢光滑金属球体作为标准定标体验证了系统的可靠性,测量结果与理论值误差小于 3 dBsm,系统的信噪比优于 24 dB。首次利用该系统开展了 3.11 THz 频点处不同材料及涂覆层圆形金属平板及不同底面直径圆锥体 RCS 的测量。通过比较分析发现了表面阳极氧化和喷漆处理的航空铝及 P304 不锈钢与纯航空铝平板的 RCS 区别,以及不同底面直径的圆锥体 RCS 差异,为太赫兹频段复杂目标体 RCS 的研究奠定基础。

关键词: 太赫兹 (THz) 波, 雷达散射截面 (RCS), 标准体, 定标

47. 迈克尔逊干涉法精确测量太赫兹频谱及目标速度

刘楚, 钟凯, 史杰, 靳硕, 葛萌, 李吉宁, 徐德刚, 姚建铨

红外与激光工程, 2018, 47(11): 1117006.

摘要: 搭建了一套迈克尔逊干涉仪,对 CO_2 激光的 9P36 和 9R10 谱线泵浦 CH_3OH 气体所产生的频率分别为 2.52 THz 和 3.11 THz 的太赫兹激光器输出频谱进行了精细测量。测量系统频率分辨率约为 1 GHz,测量结果显示 CO_2 激光泵浦的太赫兹源为单色源并具有极窄的线宽,波长与激光器标称值进行对比具有很好的一致性。基于这套系统实现了对干涉仪动臂目标的运动速度准确测量,提出了两种分别适用于匀速运动和变速运动情况下的速度反演方法,反演结果与设定值均相符。结论表明,迈克尔逊干涉仪不但可以精确测量太赫兹波源的频谱,同时配合单色太赫兹源可以准确测量目标速度,为太赫兹波段光谱、成像等领域的应用奠定基础。

关键词: 太赫兹, 迈克尔逊干涉仪, 频谱, 速度

48. 基于衰减全反射式太赫兹时域光谱技术的食用油光谱特性研究

聂美彤，徐德刚，王与烨，唐隆煌，贺奕焮，刘宏翔，姚建铨

光谱学与光谱分析，2018, 38(7): 2016-2020.

摘要： 食用油是人类营养和能量的重要来源，为人体提供必需的脂肪酸，研究食用油在太赫兹波段光学特性，对食用油成分分析及品质评价具有重要价值。衰减全反射式太赫兹时域光谱技术是一种新型的太赫兹时域光谱技术，通过样品与倏逝波的相互作用，获取样品的太赫兹光谱。与透射式或反射式太赫兹时域光谱技术相比，该技术能有效地避免测量食用油等液体样品时样品池对光学参数的影响，并能获得样品的精确光学参数。分别利用透射式太赫兹时域光谱技术和衰减全反射式太赫兹时域光谱技术测量了大豆油的吸收光谱。结果表明，与透射式太赫兹时域光谱技术相比，衰减全反射式太赫兹时域光谱技术能更有效地提取大豆油的吸收系数、吸收峰分布等光学特性。进一步利用衰减全反射式太赫兹时域光谱技术研究了大豆油、核桃油、葡萄籽油在太赫兹波段的光学特性，获得了三种食用油在 1~1.8 THz 范围内的折射率谱和吸收光谱。利用密度泛函理论计算了食用油中四种主要成分（软脂酸、硬脂酸、油酸和亚油酸）在太赫兹波段的振动、转动模式，理论计算结果同实验测量结果吻合较好。研究表明，在太赫兹波段食用油的吸收峰与所含脂肪酸分子种类与含量有关，其主要来源为脂肪酸分子的低频振动和转动。研究成果对食用油成分定性定量分析及品质检测等具有指导意义。

关键词： 太赫兹光谱，食用油，密度泛函理论

49. The terahertz electromagnetically induced transparency-like metamaterials for sensitive biosensors in the detection of cancer cells

Xin Yan, Maosheng Yang, Zhang Zhang, Lanju Liang, Dequan Wei, Meng Wang, Mengjin Zhang, Tao Wang, Longhai Liu, Jianhua Xie, JianquanYao

Biosensors and Bioelectronics, 2019, 126: 485–492.

Abstract: A kind of novel biosensor based on the electromagnetic induced transparency like (EIT-like) metamaterials (MMs) have been proposed. It demonstrates that the symmetry-breaking double-splits ring resonators can realize the EIT-like plasmonic resonance, the according transparency window occurs at 1.67 THz. The coupled oscillators model illustrates that with the increase of asymmetry degree of double splits, the coupling between bright and dark mode is enhanced. Consequently, the non-radiative damping γ_2 grows from 1.45 to 1.85 THz and coupling coefficient κ from 3.46 to 4.49 THz^2, while the radiative damping γ_1 decreases from 11.5 to 9 THz. Such EIT-like MMs were evaluated in simulation as the refractive index sensors, which the theoretical sensitivity was calculated to 455.7 GHz/RIU (RIU, Refractive Index Unit) under 11 μm-thick analyte layer. Meanwhile, the dependence of full width at half maximum (FWHM) on analyte thickness was also studied. In experiments, it is found that the frequency shift Δf increases from 50 to 90 GHz when the oral cancer cells (HSC3) concentration improves from 1×10^5 to 7×10^5 cells/ml. The maximum experimental sensitivity approaches 900 kHz/cell ml^{-1} at 7×10^5 cells/ml. Additionally, the apoptosis of cancer cells under the effect of anti-cancer drug was investigated. It shows that with the increase of anti-cancer drug concentration from 1 to 15 μM and the extension of drug action duration from 24 to 72 h, the Δf changes from 140 to 70 GHz and 140–40 GHz, respectively. Besides, the corresponding FWHM also increases from 237.9 to 305.4 GHz and 237.8–337.6 GHz. The results measured by MMs biosensors and biological method exhibit a relatively good agreement, showing a great potential for cells measurement with the sensitive biosensors based on the EIT-like MMs.

Key words: Metamaterials, Electromagnetically induced transparency like, (EIT-like) Fano resonance, Biosensors, Terahertz

50. Multiple modes integrated biosensor based on higher order Fano metamaterials

Xin Yan, Zhang Zhang, Lanju Liang, Maosheng Yang, Dequan Wei, Xiaoxian Song, Haiting Zhang, Yuying Lu, Longhai Liu, Mengjin Zhang, Tao Wang, Jianquan Yao

Nanoscale, 2020, 12(3): 1719–1727.

Abstract: A multiple modes integrated biosensor based on higher order Fano metamaterials (FRMMs) is proposed. The frequency shift (Δf) of x-polarized quadrupolar (Q_x), octupolar (O_x), hexadecapolar (H_x), y-polarized quadrupolar (Q_y) and octupolar (Q_y) Fano resonance modes are integrated to detect the concentration of lung cancer cells. In experiments, the concentrations of lung cancer cells can be distinguished by the shape and distribution of integrated graphics. In addition, an anomalous response in Δf at resonant mode is surprisingly observed. As increasing the cell concentration, the Δf at the Q_x-dip, Q_y-dip and O_y-dip successively experience increasing frequency shift stage (IFSS), decreasing frequency shift stage (DFSS) and re-increasing frequency shift stage (RIFSS). The extraordinary DFSS confirmed by single-factor analysis of variance (ANOVA) means an unusual physical phenomenon in metamaterial biosensors. By introducing a new dielectric constant ε_f, we amend the perturbation theory to explain the unusual phenomenon in Δf. With the change of mode order from Q_x to H_x, the ε_f increases from -2.78 to 0.75, which implies that the negative ε_f leads to the appearance of DFSS. As a platform for biosensing, this study open a new window though multiple modes integrated perspective.

Key words: Biosensor, FRMMs, Lung cancer cells, ANOVA

51. Study of the dielectric characteristics of living glial-like cells using terahertz ATR spectroscopy

Yuye Wang, Zhinan Jiang, Degang Xu, Tunan Chen, Beike Chen, Shi Wang, Ning Mu, Hua Feng, Jianquan Yao
Biomedical optics express, 2019, 10(10): 5351–5361.

Abstract: The attenuated total reflection spectroscopy system with the Si container attached on the prism has been demonstrated as an efficient technique to obtain the dielectric properties of living cells in the THz range. We proposed a method to determine the dielectric responses of living cells based on the combination of the single-interface and two-interface ATR models without cell thickness. The experimental results for living glial-like cells (PC12, SVG P12 and HMO6) showed the dielectric responses in the THz region were related significantly to cell number, intracellular fluid, and cell structure. Moreover, the glioma cells (C6 and U87) exhibited different dielectric properties compared with the glial-like cells, which could be one reason for the glioma tissue diagnosis using THz wave.
Key words: Dielectric properties, Living cells, Terahertz ATR spectroscopy

52. Study of *in vivo* brain glioma in a mouse model using continuous-wave terahertz reflection imaging

Limin Wu, Degang Xu, Yuye Wang, Bin Liao, Zhinan Jiang, Lu Zhao, Zhongcheng Sun, Nan Wu, Tunan Chen, Hua Feng, Jianquan Yao
Biomedical optics express, 2019, 10(8): 3953–3962.

Abstract: We demonstrated that *in vivo* brain glioma in a mouse model using a continuous-wave terahertz reflection imaging system, as well as the *ex vivo* fresh brain tissues in mouse model. The tumor regions of *in vivo* and *ex vivo* brain tissues can be well distinguished by THz intensity imaging at the frequency of 2.52 THz. The THz images with high sensitivity correlated well with magnetic resonance, visual and hematoxylin and eosin stained images. Furthermore, the THz spectral difference between brain gliomas and normal brain tissues were obtained in the 0.6 THz to 2.8 THz range, where brain gliomas have the higher refractive indices and absorption coefficients, and their differences increase particularly in the high frequency range. These results suggest that THz imaging has great potential as an alternative method for the intraoperative label-free diagnosis of brain glioma *in vivo*.
Key words: Brain glioma, Terahertz imaging, Label-free diagnosis

53. Electromagnetically induced transparency-like metamaterials for detection of lung cancer cells

Maosheng Yang, Lanju Liang, Zhang Zhang, Yan Xin, Dequan Wei, Xiaoxian Song, Haiting Zhang, Yuying Lu, Meng Wang, Mengjin Zhang, Tao Wang, Jianquan Yao

Optics Express, 2019, 27(14): 19520–19529.

Abstract: A biosensor based on electromagnetically induced transparent (EIT) metamaterials (MMs) is proposed owing to the low loss and high Q-factor. The theoretical sensitivity of the biosensor based on EIT-like MMs were evaluated up to 248.8 GHz/RIU (RIU, Refractive Index Unit). In experiments, the cancer cells A549, as an analyte, are cultured on EIT-like MMs surface. The results show that when the cell concentration increases from 0.5×10^5 to 5×10^5 cells/ml, the frequency shift Δf could change from 24 to 50 GHz. Moreover, the coupled oscillators model is applied to explain the effect of the refractive index of analyte in simulations and the cell concentration in experiments on the EIT-like MMs. The fitting results exhibit that the refractive index of analyte and cell concentration significantly affect the radiative damping of the bright mode resonator γ_1. The proposed EIT-like MMs biosensors show great potentials for cell measurement because any change that results in the lineshape variation in EIT-like MMs can only be attributed to the change of external dielectric environment due to the suppression of radiative losses.

Key words: EIT-like MMs, Biosensor, Cell concentration, Cell measurement, Radiative losses

54. Feasibility of terahertz imaging for discrimination of human hepatocellular carcinoma

Feng Duan, Yuye Wang, Degang Xu, Jia Shi, Linyu Chen, Li Cui, Yanhua Bai, Yong Xu, Jing Yuan, Chao Chang

World journal of gastrointestinal oncology, 2019, 11(2): 153–160.

Abstract:

BACKGROUND

Hepatocellular carcinoma (HCC) is one of the most common malignant tumors worldwide, and novel methods for early/rapid diagnosis of HCC are needed. Terahertz (THz) spectroscopy is considered to have the potential to distinguish between normal liver tissue and HCC tissue; however, there are few reports on it. We conduct this observational study to explore the feasibility of THz imaging for the diagnosis of HCC.

AIM

To evaluate the feasibility of THz for discriminating between HCC and normal liver tissues using fresh tissue specimens obtained from HCC patients who had undergone surgery.

METHODS

Normal liver tissue and HCC tissue were cryosectioned into 50 μm-thick slices and placed on cover glass. Two adjacent tissue sections were separated subjected to histopathological examination by hematoxylin and eosin staining or THz transmission examination, and THz images were compared with pathologically mapped images. We determined the typical tumor and normal liver tissue regions by pathological examination; the corresponding areas of adjacent sections were examined by THz transmission.

RESULTS

The transmission rate of HCC tissue was 0.15-0.25, and the transmission rate of typical HCC tissue was about 0.2. THz transmittance in normal liver tissue is slightly higher than 0.4, but there were many influencing factors, including the degree of liver cirrhosis, fat components, ice crystals in frozen sections, and apoptosis.

CONCLUSION

In conclusion, this study shows that THz imaging can detect HCC tissue. Further research will yield more detailed data of the THz transmission rates of HCC tissue with different degrees of differentiation.

Key words: Terahertz imaging, Hepatocellular carcinoma, Hematoxylin and eosin staining, Terahertz transmittance, Liver pathology

55. Super-resolution reconstruction for terahertz imaging based on sub-pixel gradient field transform

Youdong Guo, Furi Ling, He Li, Siyan Zhou, Jie Ji, Jianquan Yao

Applied optics, 2019, 58(23): 6244–6250.

Abstract: This paper presents the gradient-guided image super-resolution reconstruction for terahertz imaging to improve the image quality, taking advantage of super-resolution reconstruction based on adaptive super-pixel gradient field transform. Moreover, spatial entropy-based enhancement and a bilateral filter are introduced to ensure better performance of the reconstruction. Furthermore, we compare the performance of reconstruction operated on terahertz images with frequencies of 0.1 THz, 0.3 THz, 0.5 THz, and 0.7 THz. Experimental results demonstrate that this method successfully improves the image quality and reconstruct high-resolution images from low resolution images with the peak signal-to-noise ratio and structural similarity index improved. In addition, the signal frequency and intensity are demonstrated to affect the performance of reconstruction.

Key words: Super-resolution reconstruction, Terahertz imaging, Sub-pixel gradient, Spatial entropy-based, Bilateral filter

56. Dual-wavelength terahertz sensing based on anisotropic Fano resonance metamaterials

Yuying Lu, Maosheng Yang, Zhang Zhang, Lanju Liang, Jining Li, Jianquan Yao
Applied optics, 2019, 58(7): 1667−1674.

Abstract: Higher order Fano resonance metamaterials provide a desirable platform for biosensing applications. In this work we exhibit higher order Fano modes by designing an elliptical metallic ring. The simulation results show that for Fano resonant metamaterials, the higher order modes lead to improved sensitivity to refractive index change and larger frequency shifts. Numerically, the sensitivity of dip A (quadrupole mode) is 112 GHz/RIU, whereas the sensitivity of dip B (octupole mode) is 234 GHz/RIU, over 2 times that of dip A. According to our experimental results, the Fano resonant frequencies of dip A and dip B exhibit redshift as the concentration of the anti-cancer drug methotrexate decreases from 120 to 90 μM, with the cell analyte concentration of A549 cells at 5×10^5 cell/ml. For dip A (quadrupole mode), there is a frequency shift of 0.84 GHz for drug concentration of 120 μM and a frequency shift of 22 GHz for a 90 μM drug-treated sample. For dip B (octupole mode), there is a frequency shift of 6.767 GHz for the drug concentration of 120 μM treated sample and a frequency shift of 51.815 GHz for the drug concentration of 90 μM treated sample. Furthermore, the frequency shift of dip A is always smaller than that of dip B for both 90 μM and 120 μM drug concentrations. Such phenomena indicate that dip B is much more sensitive than dip A. The enhanced sensitivity of higher order Fano metamaterials makes it possible to realize high-performance terahertz sensing for biomedical applications.

Key words: Fano modes, Refractive index, Frequency shifts, Terahertz sensing, Biomedical applications

57. Sensitive detection of the concentrations for normal epithelial cells based on Fano resonance metamaterial biosensors in terahertz range

Maosheng Yang, Zhang Zhang, Lanju Liang, Xin Yan, Dequan Wei, Xiaoxian Song, Haiting Zhang, Yuying Lu, Meng Wang, Jianquan Yao

Applied optics, 2019, 58(23): 6268–6273.

Abstract: In this paper, we have cultured normal epithelial cells (HaCaT) as analytes to detect the sensitivity of a biosensor based on Fano resonance metamaterials (FRMMs). The frequency shift Δf of the transmission spectrum was experimentally measured at three different concentrations (0.2×10^5, 0.5×10^5, and 5×10^5 cell/ml) of HaCaT cells. By employing the FRMMs-based biosensor, the detection concentration of HaCaT cells can approximately arrive at 0.2×10^5 cell/ml; further, the corresponding Δf is 25 GHz, which reaches the measurement limit of the THz–TDS system. Additionally, the increase of HaCaT cell concentration causes a different redshift of Δf from 24–50 GHz, and the maximum of Δf can reach 50 GHz when the HaCaT cell concentration is at 5×10^5 cell/ml. Similarly, the simulated results show that the Δf depends on the numbers of analytes with a semiball shape and the refractive index of analytes. The theoretical sensitivity was calculated to be 481 GHz/RIU. The proposed FRMMs-based biosensor paves a fascinating platform for biological and biomedical applications and may become a valuable complementary reference for traditional biological research.

Key words: HaCaT, FRMMs, Biosensor, Sensitivity, Terahertz

58. The biosensing of liver cancer cells based on the terahertz plasmonic metamaterials

Maosheng Yang, Zhang Zhang, Xin Yan, LanJu Liang, Dequan Wei, Longhai Liu, Yunxia Ye, Yunpeng Ren, Xundong Ren, Jianquan Yao

Proceedings of SPIE, 2019, 11196: 111960E.

Abstract: Real-time detection for living cells in vitro is essential for cell physiology, leading to a strong requirement of low cost and label free biosensors. At present, the terahertz plasmonic metamaterials (TPMMs) are an especially attractive application for biosensing owing to their sharp resonances respond. Compared with traditional biosensors, such as flow cytomertry, the TPMMs biosensors have many unique advantages, containing real-time monitoring, free label and high sensitivity. In this paper, we proposed a TPMMs which is designed by digging out periodically arranged regular hexagonal holes on the metal plate with the thickness of 200 nm. The samples of the TPMMs is used as a platform for detecting liver cancer cell GEP2 concentration at five levels (1×10^4, 5×10^4, 1×10^5, 3×10^5 and 5×10^5 cells/ml). The results show that The THz PMMs biosensor cannot distinguish cell concentrations within the orders of magnitude between 1×10^4 and 5×10^4 cells/ml, however, it can distinguish cell concentrations within the orders of magnitude between 1×10^4 and 1×10^5 cells/ml based on the *x*-polarized reflection spectrum TPMMs biosensor. On the other hand, the transmission spectrum TPMMs biosensor has a significant detectability of the orders of magnitude cell concentration between 10^4 and 10^5 cells/ml. The proposed TPMMs biosensor paves a fascinating platform for have been widely applied for cell detection, biotechnology.

Key words: Real-time, Electromagnetic wave, Terahertz, Plasmonic, Metamaterials, Biosensor

59. Biosensor platforms of the polarization-dependent metamaterials for the detection of cancer-cell concentration

Z. Zhang, M. Yang, Lanju Liang, X. Yan, Y. Lu, D. Wei, J. Yao

Proceedings of SPIE, 2019, 11196: 111960X.

Abstract: The rapid detection of cancer cells is crucial for clinical diagnosis in biomedical field. The traditional flow cytometry (FC) in visible band, a fluorescence-labelling detection, gives rise to the complicated sample preparation and the irrecoverable antibody consumption; it blocks the development toward a convenient detection platform with fast, inexpensive and non-labelling. Here, a specifically designed metamaterial based on split ring resonators (SRRs) is proposed. Such metamaterial operating in terahertz (THz) range exhibits polarization-dependent resonances, which are observed both in experiments and simulations. Additionally, the biosensing property of the metamaterial is investigated. On metamaterial surfaces, the lung cancer cells A549 are cultured. Under the irradiation of x-polarized THz waves, it is found that for the cell concentrations from 1×10^5 cells/ml to 5×10^5 cells/ml, the maximum frequency shift Δf (the frequency difference between measured sample and bare one) at 2.24 THz increases from 15 GHz to 137 GHz, respectively. Such results also imply that a larger cell concentration leads to a higher frequency shift. Subsequently, the samples are further measured at different polarization angles. The results show that for cell concentration of 5×10^5 cells/ml, the Δf exhibits the same value of 130 GHz when polarization angle equals 30° and 150°, and 15 GHz when polarization angle equals 60° and 120°. Our proposed metamaterial may supply a potential biosensing method for the detection of cancer cells, exhibiting a new insight toward the cancer cell biosensing with certain information of polarization response.

Key words: Terahertz, Metamaterial biosensors, Cancer cells, Polarization-dependent

60. 太赫兹波三维成像技术研究进展

王与烨, 陈霖宇, 徐德刚, 石嘉, 冯华, 姚建铨

中国光学, 2019, 12(1): 1-18.

摘要：太赫兹波具有良好的光谱特性、非电离性和对许多非极性材料具有穿透性, 在无损探伤、安检、生物医学诊断、艺术品鉴别等领域表现出许多独特的优点。特别是, 太赫兹波三维成像技术能够实现样品内部信息探测, 逐渐成为当前的研究热点, 并展现出广阔的发展前景。本文重点介绍了太赫兹波三维成像的几种常用技术, 包括其基本原理和对应的研究进展, 并分析了存在的问题和发展趋势。

关键词：太赫兹波, 三维成像, 计算机辅助层析, 光学相干层析, 飞行时间层析

61. 基于太赫兹时域光谱系统的脑缺血检测

王与烨, 孙忠成, 徐德刚, 姜智南, 穆宁, 杨川燕, 陈图南, 冯华, 姚建铨

光学学报, 2020, 40(4): 0430001.

摘要：脑缺血是神经外科的常见疾病之一, 目前急需一种快速、准确的脑缺血检测方法。本文利用透射式太赫兹时域光谱技术对新鲜和石蜡包埋的脑缺血组织进行检测, 通过计算左脑与右脑的太赫兹吸收系数的相对差异来研究脑组织的缺血程度, 这可以减小个体差异的影响。实验发现随着缺血时间的延长, 新鲜组织的左右脑吸收系数的相对差异先变大后减小, 这主要是缺血区域的水含量升高和细胞密度下降所引起的。而石蜡包埋的脑缺血组织的左右脑吸收系数的相对差异随缺血时间的延长逐渐升高, 这主要归因于缺血区域的细胞密度逐渐下降。实验结果表明, 利用太赫兹时域光谱系统可以实现最早缺血两小时的脑缺血组织的检测, 这为脑缺血的早期无标记快速诊断提供了有效的技术手段。

关键词：太赫兹时域光谱系统, 脑缺血, 吸收系数, 相对差异, 新鲜组织, 石蜡包埋组织

62. 基于分块压缩感知理论的太赫兹波宽光束成像技术

王与烨,任宇琛,陈霖宇,李长昭,张超,徐德刚,姚建铨

光学学报, 2019, 39(4): 0407001.

摘要:提出了一种基于分块压缩感知理论的太赫兹波宽光束成像技术。模拟结果显示,该技术可以实现高分辨率、高质量的快速成像。采用连续太赫兹波 CO_2 气体激光器光源,基于宽光束矩阵调制采样,对不同物体进行了分块压缩感知成像,并与基于单像素随机采样的分块压缩感知方法进行了对比。结果表明,所提技术具有更高的成像稳定性,且其采样过程对不同成像物体的普适性更好。

关键词:成像系统,太赫兹成像,分块压缩感知,矩阵调制,采样技术

63. 基于太赫兹波成像的鼠脑创伤三维重构

王与烨,陈霖宇,徐德刚,陈图南,冯华,姚建铨

光学学报, 2019, 39(3): 0317002.

摘要:采用 Feeney's 方法制备了鼠脑创伤模型,利用太赫兹波透射式成像系统对鼠脑的组织切片进行成像检测,结果表明:鼠脑创伤区域相比于正常区域具有更低的透过率。采用三维重构技术实现了鼠脑的三维建模,该模型可以清楚地反映鼠脑内部创伤区域的空间分布。基于太赫兹波多深度切片成像的三维重建技术在生物医学精确诊断方面具有巨大的应用潜力。

关键词:医用光学,生物医学成像,外伤性脑损伤,太赫兹波,三维重建

64. 基于梯度变换的太赫兹图像超分辨率重建

郭佑东, 凌福日, 姚建铨

激光技术, 2019: 1–10.

摘要： 为了提高太赫兹图像的质量，克服边缘模糊的缺陷，采用有理分形插值和基于梯度变换的图像超分辨率重建算法相结合的方法对 0.25 THz, 0.50 THz 和 0.75 THz 图像进行超分辨率重建实验，并对实验结果进行了定量分析，并利用基于空间信息熵的直方图匹配技术和双边滤波器对重建算法进行优化，增强了该方法的适用性。结果表明，对经过插值的太赫兹图像采用基于梯度变换的超分辨率重建方法处理之后，0.25 THz, 0.50 THz 和 0.75 THz 图像的边缘强度分别提高了 169%，116% 和 104%；平均梯度分别提高了 16%，28% 和 24%；同时，成像信号频率和强度将对重建性能产生影响。该方法可以有效恢复太赫兹图像当中的细节信息，锐化图像边缘，提高图像质量且不会出现振铃现象，具有较好的实用价值。

关键词： 图像处理, 太赫兹, 超分辨率重建, 插值, 梯度变换, 边缘锐化

65. The antibody-free recognition of cancer cells using plasmonic biosensor platforms with the anisotropic resonant metasurfaces

Zhang Zhang, Maosheng Yang, Xin Yan, Xinyue Guo, Jie Li, Yue Yang, Dequan Wei, Longhai Liu, Jianhua Xie, Yufei Liu, Lanju Liang, Jianquan Yao

ACS Applied Materials & Interfaces, 2020, 12(10): 11388–11396.

Abstract: It is vital and promising for portable and disposable biosensing devices to achieve on-site detection and analysis of cancer cells. Although traditional labeling techniques provide an accurate quantitative measurement, the complicated cell staining and high-cost measurements limit their further development. Here, we demonstrate a nonimmune biosensing technology. The plasmonic biosensors, which are based on anisotropic resonant split ring resonators in the terahertz range, successfully realize the antibody-free recognition of cancer cells. The dependences of Δf and the fitted phase slope on the cancer cell concentration at different polarizations give new perspective in hexagonal radar maps. The results indicate that the lung cancer cell A549 and liver cancer cell HepG2 can be distinguished and determined simply based on the enclosed shapes in the radar maps without any antibody introduction. The minimum concentration of identification reduces to as low as 1×10^4 cells/mL and such identification can be kept valid in a wide range of cell concentration, ranging from 10^4 to 10^5. The construction of two-dimensional extinction intensity cards of corresponding cancer cells based on the wavelet transform method also supplies corresponding information for the antibody-free recognition and determination of two cancer cells. Our plasmonic metasurface biosensors show a great potential in the determination and recognition of label-free cancer cells, being an alternative to nonimmune biosensing technology.

Key words: Metasurfaces, Terahertz, Antibody-free biosensing, Cancer cells, Continuous wavelet transform

后记

"光"是人类认识世界之源头。当今世界五"光"十色，绚丽夺目。"光"纤织就的通信网络把地球变成了一个"村"，实现了世界的互联、互通，引领人类社会向着高效、和谐及智慧的方向前进。

"追光"是我毕生之追求，我和我的团队从事的科研道路是一条"追光"之路：激"光"、非线性"光"学、可调谐激"光"器、全固态激"光"器、激"光"倍频、绿"光"、红"光"、蓝"光"、激"光"全色显示、激"光"测距仪、激"光"高速流场检测、激"光"医疗仪、激"光"致盲研究、"光"通信、"光"纤技术、"光"纤传感技术、太赫兹激"光"技术、微纳"光"电子技术、超材料"光"学、海洋"光"学及"光"声技术、涡旋"光"学、拓扑"光"学、量子"光"学等。我毕生所学、所思、所做都是以"光"和"光电子"为载体，围绕着辐射与物质相互作用来进行的。

我期望深入认识世界、认识宇宙，用"光"照亮未来，为人类及社会服务。"光"是多彩之光、温暖之光、亮丽之光、力量之光、希望之光、探索之光。如今，这些"光"已经聚合成一个较大的"光团"，我们团队中的老师及研究生们均以自己巨大的努力和奋斗，发出了不同谱线的"光芒"。本书中所有的论文及研究成果都是团队共同的成果，大家共同完成的对"光团"的反映。在这里，我还必须感谢国内外激光与光学界的新老同人及朋友们，他们给我的鼓励、支持及协助，是绝对不能缺少的，正像旁轴"光"对我们的"光团"起到的调色、增强与和谐的作用一样。

孔子在《论语》中说过："不知命，无以为君子也；不知礼，无以立也；不知言，无以知人也。"在50余年追"光"的科研、教学实践中，我认识了"光"的本质，深入到"光"科技发展的前沿，更重要的是学会了用"要学会做事更重要的是先要学会做人"这个历史早已证明的真理教育和鞭策自己。我希望自己和团队的成员应该学会做人，光明磊落，具有谦虚、诚恳、正直、坦荡、好学、宽容的品德修养。

在追"光"的过程中，我也知人甚多。我始终要感谢在追"光"道路上支持和帮助过我的老师、学生、同人、朋友及家人，是"光"的纽带把我们联系在一起。我还要感谢钟凯副教授、梁兰菊教授、吴亮副教授、张雅婷副教授，感谢陈治良、杨茂生等博士生、硕士生，感谢徐棪等人在本书资料整理及出版过程中付出的辛勤的努力。

科学技术的广泛应用，深刻地影响着世界，促进了经济繁荣，给人类带来福祉。科技创新作为引领发展的第一动力，已呈现出多点群发和持续突破的态势，新兴技术、颠覆性技术还在不断涌现。我们要紧跟时代的步伐，瞄准国家重大需求，加强基础研究，注重学科交叉，发挥信息科技在经济、社会与人民生活中的主导作用。

世界的历史、人类的命运、科技的发展正进入崭新的机遇期。让我们以"激光"为利器，让"海"容纳百川，让"智慧"开启未来，让"光"照亮世界，照亮人类前进的步伐。

以上是我在病房里随想随写的一点字句，就作为后记吧！

姚建铨

2020年10月写于

天津医科大学总医院第二住院楼829病房